Structure and Motion:
Membranes, Nucleic Acids & Proteins

Structure and Motion: Membranes, Nucleic Acids & Proteins

Proceedings of the International Symposium on Structure and Dynamics of Membranes, Nucleic Acids & Proteins held in Rome, Italy, April 23-27, 1984 at the Accademia Nazionale dei Lincei, under the sponsorship of the Academy, National Foundation for Cancer Research and International Business Machines Corporation.

Edited by

E. Clementi & G. Corongiu

International Business Machines Corporation

and

M. H. Sarma & R. H. Sarma

State University of New York at Albany

Adenine Press, P.O. Box 355
Guilderland, New York 12084

CHEMISTRY

Adenine Press
Post Office Box 355/340
Guilderland, New York 12084

Cover illustration: The cover illustration depicts a complex between the B-DNA dodecamer of sequence C-G-C-G-A-A-T-T-^{Br}C-G-C-G and the antitumor antibiotic netropsin. This single crystal x-ray analysis is described in this volume and in Kopka *et al.* (1985), *Proc. Natl. Acad. Sci. U.S., 82.* The multi-ring netropsin molecule sits in the center of the minor groove, displacing the water molecules of the spine of hydration and making hydrogen bonds to A·T base pairs. This drawing was made on a Nicolet Zeta plotter drive by a VAX 11/780 computer, using the program SPACEFILL by David Goodsell.

Library of Congress Cataloging in Publication Data

International Symposium on Structure and Dynamics of
    Membranes, Nucleic Acids & Proteins (1984:
    Accademia nazionale dei Lincei)
    Structure and motion.

    Bibliography: p.
    Includes index.
    1. Molecular dynamics—Congresses.   2. Nucleic acids—
Congresses.   3. Membranes (Biology)—Congresses.
4. Proteins—Congresses.   5. Molecular structure—
Congresses.   I. Clementi, Enrico.   II. Accademia
nazionale dei Lincei.   III. National Foundation for
Cancer Research.   IV. International Business Machines
Corporation.   V. Title.
QP517.M65157   1984   574.87'328   85-5993
ISBN 0-940030-12-8

*Set in type in the United States by Word Management Corporation.
Printed in the United States by Hamilton Printing Company.*

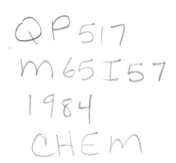

Preface

These are the Proceedings of the International Symposium: Structure & Dynamics: Nucleic Acids, Proteins & Membranes, held in Rome, Italy, April 23 through 27, 1984. The conference was held at the Accademia Nazionale dei Lincei, and was sponsored by the Academy, the International Business Machines Corporation and the National Foundation for Cancer Research.

In the opening address, Prof. Giuseppe Montalenti, President of the Accademia dei Lincei, observed that the above three chemical systems, around which the Symposium was centered, are so important for the existence and support of living organisms as to have become the primary topics for most of today's frontier problems in physical and natural sciences. Indeed, the Symposium did address a large number of issues, from physics and chemistry to biology and medicine. Thus, it was befitting to hold the Symposium at the Lincei, the oldest academy in Europe, where inquisitive minds, since the time of Galileo Galilei, often gather to discuss the status of our understanding of natural phenomena.

The laws governing motion and their manifestations in nature, a crucial curiosity for the giant mind of Arcetri, once more were under scrutiny either in the quantized or in the classical form, either for a single or a collection of particles. Old tools and probes, for instance, infrared radiations and X-rays, were complemented by relatively new ones, such as two dimensional N.M.R. and neutron beams. In this regard, we note with interest that the days of the Symposium fell during the half-centennium celebration of Fermi's and his Roman school's discovery of the neutrino. It also should be noted that at the Symposium the probes were either experimented with or simulated on more and more powerful computing instruments.

Studies on cell's growth, on molecular genetics and evolution of matter, as well as molecular diseases, as for example, cancer, were in the Symposium's agenda. This shift of attention from nuclei and electrons to problems of life and death, or, more simply, from the "inorganic" to "living" matter leads to questions on man and society. In this spirit, the visit and the private audience in the Sala del Concistoro with Pope John Paul II was a highlight of the symposium. For the record, we reproduce the address by one of us (E.C.) to Pope John Paul II and the address of the Pope to the symposium delegates in the following pages. We thank Mons. G. B. Re, Assessore, Segreteria Di Stato, for giving us permission to reproduce the symposium related material from Vatican. We thank Mrs. Alberta Martino for providing proper coordination for the volume and feeding most of the manuscripts to the computer. We thank Prof. R. E. Dickerson and his group for the design of the cover illustration. We thank and congratulate the various authors of this volume for their notable contributions.

E. Clementi & G. Corongiu
IBM Corporation
Kingston, New York

M. H. Sarma & R. H. Sarma
State University of New York
Albany, New York

Professor Clementi and Pope John Paul II

Prof. Enrico Clementi's Address to Pope John Paul II.

Your Holiness,

We are most grateful and moved to be in your presence and pleased to acknowledge Monsignor Angelini for his assistance. We have come from the West and the East to discuss and learn from each other.

Your Holiness, we are mathematicians, physicists, chemists, biologists and computer experts. We study how electrons, atoms, and molecules do play an important role in self-organization of matter in genetics, in cancer and in brain mechanism. Our common bond is in learning and in attempting to apply our most limited and very rudimental knowledge to specialized problems which might be relevant to much broader questions on life, aging, sickness, and death.

Among us are Mr. and Mrs. Franklin and Tamara Salisbury, founders and Directors of the National Foundation for Cancer Research, who have helped many scientists throughout the world by financing research against cancer with money collected from the generosity of the American people. Among us, also, are Dr. and Mrs. Ivar Giaever, Nobel winner in physics, Prof. Wlodzimierz Kołos from the Polish Academy of Sciences, and Prof. and Mrs. Michael Kasha of the U.S. Academy of Science. Their fundamental studies in physics, chemistry, and biochemistry are representative of our common goals and interests.

Your Holiness, we know of your concerns for peace, freedom, social justice, human dignity, and for poverty, sickness and ignorance. We are most grateful for this audience in the bonds of these hopes and ideals.

Enrico Clementi

Chairman, Organizing Committee
Rome Symposium on Structure & Dynamics:
Nucleic Acids, Proteins & Membranes

Vatican City, April 27, 1984
Sala del Concistoro

Pope John Paul II and the delegates at the Rome Gathering in Membranes, Proteins and Nucleic Acids

Pope John Paul II's Welcome Address

Dear Friends,

It is an honour and a pleasure for me to welcome to the Vatican those taking part in the International Symposium sponsored by the Accademia Nazionale dei Lincei, the International Business Machines Corporation and the National Foundation for Cancer Research. As I greet you today, I wish to express my <u>deep appreciation of the important contribution</u> you are making to the health and happiness of the human family. Through your generosity and self-sacrifice in the disciplined and often tedious task of research, you have greatly increased our understanding of the causes and nature of cancer and of the best methods for treating it effectively. This Symposium is one more example of your untiring efforts in this regard and of your dedicated concern for those throughout the world who suffer from this dreaded disease.

Several months ago, I issued an Apostolic Letter on the Christian meaning of human suffering. In that document I sought to bring the light of Christ to that experience which is an essential part of every person's life. In addition to my desire to help people find meaning in the mysterious face of suffering, I also wished to draw attention in <u>gratituded</u> to those, like yourselves, who are particularly sensitive to the sufferings of others, and who strive to offer not only sympathy and compassion but concrete efforts to alleviate those sufferings. In this regard, I stated: "How much there is of 'the Good Samaritan' in the profession of the doctor, or the nurse, or others similar! Considering its 'evangelical' content, we are inclined to think here of a vocation rather than simply a profession. And the institutions which from generation to generation have performed 'Good Samaritan' service have developed and specialized even further in our times. This undoubtedly proves that people today pay ever greater and closer attention to the sufferings of their neighbour, seek to understand those sufferings and deal with them with ever greater skill. They also have an ever greater capacity and specialization in this area. In view of all this, we can say that the parable of the Samaritan of the Gospel has become <u>one of the essential elements of moral culture and universally human civilization.</u> And thinking of all those who by their knowledge and ability provide many kinds of service to their suffering neighbour, we cannot but offer them words of thanks and gratitude" (<u>Salvifici Doloris</u>, 29).

For years, medical research has required the adoption of methods of advanced specialization in order to pursue new discoveries. More recently, an <u>interdisciplinary approach</u> has been increasingly needed, one which encompasses the insights provided by various fields of knowledge, such as medicine, biology, chemistry, physics, mathematics and so forth. All of this points to a need for greater dialogue and collaboration among the men and women of the different sciences. Together with this, more and more scientists and researchers feel the importance of placing the results of their research within a wider social and cultural context and of giving due consideration to the moral principles and spiritual values which are associated with new discoveries. Your desire to meet the Pope on the occasion of this International Symposium shows your own sensitivity to these dimensions.

Medical research, and indeed all scientific study, needs the support and guidance of spiritual and moral values. For such research is ultimately intended for the good of the whole person, even if the immediate aim is the treatment of certain tissues or organs of the body. A deep unity exists between the body and the spirit, a unity which is so substantial that the most spiritual activity is affected by the bodily condition, and the body itself achieves its proper and final purpose only when directed by the spirit. I would therefore like to offer encouragement to all those who are promoting an interdisciplinary approach to cancer research and other medical problems, and I would urge that there be included in this approach the wisdom which is found in the spiritual heritage of the human family. I assure you, in this regard, of the great interest of the Catholic Church in your research, and of a readiness to dialogue and to share with you the spiritual and ethical traditions of the Christian faith.

The interdisciplinary nature of science today has also involved an internationalization among those carrying out research. This is clearly seen in your present Symposium. And it is a hopeful sign of an increasing spirit of brotherhood and fruitful cooperation among men and women of good will from all nations. I pray that your successes and achievements will, in a similar way, bring hope and assistance not only to a fortunate few but to people throughout the world.

Ladies and gentlemen, yours is indeed a noble and vitally important task. May God grant you joy and strength in your work, and may he bless you and your families with his rich and abiding peace.

Structure & Motion: Membranes, Nucleic Acids & Proteins,
Eds., E. Clementi, G. Corongiu, M. H. Sarma & R. H. Sarma,
ISBN 0-940030-12-8, Adenine Press,   Copyright Adenine Press, 1985.

CONTENTS

Part I. Conceptual & Theoretical Background

Part II. Structure & Motion: Cells, Membranes & Proteins

Part III. Structure & Motion: Nucleic Acids

Structure & Motion: Membranes, Nucleic Acids & Proteins,
Eds., E. Clementi, G. Corongiu, M. H. Sarma & R. H. Sarma,
ISBN 0-940030-12-8, Adenine Press,   Copyright Adenine Press, 1985.

Intermolecular Forces –
The Glue for any Biomolecular Structure

A. D. Buckingham
University Chemical Laboratory,
Lensfield Road, Cambridge, CB2 1EW, England

Abstract

The nature and strength of intermolecular forces are examined, with particular reference to materials of biological significance. The forces may be of long or short range, additive or non-additive, and enthalpic or entropic. The long-range interactions are subdivided into electrostatic, induction and dispersion contributions; these normally give a good description of the attraction almost as far in as the equilibrium separation. The electrostatic interaction is that of the charge distributions of the free molecules and is commonly dominant. The short-range forces can generally be represented adequately by self-consistent-field calculations, or crudely but usefully by hard-spheres located on the atoms. The effects of a medium are considered and it is shown that it may enhance or diminish the strength of the interaction.

Introduction

Molecules attract one another when they are far apart, since liquids and solids exist. They repel one another when close, since densities are finite and have their familiar magnitudes in liquids and solids. This well-known truth is illustrated in figure 1 which shows a typical interaction energy U of two spherical atoms. For two argon atoms, the well-depth ϵ is 0.198 x 10^{-20}J ($\epsilon/k = 143$K) and the equilibrium separation R_e is 3.76 x 10^{-10}m (1,2).

The generalization of this simple one-dimensional function $U(R)$ to include relative orientations, internal coordinates, and the presence of additional molecules making up the medium, is central to both structure and dynamics and hence to this Symposium.

It is normal, and indeed necessary if potential energy functions are to be used, to invoke the Born-Oppenheimer separation of the nuclear and electronic motion and to represent the interaction energy as a function of the positions of the nuclei (3). This approximation is not a limitation unless we are concerned with systems with very low-lying excited electronic states. So we represent the intermolecular energy U as a function of the relative positions of all the nuclei of the interacting molecules. And the number of variables increases rapidly with the size of the molecules—it is 1

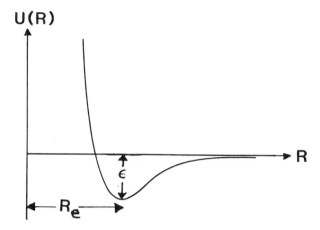

Figure 1. The interaction energy $U(R)$ of two spherical molecules as a function of their separation. The well depth is ϵ and R_e is the equilibrium separation.

for two atoms (i.e., R); 3 for an atom in interaction with a diatomic (R, r, θ where R is the distance between the atom and some convenient origin in the diatomic, such as the centre of mass, r is the separation of the nuclei in the diatomic, and θ is the angle between r and R); 6 for two diatomics ($R, r_1, r_2, \theta_1, \theta_2, \phi$, where ϕ is the angle between the planes formed by the molecules with R); and it is 12 for two water molecules. In general, for molecules containing N_1 and N_2 nuclei, there are $3(N_1+N_2)-6$ independent variables, of which $3N_1-6$ and $3N_2-6$ are vibrational coordinates in each molecule and the remaining 6 ($R, \theta_1, \chi_1, \theta_2, \chi_2, \phi$, where χ_1 and χ_2 determine the orientation of molecules 1 and 2 about their axes at angles θ_1 and θ_2 to R) give the relative positions and orientations of the two molecules. So if molecules were to interact as rigid bodies, we need consider only 6 variables. But molecules are non-rigid and there may be some motions (such as the inversion motion in NH_3, and internal rotation in CH_2Cl-CH_2Cl or in a polypeptide) for which there is only a small change in energy with a large change in some internal coordinate. In such cases, the influence of the environment in producing changes in the energy surface involving this coordinate is of interest.

The six translational and orientational degrees of freedom of an interacting pair of non-linear polyatomic molecules generally fluctuate slowly compared to the intramolecular vibrations. For some purposes, such as rotational relaxation, it may be sufficient to average U over the vibrational motion, thereby reducing the number of variables upon which U depends to just 6. And for vibrational relaxation of a particular mode, it may be reasonable to average over other vibrational modes, thus reducing the effective dimensionality of the problem.

Since we are concerned with intermolecular forces, we should consider what we mean by a *molecule* and by a *force* (4). Two argon atoms, with an interaction energy as in figure 1, can form a bound diatomic Ar_2 but we do not normally consider the species Ar_2 as a molecule, since the binding energy is only about $\frac{1}{2}kT$ at room temperature. Collisions may easily dissociate Ar_2, and there are many thermally populated vibration-rotation states $|vJ\rangle$, each with a different mean bond length $\bar{R} = \langle vJ|R|vJ\rangle$ and a large uncertainty $\langle vJ|(R-\bar{R})^2|vJ\rangle^{\frac{1}{2}}$ in R. We prefer to speak of Ar_2 as a dimer of argon atoms. Similarly, H_4, formed on cooling gaseous hydrogen to about 20K at 1 atmosphere, is an infrared-active species in which two bonds are very similar to that in H_2 (5); we prefer to think of H_4 as $(H_2)_2$, i.e. as a hydrogen molecule dimer. By a molecule we mean a group of atoms (or a single atom) whose binding energy is large compared to kT, so a molecule interacts with its environment without losing its identity, although its structure and properties may be changed by the environment.

And what do we mean by a force? In figure 1, the force is $-dU/dR$ and there is no difficulty. This concept may be extended to a many-dimensional surface when the forces and torques acting on the molecules are determined by the gradients of U. But what is the effective force on two ions in aqueous solution? It is convenient to consider the potential of average force $A(R)$ which is a Helmholtz free energy and is the mean interaction energy of the two ions at a fixed separation R, averaged over all configurations of all the other molecules and ions in the solution. $A(R)$ is the sum of $U(R)$ and $-TS(R)$ where both $U(R)$ and the entropy $S(R)$ are functions of the temperature T. The entropic contribution may be thought of as arising from the change in the order in the environment resulting from the interaction of the pair. The attractive force in a stretched rubber band is due to a decrease in entropy on stretching (6); and the hydrophobic effect (7), that appears to produce an attractive force between hydrocarbon chains in aqueous media, depends on $S(R)$, since the decrease in S in forming a cage of water molecules is less in the case of a close pair of chains than when they are far apart.

Intramolecular vibrational motion is generally too rapid to permit adjustment of the relative positions and orientations of neighbouring molecules, so their contribution to the entropic force $T\partial S(R)/\partial r_i$ may normally be neglected.

The Origin and Magnitude of Intermolecular Forces

It is often convenient to divide intermolecular forces into long-range and short-range forces. Long-range forces act when the molecules are sufficiently far apart that their electron clouds overlap insignificantly; then U varies as a power series in $1/R$. Short-range forces result from the overlap; they originate in the coulomb and exchange interactions and increase exponentially with decreasing R (2,8).

Long-range forces can be subdivided into a number of distinct types, as shown in table 1 (9). The electrostatic energy is the interaction of the permanent charge distribution of the molecules. If the interaction is treated as a perturbation to the non-interacting and non-overlapping molecules, the electrostatic energy is the first-order perturbed energy. It may give an attractive or a repulsive force, and is

Table I
A classification of molecular interaction energies

Range	Type	attractive ($-$) or repulsive ($+$)	additive or non-additive
short	overlap (Coulomb and exchange)	$-$ $+$	non-additive
long	electrostatic	$\mp$	additive
	induction	$-$	non-additive
	dispersion	$-$	nearly additive
	resonance	$\mp$	non-additive
	magnetic	$\mp$	(weak)

additive (i.e., the electrostatic energy of three molecules 1,2,3, is the sum of the three pairs 12,23,31).

The induction energy is due to the distortion of the charge distribution of one molecule by the permanent charge distribution of its neighbours. At large separations it is equal to $-\frac{1}{2}\sum_i \alpha_i F_i^2$, where α_i is the static polarizability of molecule i and F_i the electric field strength at the ith molecule resulting from the charge distribution of its neighbours (9). Since α_i is positive for molecules in their ground electronic states, the induction energy is negative, and this remains true at shorter range when induced quadrupole and higher moments are significant, since the distortion of each molecule must be such as to lower the total energy.

The dispersion energy results from a correlation in the fluctuations of the electron positions and is therefore negative. At large separations it can be represented as $-CR^{-6}$ but there are shorter-range terms, including an angle-dependent term in R^{-7} if either molecule lacks a centre of symmetry (9). Rigorous formulae exist relating the dispersion energy to the dynamic polarizabilities of the free molecules (10), but in practice we are generally content to use London's approximate formula (9) for C:

$$U_{dispersion} = -CR^{-6} = -\frac{3\,I_1 I_2}{2(I_1 + I_2)}\,\frac{\alpha_1 \alpha_2}{(4\pi\epsilon_o)^2 R^6}$$

where ϵ_o is the permittivity of free space ($4\pi\epsilon_o = 1$ e.s.u. $= 1.11265 \times 10^{-10}$ CV^{-1}m^{-1}).

Resonance energy is the additional energy that results from the lifting of degeneracy by the interaction. The degeneracy may arise because one of the molecules is in a degenerate state, as in the interaction of a H atom with principal quantum number 2 with an ion or polar molecule. The degeneracy might also result from the exchange of excitation between identical molecules, as in the case of a vibrating molecule having one quantum of excitation in its ith mode ($v_i = 1$) near an identical molecule with $v_i = 0$. The lifting of the degeneracy by the interaction produces two or more potential surfaces which lie above and below zero, according to the quantum numbers describing the state of the pair.

Magnetic interactions are weak relative to electric ones, since a magnetic dipole may be of the order of 1 Bohr magneton $= 0.9274 \times 10^{-20}$ e.m.u. $= 0.9274 \times 10^{-23} Am^2$, while electric dipoles $\sim 1D = 10^{-18}$ e.s.u. $= 3.336 \times 10^{-30} Cm$, so that magnetostatic energies are typically $\sim 10^{-4}$ of electrostatic energies. In optically active species, where the molecules exist in right- and left-handed forms, there is a coupling of the fluctuating electric and magnetic moments, giving rise to a weak dispersion energy that is dependent on the handedness of the molecules. This weak dispersion energy varies as R^{-6} and is attractive between similar species (i.e. left with left, and right with right) (11). However, it is probable that this difference is negligible and that the important discriminatory forces are of shorter range and depend on the molecular shape (12).

When the overlap of the electron clouds is significant, the anti-symmetrization of the wavefunction with respect to exchange of all pairs of electrons leads to short-range repulsion. One important route to the evaluation of short-range interactions is through Hartree-Fock SCF theory. This approach can give useful results, but if only a single electron configuration is employed there can be no electron correlation and hence no dispersion energy. Since the dispersion energy is the sole source of attraction between inert-gas atoms, it is to be expected that the Hartree-Fock potential curve for such a system should have no minimum. Minima have sometimes been obtained, but these are artifacts resulting from a basi-set extension effect and can be eliminated by the addition of 'ghost orbitals' in the calculations on the separate molecules (13,14).

In the region of electron overlap, the identity of the interacting molecules is partly removed and they merge to form a 'supermolecule'. It is therefore unlikely to be helpful to seek a rigorous theory of short-range forces which relates the interaction to the properties of the free molecules. But since the repulsion increases rapidly as R decreases from its equilibrium value, the structure and properties of interacting molecules may not be sensitive to the precise form of the repulsive potential.

It is useful to know the approximate magnitude of the various contributions to the intermolecular potential. The relative importance of each varies from system to system. Thus electrostatic and induction energies are zero in the inert gases, whereas in hydrogen-bonded systems the electrostatic energy is predominant (15).

The interaction energy of singly-charged positive and negative ions at a separation R is $-e^2(4\pi\epsilon_o R)^{-1}$ which is $-46 \times 10^{-20} J = -280 kJ\ mol^{-1}$ for $R = 5 \times 10^{-10} m$ (this could be substantially reduced by the presence of a polar medium). The energy of two colinear dipoles $\mu(\mu_z = \Sigma e_i z_i)$ of magnitude 1D (1D $= 3.336 \times 10^{-30} m$) separated by $R = 5 \times 10^{-10} m$ is $-2\mu^2(4\pi\epsilon_o R^3)^{-1} = -0.16 \times 10^{-20} J = -0.98\ kJ\ mol^{-1}$, and that of two perpendicular linear quadrupoles (9) $(\Theta_{zz} = \frac{1}{2}\Sigma e_i(3z_i^2 - r_i^2))$ of magnitude 1B $= 10^{-26}$ e.s.u. $= 3.336 \times 10^{-40} Cm^2$ is $-3\Theta^2(4\pi\epsilon_o R^5)^{-1} = -0.010 \times 10^{-20} J = -0.058\ kJ\ mol^{-1}$. The dispersion energy between a pair of $-CH_2-$ groups separated by $5 \times 10^{-10} m$ is approximately $-0.060 \times 10^{-20} J = -0.36\ kJ\ mol^{-1}$. For two long parallel linear chains, each containing n CH_2 groups, at a separation d, the total dispersion energy varies as nd^{-5} and, for $d = 5 \times 10^{-10}$m, is equal to $-0.3n \times 10^{-20}$J $= -1.7n$ kJ mol^{-1} (16). These forces provide a simple explanation of differences in the cohesive energy of cis-unsaturated fatty acids as compared to trans-unsaturated or saturated fatty acids (16).

The sublimation energy of crystalline argon at 0K is 7.74 kJ mol^{-1}, and of this 107% comes from the pairwise interaction of the atoms (97% from the 12 nearest neighbours and 10% from more distant neighbours), leaving -7% for the many-body contribution (1). The sublimation energy of CO_2 at 0K is 27 kJ mol^{-1} and of this approximately 52% is due to the quadrupole-quadrupole interaction ($\Theta = -15 \times 10^{-40}$Cm^2 (17)) and 48% to the dispersion forces.

The Effect of a Medium

A medium of relative permittivity, or dielectric constant, ϵ_r reduces the electro-static energy of interaction of two molecules immersed in it by ϵ_r. The effects of a medium on dispersion energy have been examined (18,19), and it can be helpful to introduce an 'effective' or 'excess' polarizability α^* which may be used to give the interaction free energy. The dispersion force between any two similar spherical systems is always attractive, regardless of the nature of the medium, so that two bubbles or two colloidal particles attract one another.

The presence of polarizable matter between interacting molecules may increase their mutual potential energy. For example, if a sphere of polarizability α is at the point midway between a pair of charges q and $-q$ at a separation R, the interaction energy is $U(R) = -q^2(4\pi\epsilon_o R^3)^{-1}(1+32\alpha(4\pi\epsilon_o R^3)^{-1})$. However, the sphere would not change the potential energy of two charges of the same sign, for which $U(R) = q^2(4\pi\epsilon_o R)^{-1}$. If spheres of polarizability α are at a fixed distance d beyond each of the charges q and $-q$, the magnitude of the force between the charges is reduced, and takes the value

$$-q^2(4\pi\epsilon_o R^2)^{-1}[1-4\alpha d^{-2}R^{-1}(1+2dR^{-1})(1+dR^{-1})^{-5}(4\pi\epsilon_o)^{-1}].$$

If the two charges have the same sign, the force of repulsion is *enhanced* to

$$q^2(4\pi\epsilon_o R^2)^{-1}[1+4\alpha d^{-2}R^{-1}(1+2dR^{-1}+2d^2R^{-2})(1+dR^{-1})^{-5}(4\pi\epsilon_o)^{-1}].$$

Conclusions

We have a good understanding of fundamental aspects of intermolecular forces, but systems of practical interest are generally too complicated to permit definitive analysis. Even the origin of the observed structures of Van der Waals molecules is contentious (20,21), although it seems clear that in hydrogen-bonded species, such as HF . . . HF, HF . . . H$_2$O, H$_2$O . . . H$_2$O and HF . . . H$_2$CO, the electrostatic energy, coupled with a simple hard-sphere repulsion for each atom, gives a good description of the structure (20). However, for an accurate account of the electrostatic interaction at separations as close as the equilibrium point, it is necessary to use a 'distributed multipole analysis' (22) which overcomes the convergence problems of the usual molecular multipole expansion provided the charge distribution in the free molecule is known. Similarly, assignment of polarizabilities to atoms, or groups of atoms, should permit accurate descriptions of the induction and dispersion interactions.

References and Footnotes

1. J. A. Barker in *Rare Gas Solids*, Ed., M. L. Klein and J. A. Venables, Academic Press, New York, chap.4 (1976).
2. G. C. Maitland, M. Rigby, E. B. Smith and W. A. Wakeham, *Intermolecular Forces. Their origin and determination.* Oxford University Press (1981).
3. J. O. Hirschfelder and W. J. Meath, *Adv. Chem. Phys., 12,* 3 (1967).
4. H. C. Longuet-Higgins, *Discussions Faraday Soc., 40,* 7 (1965).
5. A. Watanabe and H. L. Welsh, *Canadian J. Phys., 43,* 818 (1965); *45,* 2859 (1967). A. R. W. McKellar and H. L. Welsh, *Can. J. Phys., 52,* 1082 (1974).
6. See *A Discussion on Rubber Elasticity, Proc. Roy. Soc. A, 351,* 295 (1976).
7. See *The Hydrophobic Interaction, Faraday Symposia, 17* (1982).
8. H. Margenau and N. R. Kestner. *Theory of Intermolecular Forces,* 2nd ed., Pergamon Press, Oxford (1970).
9. A. D. Buckingham in *Molecular Interactions: From Diatomics to Biopolymers,* Ed., B. Pullman, Wiley, New York, chap.1 (1978).
10. A. Dalgarno, *Adv. Chem. Phys., 12,* 143 (1967)
11. C. Mavroyannis and M. J. Stephen, *Molec. Phys., 5,* 629 (1962).
12. S. F. Mason, *Int. Rev. Phys. Chem, 3,* 217 (1983).
13. S. F. Boys and F. Bernardi, *Molec. Phys., 19,* 553 (1970).
14. N. S. Ostlund and D. L. Merrifield, *Chem Phys. Letters, 39,* 612 (1976).
15. H. Umeyama and K. Morokuma, *J. Amer. Chem. Soc., 99,* 1316 (1977).
16. L. Salem, *Can. J. Biochem. Physiol., 40,* 1287 (1962).
17. M. R. Battaglia, A. D. Buckingham, D. Neumark, R. K. Pierens and J. H. Williams, *Molec. Phys., 43,* 1015 (1981).
18. A. D. McLachlan, *Discussions Faraday Soc. 40,* 239 (1965).
19. J. Mahanty and B. W. Ninham, *Dispersion Forces,* Academic Press, London (1976).
20. A. D. Buckingham and P. W. Fowler, *J. Chem Phys., 79,* 6426 (1983).
21. F. A. Baiocchi, W. Reiher and W. Klemperer, *J. Chem., Phys., 79,* 6428 (1983).
22. A. J. Stone, *Chem. Phys. Letters, 83,* 233 (1981).

Structure & Motion: Membranes, Nucleic Acids & Proteins,
Eds., E. Clementi, G. Corongiu, M. H. Sarma & R. H. Sarma,
ISBN 0-940030-12-8, Adenine Press, Copyright Adenine Press, 1985.

Biomolecular Interaction and Stability: Reflections on the Role of Topological Order and Stereodynamic Disorder of Solvent

M.B. Palma-Vittorelli and M.U. Palma

Istituto di Fisica, Università di Palermo, and
Istituto Applicazioni Interdisciplinari della Fisica, Cons. Naz. Ricerche
via Archirafi 36, I 90123 Palermo, Italy

Abstract

Consideration of connectivity properties of the system of H-bonds in water has offered in the recent past the basis for a unifying approach of the main properties of water (as a pure liquid, as a solvent, and in its anomalous supercooled state). Topologic connectivity properties of H-bond pathways provide the basis for an ordered classification of the otherwise disordered and flickering stereodynamics of water molecules.

Following previous work from our laboratory, we have conceptually sorted out connectivity pathways of H-bonds of solvent water, and discussed their role in terms of the standard statistical thermodynamics of solvent-mediated solute-solute interaction. The resulting model can provide a unifying interpretation of the responses of systems as different as biomolecules in solution, supercooled water, and a class of crystalline solids containing a subsystem of interacting protons, to isotopic substitution and to disturbances in motional correlations within the set of mobile protons. The model could also have a heuristic value with respect to biomolecular pattern recognition.

The exceptionally large variety of configurations having very close Gibbs' free energies, wich characterize water as a unique solvent, acquires in this model a specific relevance to stability and function of solute biomolecules. This occurs because the thermally excited, chaotic flickering of solvent configurations within a well-defined topologic context, involves a large number of solute-solvent stereodynamic microstates, into one and the same configurational/functional biomolecular state. This possibly offers a contribution toward the understanding of reasons allowing a mechanistic description of the stability and function of biomolecules, despite their being objects on the borderline of quantum mechanics.

Introduction

A few exceptional features characterize biomolecules: i) they undoubtedly exhibit an order, but this is the unusual and loose order of strings characterized by unique sequences, and conformed 3-dimensionally so as to obtain a far from uniform filling of the overall occupied space—a condition essential to their task; further, their function requires a possibility of transition between at least two conformations, which may often be visualized as a "switching"; ii) their extraordinary properties, as the permanence of their functions, and their perpetuability which originated e.g. Leibnitz's and Schroedinger's wonder, include a great flexibility, so as to allow meaningful responses (within a certain dynamic range) to environmental changes in pH, other solutes, temperature etc.; iii) they are large enough to allow being considered mechanistically in their operation and yet, sufficiently microscopic to require a quantum mechanical treatment.

The macroscopic, mechanistic description of the operation of a machinery is possible, as it is well known, when a very large number of microscopic states correspond to one and the same functional macroscopic state. Since biomolecular operation can be described mechanistically, a large number of microscopic states must contribute to each biomolecular functional state. In other words, all stereodynamic microscopic configurations in the phase space accessible to a biomolecular system, and corresponding to its functional states, must sum up to figures large enough to justify a mechanistic description of biomolecular operation. A first set is that of all geometric and dynamic states allowed to the functional biomolecule in itself. This will already be a rather ample set, as a consequence of the non uniform filling of biomolecular volume. Further, we must consider that biomolecules operate in mechanical and thermodynamical contact with the solvent, so that all pertinent stereodynamic configurations of the solvent must somehow be taken into account (1)(2). How to sort them out is of course a conceptually relevant problem.

A long time was required for these notions to take shape. Indeed, it has taken some 25 years to digest the philosophical contents of data on hydrogen exchange times in proteins (3); of the earliest suggestion of the interest of studying biomolecular stereodynamics (4), and of that of taking into account highly non-linear effects (4)(5); of indications of the need of taking into account solvent stereodynamics (6); of indications of the existence of biomolecular conformational substates and of their functionality (7-9). A variety of experimental work has since followed along diversified, often cross-fertilizing lines (10-12).

On the whole, geometry and dynamics of biomolecules and of solvent constitute four coupled subsystems, and in principle this coupling cannot be disregarded, as it may possibly originate a variety of feedback effects (13). Nevertheless, the timescale of biomolecular dynamics is considerably slower than that of solvent dynamics. The direct coupling of solute-solvent molecular dynamics can therefore be neglected but, as we shall see later, a cross coupling may reappear through configurations. Further, it has to be borne in mind that water is a unique solvent, whose anomalous properties can be thermodynamically expressed in terms of very large fluctuations in density, local entropy and local enthalpy (14)(15). These anomalous properties are based on a great flexibility (afforded by the tetrahedral structure of the H_2O molecule and by cooperativity of Hydrogen bonding) of building structures and connections, thus allowing an exceptionally large number of stereodynamic molecular configurations, characterized by large differences in entropy and enthalpy but such that $\Delta H \approx T \Delta S$, so that $\Delta G \approx 0$. As a consequence, very ample spatio-temporal fluctuations are possible within this large set of stereodynamic configurations. From the chaoticity of this flickering among widely different geometric and dynamic configurations, however, an element of order can be sorted out. This consists in the statistical permanence of topologic properties of pathways of hydrogen bonds within the solvent (1)(2)(14)(16-19), that we shall discuss in more detail in the next section. Since an infinite number of geometrically unequivalent configurations can be grouped within one topological order, the latter concerns a higher hierarchical level.

Thermodynamics of biomolecule-solvent interaction has been widely studied and discussed and the stabilizing termodynamic role of the solvent is well known (20-24). In addition, however, one wants to envisage a microscopic model, useful for the interpretation, or prediction, of perhaps more elusive microscopic and kinetic behaviours, and for a microscopic understanding of the reasons which legitimate the mechanistic treatment of biomolecular function. This will be discussed in the next sections. We shall use the notion of H-bond connectivity pathway, as it has emerged from computer simulation (16-18) and theoretical work (14), and discuss a microscopic picture of solvent stereodynamics-biomolecule interaction (1)(2)(13), its thermodynamic relevance (2), and its interpretative/predictive power. The scenario, as it results from these introductory note is one of solute biomolecules moving and fluctuating chaotically within the set of substates pertaining to their conformation, as dictated by the physico-chemical environment (7)(25) and also allowing, if this is the case, highly non linear wavelike modes (4)(5)(26); and of solvent molecules chaotically flickering within an ample set of stereodynamic configurations with an element of order consisting in the permanence of some topologic properties of the system of H-bonds, and also, when this is the case, allowing for ampler, hydrodynamic motions.

Flickering connectivity pathways of H-bonds

Available evidence

By current consensus, liquid water is viewed in terms of a random network of H-bonds. This picture has emerged out of computer simulation (16-19), theoretical (14)(27) and experimental (28)(29) work. Computer experiments have been particulary useful in providing a host of quantitative and pictorial details at individual atomic/molecular level, that could never have been obtained from laboratory experiments. This should not prevent us from appreciating the existence of limitations, as for example those due to their basic assumptions (such as ergodicity), to approximations in intermolecular potentials (particularly at large distances (30), making it difficult the evaluation of the entropic part of Gibbs' free energy), or to restricted sampling (31-34). A few pictorial examples are shown in Fig.1. The desirable approach is then a comprehensive and complementary use of simulation and experimental data (18)(37). To our present purpose, it will suffice to recall briefly some computer simulation results, helpful to obtain a better insight in the physics of the random 3-dimensional network of hydrogen bonds in liquid water.

Uninterrupted paths of hydrogen bonded H_2O molecules were first suggested to exist, as departing from solute ions, by Monte Carlo simulation experiments by Clementi and his group. The evidence came from 3-dimensional pictures obtained by video displays of a number of specific configurations (16). The feature was noted over and over, either using different potentials, or different temperatures, or different ions, and this opened the way for a possible extension of traditional ideas on solvation water depicting H_2O molecules as evenly distributed around solute ions. In these systems, the dipole moments of water molecules tend to be oriented along the electric field set by the ion. Filamentary configurations are therefore observed (16), as a consequence of the orienting (and to some extent compressing) effect of the electric field. Note that, since the electric dipole moment is not parallel to one $O-H$ direction, hydrogen bonds will result distorted and weaker along these filaments (38), the distortion energy being compensated by that due to interaction with the electric field. The stabilizing feedback mechanism which we shall discuss shortly, may provide additional stability to these special pathways.

Molecular Dynamics studies showed that the structure of pure water should be thought of in terms of a flickering random network of irregular and distorted hydrogen bonds (19). There is no evidence for the appearance of geometrically regular patterns. Rather, uninterrupted paths (or "connectivity pathways"), closed polygons, and cages of imperfectly formed hydrogen bonds are observed. These features are therefore subjected to being analyzed in terms of topological (as distinct from geometric) properties (see Fig.1). It should be appreciated that the specific picture of connectivity eventually obtained will depend upon the prescription used for the existence of a bond (19). This notwithstanding, straightforward application of the Flory-Stockmayer-Kirkpatrick treatment of gelation-percolation (39-41) shows that for any reasonable estimate of the average number of H-bonds in water, derived either from Molecular Dynamics studies or from a variety of very different experimental observations, liquid water at biological temperatures is a percolating system of connectivity pathways of hydrogen bonds (18)(19). This conclusion, therefore, as suggested by molecular dynamics simulation, rests on broader and firmer grounds, both experimental and theoretical. Of course, as a consequence of the rapid flickering of connectivity pathways of H-bonds, the "gel" picture of liquid water is in no way conflicting with the hydrodynamic properties of the liquid. Since rapid flickering and change of shape does not affect the permanence of the percolative character of pathways, the latter is a unifying feature in the continuously changing picture of water. Within this view, H_2O molecules could be grouped according to their topological properties. Nevertheless, it is useful stressing that this grouping should not be confused with a cathegorization of molecules into species defined in terms of a "yes or not" view of H-bonding (29). Such cathegorization, and so called mixture models have in fact not been confirmed by the data, and must be discarded.

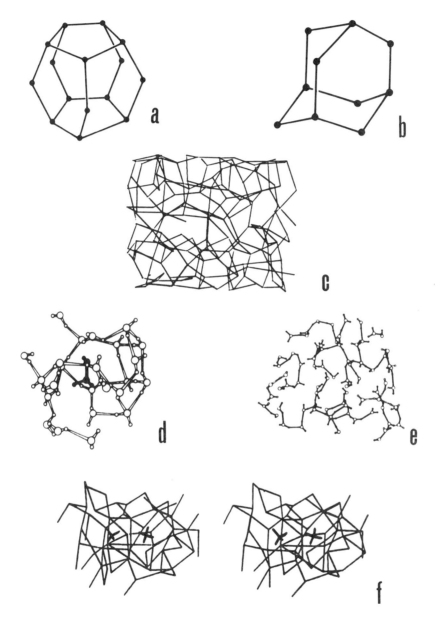

Figure 1. Examples of connectivity pathways of hydrogen bonds in water, as evidenced by computer simulations: a) and b) bulky unstrained polyhedra as individuated by oxygen atoms only, from Molecular Dynamics. External bonds to the surrounding network are not shown. (By permission, from Ref. 19) c) random network in bulk water from Monte Carlo, evidencing bonds having a statistical permanence higher than 10%. (Courtesy of Prof. S. Fornili and Dr. M. Migliore) d) cage surrounding a methanol molecule from Monte Carlo. (By permission, from Ref. 35) e) a typical Monte Carlo side view of one half ($x<0$) of the water molecules around the Agarose double helix. (By permission from Ref. 36) f) stereo view of typical cage-sharing structure corresponding to the most probable configuration in the first window of Monte Carlo umbrella sampling, for water surrounding two methane molecules. (By permission from Ref. 34)

The topological approach also allows an interpretation of the set of intriguing anomalous properties in the supercooled region (28) as schematically illustrated in Fig.2. A proposed statistical-mechanical interpretation involves here again connectivity properties (14). In this case, clustering and percolation concern 4-bonded

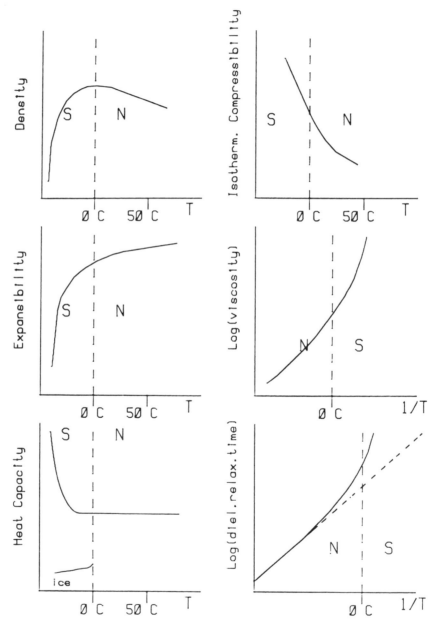

Figure 2. Schematic illustration of anomalous properties of water in the supercooled region. Vertical broken lines separate normal (N) and supercooled (S) regions. (Redrawn from Refs. 14 and 28)

water molecules. It is the latter clustering which seems to contribute to density, local entropy and local enthalpy fluctuations (14)(42) and to the anomalous behaviour of transport coefficients (14)(29)(43). This again exemplifies how topologic properties can be used for appropriately describing a physical state of water, and its cooperative onset.

In order to be useful, the topologic approach should permit viewing in a unified fashion the most remarkable properties of water, as a pure liquid and as a solvent. It is of interest to see how the percolative network of H-bonds of H_2O molecules having at least 2 hydrogen bonds in ordinary liquid water is perturbed by the presence of simple apolar solutes. It is thermodynamically well established that bringing apolar surfaces in solution implies a decrease of the entropy of the solute-solvent system (20-23). In the case of simple solutes, this was interpreted microscopically in terms of the formation of clathrates, thought as regular and more or less permanent polyhedra enclosing the solute (20). The picture obtained by computer simulation is qualitatively similar yet more diversified, visualizing the formation of irregular cages of H-bonds around apolar solutes, still flickering but having a somewhat higher stability as compared with unperturbed connectivity pathways of H-bonds in bulk water (19) (32-35) (44)(45) (see Fig.1). In this case, the topologic notion of unidimensional percolative pathways allows us visualizing a specific surface. This is the convex surface defined by the cage of H-bonds pathways surrounding the apolar surface but making no contact with it, while maintaining their percolative character throughout the solvent. This additional notion helps understanding the rich flexibility of hydrophobic interactions: thermodynamic data and statistical thermodynamic treatments (46) had suggested the existence of "contact", as well as "solvent separated" hydrophobic interaction (20)(23). Computer simulations for more than one apolar solute have visualized the possible occurence of a nice variety of solute-solvent configurations of comparable thermodynamic probabilities, having either a unique cage, or vertex-sharing cages of solvent H-bond pathways (32-35) (44)(45). These surfaces or cages will share the flickering feature of their elements, that is of connectivity pathways. The flexibility of hydrophobic interaction may be amplified by the simultaneous presence of hydrophilic and hydrofobic groups since, in a percolative model, hydrophobic and hydrophilic hydrations do not need to exhibit additive-subtractive properties (21)(30). We shall return on this point in our discussion of the effects of the addition of alcohols to the aqueous solvent.

The case of solute biomolecules is more complex, since they possess both hydrophobic and hydrophilic surfaces, in delicate balance. But of course it is here of more direct interest. For nucleic acids, thermodynamic studies of conformational equilibria in water and in non-polar solvents show that water-biomolecule interaction has a considerable role in conformational stability (22-24)(47)(48). Monte Carlo simulations of nucleic acid double helices at different degrees of hydration evidence extensive connectivity pathways of H-bonds, making in-groove, trans-groove, and interphosphate connections (49). The organization of these pathways extends outside the dimension of the B-DNA molecule (49). Features visualized by Monte

Carlo studies at different hydrations (49) are nicely found also in X-ray laboratory studies of ordered water structures in DNA oligomer crystals (50)(51), where "spines" of water molecules are observed. One such a "spine" in the minor groove of a B-DNA is formed by an alternating zig-zag array of H_2O molecules, hydrogen bonded to nitrogen and oxygen atoms on adjacent base pairs, and this apparently keeps the minor groove from opening (50)(51).

Another biomolecule studied by computer simulation in its interaction with solvent water is Agarose, a widely spread biostructural polysaccharide (36). The interest of comparing it with DNA lies in the fact that Agarose is electrically uncharged (with the exception of infrequent atypical residues interrupting the polymer chain, which at any rate were not included in the simulation study). Despite this difference, and despite the fact that interaction of B-DNA with water was found to increase dramatically by adding counter-ions to the solute (49), connectivity pathways of H-bonds were again found, as departing from/arriving at hydration sites, even if constituting a somewhat less focussed structure (36).

A remark is here in order, concerning the relation between topological structures of H-bonds in water as evidenced by X-ray experimental data on single crystals, or by computer simulations at different hydration levels, and similar structures at physiologic dilutions. We recall that a perfect crystal having neither internal motions nor crystal defects at room temperature is in a highly improbable state. The minimum of Gibbs' free energy for a crystal at equilibrium is obtained when static crystal defects (point-like defects, dislocations, etc.) and internal crystal dynamics are taken into account. The perfect and rigid crystal is but a starting point for the understanding of the actual situation of a real crystal at room temperature. Much in the same way, the situation of a biomolecule as surrounded by water in a crystal must be taken as nothing else but the starting point for the understanding of the far more complex situation of a functional biomolecule actually operating in its solvent (and capable of undergoing functional conformational transitions). In this case, the very great flexibility of H_2O molecules in building structures and connections will be expected to afford a much ampler variety of possible pathways of H-bonds connecting H-bonding sites, charged groups, etc., and of possible cages of such pathways around hydrophobic areas. This variety will appear even ampler if we consider the set of stereodynamic conformational substates of a functional biomolecule in solution (7)(8)(25).

Summing up, connectivity pathways of H-bonds as evidenced by computer simulations and X-ray data are statistically meaningful and should be taken only as a first (but important) step towards an understanding of the actual structures of pathways surrounding a functional biomolecule in solution. Also, they should be intended in a dynamic sense, as they will be expected to show a flickering among different patterns, at relatively little expense of the total energy of the system (49). If here we recall the very special property of water, of accomodating large and closely compensating enthalpy-entropy fluctuations (14)(15) this dynamic aspect will appear all the more important. The timescale will be that of structural relaxation for bulk water (10^{-11} sec. and perhaps 10^{-10}-10^{-9} around solutes (52-54)).

A stabilizing mechanism

We have just seen how statistically meaningful connectivity pathways of H-bonds are suggested to exist and how they can be sorted out as elements of order, in the thermalized state of simulated water molecules. A rapid survey of the types of molecular motions in liquid water, as known from a host of experimental data and theoretical treatments (52), is now useful for appreciating that the permanence of these elements can be reinforced by an expected collective stereodynamic behaviour of water molecules. As we shall see in the next subsection, the mechanism can be more efficient in the presence of solutes, so that connectivity pathways of H-bonds may have a marked role in solute-solvent interaction.

Molecular motions of water can be conveniently grouped in internal motions, oscillations, and slower diffusional motions. We shall neglect those of the first type (internal), as they will scarcely interfere with the pattern of H-bonds. Oscillations are distinguished in translational and orientational, and they occur on a timescale centered around 10^{-13} sec. (52). Both types of oscillations will introduce some amount of disorder along a connectivity pathway of H-bonds, but will not heavily interfere with its possible existence because no change in the average position/orientation of H_2O molecules is involved. At variance, motions implying a diffusion away of the H_2O molecules (translational or rotational) will interrupt H-bonds pathways and favour the formation of new bonds elsewhere. These motions occur on a timescale of the order of 10^{-11} sec. (52), so that this will be the expected flickering time of H-bonds connectivity pathways. The idea of the V-structure of water is here useful, and it can be thought of as described by a snapshot taken with a shutter time shorter than 10^{-11} sec. and longer than 10^{-13} sec., so as to allow vibrational averaging, while preventing diffusional averaging (52). This type of situation is illustrated in Fig.3 as a sequence of four rapid snapshots. When averaged out, these make up the V-structure (55). What we have called the flickering of connectivity pathways will therefore be the passage from one V-structure to another (equally permitted), viewed so as to sort their unifying topological feature out of a succession of such snapshots.

The intrinsically stereodynamic nature of these topological features can be easily appreciated. Let us consider a pair of H-bonded water molecules. At $0°K$ they will sit in the minimum of a potential well (in the appropriate rotational coordinate), performing zero-point librations. At non-zero temperature, thermal excitations of librational states will cause a motional broadening of the potential well, and a consequent destabilization of the H-bonding (2)(6)(27). We now consider several H_2O molecules along a connectivity pathway of hydrogen bonds. For the sake of distinguishing between geometric and dynamic effects, we consider a connectivity pathway as revealed by Monte Carlo simulations on the basis of static configuration energies (36)(49). Geometric angular correlation is evident along these pathways, and when dynamics is taken into account, this will have the double effect of reducing the thermal broadening of the potential wells and of correlating librational oscillations along pathways. In turn, this will further reduce librational amplitudes. A stereo-dynamic feedback will thus operate (1)(2)(6)(13), stabilizing the hydrogen

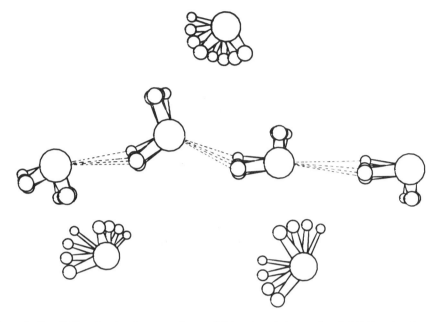

Figure 3. Four H_2O molecules along a pathway of H-bonds and three unbonded H_2O molecules, schematically shown as four superposed configurations, each seen as a snapshot taken at a shutter time shorter than 10^{-13} sec. The V-structure is obtained by averaging out the four configurations (Ref.52). Notice the angular correlations and the reduced angular amplitudes along the pathways of H-bonds. Flickering from one such structure to another (topologically equivalent) occurs when one or more molecules diffuse away, causing a rearrangement of pathways. (Courtesy of Dr. M. Migliore)

bonding in a cooperative way along the pathway, as visualized in Fig.3. We note that the cooperative character of H-bonds has been suggested since a long time (56) and the electronic contribution to cooperativity has been calculated (57) and taken into account as non-pair additivity of potentials (58). This adds to the dynamic contribution to cooperativity that we have just discussed.

The occurence of a stereodynamic feedback capable of reinforcing cooperatively the H-bonding was proposed long ago (6)(59), perhaps too long ago, when unavailability of the notion of percolative paths of H-bonds made it difficult to apply. It is now implicitly contained in Molecular Dynamics simulations. (The expected reinforcement, and the equally expected sharpening of the V-structure has in fact been found in one simulation of hydrophobic hydration (60)). Having elicited the possible mechanism of geometric-dynamic feedback along connectivity pathways of H-bonds, however, will help us to understand the thermodynamic significance of these pathways on timescales well beyond the possibilities offered presently by Molecular Dynamics. Any cause tending to reduce the H_2O libration amplitudes, will be expected to have amplified effects if the mechanism operates, and this will be true for all water molecules (if possibly more so along connectivity pathways). This expectation agrees with the increase of librational frequencies observed when temperature is decreased (52)(61). In a statistical random network model of water (27), the dependence of librational and vibrational frequencies on angular disorder was also taken into account. This study allowed a self-consistent fitting of most spectroscopic and thermodynamic data, except the anomalous behaviour in the supercooled region illustrated in Fig.2. Librational correlations as a source of cooperative effects acting preferentially along connectivity pathways of H-bonds were not taken into account in this noteworthy model. It can be expected, however, that the cooperative stereodynamic mechanism discussed above should have a particularly significant role in the percolation of 4-bonded molecules, assumed to characterize the behaviour of supercooled water (14)(28). This may be one of the reasons why the statistical random network model (27) does not allow an interpretation of the anomalous behaviour in the supercooled region . Also, we shall see that the expected role of the cooperative stereodynamic feedback along H-bond pathways is confirmed by the behaviour of D_2O and of water-ethanol mixtures at sub-zero temperatures. A similarly significant role is expected when librational amplitudes are reduced at specific solute sites as we will further discuss in the following.

The same type of stereo-dynamic mechanism of stabilization is known to operate in simpler and well defined geometric configurations which, for this very reasons are affected in a much more dramatic and often hysteretical way. This is the case, e.g., of crystals where more than one equivalent position is available to a subsystem of interacting protons, allowing them to jump from one to the other at ordinary temperatures. The case of H-bonded ferroelectrics is best known (62) but it must be understood that the onset of a ferroelectric state is neither necessary nor sufficient for the mechanism to take place (59). We may consider the case of hexammine nickel halide crystals, where the interacting NH_3 groups undergo quasi-free rotations at room temperature, but they "freeze" sharply at a critical temperature. The freezing occurs cooperatively and reversibly, and no onset of electric polarization is involved. A series of experimental and theoretical studies (59)(63) has shown it to be due to a stereo-dynamic stabilizing feedback, similar to that just discussed for water. Use of deuterium instead of hydrogen enhances the feedback mechanism and stabilizes the frozen state. Conversely, disturbances induced progressively in various ways in the propagation of the feedback throughout the crystal, destroy cooperativity in a similarly progressive way (59). Recalling these effects of disturbances will be of help when we shall discuss perturbations of the aqueous solvent by deuterium or alcohols.

A simplified description of two interacting subsystems in a mean field approximation (64) is here useful. We shall refer to Ref. 64 and to Fig. 4. The macroscopic state of the overall system is described by the two order parameters, S_1, and S_2. Each isolated sub-system obeys an equation of state:

$$S_i = F_i (V_i) \qquad (1)$$

Here V_i is the driving force acting on the i-th subsystem, as resulting from the

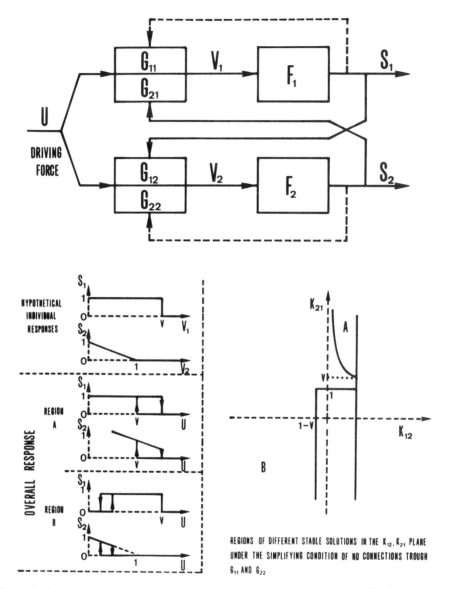

Figure 4. Upper: a system composed by two subsystems coupled by crossed feedback. Broken lines represent additional feedbacks internal to each subsystem. Lower left: an example of simple individual responses of the two subsystems, and two of the corresponding overall responses of the entire system, as per Ref. 64, neglecting internal feedbacks. Notice entrainment effects. Lower right: Regions in the (K_{12}, K_{21}) plane corresponding to different types of solutions in a mean field, linear approximation and in the absence of internal feedbacks. (Redrawn from Ref. 64 by courtesy of the Authors)

combined action of the external driving force U, and of the reacting fields:

$$V_i = U + G_{ii}(S_i) + G_{ji}(S_j) \tag{2}$$

where the G'$_s$ are the appropriate coupling functions. In the simplifying assumption of G$_{ii}$ (S$_i$) = 0, that is in the absence of feedbacks within each individual subsystem, and in a linear approximation, we have:

$$V_j = U + K_{ji} S_j \qquad (3)$$

Already in this oversimplified situation, a variety of possible behaviours exists (64). In Fig. 4, several regions in the (K$_{12}$, K$_{21}$) plane are identified, each corresponding to a different overall behaviour. In the figure two examples of possible mutual entrainments of the two subsystems are given (relative to regions A and B of the (K$_{12}$, K$_{21}$) plane), under the hypothesis that of the two individual, isolate subsystems, one shows a sharp transition and the other shows a monotonic response. This shows how even modest couplings can originate qualitatively and quantitatively relevant effects, including sizeable mutual entrainments (64). The variety of possible behaviours naturally increases when feedbacks internal to each subsystem are included. These are schematically indicated by broken lines in the upper part of Fig. 4.

The subsystem undergoing a sharp transition could be, in this example, a solute biomolecule going across a functional, conformational transition. The other subsystem could stand for solvent water, its monotonic response being justified by the great variety of allowed geometric and dynamic configurations of H$_2$O molecules. Internal feedbacks (broken lines in Fig. 4), indicate in this case the (stereodynamic) coupling between geometrical configurations and accessible dynamic states of biomolecule and water respectively. Nevertheless, we have focussed here on water only, that is on the possible occurrence of a geometric-dynamic coupling along connectivity pathways of H-bonds. (For a similar coupling within a biomolecule, one may refer to Ref. 9.) Before discussing the possible thermodynamic relevance of these pathways to biomolecular stability and function, we notice how consideration of the entire set of coupled subsystems (that is a biomolecule with its internal stereodynamic coupling, and the solvent with its own internal stereodynamic coupling), is not only a philosophical necessity, but it can add qualitatively new contributions to our understanding of biomolecular functional sability.

Solute-solvent interaction

So far, we have described connectivity pathways of H-bonds as statistically relevant topological elements providing in water at biological temperatures the percolative properties which are expected on the basis of any reasonable estimate of its extent of H-bonding (18)(19). As we have seen, these elements should be considered dynamically in two respects. The first concerns their flickering, and the second concerns the suggested stereodynamic feedback mechanism. Operation, or reinforcement of this mechanism along a path, will increase its statistical relevance. Reinforcement is expected whenever the libration amplitude is limited either generally along the path, or locally at certain locations. By way of example, we may think of electric fields around ions; or of hydrogen-bonding sites (e.g. at biomolecular surfaces) possibly originating a pathway (1)(49); or of the sterical hindrance in the vicinity of apolar surfaces; or we may think of substituting Hydrogen by Deuterium, so as to deal with smaller librational amplitudes, as we shall discuss later. In all these circumstances, stabilizing effects have been observed (1)(2)(13)(29)(32-36)(49). Interestingly, computer simulation shows the expected enhancement of water-water molecular interactions within the hydrophobic solvation shell, and the more so in the region which includes solvent taking part in the solvent-separated hydrophobic interaction (60). Conversely, the presence of solutes acting in competing or synergetic ways, may disturb connectivity paths.

To discuss the thermodynamic implications of the foregoing considerations, we consider two solute molecules in water, and the process of bringing them from a mutual distance R_o to r, in the presence of N solvent water molecules, and under pair potential approximation. The specificity of this case will not affect the applicability of our central conclusions to more complex cases, e.g. that of a functional switch between two configurations of a biomolecule (21). Let R_1 and R_2 be the positions of the two solutes, $R = R_2 - R_1$ their relative distance, and X_i the position of the i-th water molecule (including orientational coordinates). The potential energy of the system is:

$$U = U_{ss}(R_1,R_2) + \sum_i[u_{sw}(R_1,X_i) + u_{sw}(R_2,X_i)] + \sum_{i,j} u_{ww}(X_i,X_j) =$$
$$= U_{ss}(R) + U_{sw}(R) + U_w \tag{4}$$

where the $U_{sw}(R)$ and U_w indicate sums over all pairs of molecules. The three terms represent the direct solute-solute, the solute-water, and the water-water contributions, respectively.

The Gibbs' free energy change in the process and in the presence of N molecules of solvent will be:

$$\Delta G = -kT \, \Delta\ln \int[dV \, dX^N \exp(-(U+PV)/kT)]$$

$$\Delta G = \Delta U_{ss} + \int_{R_o}^{r} dR \frac{\int dX^N (\partial U_{sw}/\partial R) \exp[-(U_{sw}+U_w)/kT]}{\int dX^N \exp[-(U_{sw}+U_w)/kT]} \tag{5}$$

$$\Delta G = \Delta U_{ss} - kT \Delta\ln \langle \exp(-U_{sw}/kT) \rangle \tag{6}$$

where the weighted averaging implied in the integration is intended as carried over all possible configurations of the solvent (21). Each microscopic configuration of the solvent, therefore, brings into the averaging process a contribution which is measured by its statistical weight.

The expression above can be rewritten as the sum of a direct, solvent independent, solute-solute term, and of an indirect, water-mediated term:

$$\Delta G = \Delta U_{ss} + \Delta G_{ws} = \Delta U_{ss} + \int_{R_o}^{r} \langle \frac{\partial U_{sw}}{\partial R} \rangle \, dR \tag{7}$$

Accordingly, the generalized solute-solute driving force and the solute-solute correlation function can be written (21):

$$f_{ss} = -\text{grad } G = f_{ss}(\text{dir}) + f_{ss}(w) \tag{8}$$

$$g_{ss} = \exp(-G/kT) = g_{ss}(\text{dir}) \, g_{ss}(w) \tag{9}$$

It follows that the solvent-mediated term of solute-solute driving forces (2) and correlation functions are determined by the configurations of the solvent wich are statistically most relevant. In our model, these are the H-bond connectivity pathways, and it follows that they play the major role in solvent-mediated solute-solute interactions, and solute-solute driving forces. As the effects of hydrophilic sites and hydrophobic areas (imposing topological constraints on the pattern of H-bonds) are not expected to be additive (21)(30) and mutual reinforcements are expected to occur (65), connectivity pathways of H-bonds viewed within the foregoing plain thermodynamic framework (2)(21) allow for pattern-sensitive driving forces encompassing several water-water distances and depending upon mutual presentation and distances of solutes.

The entropy and enthalpy contributions in eqs. 5 and 6 can be derived by standard thermodynamics, and they contain large compensating terms (15)(21) (66). In actual experiments, however, linear fittings of van't Hoff plots are often used to evaluate entropy and enthalpy changes by identifying them with intercept and slope, respectively, of the fitting straight line. Conceptually, this amounts to expanding ΔG as per eq. 7, in a power series of temperature and retaining the constant and linear terms only. Considered that the basis for this is unwarranted, and also considered the existence of entropy-enthalpy compensating terms, quantities so evaluated not necessarily coincide with those obtained calorimetrically, and with those derived theoretically, although coincidences can interestingly be found (67) (68). We may conclude that the foregoing thermodynamic derivations should not be expected to provide grounds for quantitative comparisons; yet, they provide a necessary frame for qualitative evaluations, comparisons, and predictions.

Perturbing the solvent

We shall now discuss what can be expected as a consequence of some specific perturbations of the network of flickering H-bond connectivity pathways, and compare that with available results. Since a vast literature exists on the effects of solvent perturbation, any attempt of making an exhaustive review would go too far beyond the scope of the present work. Our choice is therefore bound to be arbitrary, and limited to cases thought to be significant within the present context.

Effects of Deuteration

If the stereodynamic feedback discussed in the foregoing section is operative and if connectivity pathways of H-bonds play a major role in solvent-mediated solute-solute interaction, a sizeable alteration of a dynamic parameter such as by substitution of hydrogen by deuterium, is expected to cause measurable effects, which we are now going to discuss. We have discussed elsewhere this type of effects (2) so that we shall here refer to a few cases only. We first recall that zero point and higher energy levels are lower for a deuteron with respect to a proton in the same potential well. This standard quantum mechanical fact has two simultaneous consequences: it increases the thermal population of the first excited level, and it makes deuterium bonding stronger and more localized (69). In water, librational frequencies will be reduced as the inverse square root of the moments of inertia, and motional amplitudes will also be reduced. This will cause a decrease of orientational disorder. If a stereodynamic feedback exists along connectivity pathways of hydrogen bonds, deeper potential wells, stronger H-bonding, and reinforcement of connectivity pathways are expected in D_2O. The first effect shows up in at least one librational mode (61)(70). For this mode, the frequency in D_2O is about 5% higher than the expected value, which indicates an increase of the force constant by about 10%. The second effect is also observed: for a self-consistent fitting of values of thermodynamic quantities relative to H_2O and D_2O, a stronger hydrogen bonding, in addition to contributions due to zero point differences, must be assumed (71). As to a reinforcement of connectivity pathways in D_2O, no direct evidence is available. However, we notice that C_p is 10% larger in D_2O than in H_2O, while a 3% difference only can be accounted for by the vibrational frequency change (72). The rest largely represents an increased configurational energy (72), which at least in part may reflect the operation of the stereodynamic feedback along connectivity pathways of hydrogen bonds. If so, the resulting angular amplitudes (such as in Fig. 3) will be reduced, the more so along the hydrogen-bonds pathways.

A richer phenomenology, also indicative of a reinforcement of a stereodynamic feedback mechanism, is observed when the supercooled region is approached (28), where effects of cooperativity are evidenced. In D_2O the temperature of the anomaly is about 5°C higher, but this is not the only effect of isotopic substitution. Viscosity, self diffusion coefficients, dielectric and nuclear relaxation times, all appear to follow the same exponential law, with a logarithmic slope higher than in H_2O (28). This evidences a role of librational dynamics of H_2O molecules in the cooperative onset of the anomalous behaviour. Non-pair additivity of potentials (58) can of course add to cooperativity, yet not with the observed mass dependence. Further supporting evidence comes from a comparative spectroscopic study of H_2O and D_2O, extended to the supercooled region. The study shows that in the sub-zero region, the anomalous behaviour is not observed in the number of H-bonds of given geometry, but shows up when molecular species identifiable with connectivity properties are sorted out (29).

We now return to solute-solvent interaction. As a consequence of the stronger hydrogen bonding in D_2O, the U_w term in eq. 1 will be larger. From eq. 4 we see that no enthalpic change will be expected from H—D substitution in the solvent (and therefore, at exchangeable H sites in the solute), unless the process implies a rather unlikely net change of the number of hydrogen/deuterium bonds (solute-solute vs. solute-solvent). This may no longer be true in presence of electrostatic contributions large enough to make significant the 0.5% change in dielectric constant (47)(73). We see from eq. 2 that if the U_w term is proportionally altered by deuteration for all solvent configurations, considered that $U_w \gg U_{sw}$, this effect will be equivalent to a temperature shift, that is to an entropic effect (2). The change of H-bond strength is expected to correspond to a change in U_w of not much more than 1% (52), so that the expected temperature shift will be of about 3°C. Many observations are reported of deuterium isotopic effects of this type and size (74). They agree (2) with the (least dramatic) expectations based on the model discussed above. For nucleic acids, it was possible to clarify early uncertainties on the occurrence of an effect, by a thorough self-consistent interpretation of a series of different conformational transitions of the same systems (2)(47), and by stretching accuracies to much higher limits (73). The same kind of evidence is available for a biostructural polysaccharide (75). Similarity of results, irrespective of the presence of charged groups provides additional evidence against the operation of a trivial effect of the slightly altered dielectric constant (2)(47)(73). For proteins, results are somewhat more controversial or less conclusive (74). This may be due to the fact that functional effects have been often studied. As it has been clarified only recently, D_2O may have contained heavy metal traces, and as it is well known, these may have very strong functional consequences on proteins, thus possibly causing artifacts.

If connectivity pathways have the proposed role, and if the proposed stabilization mechanism is operative, we expect more complex effects of D_2O leading to stronger stability, order and cooperativity, in addition to effects which can be compensated for by a mere temperature shift. This is e.g. what is shown by experiments with sickle cell Hemoglobin (2). This is a mutant species of Hemoglobin in which as discovered by L. Pauling, the balance between hydrophilic-hydrophobic groups of normal (Adult) Hemoglobin is tipped by a genetic mutation towards the hydrophobic side. As a result, a termoreversible gelation occurs when temperature is raised above a critical value. The thermodynamics of this transition (67)(68) shows a very interesting behaviour of entropy and enthalpy, including a measured equivalence of van't Hoff and calorimetric enthalpies (67)(68). Deuterium isotope effect in Minimum Gelling Concentration (mgc) causes (within the experimental error) a translation of the mgc curve towards remarkably lower (25%) concentrations (2). This does not affect the enthalpy term of the gelation transition, as expected. The entropic effect, however, is far from being accounted for by a proportional change in temperature, and causes a stabilization of the more ordered phase.

Similar stabilization effects are observed on the Critical Micelle Concentration (cmc) of aqueous n-Dodecyltrimethylammonium Bromide (76). Here again the cmc value decreases remarkably (15%) and the effect appears to be essentially entropic, and yet impossible to compensate for by a change of temperature. Another case of phase separation shows a differently noticeable effect. In binary isobutyric acid/ water mixtures, solvent deuteration lowers the critical concentration by 6%, and also raises the critical phase separation temperature by a remarkable 19°C (77). Effects of this type and sizes are hardly explained in terms of a uniformly distributed increase of hydrogen bonding, reflected only in a proportional change of the U_w value in eq. 5, for all configurations.

In addition to these effects at equilibrium, solvent deuteration has been observed (75)(78)(79) to have interesting qualitative and quantitative consequences in the behaviour of a thermoreversibly gelating Agarose-water system. This biostructural, uncharged, unbranched polysaccharide, exhibits very strong gelation properties when dissolved in water. Gelation is caused by molecular ordering in double helices, followed by partial supramolecular ordering, both occuring cooperatively, as evidenced by a large thermal hysteresis (80-82). Difficulties in the formulation of a universal description of gelation phase transitions (83) may reflect their sensitivity to molecular details (79). If this is a correct inference, molecular recognition might (although, it does not need to) be the initial step of the molecular/supramolecular ordering process. In this respect, the system might offer the additional interest of a model system. Recent studies of these systems have shown remarkable dynamic instabilities of self-assembly and transport properties (79), while previous work and observations up to thousands of hours had shown that the nature of the transitions (78)(84), the time required for it to occur, and the structure of gels so obtained depend strongly upon temperature and polymer concentration (78)(79). In Agarose-water systems, solvent deuteration causes effects which again cannot be treated on the simple basis of a proportional temperature change, as predicted in the case of a uniform increase of U_w for all configurations of water molecules. Deuteration accelerates transition kinetics, and so does a temperature decrease (85). However, the structural properties of the order obtained in D_2O are qualitatively different from those of the order which is obtained in H_2O at any temperature (79). This is illustrated in Fig. 5. Further, the effect of deuteration on critical instabilities at the sol-gel transition again shows complex features, on top of a general temperature shift. Instabilities in self-assembly properties are ampler, and similar instabilities in microviscosity are restricted within a smaller temperature range. These effects seem therefore to provide supporting evidence in favour of a coupling between the stereodynamics of this biopolymer (involved in the ordering process, and in the observed instabilities at the transition), and the stereodynamics of solvent, whose role would be reinforced by a cooperativity having a dynamic origin. Apart from the effects of isotopic substitution, instabilities in microviscosity possibly indicating a role of solvent dynamics at large, that is hydrodynamics (53)(54)(88).

Effects of Alcohols

As we have seen, effects predictable in the hypothesis of a cooperativity having a dynamic origin are in fact found upon deuteration of the aqueous solvent (and of exchangeable hydrogens of the solutes). Along pathways of H-bonds, if existing, these effects are expected to appear enhanced, and the experimental results are

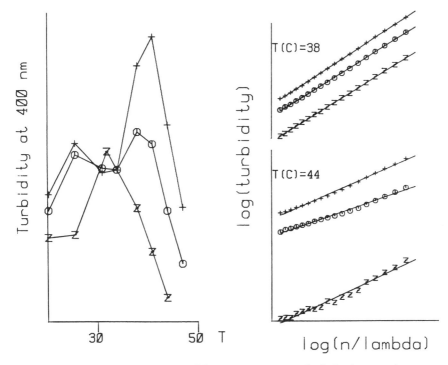

Figure 5. Effects of solvent perturbations on 0.5% Agarose aqueous gels obtained at several temperatures. Note that the structure of the gel is characterized (Ref. 79) by the total scattered light at a given wavelength and by the log-log dependence of the total scattered light upon wavelength. Left: turbidity at fixed wavelength vs. temperature of isothermal gelation. Right: log-log wavelength dependence of total scattered light in gels obtained at two temperatures. Symbols: o o o o H_2O; + + + + D_2O; z z z z ethanol at 3% molar concentration in H_2O (Redrawn from Ref. 79)

consistent with this expectation. However, in lack of direct evidence it is desirable to look for additional supporting data. These may come from the large amount of studies in current literature, concerning perturbation of the aqueous solvent with small amphiphilic molecules. The rationale for this lies in the expectation that even a small number of these molecules should perturb the "natural" arrays of connectivity pathways, as a consequence of the interplay of hydrophilic/hydrophobic sites (2).

Convenient perturbations can be obtained using alcohols in sufficiently low concentrations (a few units percent in molar fraction), so as to obtain an effective, yet non-destructive perturbation of H-bonds connectivity pathways. Thermodynamic properties of alcohol-water mixtures have been extensively studied (20)(21)(65)(89). Here we wish to mention that in the range of concentrations which concern us, all monohydric alcohols exhibit large and negative enthalpies and excess entropies of mixing (typical of hydrophobic interactions), having minima at molar fraction concentrations lying well below the normal 50% as illustrated in Fig. 6. This, together with activity and virial coefficients, and excess partial molar volumes is indicative

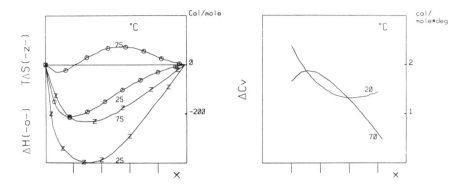

Figure 6. Selected thermodynamic data relative to ethanol-water systems at several molar fraction of ethanol, showing how the most interesting effects occur in the range (0-10%) molar fraction. Left: o o o o Enthalpy of Mixing; z z z z Excess Entropy of Mixing. Right: Excess Heat Capacity. Note that by definition $\Delta C_v = 0$ at both extrema, so that at least two maxima should occur in the 20°C case. (Redrawn from Ref. 20)

of an associative tendency of alcohols, increasing with the size of alkyl groups (20)(21)(23)(30).

Indications as to the type of perturbation induced on water structure by alcohols come from the existence of this associative tendency, as further precised by heat capacity, volume and expansibility data of alcohol-water binary systems (65). In particular, these data suggest a reinforcement of hydrophobic hydration due to strong hydrogen-bonding interactions of polar groups with water (65), just as it could have been predicted on the grounds of the percolative model of H-bonds. A Monte Carlo study of methanol (35) seems to confirm this view. We shall call "amphiphilic hydration" the result of the mutual hydrophobic-hydrophilic reinforcement not limitedly to alcohols, and both in the intra- and intermolecular case. Altogether, the interplay of the hydrophobic and hydrophilic sites (30) of a solute species will be expected to perturb in various (and perhaps to some purposes more efficient) ways the structure of H-bond pathways, and in some instances its very percolative properties.

Breakage of percolative properties is suggested by studies of the behaviour of sound velocity (90) and density fluctuations (91) in supercooled water, as disturbed by the addition of small quantities of ethanol. In particular, the structure factor S(q) as measured by the angular dependence of the scattered light, shows for pure water in the supercooled region a minimum and the pronounced rise at vanishing q-values due to an extended correlation length, typical of critical behaviour. Upon addition of small quantities of ethanol, a progressive reduction of the correlation length is observed and the anomalous behaviour disappears for a 5.8% molar fraction, thus indicating that at this concentration (corresponding to an average ethanol-ethanol separation of five H_2O molecules) and at the temperature of the experiment (253°K), long range correlation is no longer observed (91). Accordingly, a normal behaviour of the velocity of sound is restored at about the same ethanol concentration (90), as

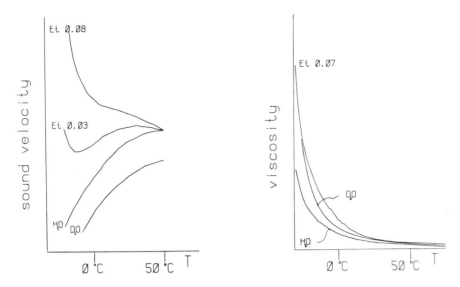

Figure 7. Perturbation by ethanol and by isotopic H-D substitution of anomalous transport properties of water in the supercooled region. Left: between 0.03 and 0.08 ethanol the anomalous behaviour (and so also percolation of four H-bond pathways) is seen to disappear, while D_2O enhances it. Right: despite the disappearance of the anomaly (confirmed by measurements of the Structure factor in light scattering as discussed in the text), viscosity values show a remarkable increase at 0.07% ethanol. This is expected if significant (and flickering) cages of H-bond pathways surround ethanol molecules by amphiphilic hydration. The effect of isotopic substituion is also shown, to illustrate that the two disturbances do not need to cause effects of opposite signs. (Data from Refs. 28, 90, 92).

shown in Fig. 7. It would not be correct, nevertheless, to infer that in these conditions topologic connectivity properties have all disappeared. We recall that if the model of percolation along pathways of 4-H-bonded molecules (14) is correct, disappearance of the anomaly evidences the disappearance of this type of percolation only. On the contrary, open and closed pathways connecting not necessarily 4-bonded water molecules may well exist, even if strongly perturbed, and certain geometries are even expected to be established/stabilized by the operation of amphiphilic hydration (2)(13)(35)(65) and of solvent-separated interaction (20)(23) (34)(46) among amphiphilic Et−OH molecules. Supporting evidence for the occurrence of longer-living connectivity pathways arranged in cages, if not percolating, comes from the abnormally high viscosity of diluted water-ethanol mixtures at temperatures between 0 and −30°C (92) shown in Fig. 7. This again is in favour of the operation of a stabilizing amphiphilic interaction of Et−OH molecules, as suggested by thermodynamic data (65)(89). These data show how not necessarily deuterium and Et−OH perturbations should be expected to cause effects of opposite signs.

From the foregoing discussion we infer that interesting effects can be expected on biomolecules by adding to the aqueous solvent amphiphilic perturbants (alcohols in the simplest case), in proportions as low as a few percents in molar fraction. A host of data is available in the literature in this field. To our purpose, we shall best refer to an extended series of studies by Cordone and collaborators (93-99) on the effects of monohydric alcohols. These studies are of particular interest to the present discussion, for two reasons. First, they offer a thorough and systematic set of data on amphiphilic perturbations of the aqueous solvent, obtained by maintaining unchanged the hydrophilic part and systematically varying the hydrophobic moiety of perturbants. Second, the authors have studied mostly Hemoglobin (adult A), and this protein constitutes a paradigmatic model system of regulatory enzyme (100)(101). So, we shall focus our attention on their results and conclusions on the effects of amphiphilic perturbants on the transition between the two functional states T and R of Hemoglobin, which governs the oxygen uptake of the latter (100)(101).

Specifically, the authors measure the change in the difference of free energy between the T and R states. They make it clear that the quantity they measure is not altered by alcohols binding to Hemoglobin as allosteric effectors and that rather, the allosteric constant is altered through general solvent effects. Two contributions are distinguished. One is linearly related to the variation of the inverse bulk dielectric constant of the solvent, as illustrated in Fig. 8 and is considered to be due to the altered interaction energy of charge densities at protein surface. The remaining and most interesting contribution, also shown in Fig. 8 is attributed to a smaller free energy needed to expose to the perturbed solvent hydrophobic surfaces of the protein, as a consequence of the T−R transition. These effects are observed to increase systematically and more than linearly, both with alcohol concentration and with the size of alkyl groups. A thermodynamic analysis also shows that the change in the free energy difference between T and R states is the result of much larger (at least one order of magnitude) and quasi-compensating ΔH and $T\Delta S$ terms. This effect results from the very unique properties of water, discussed in the indroduction (15). The overall line of interpretation in terms of altered hydrophobic interactions is in quantitative agreement with the estimated free energy of transfer of the hydrophobic groups involved, from pure water to the perturbed solvent. Further support comes from a study of denaturation kinetics of Met-Hemoglobin, as affected by the same solvent perturbations (98). Again the two contributions were present and they were much larger, as expected. Hydrophobic contributions in this case, and those relative to the T−R transition, were found to be linked by one and the same correlation, irrespective of the specific alcohol (98).

For the entire series of monohydric alcohols studied, these interesting effects occur in the range 0.4-6% molar fraction (93-99), corresponding to inter-alcohol separations from 5 to 12 water molecules. As we have already noted, the anomalous behaviour of water at sub-zero temperatures is observed to be altered and to disappear in the same range of ethanol concentrations (90)(91). We infer that the same qualitative and quantitative perturbation is responsible for the two effects. We have seen that the anomalous behaviour of supercooled water is well interpreted by a percolative model (14), and that its disappearance in the presence of small proportions of ethanol is attributable to amphiphilic hydration and interaction of the perturbant molecules. It follows that in this range of concentrations, patterns of

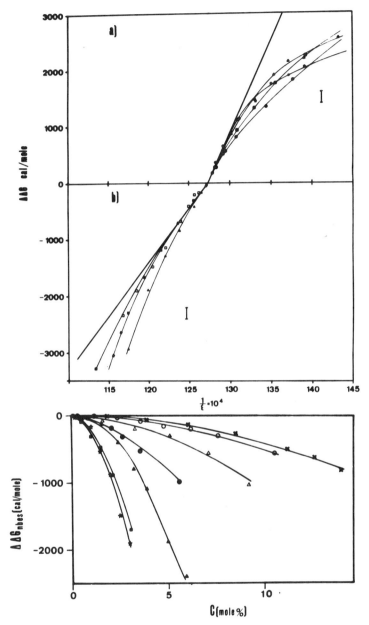

Figure 8. Effects of alcohols and of amides on human Hemoglobin (Adult A), as measured by the alteration $\Delta\Delta G$ of Gibbs' free energy difference ΔG between T and R states. Upper: plotting $\Delta\Delta G$ vs. the reciprocal dielectric constant of the perturbed solvent allows identification of bulk electrostatic (bes) contributions, shown by broken straight lines. These have, as expected, opposite signs for alcohols and amides. Note that the slight slope difference of the two broken straight lines disappears when pH corrections are accounted for (L. Cordone, private comm.). Lower: non-bulk electrostatic (nbes) contributions to $\Delta\Delta G$, as obtained upon subtraction of bes contributions from the upper plot. Notice departure from linearity. Symbols: ● methanol; ▲ ethanol; ■ *iso*-propanol; ★ *n*-propanol; □ formamide; ✕ acetamide; △ N-methylformamide. (By permission, from Ref. 98)

H-bond connectivity pathways are essentially dictated by alcohol and interalcohol amphiphilic hydration. This remains probably true at the temperatures used for experiments with Hemoglobin (93-99). If so, exposure of apolar groups of the protein, consequent to the $T-R$ transition, is expected to occur at a lesser free energy cost: in the perturbed solvent room for exposing the apolar groups would be obtained only in part by creation of cages of H-bonds around them, and in part by a plain dislocation of the perturbed solvent and of its pattern of amphiphilic hydration as a whole. This is the microscopic view complementing the thermodynamic description (93-99) in terms of a weaker hydrophobic interaction.

An interpretation of this type is bound to be qualitative, as long as no hints will be available (e.g. from computer experiments) to the microscopic taxonomy of multiple amphiphilic hydration. If, out of this taxonomy, we consider in a first coarse approximation pairs or multiple configurations, we may already expect the observed departure of the discussed effect (93-99) from a linear dependence upon concentration of alcohols and size of alkyl groups, as evidenced in Fig. 8.

Conclusions

We have reviewed how the extraordinary variety of possible, geometrically disordered and seemingly uncorrelated structures of bulk water, flickering within 10^{-11} sec. can be viewed as orderly at a higher, topological rather than geometric level (18)(19).

The foregoing discussion concerned the possible relevance of H-bond connectivity pathways in water to the stability and functionality of solute biomolecules. These topologic structures are, as we have seen, visualized by computer experiments and X-ray data on single crystals. We have conceptually sorted them out, and discussed their role in terms of the thermodynamics of solvent-mediated solute-solute interaction. This has provided a model, useful in pointing out explicitly that: i) pathways of H-bonds appear to have a relevant statistical role in solvent-mediated solute-solute interactions and thus also in solute conformational stability; ii) the flickering character of H-bond pathways in solutions (as contrasted by their stability in crystals) might be thought to weaken their role, but as a consequence of the timescales involved, the inverse can be expected. This is so because the lifetime of each pathway is 10^{-11}-10^{-9} sec. (depending on how much it is influenced by the solute (53)(54)), in any case long enough to give it a relevant statistical weight (see eq.9), while the very flickering implies a much larger participation of microscopic configurations of the solvent to biomolecular stability; iii) the stabilizing stereo-dynamic feedback surmised to occur along connectivity pathways of H-bonds is also, if in fact occurring, implicit in computer simulations. However, having explicited it, helps further (through the mechanism of crossed coupling discussed with reference to Fig.4) in uniting and cross-coupling geometric and dynamic aspects of biomolecules and of solvent; and iv) the same feedback mechanism helps visualizing how a polar site at a solute surface is expected to be the preferential origin of a connectivity pathway, and visualizing hydrophobic-hydrophilic correlations giving rise to what we have called "amphiphilic hydration". It should be understood that by preferential origin or amphiphilic reinforcement of connectivity pathways we refer to the statistical properties of pathways and not to a preferential and permanent involvement of specific solvent molecules. These are in fact known to have lifetimes not longer than 10^{-9} sec. at any specific biomolecular site. Within this frame, the influence of solute biomolecules on water structure is viewed at a microscopic level with no need of cathegorizing "bound" and "unbound" water, so that "water remains water". This influence, however, can all the same extend for a number of steps and, through the mechanism of amphiphilic hydration, be e.g. "focussed" and concur in pattern recognition. This illustrates a possible heuristic value of the model.

The model also allows the prediction/understanding of widely different phenomena, such as the effects of isotopic and amphiphilic perturbations on the properties of supercooled water, on biomolecular stability, and on biomolecular functions. Also, it suggests the possibility of occurrence of subtle effects on viscosity at a microscopic level (102). It is supported as we have seen, by the similarity of responses exhibited by systems as different as a class of crystalline solids, biomolecules in solution, and supercooled water, to reinforcement by isotopic substitution and disturbances in motional correlations induced progressively in various ways. A deeper and exhaustive work of comparison with available data remains to be done. In particular, we have not attempted to include a comparison with data available from the wide field of Proton Transfer experiments (103). Interestingly, and despite its oversimplifications, the model is capable of generating a rich (and again complex) picture. In it, the exceptionally large set of microscopic stereodynamic configurations which are accessible to water at $\Delta G \simeq 0$, and characterize its unique properties as a solvent, acquires a specific relevance to stability and function of solute biomolecules. Specifically, the thermally excited, chaotic flickering among these configurations is seen to involve a vast number of solute-solvent stereodynamic microstates and their (statistically relevant) H-bond pathways, in one and the same biomolecular functional state, and in functions such as conformational transitions and pattern recognition. This extension of our understanding of biomolecular stability to a much ampler volume of the phase space may explain at least in part why the operation of such a microscopic object as a biomolecule, on the borderline of quantum physics, can be described mechanistically, as if it were a macroscopic machine. If confirmed, the model would illustrate the great efficiency and flexibility of a loose order (chaotically flickering within a strict topologic context) with respect to life processes. Even a provisional conclusion, however, must wait for further work.

Acknowledgments

We thank Professors L. Cordone, S. Fornili, F. Madonia and Drs. P.L. San Biagio and M. Migliore for several discussions and access to unpublished data. We also thank Mr. A. La Gattuta, Mr. M. Lapis and Mr. S. Pappalardo for technical help and Ms. M. Giannola, Ms. M. Genova-Baiamonte and Ms. V. Paladino for producing the processed typescript. The present work was supported by Consiglio Nazionale delle Ricerche (Institute for Interdisciplinary Applications of Physics), by CRRNSM (Palermo) and by M.P.I. local fundings.

References and Footnotes

1. M.U. Palma, in *Structure and Dynamics: Nucleic Acids and Proteins,* Eds. E. Clementi and R.M. Sarma, Adenine Press, New York, p.125 (1983).
2. S.L. Fornili, M. Leone, F. Madonia, M. Migliore, M.B. Palma-Vittorelli, M.U. Palma and P.L. San Biagio, *J. Biomol. Str. Dyn. 1,* p. 473 (1983).
3. K.U. Linderstrom-Lang, and J.A. Schellman, in *The Enzymes,* Eds. D. Boyler, H. Lardy and K. Myrbach, vol. 1, Academic Press, New York, p. 12 (1959).
4. H. Frohlich, in *Theoretical Physics and Biology,* Ed. M. Marois, North-Holland Publishing Co., Amsterdam (1969). H. Frohlich and K. Myrbach, *Adv. Electronics and Electr. Phys. 53,* 85 (1982).
5. A.S. Davydov, *Biology and Quantum Mechanics,* Pergamon Press, Oxford (1982).
6. G. Aiello, M.S. Giammarinaro, M.B. Palma-Vittorelli and M.U. Palma, in *Cooperative Phenomena,* Eds. H. Haken and M. Wagner, Springer-Verlag, Berlin p. 395 (1973). M.U. Palma, in *From Theoretical Physics to Biology,* Ed. M.Marois, S. Karger, Basel, p. 21 (1973).
7. G. Careri, in *Quantum Statistical Mechanics in the Natural Sciences,* Eds. B. Kursunoglu, S.L. Mintz and S.M. Widmayer, Plenum Press, New York (1974).
8. B.H. Austin, K.W. Beeson, L. Eisenstein, H. Frauenfelder and I.C. Gunsalus, *Biochemistry 14,* 3555 (1975). H. Frauenfelder, in *Mobility and Function in Protein and Nucleic Acids* (CIBA Foundation Symposium 93), Pitmann, London (1982).
9. P.L. San Biagio, E. Vitrano, A. Cupane, F. Madonia and M.U. Palma, *Biochem. Biophys. Res. Communications 77,* 1158 (1977).
10. E. Clementi and R.M. Sarma, Eds., *Structure and Dynamics: Nucleic Acids and Proteins,* Adenine Press, New York (1983).
11. Various papers in *J. Biomol. Str. Dyn. 1,* Issues 1 and 2 (1983).
12. H. Frohlich and F. Kremer, Eds., *Coherent Excitations in Biological Systems,* Springer-Verlag, Berlin-Heidelberg (1983).
13. M.U. Palma, in ref. 12, p. 84.
14. H.E. Stanley and J.Teixeira, *J. Chem. Phys. 73,* 3404 (1980). H.E. Stanley, J. Teixeira, A. Geiger and R.L. Blumberg, *Physica 106 A,* 260 (1981).
15. R. Lumry, E. Battistel and C. Jolicoeur, *Farad. Symp. Chem. Soc. (London) 17,* 93 (1982).
16. R.O. Watts, E. Clementi and J. Fromm, *J. Chem Phys. 61,* 2550 (1974). J. Fromm, E. Clementi and R.O. Watts, *J. Chem. Phys. 62,* 1388 (1975).
17. E. Clementi, *Lecture Notes in Chemistry, vol. 2,* Springer-Verlag, Berlin-Heidelberg New York, (1976).
18. A. Geiger, F.H. Stillinger and A. Rahman, *J. Chem. Phys. 70,* 2585 (1979).
19. F.H. Stillinger, *Science 209,* 451 (1980).
20. F. Franks and D.S. Reid, in *Water: a Comprehensive Treatise,* Ed. F. Franks, Plenum Press, New York *vol. 2,* ch. 5 (1974). F. Franks, ibid., *vol. 2,* ch. 1 and *vol. 4,* ch. 1.
21. A. Ben-Naim, *Hydrophobic Interactions,* Plenum Press, New York (1980). A. Ben-Naim, *Biopolymers 14,* 1337 (1975).
22. D. Eagland, in *Water: a Comprehensive Treatise,* Ed. F. Franks, Plenum Press, New York (1975) *vol. 4,* ch. 5.
23. J.I. Edsall and M.A. Mc Kenzie, *Adv. Biophys. 16,* 53 (1983).
24. J. Texter, *Progr. Biophys. Molec. Biol. 33,* 83 (1978).
25. H. Frauenfelder, in ref. 10, p. 369.
26. S. W. Englander, N.R. Kallenbach, A.J. Heeger, J.A. Krumhansl and S. Litwin, *Proc. Natl. Ac. Sci. (USA) 77,* 7222 (1979).
27. M.G. Sceats and S.A. Rice, in *Water: a Comprehensive Treatise,* Ed. F. Franks, Plenum Press, New York, (1982) *vol. 7,* ch. 2; and *J. Chem. Phys. 72,* 6183 (1980).

28. C.A. Angell, in *Water: a Comprehensive Treatise,* Ed. F. Franks, Plenum Press, New York, (1982) *vol. 7,* ch. 1.
29. G. Andaloro, M. Leone and M.B. Palma-Vittorelli, *Nuovo Cimento 2 D,* 1239 (1983).
30. L.R. Pratt and D. Chandler, *J. Soln. Chem. 9,* 1 (1980).
31. M. Rao, C. Pangali and B.J. Berne, *Molecular Phys. 37,* 1773 (1979).
32. D.C. Rapaport and H.A. Scheraga, *J. Phys. Chem. 86,* 873 (1982).
33. D.L. Beveridge, G. Ravishanker and M. Mezei, in ref. 10, p. 477.
34. G. Ravishanker, M. Mezei and D.L. Beveridge, *Farad. Symp. Chem. Soc. (London) 17,* 79 (1982).
35. G. Bolis, G. Corongiu and E. Clementi, *Chem. Phys. Letters 86,* 299 (1982). G. Bolis and E. Clementi, *Chem. Phys. Letters 82,* 147 (1981).
36. G. Corongiu, S.L. Fornili and E. Clementi, *Int. J. of Q. Biol. 10,* 277 (1983).
37. K.S. Kim, G. Corongiu and E. Clementi, in ref. 11, p. 263.
38. We thank Prof. D.L. Beveridge for this remark.
39. P.J. Flory, *J. Am. Chem. Soc. 63,* 3083, 3091, 3096 (1941).
40. W.H. Stockmayer, *J. Chem. Phys. 11,* 45 (1943).
41. K. Kirkpatrick, *Rev. Mod. Phys. 45,* 574 (1973).
42. G. D'Arrigo, *Nuovo Cimento, 61 B,* 123 (1981).
43. J. Teixeira and J. Leblond, *J. Phys. (Paris) Lett. 39,* 83 (1978). O. Conde, J. Leblond, and J. Teixeira, *J. Phys. (Paris), 41,* 997 (1980).
44. A. Geiger, A. Rahman and F.M. Stillinger, *J. Chem. Phys. 70,* 263 (1979).
45. C.S. Pangali, R. Rao and B.J. Berne, *J. Chem. Phys. 71,* 2975 (1979).
46. L.R. Pratt and D. Chandler, *J. Chem. Phys. 67,* 3683 (1977).
47. A. Cupane, E. Vitrano, P.L. San Biagio, F. Madonia and M.U. Palma, *Nucl. Acid Res. 8,* 4283 (1980).
48. J. Massoulié, *Eur. J. Biochem. 3,* 428 (1968).
49. E. Clementi and G. Corongiu, in *Biomolecular Stereodynamics I,* Ed. R.H. Sarma, Adenine Press, New York, p. 209 (1981). E. Clementi, in ref. 10, p. 321.
50. M.L. Kopka, A.V. Fratini, M.R. Drew and R.E. Dickerson, *J. Mol. Biol. 163,* 129 (1983).
51. R.E. Dickerson, this volume.
52. D. Eisenberg and W. Kauzmann, *The Structure and Properties of Water,* Oxford University Press, Oxford (1969).
53. S.H. Konig, in *Water in Polymers,* Ed. S.P. Rowland, Am. Chem. Soc., Washington, D.C., p. 157 (1980).
54. S.H. Konig, K. Hallenga and M. Shporer, *Proc. Natl. Acad. Sci. U.S. 72,* 2667 (1975).
55. F. Hirata and P.J. Rossky, *J. Chem. Phys. 74,* 6867 (1981).
56. H.S. Frank and W.Y. Wen, *Disc. Farad. Soc. 24,* 133 (1957).
57. J. Del Bene and J.H. Pople, *J. Chem Phys. 52,* 4858 (1970).
58. J.L. Finney and J.M. Goodfellow, in ref. 10, pag. 81.
59. G. Aiello, M.U. Palma and F. Persico, *Phys. Letters 11,* 117 (1964). G. Aiello and M.B. Palma-Vittorelli, *Phys. Rev. Letters 21,* 137 (1968). G. Aiello and M.B. Palma-Vittorelli, *Collective Phenomena 1,* 87 (1973).
60. P.J. Rossky and D.A. Zichi, *Farad. Symp. Chem. Soc., 17,* 69 (1982).
61. D.A. Dragaert, N.W.B. Stone, B. Curnutte and D. Williams, *J. Opt. Soc. Am 56,* 64 (1966).
62. E. Fatuzzo and W.J. Merz, *Ferroelectricity,* North-Holland Publishing Co., Amsterdam (1967). R. Blinc, *Phys. Rev. 147,* 430 (1966).
63. A.R. Bates and K.W.H. Stevens, *J. Phys. C. (G.B.) 2,* 1573 (1969). A.B. Bates, ibid., 3, 1825 (1970).
64. S. Micciancio and G. Vassallo, *Nuovo Cimento 1 D,* 627 (1982).
65. G. Roux, D. Roberts, G. Perron and J.E. Desnoyers, *J. Soln. Chem. 9,* 629 (1980).
66. T.M. Benzinger, *Nature 229,* 100 (1971).
67. P.D. Ross, J. Hofrichter, and W.A. Eaton, *J. Mol. Biol. 96,* 239 (1975).
68. B. Magdoff Fairchild, W.N. Poillon, I. Li and J.F. Bertles, *Proc. Nat. Acad. Sci. (USA) 73,* 990 (1976).
69. Not enough care is usually paid to the conclusion that a stronger bond, a larger entropy and a closer localization need to coexist. This is not paradoxical, since the closer localization concerns the geometric space, and the larger entropy concerns the dynamic space.

70. G.E. Walrafen, in *Water: a Comprehensive Treatise,* Ed. F. Franks, Plenum Press, New York (1972), *vol. 1,* ch. 5.
71. G. Nemethy and H.A. Scheraga, *J. Chem. Phys. 41,* 680 (1964).
72. G.S. Kell, in *Water: a Comprehensive Treatise,* Ed. F. Franks, Plenum Press, New York, *vol. 1,* ch. 10 (1972).
73. R.D. Blacke, *Biochemistry 20,* 5735 (1981).
74. J.J. Katz and H.L. Crespi, in *Isotope Effects in Chemical Reactions,* Eds. C.J. Collins and N.S. Bowman, Van Nostrand, USA, p.686 (1970).
75. G. Vento, M.U. Palma and P. Indovina, *J. Chem. Phys. 70,* 2848 (1979).
76. M.F. Emerson and A. Holtzer, *J. Phys. Chem. 71,* 3320 (1967).
77. E. Gulari, B. Chu and D. Woermann, *J. Chem. Phys. 73,* 2480 (1980).
78. M. Leone, S.L. Fornili and M.B. Palma-Vittorelli, in *Water and Ions in Biological Systems,* Eds. V. Vasilescu, B. Pullman, L. Packer and L. Leahu, Plenum Press, London, p. 103 (to appear in 1985).
79. P.L. San Biagio, F. Madonia, F. Sciortino, M.B. Palma-Vittorelli and M.U. Palma, Proc. Workshop on Water, I.L.L., Grenoble, 1984 *J. Phys. C7,* 225 (1984).
80. P.L. Indovina, E. Tettamanti, M.S. Giammarinaro-Micciancio and M.U. Palma, *J. Chem. Phys. 70,* 2841 (1979).
81. D.A. Rees, *Adv. Carbohydr. Res. 24,* 267 (1969).
82. S. Arnott, A. Fulmer, W.E. Scott, I.C.M. Dea, R. Moorhouse and D.A. Rees, *J. Mol. Biol. 90,* 269 (1974).
83. P.G. de Gennes, *Scaling Concepts in Polymer Physics,* Cornell University Press, Ithaca, N.Y. (1979).
84. K.L. Wun, G.T. Feke and W. Prins, *Farad. Disc. Chem. Soc. 57,* 146 (1974). G.T. Feke and W. Prins, *Macromolecules 7,* 527 (1974).
85. It is order to remark that these effects on kinetics have no relation with transition state theory and isotope effects on chemical kinetics. On this latter subject see. e.g. Refs. 86 and 87.
86. C.J. Collins and N.S. Bowman, Eds., *Isotope Effects in Chemical Reactions,* Van Nostrand, U.S.A. (1970).
87. E.M. Halevi, *Israel J. Chem. 9,* 1805 (1975).
88. K. Kawasaki and S.M. Lo, *Phys. Rev. Letters 29,* 48 (1972).
89. F. Franks and D.J.G. Ives, *Quart. Rev. Chem. Soc. 20,* 1 (1966).
90. O. Conde, J. Teixeira and P. Papon, *J. Chem. Phys. 76,* 3747 (1982).
91. L. Bosio, J. Teixeira and H.E. Stanley, *Phys. Rev. Letters 46,* 597 (1981)
92. B.L. Halfpap and Sorensen, *J. Chem. Phys. 77,* 466 (1982).
93. L. Cordone, V. Izzo, G. Sgroi and S.L. Fornili, *Biopolymers 18,* 1965 (1979).
94. L. Cordone, A. Cupane, P.L. San Biagio and E. Vitrano, *Biopolymers 18,* 1975 (1979).
95. L. Cordone, A. Cupane, P.L. San Biagio and E. Vitrano, *Biopolymers 20,* 39 (1981).
96. L. Cordone, A. Cupane, P.L. San Biagio and E. Vitrano, *Biopolymers 20,* 53 (1981).
97. L. Cordone and A. Cupane, *Biopolymers 20,* 2137 (1981).
98. A. Cupane, D. Giacomazza and L. Cordone, *Biopolymers 21,* 1081 (1982).
99. L. Cordone, A. Cupane, F. D'Alia and M.G. De Stefano, *Farad. Symp. Chem. Soc. 17,* 161 (1982)
100. E. Antonini and M. Brunori, *Hemoglobin and Myoglobin in their Reactions with Ligands,* North. Holland, Amsterdam (1971).
101. M.F. Perutz, *Nature (London) 228,* 776 (1970).
102. F. Madonia, P.L. San Biagio, M.U. Palma, G. Schiliró, S. Musumeci and G. Russo, *Nature 302,* 412 (1983).
103. M. Kasha, *personal communication.*

Structure & Motion: Membranes, Nucleic Acids & Proteins,
Eds., E. Clementi, G. Corongiu, M. H. Sarma & R. H. Sarma,
ISBN 0-940030-12-8, Adenine Press, Copyright Adenine Press, 1985.

Isomerism of Molecular Complexes and Theoretical Studies of Biomolecules*

Zdeněk Slanina

The J. Heyrovský Institute of Physical Chemistry and Electrochemistry
Czechoslovak Academy of Sciences, Máchova 7, CS-121 38
Prague 2, Czechoslovakia

Abstract

A biomolecule is considered represented by a potential energy hypersurface whose theoretical study revealed several relevant stationary points, although the system was treated experimentally as a single species. The consequences thereof are analyzed with respect to comparison of the partial theoretical and the observed overall thermodynamic or kinetic data in case of the enthalpy, entropy, and heat capacity terms. The weighting treatment used uses the weight factors of the individual isomeric structures correctly evaluated in the terms of the information available from representation of the hypersurface by their stationary points only (full respecting of the vibrational-rotational motions). In case of the heat capacity a substantial link is shown between the measured overall values and the adopted type of observation technique. Interconversions of water dimers are used as an illustrative example, and substantial contributions of isomerism to the measured reactivity characteristics are pointed out. The results obtained show that the presumption of possible simulation of the overall terms by the partial values corresponding to the most stable structure cannot be applied to biomolecules.

Introduction

The recent progress in theoretical studies of chemical, biochemical, and biological activity phenomena at molecular level is connected with a distinct transition from the traditional approach in terms of structure vs potential energy to approaches fully respecting all types of motions existing in the given model of the real system studied. Although these approaches have been connected with application of rather simple models so far, they undoubtedly allow a better representation of the physical nature of the problem. So there appear an increasing number of papers in the field of quantum chemistry, biochemistry, and biology which involve entropy, temperature, or time as quite intrinsic terms even at computational, numerical level (1-8). In all molecular sciences we can observe a linking (2) of the techniques generating the potential energy hypersurfaces or their parts with the approaches of statistical

*Part XXIV in the series Multi-Molecular Clusters and Their Isomerism; Part XXIII. see (84).

mechanics which adopt these hypersurfaces for predictions of behaviour of the systems in observable terms. This higher form of application of the potential energy hypersurfaces allows simultaneously also a qualitatively higher interplay between theory and experiment.

The modern approach to structure of biomolecules abandons their rigidity concept and takes them consequently as non-rigid species. Thus soon after introduction of the concept of non-rigidity of compounds into inorganic (9-11) and organic chemistry (12-14) the phenomena of flexibility, structural fluctuations, or plasticity are also observed with biomolecules—see e.g. (15-24). These observed phenomena of biomolecules are simultaneously studied also theoretically (25-34), usually in terms of the minimum-energy structures. This increasing cooperation between theory and experiment is functional and, in some cases, even essential. So e.g. in spite of large progress in dynamical NMR spectroscopy (35) it is still true that the average structure deduced from NMR data is quite insignificant physically (36) unless the individual contributing conformations are known. Finally it should be mentioned that the concept of non-rigid structures of biomolecules, in spite of being the most sophisticated approach used at present, represents, nevertheless, only an intermediate step in transition from classical to quantum (e.g. (37-39)) concepts of molecular structure.

A particularly pregnant manifestation of structural non-rigidity of biomolecules consists in the finding of several non-equivalent, mutually sufficiently easily interconvertible local energy minima during theoretical studies of the potential energy hypersurfaces. In the case of biomolecules and their complexes such situation can be denoted as typical, the number of the minima being often quite large, and the transitions between them being relatively easy (at least at some conditions). So at least for some choices of observation techniques and conditions it must be presumed that the individual structures involved will not be resolved, a whole set of such isomeric structures being manifested as a single group. This also means that, typically, not partial properties corresponding to a single individual structure but certain overall terms corresponding to whole this group will be measured. The first step to interpretation of these overall terms consists in construction of the weight factors corresponding to the individual structures. In the field of theoretical studies of biomolecules these weight factors are determined (40-47) by approximative methods—sometimes they are reduced to mere Boltzmann factors which may be improved by an estimation of contribution of the vibrational motions at the minima. Recently, however, it was shown in another context (48-50) that the Boltzmann factors alone can completely fail as compared to the correctly evaluated weight factors. Therefore, the latter factors are used in the present communication to study the differences between the partial terms belonging to the most stable structure and the overall terms corresponding to the whole set of the isomeric structures for the basic thermodynamic and kinetic terms. In this way it will be possible to reexamine the frequently accepted idea concerning possible representation of an isomeric mixture corresponding to a biomolecule or its complex by only the most stable (or any other) member of this mixture. The required extent of theoretical information about the biomolecule will be that resulting from the representation of the hypersurface by means of localization and characterization of its stationary points.

General Methodology

Many biologically important processes can be represented in the theoretical study by the following association reaction

$$A + B \rightleftharpoons C, \tag{1}$$

and analysis of the corresponding potential energy hypersurface will show the presence of several non-equivalent local energy minima for the association complex C. In this way the species C becomes a mixture of—let us say—n isomeric structures C_i, and the overall process (1) is decomposed into a partial association reactions

$$A + B \rightleftharpoons C_i. \qquad (i = 1, 2, ..., n) \tag{2}$$

If the observation technique used is unable to differentiate between these isomeric structures, it will provide overall values composed of contributions of all the structures in play. Thus the overall values of change of standard molar enthalpy ΔH°, entropy ΔS°, and heat capacity at constant pressure ΔC_p° are obtained, whereas the theoretical approach leads primarily to the partial values of these terms ΔH_i°, ΔS_i°, and $\Delta C_{p,i}^\circ$ connected with the realization of the process (1) by the isomer C_i. Considering establishment of complete thermodynamic equilibrium and equal proportionality between concentration of the structure and its contribution to the quantity measured primarily in the experiment (e.g. spectral intensity, ion current, pressure) for all the isomers, the relations (3)-(5) can be written between the partial and overall terms (51):

$$\Delta H^\circ = \sum_{i=1}^{n} w_i \, \Delta H_i^\circ \tag{3}$$

$$\Delta S^\circ = \sum_{i=1}^{n} w_i (\Delta S_i^\circ - R \ln w_i) \tag{4}$$

$$\Delta C_p^\circ = \Delta C_{p,w}^\circ + \frac{1}{RT^2} \left[\sum_{i=1}^{n} w_i (\Delta H_i^\circ)^2 - (\Delta H^\circ)^2 \right], \tag{5}$$

where

$$\Delta C_{p,w}^\circ = \sum_{i=1}^{n} w_i \, \Delta C_{p,i}^\circ, \tag{6}$$

and R and T stand for the gas constant and temperature, respectively. The weight factors w_i mean the molar fractions of the individual isomers in their equilibrium mixture. It is especially simple to express these factors in terms of the molecular parameters in the case of the ideal gas, where it is

$$w_i = \frac{q_i \exp(-\Delta H_{0,i}^{\circ}/RT)}{\sum_{j=1}^{n} q_j \exp(-\Delta H_{0,j}^{\circ}/RT)} , \tag{7}$$

where q_i means the partition function of the i-th isomer related to its ground state energy as the energy zero, and $\Delta H_{0,i}^{\circ}$ means the standard molar enthalpy change of the reaction (2) at the absolute zero temperature. The evaluation methods of the environment effects on values of these weight factors belong to the more general problem of medium effects on conformational equilibria (52). These methods are not fully established yet, so it is impossible to treat the situation in this case in such straightforward way as is possible in the case of gas phase.

Whereas formulae (3),(4) resemble formally the relations for a mixture of non-reacting ideal gases (although, in fact the two situations have different backgrounds—in our case the weight factors came from establishment of thermodynamic equilibrium within interconversions of isomers), the relation (5) does not. In this case, a mixture of non-reacting component would obey an analogy to relation (6). The transition to an equilibrium of chemically reacting components necessitates the addition of the second term of the relation (5), which is due to the fact that the weight factors w_i are temperature dependent in this case. However, here appears a serious problem with respect to the technique used for measurement of the heat capacity, viz. the problem of the time scale of this technique. If the measurement technique used allows full manifestation of the temperature dependence of w_i (i.e. if there is enough time to establish a new interisomer equilibrium after a temperature change), it can be recommended that the formula (5) be applied; this type of the overall heat capacity will be called relaxation heat capacity. If, on the contrary, the time scale of the measurement technique used allows no change in w_i towards a new equilibrium after a temperature change (the weight factors w_i retained their original values), then it would be better to use the relation (6). This type of the overall heat capacity will be denoted as isofractional. The relaxation and isofractional terms represent, of course, limiting situations. In a real case, a relevant state can lie also between these limits. Evaluation of any particular case would necessitate to take into account the respective kinetic factors determining the establishment of a new equilibrium and to compare them with the time scale of the technique used for the heat capacity measurements. For methodical studies it is more advantageous, however, to use an alternative expression of the formulae (3)-(6). Let us introduce the isomerism corrections (53) $\delta X_i^{(iso)}$ as the terms which must be added to the standard molar change of the thermodynamic function ΔX_i° of the process (2) connected with the isomer labelled conventionally as 1 in order to obtain the overall value ΔX°. The label 1 could be assigned to any member of the isomeric set, nevertheless, it is useful to assign this label to the most stable isomer in the given temperature range. Such isomerism corrections enable direct testing of applicability of the partial terms corresponding to the most stable structure for the simulation purposes of the overall terms. For the isomerism corrections it follows:

$$\delta H_1^{(iso)} = \sum_{i=1}^{n} w_i (\Delta H_i^\circ - \Delta H_1^\circ) \tag{8}$$

$$\delta S_1^{(iso)} = \sum_{i=1}^{n} w_i (\Delta S_i^\circ - \Delta S_1^\circ - R\ln w_i) \tag{9}$$

$$\delta C_{p,1}^{(iso)} = \delta C_{p,w,1}^{(iso)} + \frac{1}{RT^2} \left[\sum_{i=1}^{n} w_i (\Delta H_i^\circ - \Delta H_1^\circ)^2 - (\delta H_1^{(iso)})^2 \right] \tag{10}$$

$$\delta C_{p,w,1}^{(iso)} = \sum_{i=1}^{n} w_i (\Delta C_{p,i}^\circ - \Delta C_{p,1}^\circ). \tag{11}$$

It must be mentioned that these isomerism corrections (8)-(11) are more general quantities than the overall terms (3)-(6). They are not bound (presuming the ideal behaviour) to the choice of reactants in reaction (2), nor do they depend on the choice of the standard state.

So far we have considered the association equilibrium and the standard thermodynamic terms corresponding thereto. However, a deeper analysis shows (54) that— within the concept of the activated complex—the above formulae can be also applied quite straightforwardly to the case of the rate process giving the product(s) E from the reactant(s) D:

$$D \Rightarrow [D]_i^\neq \Rightarrow E \quad (i = 1,2 ..., n) \tag{12}$$

and exhibiting an n-fold parallel isomerism of the activated complex $[D]^\neq$. If the construction of the partition functions in relation (7) respects the requirements of the concept of the activated complex, this relation can also be used for calculation of the weight factors $w_i^\neq$ of the individual isomeric structures of the activated complex. *Per analogiam* we can then apply the formulae (3)-(6) and (8)-(11) to the calculation of overall activation characteristics $\Delta H^\neq$, $\Delta S^\neq$, $\Delta C_p^\neq$, and $\Delta C_{p,w}^\neq$ as well as of the corrections for the parallel isomerism of the activated complex $\delta H_1^{\neq(iso)}$, $\delta S_1^{\neq(iso)}$, $\delta C_{p,1}^{\neq(iso)}$, and $\delta C_{p,w,1}^{\neq(iso)}$.

The Methods and Model Used for Illustrative Example

For an illustrative example we must choose a system whose energetics is known so well that the above relations can be used without additional simplifications. The water dimer, $(H_2O)_2$, was chosen for the following three reasons: (i) thanks to the classical work of Matsuoka, Clementi & Yoshimine (MCY) (55) the corresponding CI potential hypersurface is known, (ii) in the water dimer there exist the interaction types typical for numerous biomolecules and their complexes, and (iii) study of the water dimer (55-70) is constantly typical for understanding of properties of gaseous and liquid water which represent the most important media both in biology and in chemistry.

Recently a new fitting of the MCY potential was carried out (71) referred to as MCY-B which led to a decrease in the mean standard deviation. From this partial point of view the MCY-B represents, so far, the best analytical representation of the original CI data (55). Therefore, within the terms of our methodical study we retained this new fit (71), because it allows us to be in the closest contact with the original quantum-chemical data (55). This, however, does not imply that the MCY-B potential represents the best potential for the water dimer at present. This problem has been investigated only partially so far. Although the MCY-B was found, e.g., to give worse results for liquid water (71) than the MCY potential does, this fact alone represents no satisfactory test (it is impossible to reduce the interactions in liquid water to a purely dimer problem). Therefore, evaluation of this problem directly on some observable properties of the $(H_2O)_2(g)$ itself will represent also a side product of our study.

The first step necessary for our purposes is representation of the MCY-B hypersurface in terms of relevant stationary points. For this purpose we need (72) the analytically constructed first derivatives of potential energy. At the localized stationary points we carried out vibrational analysis using the analytically constructed second derivatives of the potential energy and the experimental force field (74) for the monomer units and proper treatment of the coordinate redundancy (75). In accordance with the previous results (72,73), only one energy minimum was found in the relevant part of the configurational space, besides three stationary points of the type of activated complex (one imaginary vibrational frequency) corresponding to the point groups of symmetry C_s, S_2, and C_{2v}, which are denoted (in accordance with the previous usage (72)) by the terms planar linear,

Table I

Survey of Structural, Energetical[†], and Vibrational Characteristics of Three Activated Complexes Found on MCY-B Potential Energy Hypersurface

Term(s)	Planar linear	Closed	Bifurcated
Point group of symmetry	C_s	S_2	C_{2v}
Distance of oxygen atoms (Å)	2.984	2.852	3.006
ΔE_i^* (kJ/mol)	0.949	3.088	7.473
$\Delta H_{0,i}^{\ddagger}$ (kJ/mol)	0.368×10^{-1}	3.397	5.339
Intermolecular vibrational frequencies (cm^{-1})	93i; 100; 126; 166; 384; 459	122i; 102; 127; 166; 384; 668	266i; 85; 131; 179; 211; 456

[†]Potential barrier ΔE_i^* and activation enthalpy at absolute zero $\Delta H_{0,i}^{\ddagger}$ for the activation process $(H_2O)_2(g) \Rightarrow [(H_2O)_2(g)]_i^{\ddagger}$ are presented.

closed, and bifurcated structure. Main data on energetics, structure, and intermolecular harmonic vibrational frequences are given in Table I. These molecular parameters were used for construction of the partition functions in the usual approximation (76) of rigid rotor and harmonic oscillator (RRHO).

Results and Discussion

Using the information of Table I we carried out calculation of the activation parameters of autoisomerizations of the water dimer:

$$(H_2O)_2 \text{ (g)} \Rightarrow (H_2O)_2 \text{ (g)} \tag{13}$$

both via the individual realization through the individual activated complexes found and for the overall rate process to which each of the found elementary reaction paths contributes. Table II shows the temperature evolution of the weight factors $w_{C_s}^{\ddagger}$, $w_{S_2}^{\ddagger}$, and $w_{C_{2v}}^{\ddagger}$. Whereas at the lowest temperature the energetically lowest structure C_s is predominant, at about 200K the weight factors $w_{C_s}^{\ddagger}$ and $w_{S_2}^{\ddagger}$ are of the same order of magnitude, and at the end of the temperature interval studied the fraction of the activated complexes of C_s symmetry becomes comparable with the fraction of the remaining two symmetries. These results alone indicate that the representation of the overall rate process (13) via only the activated complex of lowest potential barrier could not be sufficient. For comparison, Table II also includes the values of simple Boltzmann factors $w_i^{\ddagger\cdot}$ obtained from the relation (7) in which all the partition functions are considered to be equal to 1, and the ground-state energy terms are replaced by mere potential energy. Although the differences between the $w_i^{\ddagger}$ and $w_i^{\ddagger\cdot}$ factors are not so dramatic in our particular case as in other situations reported earlier (48-50), it is, nevertheless, confirmed also here that the $w_i^{\ddagger\cdot}$ values are not well applicable to simulation of the correctly determined $w_i^{\ddagger}$ terms.

Table II
Temperature Dependence of Weight Factors[†] $w_{C_s}^{\ddagger}$, $w_{S_2}^{\ddagger}$, and $w_{C_{2v}}^{\ddagger}$ of Planar Linear (C_s), Closed (S_2), and Bifurcated (C_{2v}) Activated Complexes in the Water Dimer Interconversion

T (K)	$w_{C_s}^{\ddagger}$	$w_{C_s}^{\ddagger\cdot}$	$w_{S_2}^{\ddagger}$	$w_{S_2}^{\ddagger\cdot}$	$w_{C_{2v}}^{\ddagger}$	$w_{C_{2v}}^{\ddagger\cdot}$
100	0.983	0.929	0.158×10^{-1}	0.709×10^{-1}	0.101×10^{-2}	0.363×10^{-3}
200	0.873	0.772	0.102	0.213	0.251×10^{-1}	0.153×10^{-1}
298.15	0.765	0.669	0.166	0.282	0.692×10^{-1}	0.482×10^{-1}
300	0.763	0.668	0.167	0.283	0.700×10^{-1}	0.488×10^{-1}
400	0.685	0.600	0.201	0.315	0.114	0.844×10^{-1}
500	0.631	0.554	0.218	0.331	0.151	0.115

[†]$w_i^{\ddagger}$—weights from Eq. (7); $w_i^{\ddagger\cdot}$—simple Boltzmann factors.

Table III gives a comparison of the partial and overall activation parameters of the process (13), whereas Table IV separates the pure contributions of the parallel

Table III

Temperture Dependence of Partial Activation Parameters[†] $\Delta X_i^{\neq}$ of $(H_2O)_2$ Interconversion for Planar Linear (C_s), Closed (S_2), and Bifurcated (C_{2v}) Activated Complexes and of Overall Values[†] $\Delta X^{\neq}$

T (K)	$\Delta H_{C_s}^{\neq}$	$\Delta H_{S_2}^{\neq}$	$\Delta H_{C_{2v}}^{\neq}$	$\Delta H^{\neq}$
100	−0.259	3.083	5.166	−0.201
200	−1.006	2.195	4.701	−0.535
298.15	−1.758	1.243	4.121	−0.854
300	−1.773	1.226	4.109	−0.861
400	−2.555	0.278	3.433	−1.304
500	−3.353	−0.643	2.708	−1.847

	$\Delta S_{C_s}^{\neq}$	$\Delta S_{S_2}^{\neq}$	$\Delta S_{C_{2v}}^{\neq}$	$\Delta S^{\neq}$
100	−4.698	−5.611	−7.651	−3.974
200	−9.842	−11.673	−10.818	−6.360
298.15	−12.903	−15.546	−13.164	−7.642
300	−12.951	−15.606	−13.203	−7.666
400	−15.202	−18.335	−15.144	−8.933
500	−16.980	−20.390	−16.761	−10.142

	$\Delta C_{p.C_s}^{\neq}$	$\Delta C_{p.S_2}^{\neq}$	$\Delta C_{p.C_{2v}}^{\neq}$	$\Delta C_{p.w}^{\neq}$	$\Delta C_p^{\neq}$
100	−6.801	−7.331	−3.900	−6.806	−4.365
200	−7.622	−9.644	−5.342	−7.771	−2.832
298.15	−7.740	−9.633	−6.400	−7.961	−3.811
300	−7.743	−9.628	−6.415	−7.965	−3.834
400	−7.906	−9.332	−7.051	−8.095	−4.988
500	−8.029	−9.090	−7.419	−8.168	−5.823

[†]$X \equiv H$, S, C_p, or $C_{p.w}$; H terms in kJ/mol, otherwise in J/K/mol.

Table IV

Temperature Dependence of Isomerism Corrections[†] $\delta X_{C_s}^{\neq(iso)}$ to Activation Paramaters of $(H_2O)_2$ Interconversion Related to Planar Linear (C_s) Activated Complex as Reference Structure

T (K)	$\delta H_{C_s}^{\neq(iso)}$ (kJ/mol)	$\delta S_{C_s}^{\neq(iso)}$ (J/K/mol)	$\delta C_{p.w.C_s}^{\neq(iso)}$ (J/K/mol)	$\delta C_{p.C_s}^{\neq(iso)}$ (J/K/mol)
100	0.583×10^{-1}	0.724	-0.545×10^{-2}	2.436
200	0.470	3.482	−0.149	4.790
298.15	0.905	5.261	−0.221	3.930
300	0.912	5.286	−0.221	3.909
400	1.251	6.268	−0.188	2.919
500	1.505	6.838	−0.140	2.206

[†]$X \equiv H, S, C_p$, or $C_{p.w}$.

isomerism of the activated complex related to the structure of C_s symmetry. At the lowest temperatures, of course, the overall enthalpy term $\Delta H^{\neq}$ lies near the $\Delta H_{C_s}^{\neq}$ value, but at the end of the temperature interval studied it is already closer to the $\Delta H_{S_2}^{\neq}$ value. The $T\Delta S^{\neq}$ term is affected more distinctly by the isomerism, moreover the overall entropy term lies (within the whole temperature interval studied) outside of the region limited by the lowest and the highest partial entropy terms. With the heat capacity term, there is a distinct difference between the isofractional and relaxation terms in the whole temperature interval. Thereby the fact is stressed that the nature of measured data must be properly elucidated prior to comparison of theoretical data with the values observed. (In our particular case, of course, it can be presumed, with respect to the easy interconversions of the individual activated complexes which can be well expected, that the measurements will give rather the relaxation term.) The isomerism correction values (Table IV) show convincingly that the rate process (13) must, in most situations, be taken as a resultant of three partial processes and its simulation by the path through the activated complex of C_s symmetry is unsatisfactory. So e.g. the isomerism correction of the TS term exceeds 1.5 kJ/mol at room temperature. Hence it follows that, at least at some temperatures, the isomerism correction can be more significant than the conventionally discussed corrections for the deviations from the RRHO model. Especially interesting is the temperature course of the relaxation correction term $\delta C_{p,C_s}^{\neq(iso)}$. Besides the already mentioned substantial difference from the isofractional term $\delta C_{p,w,C_s}^{\neq(iso)}$, this course is characterized by a maximum in low temperature region. This is connected with the temperature course of the weight factors of molecular complexes of the considered type which allows a convergency to a high-temperature limit at higher temperatures, i.e. slowing down of the temperature changes. Hence, especially in low temperature region, the relaxation overall heat capacity can be considered a sensitive indicator or test of the presence of isomerism. This conclusion agrees with the sensitivity of the activation heat capacity to the reaction mechanism discussed in organic chemistry (e.g. (77-79)). The main aim of the present study is the analysis of the weighting treatment with a suitable model case. A possible prediction of the activation parameters of the process (13) is of marginal importance only. In this context let us mention, however, that if these parameters should be measured, the reactant and the product of the process (13) would have to be differentiated by suitably isotopically labelled structures. The corresponding changes in atomic masses, of course would have to be respected in the partition functions. Besides that it must not be forgotten that the theoretical values could be made more accurate by introduction of corrections for deviations from the RRHO behaviour. From the methodological point of view it is also noteworthy that the weighting treatment used (constructed so as to match with the representation of the potential energy hypersurface by its stationary points only) should not be considered to be a limit case of the Monte Carlo approach to this potential hypersurface, but rather a limit case of the Monte Carlo technique applied to the (hypothetical) Gibbs energy hypersurface. This fact is due to the circumstance that the rotational-vibrational motion is fully respected in our weighting.

Any extrapolation of the results of this study of the model case of water dimer to typical biomolecular problems must take into account that such transition would be connected with an increase in number of the relevant isomeric structures by several orders of magnitude (cf.,e.g., (42)). This is, of course, substantially manifested in the values of corrections for the isomerism ($\delta S_i^{(iso)}$ increases (51) approximately with logarithm of total number of isomers). Increasing accuracy of calorimetric experiments (80), therefore, stresses the necessity of proper attention to biomolecular isomerism in interpretation of such data, in their confrontation with theoretical data and, finally, in their theoretical prediction. The results of this report apply not only to thermodynamic or kinetic data but can concern each of structure-dependent characteristics (81).

Table V
Theoretical and Experimental Thermodynamics of the Water Dimer Formation[†]

	ΔH° (kJ/mol)	ΔS° (J/K/mol)
MCY (Ref. 72)	−13.24	−74.48
MCY-B	−12.48	−69.36
Experimental (Ref. 82)	−15.7	−74.9

[†]Standard enthalpy and entropy changes for reaction (14) at T = 573.15K; Standard state—ideal gas at 101 325 Pa.

At the end of this report let us examine the problem of the prediction power of the MCY and MCY-B fits. For this purpose useful data are given in Table V on thermodynamics of the equilibrium process:

$$2 \text{ H}_2\text{O (g)} \rightleftharpoons (\text{H}_2\text{O})_2 \text{ (g)}. \tag{14}$$

It can be seen that the MCY-B fit—as compared with the MCY values(72)—does not lead to an improvement in the agreement with experimental data (82) on the thermodynamics of the process (14). This finding, along with the worse quality (71) of the second virial coefficient from the MCY-B fit as compared with the MCY result, allows the conclusion that the original MCY fit (55) continues to be (for the predicative purposes) the most reliable quantum-chemical potential of water dimer. Clementi and Corongiu (83) stressed recently that a good (that is, small mean standard deviation) fit is needed but is in itself far from being sufficient.

References and Footnotes

1. Clementi, E., *Determination of Liquid Water Structure, Coordination Numbers for Ions, and Solvation for Biological Molecules*, Springer-Verlag, Berlin (1976).
2. Clementi, E., *Bull. Soc. Chim. Belg. 85*, 969 (1976).
3. Löwdin, P.-O., *Int. J. Quantum Chem. 12, Suppl. 1*, 197 (1977); *20*, 775 (1981).
4. Linderberg, J., *Int. J. Quantum Chem. 12, Suppl. 1*, 267 (1977).
5. Crotov, S. S., Ischenko, A. A. and Ivaskevich, L. S., *Int. J. Quantum Chem. 16*, 973 (1979).
6. Clementi, E., *Computational Aspects for Large Chemical Systems*, Springer-Verlag, Berlin (1980).
7. Slanina, Z., *Advan. Quantum Chem. 13*, 89 (1981).
8. *Int. J. Quantum Chem., Quantum Chem. Symp. 16*, (1982).
9. Dennison, D. M. and Uhlenbeck, G. E., *Phys. Rev. 41*, 313 (1932); Rosen, N. and Morse, P. M., *ibid. 42*, 210 (1932).
10. Berry, R. S., *J. Chem. Phys. 32*, 933 (1960).

11. Muetterties, E. L., *Inorg. Chem. 4*, 769 (1965).
12. Kilpatrick, J. E., Pitzer, K. S. and Spitzer, R., *J. Am. Chem. Soc. 69*, 2483 (1947).
13. von Doering, W. E. and Roth, W. R., *Angew. Chem. 75*, 27 (1963).
14. Cotton, F. A., *Acc. Chem. Res. 1*, 257 (1968).
15. Ovchinnikov, Yu. A. and Ivanov, V. T., *Tetrahedron 30*, 1871 (1974).
16. Sundaralingam, M. and Westhof, E., *Int. J. Quantum Chem., Quantum Biol. Symp. 6*, 115 (1979).
17. Bolton, P. H. and James, T. L., *J. Am. Chem. Soc. 102*, 25 (1980).
18. Mitra, C. K., Sarma, M. H. and Sarma, R. H., *J. Am. Chem. Soc. 103*, 6727 (1981).
19. Gupta, V. D., *Int. J. Quantu, Chem. 20*, 9 (1981).
20. Ramachandran, G. N. in *Biomolecular Stereodynamics II*, Ed., Sarma, R. H., Adenine Press, New York, p. 1 (1981)
21. Cowburn, D., Live, D. H. and Agosta, W. C. in *Biomolecular Stereodynamics II*, Ed., Sarma, R. H., Adenine Press, New York, p. 345 (1981).
22. Bondarev, G. N., Isaev-Ivanov, V. V., Isaeva-Ivanova, L. S., Kirillov, S. V., Kleiner, A. R., Lepekhin, A. F., Odinzov, V. B. and Fomichev, V. N., *Nucl. Acids Res. 10*, 1113 (1982).
23. Olson, W. K. and Sussman, J. L., *J. Am. Chem. Soc. 104*, 270 (1982).
24. Käivaräinen, A. I., *Solvent-Dependent Flexibility of Proteins and Principles of Their Function*, D. Reidel, Dordrecht (1983).
25. Levitt, M. and Warshel, A., *J. Am. Chem. Soc. 100*, 2607 (1978).
26. Butt, G., Walter, R., Renugopalakrishnan, V. and Druyan, M. E., *Int. J. Quantum Chem., Quantum Biol. Symp. 6*, 453 (1979).
27. Langlet, J., Claverie, P., Caron, F. and Boeuve, J. C., *Int. J. Quantum Chem. 19*, 299 (1981).
28. Paterson, Y., Rumsey, S. M., Benedetti, E., Némethy, G. and Scheraga, H. A., *J. Am. Chem. Soc. 103*, 2947 (1981).
29. Sasisekharan, V., Bansal, M., Bramachari, S. K. and Gupta, G. in *Biomolecular Stereodynamics I*, ed., Sarma, R. H., Adenine Press, New York, p. 123 (1981).
30. Karplus, M. in *Biomolecular Stereodynamics II*, Ed. Sarma, R. H., Adenine Press, New York, p. 211 (1981).
31. Venkatachalam, C. M., Khaled, M. A., Sugano, H. and Urry, D. W., *J. Am. Chem. Soc. 103*, 2372 (1981).
32. Tosi, C. and Saenger, W., *Chem. Phys. Lett. 90*, 277 (1982).
33. Olson, W. K., *Nucl. Acid Res. 10*, 777 (1982).
34. Chou, K. C., Némethy, G. and Scheraga, H., *J. Phys. Chem. 87*, 2869 (1983).
35. Sandström, J., *Dynamic NMR Spectroscopy*, Academic Press, New York (1982).
36. Jardetzky, O., *Biochim. Biophys. Acta 621*, 227 (1980).
37. Woolley, R. G. and Sutcliffe, B. T., *Chem. Phys. Lett. 45*, 393 (1977).
38. Woolley, R. G., *Isr. J. Chem. 19*, 30 (1980).
39. Claverie, P. and Diner, S., *Isr. J. Chem. 19*, 54 (1980).
40. Gō, N. and Scheraga, H. A., *J. Chem. Phys. 51*, 4751 (1969).
41. Scheraga, H. A., *Chem. Rev. 71*, 195 (1971).
42. Lewis, P. N., Momany, F. A. and Scheraga, H. A., *Isr. J. Chem. 11*, 121 (1973).
43. Flory, P. J., *Macromolecules 7*, 381 (1974).
44. Mattice, W. L., *J. Am. Chem. Soc. 99*, 2324 (1977).
45. Hagler, A. T., Stern, P. S., Sharon, R., Becker, J. M. and Naider, F., *J. Am. Chem. Soc. 101*, 6842 (1979).
46. Madison, V. and Kopple, K. D., *J. Am. Chem. Soc. 102*, 4855 (1980).
47. Ikegami, A., *Advan. Chem. Phys. 46*, 363 (1981).
48. Slanina, Z., *Chem. Phys. Lett. 83*, 418 (1981).
49. Slanina, Z., *J. Mol. Struct. 94*, 401 (1983).
50. Slanina, Z., *Chem. Phys. Lett. 105*, 531 (1984).
51. Slanina, Z., *Collect. Czech. Chem. Commun. 40*, 1997 (1975).
52. Jorgensen, W. L., *J. Phys. Chem. 87*, 5304 (1983).
53. Slanina, Z., *Advan. Mol. Relax. Interact. Process. 14*, 133 (1979).
54. Slanina, Z., *Collect. Czech. Chem. Commun. 42*, 1914 (1977)
55. Matsuoka, O., Clementi, E. and Yoshimine, M., *J. Chem. Phys. 64*, 1351 (1976).
56. Braun, C. and Leidecker, H., *J. Chem. Phys. 61*, 3104 (1974).
57. Kistenmacher, H., Lie, G. C., Popkie, H. and Clementi, E., *J. Chem. Phys. 61*, 546 (1974).

58. Curtiss, L. A. and Pople, J. A., *J. Mol. Spectr. 55,* 1 (1975).
59. Diercksen, G. H. F., Kraemer, W. P. and Roos, B. O., *Theor. Chim. Acta 36,* 249 (1975).
60. Owicki, J. C., Shipman, L. L. and Scheraga, H. A., *J. Phys. Chem. 79,* 1794 (1975).
61. Huler, E. and Zunger, A., *Chem. Phys. 13,* 433 (1976).
62. Fredin, L., Nelander, B. and Ribbegård, G., *J. Chem. Phys. 66,* 4065, 4073 (1977).
63. Odutola, J. A. and Dyke, T. R., *J. Chem. Phys. 72,* 5062 (1980).
64. Coker, D. F., Reimers, J. R. and Watts, R. O., *Aust. J. Phys. 35,* 623 (1982).
65. Park, Y. J., Kang, Y. K., Yoon, B. J. and Jhon, M. S., *Bull. Kor. Chem. Soc. 3,* 50 (1982).
66. Wormald, C. J., Colling, C. N. and Smith, G., *Fluid Phase. Equi. 10,* 223 (1983)
67. Clementi, E. and Habitz, P., *J. Phys. Chem. 87,* 2815 (1983).
68. Swanton, D. J., Bacskay, G. B. and Hush, N. S., *Chem. Phys. 82,* 303 (1983).
69. Vigasin, A. A., *Zh. Strukt. Kh. 24,* 116 (1983).
70. Szczęśniak, M. M. and Scheiner, S., *J. Chem. Phys. 80,* 1535 (1984).
71. Bounds, D. G., *Chem. Phys. Lett. 96,* 604 (1983).
72. Slanina, Z., *J. Chem. Phys. 73,* 2519 (1980).
73. Slanina, Z., *Advan. Mol. Relax. Interact. Process. 19,* 117 (1981).
74. Mills, I. M. in *A Specialist Periodical Report, Thereoretical Chemistry, Vol. 1,* Ed., Dixon, R. N., The Chemical Society, London, p. 151 (1974).
75. Gwinn, W. D., *J. Chem. Phys. 55,* 477 (1971).
76. Janz, G. J., *Thermodynamic Properties of Organic Compounds,* Academic Press, New York (1967).
77. Kohnstam, G., *Advan. Phys. Org. Chem. 5,* 121 (1967).
78. Albery, W. J. and Robinson, B. H., *Trans. Faraday Soc. 65,* 980 (1969).
79. Blandamer, M. J., Robertson, R. E., Golding, P. D., MacNeil, J. M. and Scott, J. M. W., *J. Am. Chem. Soc. 103,* 2415 (1981).
80. *Biochemical Microcalorimetry,* Ed. Brown, H. D., Academic Press, New York (1969).
81. Davies, R. H., Smith, D. A., McNeillie, D. J. and Morris, T. R., *Int. J. Quantum Chem., Quantum Biol. Symp. 6,* 203 (1979).
82. Kell, G. S. and McLaurin, G. E., *J. Chem. Phys. 51,* 4345 (1969).
83. Clementi, E. and Corongiu, G., *Int. J. Quantum Chem., Quantum Biol. Symp. 10,* 31, (1983)
84. Slanina, Z., submitted for publication.

Structure & Motion: Membranes, Nucleic Acids & Proteins,
Eds., E. Clementi, G. Corongiu, M. H. Sarma & R. H. Sarma,
ISBN 0-940030-12-8, Adenine Press,   Copyright Adenine Press, 1985.

Parallelism in Computational Chemistry: Applications in Quantum and Statistical Mechanics

E. Clementi, G. Corongiu, J. H. Detrich, H. Khanmohammadbaigi, and S. Chin, L. Domingo, A. Laaksonen, H. L. Nguyen
IBM Corporation IS&TG
Department D55/Building 701
P. O. Box 390
Poughkeepsie, New York 12602
and
National Foundation for Cancer Research
Poughkeepsie Laboratory

Abstract

Often very fundamental biochemical and biophysical problems defy simulations because of limitation in today's computers. We present and discuss a distributed system composed of two IBM-4341's and one IBM-4381, as front-end processors, and ten FPS-164 attached array processors from Floating Point Systems, Inc. This parallel system—called LCAP—has presently a peak performance of about 120 MFlops; extensions to higher performance are discussed. Presently, the system applications use VM/SP as the operating system. Three applications programs have been migrated from sequential to parallel; a molecular quantum mechanical, a Metropolis-Monte Carlo and a Molecular Dynamics program. Descriptions of the parallel codes are briefly outlined. As examples and tests of these applications we report on a study for proton tunneling in DNA base-pairs, very relevant to spontaneous mutations in genetics. As a second example, we present a Monte Carlo study of liquid water at room temperature where not only two- and three-body interactions are considered but—for the first time—also four-body interactions are included. Finally we briefly summarize a Molecular Dynamics study where two- and three-body interactions have been considered. These examples, and very positive performance comparison with today's supercomputers allow us to conclude that parallel computers and programming of the type we have considered, represent a pragmatic answer to many compute intensive problems.

I. Parallelism in Scientific Computations, and in Computational Chemistry in Particular

Since the late 1950's (1) an increasing number of chemists have realized that many aspects of chemical research can be simulated on digital computers. More fundamentally, chemists are becoming aware that the simulations need not be limited to empirical correlations, but can be derived directly from quantum theory, without empirical parametrization; in short, *"ab-initio"* computational chemistry. We recall that since the early 1960's it was known how to select extended basis sets

49

with polarization functions and the importance of correlation energy corrections was well understood (2). By 1967, R. S. Mulliken, accepting the Nobel Prize for his work in quantum chemistry, would write (3), ". . . In conclusion, I would like to emphasize strongly my belief that the era of computing chemists, when hundreds if not thousands of chemists will go to the computing machine instead of the laboratory for increasingly many facets of chemical information is already at hand."

Of course, there were and there are obstacles: some are related to our expanding expectations and needs and some to the limits in today's computer performance. However, new avenues are being explored. In this work we shall present computations which deal with molecular systems somewhat larger than customary (the energy surfaces we shall report on are for DNA's fragments with about 90 atoms) or with unusually accurate techniques (the liquid water statistical mechanical computations described below make use not only of two-body potentials, but include also three- and four-body corrections). These computations have been carried out using *a new computer system* which we have recently programmed in our department and for which we have coded a skeleton-software system. The performance of our new system is somewhat higher, in general, than the performance of a CRAY-1, the deservingly well-known—even if by now rather old—supercomputer in the scientific/engineering fields. As we shall report below in detail, this first step in hardware is now being extended and we expect soon to increase the performance by about a factor of four to five, while keeping the cost nearly constant.

The main goal of this paper is to present our experimental computer hardwares and softwares, to review modified application programs in quantum chemistry and statistical mechanics, and to lend credibility to the thesis that "parallel computers and computations represent the most *natural choice* for scientific simulation in general and for computer simulations in *materials sciences in particular*".

We shall not comment on parallelism in computer architecture and related system and/or application programming since this is the subject of a very vast literature (4a-4e).

We recall that considerable effort has been made to exploit inherent parallelism of programs and to reduce their execution times throughout the last decade (see for example, Refs. 4f-4h). These efforts, except for a few recent implementations (4i), have been concerned with parallelism at the instruction level. Furthermore, when supporting hardware has been available, or suggested, it has been in the form of a network of low precision (usually in 8 to 16 bit logic) microprocessors.

We shall restrict ourselves to *our home brand* of parallelism which can be characterized as follows: 1) not thousands, nor hundreds of CPU's are considered, but very few, less than 20; 2) each CPU is a *well-developed engine* which executes asynchronously or even independently as a stand alone from the others; 3) the system softwares needed to execute application programs are as much as possible those commonly available for sequential programming; 4) to start with we constrain ourselves to FORTRAN, since this is the most widely used scientific application language and 5) we restrict ourselves to the minimum level of programming variations relative to the old sequential codes; 6) because of the applications we are interested in, 64-bit hardware precision is required. With respect to point 1) we wish to implement a *pragmatic approach* which is in no way critical of more ambitious attempts that are likely to be standardly available for general application programming only in the next decade. Concerning point 2) we use IBM host or hosts (for example, IBM-4341 or IBM-4381) with several FPS-164 attached *array processors* (AP). Since the latter could in principle be either standard scalar CPU's (Central Processing Units) or AP's, in the following we shall refer either to "slave CPU's" or to "attached AP's" as equivalent approaches to obtain parallelism. Concerning point 3) we note that our programs operate under VM/SP system (Virtual Machines/System Product) which has been supplemented by communication routines to allow standardized communication of commands and files between Host and AP (5). We note that we are actively working on a MVS alternative. Concerning point 4) clearly one would eventually like to have a compiler and/or an optimizer which would include the above communication facilities. As a first step, we are presently considering the possibility to write a simple and rather rudimentary pre-compiler to simplify the migration of codes from sequential to parallel.

In our configuration we can execute one or many jobs on the hosts and from 1 to 10 jobs on the AP's. Presently, when we run a given job on two or more AP's (parallel mode) we attempt to ensure that the amount of data flowing from one AP to another AP, via the host processor, is kept to a minimum. Indeed, jobs requiring transfers of long files from AP to AP are not optimal on our configuration because of channel limitations. Alternatively stated, the parallel sections of a given task need at present to be rather independent and not tightly coupled to one another. This characterization is important and to stress it we have named our experimental machine "LCAP", an acronym for "Loosely Coupled Array of Processors". Steps are now being taken not only to improve the performance of LCAP by upgrading the AP's, but also to allow jobs with heavy AP to AP data transfer; a move towards a "CAP" configuration.

In general, we operate with a "master" system which parcels out tasks to "slave" systems, the latter working in parallel but not synchronized. At times the task of the master might be essentially equal to the subtask of the slaves; under this conditions the master too works in parallel with the slaves. In general, *the total execution time,* T, of a given application program flows through two different types of time intervals; one when only the master system works, and one when the slaves are active. Generally, the sum of the former defines the sub-total time, T(s), for the *sequential* part of the execution, while the remaining execution time is the sub-total time for the *parallel* execution, T(p). With obvious notation: $T = T(s) + T(p)$ and $T(s) = \Sigma_i t_i(s)$ and $T(p) = \Sigma_j t_j(p)$, where i and j are indices related to successive time intervals. If we consider a system with slaves, then in the time interval $t_j(p)$ the slaves are active for time intervals $t_j(\alpha),.....,t_j(\omega)$; only under optimal conditions do $t_j(\alpha) = t_j(\beta) = ... = t_j(\omega)$. Otherwise, it is assumed that $t_j(p)$ is the longest time interval within the set $\{t_j(\alpha),....t_j(\omega)\}$. One seeks to have $t_j(p)$ much larger than the system overhead and communication time; further one seeks to have $T(p)$ much larger than $T(s)$. Those application programs either fulfilling or approaching the above conditions are *ideal* for parallelism; for additional details see Clementi, et al (5).

The asynchronous character of our system is due to the fact that each of the ω subtasks performed in parallel are similar but not exactly equal. For example, if each subtask is the computation of a list of two electron integrals (see below), then depending on how many integrals are neglected (because they are smaller than a certain threshold), one slave may finish before another slave. Another reason for the asynchronization is that the master "talks" sequentially to the slaves. For optimal parallelism, a given parallel subtask should last as long as possible (granularity of the problem) and its execution should be as independent as possible from the execution of the remaining subtasks.

For example, in a molecular wavefunction program the computation of each molecular integral (for example in the two-electron integral files) is independent from the computation of any other integral. Therefore, one would expect quantum chemical integral packages generally to be particularly suited for parallelism. The gathering of the above integrals to construct matrix elements is another obvious parallel-oriented task. *But "parallelism" is a technique of broad applicability in physical sciences.* For example, the computation of the energy at different values of the K-vectors provides us with one more example of a parallel application, this time from solid state. In this work we shall restrict ourselves to quantum chemistry, Metropolis-Monte Carlo (6) and Molecular Dynamics (7).

Note that there are problems, such as the theoretical prediction of the tertiary structure of proteins, which are so highly parallel that one needs only a few hours to program the migration from sequential to parallel.

Our system is essentially a Multi Instruction Stream Multi Data Stream (MIMD) system (4j), in the form of a distributed network of nodes. Each node of the network is a full processor capable of executing complete programs. Therefore, the implementation of a parallel code can be done at the FORTRAN level. This leads to a high-level parallelism on a distributed system, where *global parallelism* of a given code is revealed. The distributed nature of the system, on the other hand, allows appropriate utilization of the parallelism of the code; i.e. the "degree of parallelism" (4b) of the software is matched by the hardware. An important advantage of high-level parallelism, as in high-level programming languages, is its portability. A high-level parallel code may be run on any system with a rudimentary communication protocol. In addition, improvements at the instruction level can be independently pursued without disturbing the setup of the parallel algorithm.

The data transfer aspect of a parallel program seems to be of importance regardless of the actual breakup of the code and the particular communication protocol (4k).

Therefore, a proper design strategy for the breakup of a program must take into account a few factors concerning communication performance from the early stages. These may be characterized as follows: 1) transfer time vs. residence time, 2) transfer count, and 3) uniformity of subtask allocation.

Assume that a master processor, $P(i)$, is to send some data, $D(i,j)$, to a slave processor, $P(j)$. *Transfer time,* $Tt(i,j)$, is measured, ideally, from the moment the receiving processor needs the data, to the moment the data is usefully accessible to the receiver. $Tt(i,j)$ is all the overhead associated with a data transfer.

The elapsed time between the receipt of some data, $D(i,j)$, by a slave processor, $P(j)$, and the time when another data transaction is made by $P(j)$, is the *residence time, Tr(i,j), of D(i,j) in P(j)*.

In a typical multitasked code, there are many such data transfers. From a global viewpoint, therefore, the sums Tt and Tr of all $Tt(i,j)$'s and $Tr(i,j)$'s would be more meaningful.

The ratio of residence time to transfer time $E=(Tr/Tt)$, is a measure of the efficiency, E, of the parallel code. The higher the ratio, the more efficient the multitasked process is. The ratio E decreases with decreasing the hardware transfer rate, by increasing the amount of transferred data, increasing processor speed, and often, but not necessarily, by increasing the number of processors.

Typically, program execution comes to a halt as a processor waits to receive refreshed or new data, thus increasing the transfer time. Asynchronous data transfer and/or shared memory implementations have been major methods of shortening data transfer time. Asynchronous data transfer overlaps Tt with Tr, while a shared memory causes a genuine decrease in Tt.

The amount of data to be transferred has a direct impact on Tt. The dependency is linear as long as the transferable data is small enough as to fit in the main memory of the processor. It is sometimes possible to avoid transferring certain data in a given program. For example, *subtask-invariant* data (data that is used, but not modified by a subtask) can always be transferred to a subtask once, and held for subsequent use. Similarly, *main task-invariant* data (data that is repeatedly modified by a subtask but is used by the main task only after a final modification) can be created by subtasks, and kept, to be sent upon request. Using standard features provided by almost all high-level programming languages, including FORTRAN, such "factoring-out" techniques can be easily employed in an IBM/AP environment, and may result in considerable time savings.

Transfer count, Nt, is the number of data transfers initiated in the course of the execution of a multitasked code. It is a global quantity. There exists a constant "fixed cost" associated with data transfer initiation. It is, therefore, desirable to initiate as few data transfers as possible. If two or more subtasks can be collapsed into one larger subtask, with the same total transferable data, the time saved by foregoing the otherwise necessary communication after the first subtask, and before the second, could be appreciable. More efficient data transfer schemes, such as a shared memory, decrease the importance of Nt.

Concerning the *uniformity of subtask allocation,* we note that it is important that a task be distributed as equally as possible among the processors. This is equivalent to a constant Tt+Tr for every processor. Thus, a slower processor should have less to work on, and a processor with a different data transfer time (due to a lower/higher capacity channel interface) should get an appropriately adjusted share of the task.

Concerning the applicability of the above parallel machine to physics and chemistry we conclude as follows. In general, given a system of n particles, either classical or quantized, the computation of the interactions between the i-th particle and the n-1 remaining particles *is independent* from the computation of the interactions of particle $j \neq i$ with the n−1 remaining particles; *this computational independence leads to parallelism in simulations* in physical sciences. Therefore, one would expect parallel programming to be widely applicable in computational physics and chemistry, in statistical mechanics and in much of bio-physics due to the intrinsic granularity of these computations. Even if outside the scope of this paper, we note that parallel computations can be used most efficently in other fields including econometrics, graphics, theory of games, etc. etc.

II. LCAP, a New Parallel Computer System: Present Configuration

The computer configuration that we have assembled and tested consists of ten FPS-164 attached processors; seven are attached to an IBM-4381 host and the remaining three to an IBM-4341 host; see Fig. 1. The FPS-164 processors are attached to the IBM hosts through IBM 3Mbyte/sec channels. A third IBM-4341, connected to graphic stations, completes the host processor pool. The three IBM systems are interconnected, channel to channel, via an IBM-3880 connector. One attractive feature of the above system is the possibility to switch one, two or three FPS-164's from the IBM-4341 host to the IBM-4381 host. This gives us the flexibility of a "production system" with seven AP's and an "experimental" system with three AP's. The latter system is used during "prime time" for program development, experimentation of new hardware, debugging of system and/or application programs, etc., etc. During off-prime time and weekends, we typically work with two systems of four AP's each, and a third system with two AP's. Clearly depending upon user demand, we can use all ten AP's for a single job, or at the opposite extreme, each one of the ten AP's independently on ten different jobs.

Concerning random memory real storage, each FPS-164 has 4 Mbytes of storage. For two AP's the memory is larger (8 and 10 Mbytes, respectively); this feature is very convenient for special applications where one of the AP's is used as a "master". The memory on the IBM-4341 model M2 and model P2 are 8 Mbytes and 16 Mbytes, respectively; for the IBM-4381, 16 Mbytes are available.

loosely Coupled Array of Processors (LCAP)
Scientific/Engineering Computations, Dept. 48B, IBM–Kingston, DSD

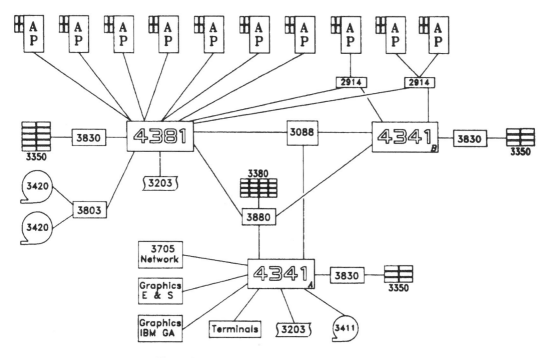

Figure 1. Configuration of LCAP as of June 1984.

In addition to the ten sets of four disks per AP (for a total of 5.4 Giga bytes) there are banks of IBM-3350 and IBM-3380 disks giving a total of about 25 Giga bytes. Tape drives, printers and network communication hardware complete our configuration.

We have *not* yet implemented in our applications the use of the TRAM (Table random access memory) on the FPS-164, *nor* the recently announced 164/MAX boards, *nor* the FPS Math-Library. Notice that under these conditions our system is about a 60 MFlops peak-performance system. The use of the TRAM feature brings the system up to about 120 MFlops peak performance. Recalling that each 164/MAX board was announced as a *"vector"* feature which increases the performance by up to 22 MFlops, our system with two 164/MAX boards per AP will have a peak performance of $10(22 \times 2 + 11) = 550$ MFlops. It is noted that since a *maximum* of fifteen 164/MAX boards can be added onto each AP, our system can grow to a maximum of about 3410 MFlops. Clearly, we wish to *first* explore the gain one can realistically obtain with only a few 164/MAX boards per AP, thus we shall start at 550 MFlops.

Presently, we have adapted for parallel execution application programs related to chemistry and biochemistry, but we wish to extend to physics, engineering, and other scientific applications. We are interested in performing these experiments not only to learn how to improve our system, but also to explore the applicability of parallelism *in general. For this reason, we are implementing a "visiting scientists" program where scientists from Universities and other Research Institutions would be able to investigate at our laboratory the possibility of adapting parallelism to other scientific problems. Scientists interested in learning more about this program can contact one of us (E.C.) for additional information.*

The computational timing of our system is discussed elsewhere (5). Here we report only a few examples obtained with our system, using the IBM-4341 model M2 (or the IBM-4381 model II) and without using any of the added features discussed above (TRAM, 164/MAX and Math.-Lib.). In Table 1 timing results are reported for four different application programs running with various numbers of AP's. These results are compared with the same computer programs running on a CRAY-1S. Note that for the integral, Monte Carlo and Molecular Dynamics programs, the execution time on our system with six AP's almost equals the CRAY-1S. For the SCF, the results are not as good. However, as mentioned in Section III, we have not yet optimized the sequential part of the code in the SCF. Once this will be accomplished, we expect that the performance for the SCF will approach the relative performance obtained for the integral, Monte Carlo and Molecular Dynamics routines. Note from Table 1 that on increasing from one AP to three AP's, to six AP's the improvement in execution times degrades slightly. This is due, in part, to system and communication overhead; we presently are investigating hardware modifications which should reduce this overhead.

The "peak performance" expectations previously reported are—in general—not too interesting, since the corresponding "sustained" performance can easily be notably less "spectacular". Therefore, it might be of interest to present an estimate—even if provisional and very rough—on the expected performance with reference to some of the data in Table 1. Let us consider the SCF timing of 34.1 minutes obtained with 6 AP's in parallel for the example "SCF, 42 atoms" (see Table 1, 4th line, 4th column). By using the *TRAM memory* we expect a speedup up to 22 or at best up to 15 minutes; by using, in addition, *one 164/MAX board,* we expect to need from 17 to 9 minutes; finally by using a *second 164/MAX board* we expect to further reduce the original timing to within the values of 15 and 7 minutes. As to be expected, the original sequential code itself can be improved; presently, two of us (G.C. and L.D.) are working on this aspect.

We conclude this section by stressing the rather exceptional flexibility of our system (scalar, vector and parallel), its high performance and "user friendliness", indirectly demonstrated by the existence of several large-scale applications already

Table I
Comparison of execution times for our system with different numbers of AP's
(without use of TRAM, 164/MAX and Math.-Lib) versus the time needed on a CRAY-1S.
ELAPSED TIME FOR STANDALONE CONFIGURATIONS (MINUTES)

JOB	1 AP	3 AP's	6 AP's	CRAY-1S	10 AP's	HOST
Integrals (27 atoms)	71.7	24.0	12.3	10.6	7.8	4341
SCF (27 atoms)	46.7	21.0	17.5	8.6	12.0	4341
Integrals (42 atoms)	203.4	68.9	38.3	32.3	21.2	4341
SCF (42 atoms)	108.5	44.9	34.1	19.6	22.0	4341
Integrals (87 atoms)	2163.0	730.0	380.0	309.0	247.0	4341
Monte Carlo	162.1	57.8	32.0	28.4	22.0	4341
Mol. Dyn.	87.1	38.3	23.3	20.1	—	4381

working in parallel on our system. In addition, we stress the high reliability: any one of the hosts can be used to enter our system and the event of *all* ten FPS-164's being *"down" at the same time* is most unlikely. Finally, the realistic possibility to increase the performance, even if "peak" rather than "sustained", from about 500 MFlops to about 3000 MFlops, opens the door to computational "experiments" previously "unthinkable". However, this optimism should be tempered by the realization that we are only at the beginning in our exploration and many aspects, both in hardware and in software, might prove to be more of an obstacle than presently forecasted.

III. Parallelism in Quantum Chemistry

Here we report on a computation performed on LCAP; we shall deal with the possibility that a proton can tunnel from one base in a DNA base pair to another, thus altering the genetic code. We recall that the first *all-electron, ab initio* calculation concerning nucleic acids was a computation for the separated bases (8); it was performed in 1968 on an IBM 360/50 computer (the largest base consists of 16 nuclei and 78 electrons). The first *all-electron, ab initio* computation of a base pair (9) was obtained using an IBM 360/195 computer (the computation dealt with a system of 29 nuclei and 136 electrons). Further, we note that the first *all-electron, ab initio* computation for a rather large fragment of DNA (10) was performed on an IBM 370/3033 (83 nuclei and 392 electrons). The latter required approximately five days of standalone computer time in order to obtain the wave functions and the energies for only two DNA fragments in the B and the Z conformations. *In the next section* of this paper we shall present the preliminary results of a computer application where cross sections of energy surfaces for systems consisting of 87 to 94 nuclei and 408 to 424 electrons have been considered. The study deals with the hydrogen bonds in base pairs, specifically the energy variations which are concomitant to displacement of protons along the hydrogen bond direction in the Guanine$-$Cytosine (G$-$C) pair.

It is noteworthy that by plotting the number of atoms versus the year of the computation one obtains a nearly linear plot for a problem with n^3-n^4 type dependency. This rather pleasant feature, however, should be put into proper perspective. The above computational examples (8-10) were not presented as "routine" work, but rather as "exceptional" efforts. In contrast, the computations presented in the next section of this paper are presented as "routine" ones, at least for *the LCAP computer system and the application programs as described below.* Indeed, we shall present not only a *few points* on a potential curve, but many points on *many surfaces.*

The Quantum Chemical molecular program, which we shall use and discuss, approximates solutions to the Schroedinger equation within the one-electron model (Hartree-Fock (11) approximation). In particular, the program is based on the SCF-LCAO-MO concept (12) using an analytical basis set of cartesian gaussian type functions (13). The gaussian type functions can be contracted following E. Clementi et al. (14,15). Several versions of this program have been written in the last twenty years (16), both to keep up with computer evolution and to introduce and test new concepts and algorithms. For example, the possibility to *"add"* or *"subtract"* atoms from a previous molecular computation was considered since 1966 (17a); the possibility to change the coordinates of one or more atoms while keeping the integral file for the unmoved atoms ("move" option) and the possibility to merge files related to two separated molecules ("merge" option) to compute intermolecular energies were also considered and implemented (see 17 b) and h)). In addition, simple numerical techniques to estimate the electronic correlation energy as a functional of the density matrix were proposed and tested (17,18). Further, algorithms to save computer time by taking advantage of *local symmetry* were proposed and implemented (16d,16e). Some of these computer programs and concepts have been adopted and modified and are present in most of the computer programs used by the quantum chemistry community. For information concerning other computer programs for the computation of molecular wave functions we refer the reader to the papers by M. Dupuis et al. (19a), V. R. Saunders and M. F. Guest (19b), J. Almlof et al. (19c), W. J. Hehre, et al, (19d), L. E. McMurchie and E. R. Davidson (19e), P. S. Bagus et al. (19f) and K. Ohno (19g). From these references the interested reader can obtain information on other equivalent computer programs that we have not listed above. *In this work we are not interested in "sequential" computer* programs which use either "scalar" or "vector" computers, *but rather our emphasis is on using "parallel" computers.* Indeed, these programs (12-19) have one aspect in common, namely the use of only *one* CPU. In this work we shall depart from the past and consider the use of several CPU's *in parallel.*

The version of the molecular program used in this work is the one described in Ref. 16c. We do not report here the algorithms of the molecular program, since they are extensively documented elsewhere (16b). We recall that the above computer program is subdivided into two tasks: the first computes the one and two electron integrals for a given set of nuclei and a given basis set and stores the integrals into two files (the first for the one-electron and the second for the two-electron integrals); the latter task retrieves iteratively the two integral files to build the Fock matrices and solves the pseudo-eigenvalue problem (11,12).

Let us comment on the first task. A *flow-chart* of the program is given in Fig. 2. The first subroutine called is the control subroutine, CNTL, which prepares an area in the main storage for the dynamic allocation. The size of the area is specified in input. The control is then passed to the INTGRL subroutine which calls the appropriate routines in the order from left to right in Fig. 2 (computation of the one and two electron integrals). In the subroutine INPUT, the geometry and the basis set for the computation are loaded. The input has the same characteristics as those described in Ref. 16c; in addition the *number of CPU's* available to the job for parallel execution has to be specified. Notice that this parameter is a variable, and presently the only limit is that the number of CPU's to be used must be smaller (or at most equal) to the number of atoms given in input. Partial rewriting of the original

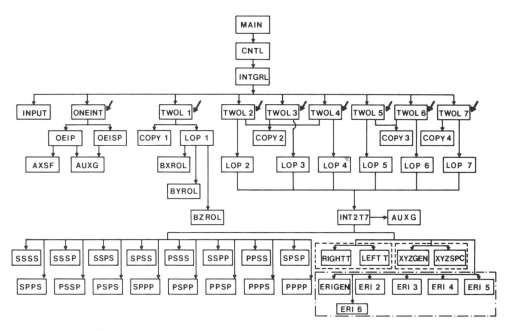

Figure 2. Flow chart for the one and two electron integral programs.

molecular program was needed for those subroutines distinguished by an arrow (see Fig. 2.) The subroutine, ONEINT, loops on the basis set centers to compute the one-electron integrals. The subroutines TWOL1,...,TWOL7, loop on the centers to compute one-center, two-center, three-center and four-center integrals. The main modification (needed to migrate from sequential to parallel code), which is, however, *very similar* in each one of the above subroutines, consists in identifying the most external loop, which runs over the atomic centers, and stepping the DO loop in the following exemplified way:

             (sequential)   (parallel)

                                        IBEGIN = ICPU
       DO 10 I = IBEGIN,NCENTERS   DO 10 I = IBEGIN,NCENTERS,NCPU

Here ICPU designates the i-th CPU out of the NCPU's available for the specific job. In this way, each CPU executes part of the integrals (parallelism) and writes into *its own* dedicated disk the integrals it generates. The NCPU integral files (NCPU = number of CPU's used) are such that 1) they can be restarted at any time in the computation, 2) integrals generated by introducing additional atomic centers can be added to the old file, 3) integrals generated for selected atomic centers can be subtracted, 4) integrals smaller than a pre-assigned threshold (given in input) are not stored; therefore, it is necessary to store also the four indices corresponding to each integral (notice that the larger the molecule, the larger the number of integrals nearly vanishing), 5) the four indices corresponding to each integral are stored in one word (64 bits on the AP, 4 bytes on the IBM) to save disk space.

To summarize, the characteristics and limits of the present "parallel" integral package are: a) the main storage is dynamically allocated, b) the program allows overlay, c) if computing molecular complexes and evaluating the basis set superposition error, only the one-eletron integrals are recomputed, d) the basis set can include s, p, d and f cartesian functions, e) the current version allows up to 200 centers, 104 primitives per center, 48 contracted functions per center, and up to 2000 point charges (these limits can be easily extended).

Let us now describe the modifications to the program in Ref. 16c introduced in the SCF package in order to migrate from sequential to parallel mode. Again, as an input parameter, the number of slave CPU's (where the integral files are stored) is given. We must point out that, at present, the routines working in parallel are those which prepare the Fock matrices; the diagonalization and normalization routines are still running in only one processor (sequential rather than parallel code). work is in progress to parallelize these subroutines and thus notably reduce the sequential time, T(s) for the S.C.F. program. A brief sketch of the computer program for the parallel version is summarized below:

   a) each slave CPU prepares the H and S sub-matrices and sends it to the host processor (master)

b) the host sends the D matrix to each CPU

c) each slave CPU reads the integrals from the disk storage, prepares the P sub-matrix and sends it to the host (parallelism in the I/O!)

d) the host generates, using the NCPU sub-matrices, the Fock matrix and diagonalizes it.

The tasks from b) to d) are repeated until convergency is reached. Notice that all the I/O *is done in parallel* and that the most time consuming part of each iteration is the construction of the Fock matrix. This suggests the advisability to recompute the integrals at each iteration *especially* in the future for eventual array processors of much higher performance than the present FPS-164. In this context we should point out that *parallelism in the I/O* could be of great importance in those C.I. type programs where the elapsed time is a bottleneck. We must point out that a bit of care has to be taken when changing a program from sequential to parallel mode. Once the host sends work to the n-CPU's it can continue the computation until it requires data from the n-CPU's; interrupts are inserted into the program to check if all the data have arrived and if the n-CPU's are free for executing new tasks. Presently the above scheduling between sequential and parallel tasks of a given application program is decided *by the programmer.* This requires a detailed knowledge of the Fortran code and of the algorithms used. We stress, however, that the overall approach is *general* in the sense that the majority of the programs in quantum chemistry can be converted from sequential to parallel mode. We wish to point out that "parallel" does not exclude "vector"; indeed it should be obvious that at least in principle each CPU can be a vector machine. For this reason we are considering to add to each one of our FPS array processors two 164/MAX boards.

In the *next section* we report on an application where we have made use of the parallel program sketched above. At present the application is still in progress but the data reported are sufficient to clearly demonstrate that the computer program is working properly; in addition the execution time is most attractive, pointing out that *parallelism is the proper avenue for large chemical applications.*

With the program structured as described above, and with our computer system we have opened the realistic possibility to compute *routinely molecular systems* in the range of 50-150 atoms at the *ab initio* level either using *minimal* (or near-minimal) basis sets or *extended* basis sets (with the pseudo-potential technique). From the above, it should be obvious that many crystal-orbital and solid state computer applications can also gain substantially with migration from sequential to parallel. We are now extending, by migrating to parallel codes, C.I. and M.C.S.C.F. molecular programs; however, presently we are not in a position to present timing data.

IV. Energy Surfaces for Base-Pair Proton Tunneling in DNA

In this section preliminary results obtained from a study on the hydrogen bridges in the $G-C$ pair are reported. First, recall that the aim of a previous and separate study (9) was to verify the existence of a double well in the interaction potential for a base pair; according to Löwdin (20) a double well is generated by displacing a proton(s) along its hydrogen bond(s). From this previous study (9) however, no double well was found, but this result was presented as a very preliminary one and possibly even as an "artifact" due to the use of a minimal basis set, neglect of electron correlation, uncertainty in relative positions of the two bases forming a pair, neglect of geometry optimization and neglect of the field effects originated either by the stacks of base pairs and/or by the sugar and phosphate groups, or more importantly the neglect of the displacement of two protons *simultaneously*.

In addition to the study on the base pair, Ref. 9 reported on an analogous but more elaborate study on the hydrogen bond bridges in the formic acid dimer (a more feasible system for the computers of the 1970's). In that study two main effects were investigated: a) the effect of the basis set on the shape of the hydrogen bond potential and b) the shape of the potential for the coupled motion of two hydrogens. With regard to point a) it was concluded that the results obtained with a 7/3 basis set, a 9/5 basis set, or with a 9/5 basis set including polarization functions were all *qualitatively* similar. However, for reliable *quantitative* results the larger basis set is needed. With regard to the latter point the following conclusions were made (9): "we would like to note that the assumption of a double-well potential as typical for the h-b is likely to be incorrect for the motion of a single H atom. Our conclusion is obtained independently of basis set size and seems to be corroborated by other independent computations on molecules containing a single h-b. However, for systems with multiple h-b's there is a higher probability to have a double-well potential than for systems with a single h-b, if one considers coupled motions."

Therefore, the present paper investigates more thoroughly the possibility of a double potential well in the $G-C$ system by addressing points a) and b) above. In addition, the influence of the field effects due to the stacking of additional base pairs and of the sugar-phosphate groups are explored.

We have considered three chemical systems, a $G-C$ pair, three stacked $G-C$ pairs, and a $G-C$ pair with its sugar and phosphate groups. The system with three base pairs ($(G-C)_3$, for short) has the sequence $C-G$, (top of stack), $G-C$ (middle of stack), and $C-G$ (bottom of stack), geometrically to one another as in B-DNA: the distance between two consecutive planes (containing a base-pair) is 3.34A and one base pair is rotated by $36°$ relative to the next one. For each base pair, the relative positions of the atoms are as given in Ref. 9.

In Fig. 3 we present two views of the $(G-C)_3$ system; on the top left is a projection into the x-y plane and on the top right is a side view. The complex under consideration consists of 87 nuclei and 408 electrons; the computations have been performed using the 7/3 basis set for the C, N, O atoms (18) and 4 primitive functions for the H atoms, yielding a total of 1032 primitive functions contracted to 315 functions. In the bottom of Fig. 3, we report the fragment with the $G-C$ pair, the two sugar-phosphate groups and two additional sugar units. Thus the system to be simulated, sugar(1)$-$phosphate(1)$-$sugar(2)$-$G$-$C$-$sugar(3)$-$phosphate(2)$-$sugar(4), is com-

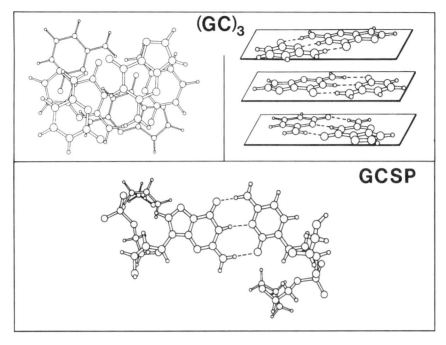

Figure 3. Pictorial representation of the (G-C)$_3$ system. On the top left is a projection into the X-Y plane, and on the top right is a side view showing the stacking of the three base pairs. On the bottom is a projection into the X-Y plane of the GCSP system.

posed of 94 nuclei and 424 electrons. The computations performed for this system use a 7/3 basis set for C,N,O and a 4 basis set for H as described above, yielding a total of 1046 primitive functions contracted to 318 functions. This computation is among the largest quantum chemical computations in the literature (the system has no point group symmetry); however works like the one in Ref. 10 are comparable.

As it is well known, the G−C base pair has three hydrogen bridges; following the notation used in Ref. 9 we designate as h-b-1 the bridge between the O6 of guanine and the NH$_2$ of cytosine, h-b-2 the bridge between N1H of guanine and N3 of cytosine and h-b-3 the bridge between NH$_2$ of guanine and O2 of cytosine.

In Fig. 4 (left, middle and right insets) the potential energy curves for the displacement of a *single* proton along its hydrogen bond are given for h-b-1, h-b-2 and h-b-3, respectively. In each inset curves A and B refer to the results obtained for the G−C system using either the 7/3 basis set (curve A) or the 9/5 basis set (curve B), while curve C refers to the (G−C)$_3$ system. The energies are expressed in a.u., while the distances (ΔR) are given as displacements relative to the standard N−H bond length for an isolated base (ΔR=0). Negative values for ΔR correspond to a shortening of the N−H bond. In each inset the three curves yield qualitatively similar results: namely a very well defined minimum for values of ΔR close to zero and an inflection point near the position corresponding to the formation of a bond in the

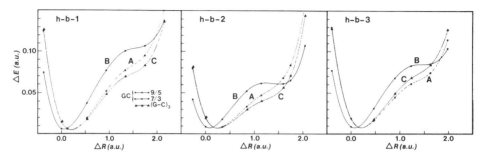

Figure 4. Inset A: Potential energy curves for the displacement of h-b-1 from Cytosine to Guanine. Curve A, denoted with squares (■), corresponds to the G−C system computed with the 7/3 basis set. Curve B, denoted with dots (●), corresponds to the G−C system with the 9/5 basis set. Curve C, denoted with triangles (▲), corresponds to the (G−C)$_3$ system. The ordinate is given as ΔE, where E is in a.u., and the minima for each curve have been aligned by adjusting the energy scales. The abscissa is given as ΔR, where ΔR represents the displacement (in a.u.) of the standard N−H bond length from its equilibrium value. Insets B and C: same for the displacement of h-b-2 and of h-b-3 respectively.

complementary base. Note, however, that for the curves computed with the 9/5 basis set there is a relatively "flat" region near the inflection point. These calculations have all been performed assuming the rigid equilibrium structure for the G−C pair, and displacing only the hydrogen atom. These "flat" regions indicate that if computations were performed with the 9/5 basis set and *optimized geometries* for the base pair system, then *perhaps* single proton transfers may yield a double potential well, as originally postulated by Löwdin (20).

Additional information concerning the effects of the size of the basis set and the effect of the field induced by the stacking of the G−C pairs can be obtained by examining more carefully the relative positions and shapes of curves A, B and C in Fig. 4. As expected, the equilibrium N−H bond length in the base pair is somewhat longer than the originally assumed standard bond length for the separated bases ($\Delta R=0$). In addition, the minimal basis set predicts a slightly longer bond length compared to the double zeta basis set, and a less repulsive curve as the displaced hydrogen begins to experience the complementary base ($\Delta R = 1.5$). These effects are evident in all three insets. However, the presence of the two stacked base pairs in (G−C)$_3$ (curve C) introduce seemingly different effects in the three hydrogen bonds. For h-b-1 (left inset) curve C is more repulsive than curve A, whereas for h-b-2 (middle inset) and h-b-3 (right inset) the opposite is observed. The reversal of the relative positions of curves A and C in h-b-2 and h-b-3 compared to h-b-1 can be rationalized with a "geometrical" argument. Likely, as the h-b-2 and h-b-3 protons are displaced they experience more and more the field of the rings of the stacked bases (see top left side of Fig. 3), whereas for h-b-1 this is less so. It would appear that the stacking favors the proton transfer for a hydrogen which experiences the field of aromatic rings in the stacked bases.

As previously mentioned, another goal of this section is to explore the potential surface for the *coupled motion* of two hydrogens. Here, three different possibilities exist: 1) the coupled motion of h-b-1 and h-b-2, 2) the one involving h-b-1 and h-b-3, and 3) the simultaneous motion of h-b-2 and h-b-3. In the first two cases the two hydrogens involved belong to different bases, however in the last case the two hydrogens are both being displaced from Guanine towards Cytosine, yielding the system G^{--}--------C^{++}. Both a two dimensional (bottom) and a three dimensional (top) representation of the energy surfaces for cases 1) (Fig. 5) and, 2) and 3) (Fig. 6) are reported. The right side of Fig. 5 reports on the surface obtained by displacing h-b-1 and h-b-2 with a partially optimized geometry for the G$-$C system using the minimal basis set. It is evident from these figures that a *very well* defined double minimum occurs only in case 1). It must be pointed out that the computations for the left inset of Fig. 5 and in Fig. 6 were carried out by moving the two hydrogen atoms of the h-b's *without* altering the geometry of either Guanine or Cytosine. During, and after completion of the hydrogen motions, the geometry of the rings will surely change. In the cases of h-b-1 and h-b-3, C=O becomes C$-$O$-$H and

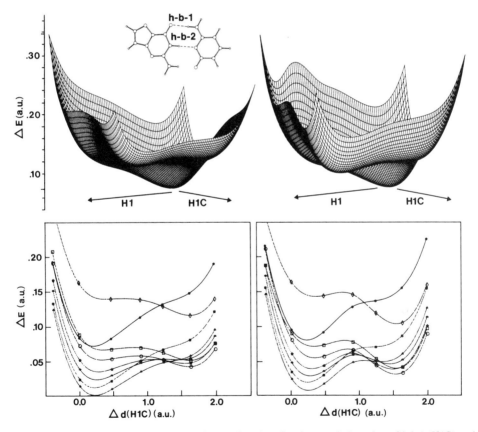

Figure 5. The potential energy cross sections and surface for the coupled motion of h-b-1 (H1C) and h-b-2 (H1), computed for the G-C system with the minimal basis set. Each cross section is given at a different displacement (ΔR(H1C)) of h-b-1 according to the following legend (a.u.); ● $=-0.3270$, ■ $=0.0$, ▲ $=0.4667$, ◊ $=0.8982$, ★ $=1.2300$, ○ $=1.6218$, □ $=1.9832$. Insets A and B correspond to h-b-1 and h-b-2 with and without partial geometry optimization respectively.

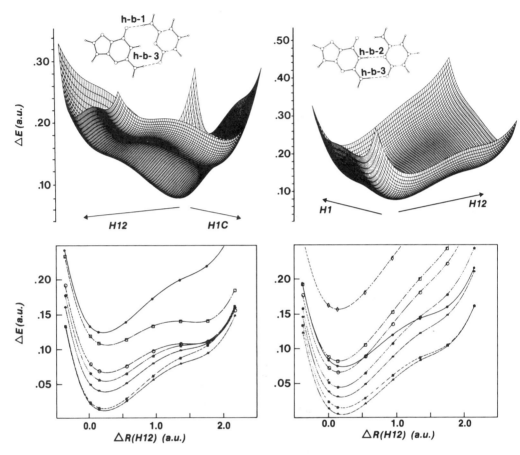

Figure 6. Left side: the potential energy surface for the coupled motion of h-b-1 (H1C), h-b-3 (H12)—See legend in Figure 6. Right side: Same for h-b-2 and h-b-3 but with the following displacements ($\Delta R(H1)$) of h-b-2 ● = −0.2631, ■ = 0.0000, ▲ = 0.3409, = 0.8418, ★ = 1.1441, ○ = 1.5973, □ = 1.7476, ◇ = 2.0487.

$C−N\!<^H_H$ becomes $C=N−H$. In these computations the $C=O$ bond length in the newly formed $C−O−H$ group is kept fixed at the same bond length as originally assumed for $C=O$. In addition, the $C=N$ bond length in the $C=N−H$ group is assumed to remain equal to the $C−N$ bond length in the $C−N\!<^H_H$ group. Certainly a geometry optimization will alter the shape of the potential surfaces, probably resulting in deeper minima and lower barriers, or perhaps even new minima. However, we expect that the double well of Fig. 5 will remain, thus providing the necessary base for Löwdin's hypothesis that the simultaneous motion of two protons resulting in the formation of tautomeric forms is a possible event. The right inset of Fig. 5 indeed shows—as expected—that geometry optimization does not qualitatively alter the above conclusion.

We note that our findings agree with results obtained in previous investigations; for example, in the study on the tautomerization energy of uracil (22,23) a geometry relaxation (especially in the region of the proton transfer) gives—as expected—a significant contribution to the computed energy. Zielinski et al. (22) have shown that in order to obtain reliable numerical data on either the tautomerization barrier or the energy difference between the tautomeric forms, geometry relaxation is needed. In the formic acid dimer, a relaxation of the heavy atoms resulted in deeper wells (more so for the tautomeric form) and a lowering of the potential barrier (24). From the curves in Fig. 5, the partial geometry relaxation does indeed affect the tautomerization barrier and the energy difference between the tautomeric forms.

In order to further assess the reliability of the above computations, a cross section of the potential energy surface of Fig. 5 has been computed for the single base pair using the 9/5 basis set and also for the larger $(G-C)_3$ system; see top left inset of Fig. 7. This cross section corresponds to first attaching the h-b-2 proton to Cytosine,

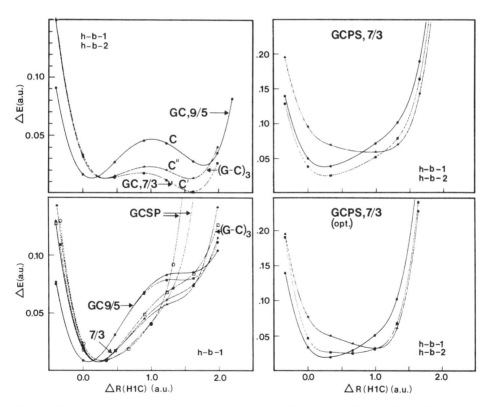

Figure 7. Top left inset: Potential energy curves for the coupled motion of h-b-1 (H1C) and h-b-2 (H1), taken as the cross sections which contains two minima. Curve C and C′ represent the $G-C$ system with the minimal basis set and the 9/5 basis set respectively. Curve C″ is for the $(G-C)$ system. For the three curves, the energy scales have been shifted to align the minima which are in the vicinity of $\Delta R=0.0$. Bottom left inset: Comparison of a cross-section obtained by moving only h-b-1. The energy scales have been shifted to align the minimum. Right insets : Three cross-sections obtained by moving h-b-1 and h-b-2 in the GCSP fragment with the 7/3 basis set without (top) and with (bottom) partial geometry optimization.

and then sequentially displacing the h-b-1 proton from Cytosine to Guanine. As expected, the three curves are all qualitatively similar, namely there is most definitely a double minimum. Quantitatively, by comparing curves C and C′ it is clear that the larger 9/5 basis, set gives deeper minima (as a consequence the barrier is higher) and the relative depths of the two wells are reversed compared with the calculations done with the 7/3 basis set. Finally, the effect of the induced field due to the presence of the two stacked pairs can be explored by comparing curves C and C″. From these curves it can be observed that in the presence of the field a double minimum still exists, but now both minima have approximately the same energy.

For the double proton transfer of h-b-1 and h-b-2 a double minimum was found; this is in agreement with the results of Schiener and Kern (25). However our conclusions differ slightly from theirs; whereas they investigated only the potential *curve* for the simultaneous motion of two protons, we have calculated the entire potential energy *surface*. The potential curve of Ref. 25 predicts a very high barrier for double proton transfer (61.6 kcal/mol). This result is due partly from the method (semi-emperical PRDDO), but more importantly is a consequence of having considered only one cross-section for the simultaneous motion of the two protons. By scanning the entire potential energy surface we have found that the optimum path for the double proton transfer does not correspond to a concerted symmetrical displacement of the two protons (see Figs. 5, 7). By following this path, the large barrier predicted by Ref. 25 is avoided.

The results on $G-C$ system with the larger basis set and on the $(G-C)_3$ system also exhibited a double minimum. This reinforces the notion that the double minimum is indeed real, and not an artifact of our calculation.

In conclusion, a double well for the double proton transfer of h-b-1 and h-b-2 in the Guanine−Cytosine base pair has been found. However, *this conclusion presently should not be extended to DNA.* Indeed, in order to analyze quantitatively the characteristics of the double proton transfer in DNA one should consider aspects such as a more complete geometry optimization, the field induced by the stacked base pairs, a larger basis set with polarization functions, the effects of the sugar and phosphate groups, and of the medium (counter ions and water). The computations in Figure 7 are very revealing in the context of the above list of "improvements". At the top right inset, we present three cross-sections of a surface obtained by moving the protons h-b-1 and h-b-2, this time for the large fragment GCSP (see Figure 3, bottom inset) with 94 atoms. At the bottom right inset we present the same computation, but this time we have partially optimized the geometry for the C=O, $C-NH_2$ COH and CNH groups as previously discussed. The basis sets are 7/3 for C,N,O atoms and 4s for the H atoms. *The double well has disappeared* because of base to base geometry optimization. Notice that in previous papers, as summarized in Ref. 10, we have insisted on the need to include the phosphate groups in any discussion dealing with the base pairs in DNA or RNA. Indeed there is a *notable charge transfer from the phosphates to the sugar to the bases.* In this context, once more we wish to express our concern on popular electrostatic potential approximations, repeatedly published in the last few years, where, somewhat conveniently, one has glossed over the possibility of large charge transfers between phosphate and sugars while smaller electronic details are emphasized.

The importance of the phosphate field brings about as a direct consequence the need to extend this work to inclusion of counter-ions and of water molecules, especially those hydrating the base pairs.

Possibly, inclusion of counter-ions will re-introduce the double well, especially when one of the counter-ions is located near one of the bases. Work is now in progress on these aspects. However, comparing Fig. 7 with Fig. 5 it is obvious that one should be very careful in extrapolating from the isolated base-pairs to a DNA or a RNA in vacuum, or from a "nucleic acids in vacuum" to a DNA double helix or a RNA in solution.

The bottom left inset of Fig. 7 summarizes the above conclusion for the case of the potential curve for h-b-1 (single proton). The $G-C$ and even the $(G-C)_3$ system either with partial geometry optimization or without, either with small (7/3) or larger basis set (9/5) show a minimum and a more or less pronounced plateau, which is the "seed" for the double potential well in considering *two* protons. However, the inclusion of the sugar and phosphate groups—with base to base partial geometry optimization—brings about only a single well. Thus *qualitatively different* answers have been obtained by increasing the "realism" in the simulation.

This "test study" would have been nearly impossible with more conventional computers. Note, in addition, that in a following study we shall select the water molecules' position and orientation and the counter-ions' position from our previous Monte Carlo studies on DNA hydration. This is possibly the first time that we experienced a *feedback* between simulations: from quantum chemistry (interaction potentials) to statistical mechanics (counter-ions and hydration in DNA) back to quantum chemistry (medium effect on the proton transfer interaction surface). Likely, in the future, computational chemistry will experience more and more such feedbacks, especially in attempting to simulate chemically complex systems; this brings about a very sharp increase in computational needs and also advises to broaden our competency so as to use quantum chemistry and statistical mechanics interchangeably.

V. Parallelism in Monte-Carlo Simulations

For some time now we have been deriving intermolecular potentials from *ab initio* computations and using these potentials in statistical mechanical computer experiments on pure solvents or solutions (26), with the goal of deriving an ordered set of approximations leading to an increasingly realistic description of the *ab initio* liquid. An important step along this road is the recent work of Clementi and Corongiu (27), which obtains a three-body interaction potential for water from ab initio calculations and applies this in Metropolis-Monte Carlo simulations of liquid water. In the last section of this paper we shall consider this problem once more, this time using Molecular dynamics.

Here we briefly review this potential, and its generalization to include a four-body intermolecular interaction potential as well, (details are available in a recent paper by J. H. Detrich, et al (28)). This lays the concrete groundwork for our main concern here, which is the development of a practical computational scheme to meet the very heavy demands made, for example, by our increasingly realistic computer simulations of liquid water.

For an N body system we can represent the total intermolecular interaction energy, U, in terms of a series of the form

$$U = \Sigma_i \, \Sigma_{j \neq i} \, U_{ij}^{(2)} + \Sigma_i \, \Sigma_{j \neq i} \, \Sigma_{k \neq i,j} \, U_{i,j,k}^{(3)} +$$
$$+ \Sigma_i \, \Sigma_{j \neq i} \, \Sigma_{k \neq i,j} \, \Sigma_{l \neq k,i,j} \, U_{i,j,k,l}^{(4)} + \dots \tag{1}$$

where the summations range from 1 to N, and the various $U^{(k)}$ are defined so that they are necessarily zero in case $N < k$.

Clementi and Corongiu base their expression for $U^{(3)}$ on a classical bond polarizibility model. The water molecules are taken to be rigid, but polarizible, so that a dipole $\vec{\mu}_{bi}$ is induced at the midpoint of each OH bond according to the relation

$$\vec{\mu}_{bi} = \alpha \vec{E}_{bi} + \delta \vec{e}_{bi} \, (\vec{e}_{bi} \cdot \vec{E}_{bi}) \tag{2}$$

here α and δ are the polarizibility constants for the water OH bonds, i is the index denoting the i-th water molecule, b is the index used to distinguish the two bonds in a given molecule, $\vec{e}_{bi}$ is the unit vector in the direction of the bond bi, and $\vec{E}_{bi}$ is the electric field at the midpoint of the bond bi. The electric field $\vec{E}_{bi}$ originates from the assumed permanent electrostatic moments of the surrounding water molecules, and is naturally expressed as a sum

$$\vec{E}_{bi} = \Sigma_{j \neq i} \, \vec{E}_{bi,j} \tag{3}$$

where $\vec{E}_{bi,j}$ is the electric field component at the bond midpoint bi originating from the permanent electrostatic moments of the j-th water molecule.

To first order in the polarizibility, this model yields the energy given by

$$U_{p1} = -\tfrac{1}{2} \, \Sigma_{bi} \, \vec{\mu}_{bi} \cdot \vec{E}_{bi} \tag{4}$$

If we substitute Eqs. (2) and (3) in this expression, we obtain a triple sum which shows the three-body character of our interaction energy.

Continuing now to second order in the polarizibility, we find an interaction energy which is conveniently expressed in terms of dipole-dipole interactions, namely

$$U_{p2} = \Sigma_{bi} \, \Sigma_{b'j,j \neq i} \, [\vec{\mu}_{bi} \cdot \vec{\mu}_{b'j}/r^3_{bi,b'j} +$$

$$-3(\vec{r}_{bi,b'j} \cdot \vec{\mu}_{bi})(\vec{r}_{bi,b'j} \cdot \vec{\mu}_{bj})/r^5_{bj,b'j}]$$

(5)

where $\vec{r}_{bj,b'j}$ is the distance vector pointing from bond midpoint bi to bond midpoint b'j. We can again substitute Eqs. (2) and (3) here to obtain an expression with a four-fold summation which clearly shows the four-body character of the interaction energy in Eq. (5). In practice, the interaction terms are computed by first evaluating Eq. (3) and then Eq. (2). Hence our four-body term requires a single loop followed by a double loop, instead of a four-fold loop; this economy is very important. Even so, liquid water simulations incorporating the four-body term are far more demanding than those with only two- and three-body interactions.

The actual parameters used by Clementi and Corongiu in realizing their expression for the three-body interaction term are derived from a fit to *ab initio* SCF calculations on the water trimer at 173 different geometries; to ensure accuracy, each of these calculations uses a large basis set and corrects for superposition error by means of the counterpoise technique (29). We have seen that the four-body term we propose is already implicit in the model adopted by Clementi and Corongiu, so we are obliged to carry over their parameters as well. It remains to verify that our model does in fact account for the four-body interaction, and we have carried out several *ab initio* SCF calculations for the water tetramer at different geometries as a check. These calculations support the use of our model, and also indicate that the three-body terms found in Eq. (5) are helpful in giving a good description of the three-body interaction.

The production calculations described here have a unit cell containing 512 water molecules, and we find that, even with the economies permitted by our model for the three- and four-body interaction terms, our calculations require two orders of magnitude more run time than analogous calculations with two-body interactions only.

Let us now briefly outline our programming solution for the migration of our Monte Carlo codes from sequential to parallel (on the LCAP). The practical implementation of this system requires that the main program (which runs on the host) be able to closely control the subtasks running on the AP's, periodically transferring data to them, starting them, and subsequently gathering up each set of subtask output.

Our Metropolis-Monte Carlo liquid simulation programs can all be broken down according to the following scheme:

(1) initial input and setup,
(2) generate tables for energy evaluation,
(3) evaluate initial energies,
(4) generate trial move coordinates,
(5) generate trial move energy arrays,
(6) accept or reject move,

(7) update coordinates,
(8) update energy arrays,
(9) final results and output.

Items (1), (2) and (9) occur only once for a particular run, but the remaining steps take place inside loops over the desired number of moves. This includes item (3), since the energies should be recomputed from time to time in order to avoid the accumulation of roundoff error. The number of moves in a run is typically very large (many thousands), so virtually all the run time resides in items (3-8). For a single move, handling the trial move coordinates and the calculation to accept or reject the move are not demanding; by far the heaviest computational demands come from the evaluation of the energy differences associated with the move.

From this breakdown it is fairly clear which parts of the program are worthwhile to modify so they run in parallel on the AP's. The initial input and setup and the final results and output, items (1) and (9), run best on the host, while the evaluation of the move energies should be put on the AP's in parallel; this includes not only items (3) and (5), but also the table generation, item (2).

We thus arrive at a plan for the program where input and run setup is handled on the host, after which a copy of the initial coordinates is transmitted to each AP; each AP thereupon generates its own set of tables for future use and then proceeds with the initial energy evaluation. The energy parameters required by the host are then gathered from the AP's and the loops associated with move generation are entered; these loops are under the control of the host. Within them, the generation of the trial move coordinates takes place on the host, which then transmits a copy of these coordinates to each AP to use in its portion of the generation of the trial move energy arrays. The AP's then transmit back to be gathered up by the host the energies associated with the move. On this basis, the host decides whether to accept the move, and transmits this decision to each of the AP's, so they may update their coordinate and energy arrays if necessary. Control then passes back to the host, and the loops are recycled or exited for final results and output.

We have added only a few refinements to this program plan, in order to minimize the time taken up with transmission between host and AP's (28).

The AP's must retain the old trial move coordinates in memory when control is passed back to the host. In addition to these coordinates, we require retention of the tables that were set up at the beginning of the run, all the intermediate energy arrays, and the set of coordinates for all of the molecules. It is an enormous savings in transmission overhead to retain this data in AP memory throughout the run. Fortunately, this capability is standard with the APEX software normally supplied with the FPS-164: provided a new program image is not loaded, material in the FPS-164 memory is undisturbed unless specifically altered by an APEX software call (30).

It is well worth the trouble to minimize host-AP transmission overhead. With only a two-body interaction, evaluation of a single move (on one AP) takes place on the order of several hundredths of a second, and host-AP move transmission time can be an appreciable fraction of this if one is careless.

The remaining task in migrating our Monte Carlo programs to parallel mode is the actual splitting of the move energy evaluation among the various AP's. We shall not be too concerned with details here, since these naturally vary according to the specifics of a particular program; instead, we have presented elsewhere (28) organizational features that tend to emerge in most of our programs.

We recall that many intermediate arrays are maintained for move energy evaluation. An obvious recipe for migration to parallel execution is to divide the arrays among the various AP's, with each AP responsible only for the evaluation of its particular portion of the array (28).

We note that the strategies we have implemented for handling the arrays not only support parallel use of the AP CPU's, but also the AP memories: Since the arrays are distributed over the different AP's, without much duplication, we have the sum of all the AP memories available to the run. Thus our parallel strategy not only allows practical execution of runs requiring more CPU time, but also those with larger memory requirements, or both.

VI. Liquid Water M.C. Simulation with up to Four-Body Interactions

The previous section intended to give some idea of how we actually accomplished migration of our programs to parallel execution mode. We note that the idea of parallel execution actually has a rather long history; we cannot claim much originality in this area. The distinction we do claim is that we have put these ideas to practical test in production calculations of genuine scientific interest. As a test for our scheme in this section, we discuss the Metropolis-Monte Carlo code we are using for liquid simulation, incorporating the three- and four-body interactions described above. This calculation is very heavy in its computational demands, and can hardly be handled with any normal computer resources, so that it fulfills every requirement as a good practical test of the LCAP system. Furthermore, it is of considerable interest, since the most sophisticated calculations performed heretofore (27) yield an enthalpy which is about 1.674 KJ/mole above the experimental value, 41.3 KJ/mole at room temperature and one obvious explanation of this discrepancy is the omission of four-body interactions.

The calculations we are pursuing to investigate this question are not yet to the point where we can present conclusive results: Our current runs include about 1,000,000 moves, which is enough to achieve equilibration, and nearly enough to support *good* statistics. On a preliminary basis, however, we appear to be getting an enthalpy about 3.35 KJ/mole *below* the results incorporating only the three-body terms. This suggests that the four-body interactions are indeed significant in getting a good liquid water simulation. One notices that our results actually fall below experiment; presumably this is due to our omission of the quantum correction, which is estimated to be of the right sign and magnitude (3 to 4 KJ/mole) to bring our results into better agreement with experiment (31). We shall return to this point in the last section of the paper. Here we wish to comment that this exceptional agreement obtained by using *ab initio* potentials, could in itself justify our efforts in assembling the LCAP system.

In Figure 8 (top) we report our simulated pair correlation function g(O—O) obtained with two-body (left), two- and three-body (middle) and two-, three- and four-body interaction potentials (right). The full line is for the experimental data. In the same figure we report the simulated X-ray and neutron-beams scattering intensities. The agreement with the experimental data and the "convergence" of

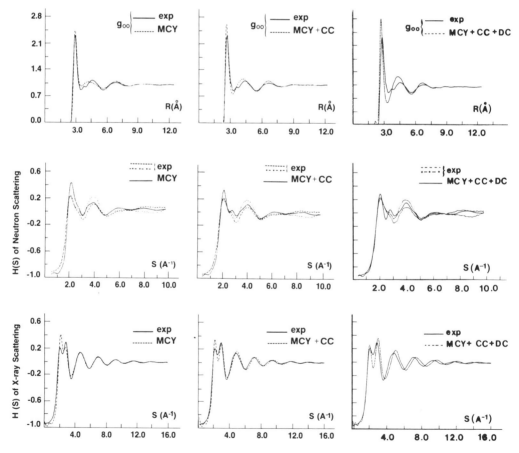

Figure 8. Simulation of liquid water at room temperature with two- (left), two- and three- (middle) and two-, three- and four-body interactions (right). Top: Pair correlation function, g(O-O). Middle: Comparison with X-ray scattering. Bottom: Comparison with neutron-beam data.

the many-body series are most gratifying. Notice, in particular, how the many-body brings more and more structure into the g(O−O) and into the beam scattering intensity. After about ten years of systematic work, the intensity in the split first peak is finally correct. These results point out that *now* it is reasonable for us to introduce vibrations into our model (of course, *ab-initio,* and without any empiricism).

We may conclude that LCAP, our experimental system, is in fact already achieving one of its principal goals even at this rather early stage. It is providing us with a means of accomplishing scientific calculations that are essentially out of reach for a more conventional computer configuration.

VII. Parallelism in Molecular Dynamics

As it is known, the Molecular Dynamics, MD, method was first developed for rigid spheres fluids by Alder and Wainwright (32). This computer modeling method was later extended to more realistic Lennard-Jones potentials for simple liquids by Rahman (33), Levesque and Verlet (34) and others. The MD method, as standardly used, has two major drawbacks, i) accurate potential-energy functions for real systems are not readily available, ii) it requires numerical solution of vast number of coupled nonlinear differential equations which must be solved many times.

The first drawback is being confronted with increasingly accurate potentials obtained from quantum mechanical computations as indicated in the previous section and elsewhere (26). This section attempts to confront the second drawback, by considering our own brand of parallelism, as put forward in the first two sections of this paper.

We recall that molecular dynamics is used to calculate the observed classical trajectory for an assembly of N (usually several hundreds to thousands) interacting molecules in a box of side L. The Hamiltonian of a N-molecule system is

$$H = H_T + V(\vec{q}_1, ..., \vec{q}_N) \tag{6}$$

$$H_T = \Sigma_i \, \vec{p}_i^2/2m \tag{7}$$

where $(\vec{q}_1, ..., \vec{q}_N)$ denotes the generalized coordinates and $(\vec{p}_1, ..., \vec{p}_N)$ denotes the generalized conjugate momenta, H_T is the total kinetic energy of the system, $V(\vec{q}_1, ..., \vec{q}_N)$ represents the total potential energy, which is the sum of two-body, three-body, ..., n-body interactions (see Eq. 1). In general, the two-body interaction energy is assumed to be a pairwise sum of spherically symmetric functions. For molecules possessing charged sites, the interaction of a charged site with the charged sites outside the simulation box constitutes another term in the potential energy, V. Using the Ewald summation method, this reciprocal space (RC) potential term, $V_{RC}^{Coulomb}$, can be conveniently expressed as

$$V_{RC}^{Coulomb} = (\tfrac{1}{2}\pi L) \sum_{\vec{k}=\vec{0}} A(\vec{k}) \left\{ [q_i \cos(\tfrac{2\pi}{L}\vec{k}\cdot\vec{r}_i)]^2 + [\Sigma_i q_i \sin(\tfrac{2\pi}{L}\vec{k}\cdot\vec{r}_i)]^2 \right\} \tag{8}$$

where $\vec{k}$ is a vector in the reciprocal lattice of the simulation box, q_i is the charge of the site i, and $A(\vec{k})$ is the Fourier component of the long-range potential (35-37).

For a 6N-dimensional phase space (Γ-phase space), the state of the system is represented by a point $(\vec{q}_1(t), ..., \vec{q}_N(t); \vec{p}_1(t), ..., \vec{p}_N(t))$ which evolves according to Hamilton's equation

$$\partial\vec{q}_i/\partial t = m^{-1} \nabla_{p_i} H \qquad\qquad i = 1, ..., N \qquad\qquad (9a)$$

$$\partial\vec{p}_i/\partial t = -m^{-1} \nabla_{q_i} H \qquad\qquad i = 1, ..., N \qquad\qquad (9b)$$

subject to a given N-molecule initial condition $(\vec{q}_1(0), ..., \vec{q}_N(0); \vec{p}_1(0), ..., \vec{p}_N(0))$.

The MD method essentially consists of solving the initial value problem of Eq.(9). There are various integration algorithms for the solution of this problem. In general, the purpose of any integration procedure is to determine the phase point Γ_{n+1} representing the state of the system at time $t_{n+1} = t_n + \Delta t * \Gamma_{n+1}$ once the phase point Γ_n has been specified. The approach employs a power series in t, which is truncated at a given term, say $(\Delta t)^k$, where k is the order of the integration method.

The organization of the sequential molecular dynamics program MABDS (for "MAny-Body Dynamics Simulator") is depicted in Fig. 9; a detailed description of the program is given elsewhere (37).

The coupled set of equations, Eq. 9, is nonlinear in character. The strong coupling effect is apparent in the detailed form of the interacting Hamiltonian of the N-molecule system, Eqs.(6-8), especially when strong force liquid systems are considered. The mathematical problem on how one can multitask the procedure to find the solution to the coupled set of nonlinear differential equations (Hamilton's or Newton's equations of motion) has been solved and discussed in detail elswhere (37).

We are now in a position to reformulate the MD program for a parallel engine. A description of how the sequential MD code is migrated to parallel mode can now be discussed.

The sequential program was first analyzed to verify the strictly sequential and the "parallelable" parts of the computation. The timing histogram of our sequential program is given in Fig. 10 for a system containing 512 water molecules. Because the CPU time (Tr) per simulation step to execute the subroutines labelled (1)-(9) and (14) listed in Fig. 9 is relatively short and the intermix between Tr and Tt is low, these subroutines run on the primary process (host). The timing histogram in Fig. 10 also indicates that the evaluation of potential energies, site forces (and site virials) represents the computational workhorse of the molecular dynamics program. Considering the pairwise potential energy term used to compute the short-range interaction, the number of pairs to be evaluated at each time step is $N(N-1)/2$. For our simulation system of 512 MCY-waters (38), the number of floating point opera-

```
                      program Molecular_Dynamics;
              files : input, restart_configurations_input,
                      output, restart_configurations_output;
              begin;
(1)              /* Initialize */
(2)              /* System_definition */
                 if ( Look_up_tables_needed )
(3),(4)            then /* Set up force_potential_virial tables */
(5)              /* Tail corrections for short range interactions */
                 if ( Restarting )
(6)                then /* Get restart configurations from
                            previous MD or MC runs */
(7)                else /* Set up starting configurations */
                 if ( New_velocities )
(8)                then /* Randomize initial momentum */
(1)              /* Record status */
                 for Each_time_step Loop
                    Loop_begin ;
(9)                 /* Update accumulators */
(10)                /* Evaluate short range two-body force,
                        potential, and virial */
(11)                /* Evaluate long range force, potential,
                        and virial, using Ewald image method */
                    if ( Three_body & Water_is_a_component )
(*)                   then /* Evaluate three-body force,
                              potential, and virial */
(12)                /* Integrate translational motion, using a
                        finite difference method */
                    if ( Rigid_bodies )
(13)                  then /* Integrate rotational motion, using quaternion
                              algebra and finite difference methods */
(14)                /* Find average quantities */
(1)                 /* Record status */
                    end;
(1)              /* Final results and output */
              end Molecular_Dynamics.
```

Figure 9. A brief organization of the Molecular Dynamics program, MABDS.

tions per pair in the FORCET routine alone is approximately 180. Note that M is equal to the number of molecules (512 in our case) times the number of interacting sites for the water model (4 for MCY); therefore $M = 512 \times 4 = 2048$. Consequently about 7.2×10^8 floating point operations are needed in each time step to evaluate the interaction pairs. The terms involving the Ewald image potential are expressed with all pairs of charged sites included in a single sum Eq. (8). Furthermore, it is adequate to represent the potential with accuracy of $1*10E-4$ if the value of k appearing in the summation index in Eq. (8) is set to be $<2*\pi*NSHELL/L$, where the number of spherical shells in k-space (NSHELL) is chosen to be 3 (39,40). As long as the value for k is held fixed, the CPU time to evaluate the terms involved in the Ewald image potential for each simulation time step is proportional to the number of charged sites. The number of triplets in the three-body interaction (27) is $N(N-1)(N-2)/3!$.

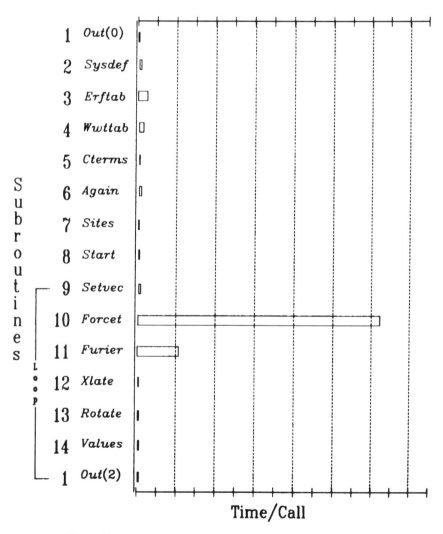

Figure 10. Relative timing histogram of the subroutines in MABDS.

Based on the timing histogram of Fig. 10 and the discussion of how to multitask the computation of a set of coupled nonlinear differential equations presented above, the calculation of the potentials, site forces, (and site virials) are the primary tasks to be broken up into subtasks. Note that the CPU time to solve the set of equations of motion grows linearly with the number of atoms and molecules present in the simulated system. The execution of the integrator modules is also done in parallel. Consequently, the routines (10)-(13) and (*) displayed in Fig. 10 were modified to run in parallel. These routines make up the calculation phase of the MD runs and are called once per time step. We have thus arrived at the following scheme for the multitasking of our "parallelable" routines.

1) At the beginning of each time step, the primary process (PP) sends to each secondary process (SP) the coordinates and velocities of the previous phase space point Γ_n.

2) The primary and the secondary processes then send these data to the AP's. Each AP evaluates its assigned portion of the short-range part of the potential, site force, and site virial. Table look-up or direct evaluation of the potential, site force, and site virial are available. Computation of the radial distribution functions can also be performed if desired. Upon completion, each AP transfers its portion of the potential, site force, and site virial back to its corresponding process. The secondary processes then send to the primary process their portions of these calculated data, where elements calculated by each SP are put together.

3) The PP sends to each SP a portion of the short-range part of the potential, site force, and site virial. These data are sent to the AP's. The AP's then evaluate the reciprocal space contribution to the Coulomb energy, site force, and site virial. Option of using real arithmetic (to evaluate Eq.(8) and the corresponding force function) or complex arithmetic is available. If three-body correction for water is required, each AP then evaluates its assigned portion of the three-body correction to the center of mass force and torque, potential, and virial. Each AP then converts the site forces into center of mass force and torque, adds to these quantities the corresponding three-body corrections, then performs the integration of the equations of motion to find the coordinates and velocities of the phase space point Γ_{n+1}. Various quantities such as mean square displacement, species potential energies, kinetic energies, etc. can also be computed.

4) The computed data in step 3 are sent from the AP's to their corresponding processes. The SP's then send back to the PP their portions of the computed data. The PP then puts these data together, computes step temperature and pressure, and performs averages for various thermodynamics quantities.

5) Periodic output may be generated at this point; and the cycle repeats from step 1 for n number of time steps. When the cycle stops, final output for average values of computed quantities is done by the primary process.

For the water system test case reported elsewhere (41) the sequential routines (10)-(13) and (*) have been migrated to two routines that execute in parallel mode. These routines mainly compute the force (and the torque) that act on the molecules, and integrate the corresponding equation of motion.

Fig. 11 describes the relationships between the CPU time and the number of simulated water molecules. The behavior, nevertheless, is clearly of a more general nature. The reduction in time as a result of using a multiprocessor enables one to consider much larger and more complex systems. Furthermore, as the size of the computation grows, there is more to be gained by using a parallel scheme, and a better incentive to use more processors.

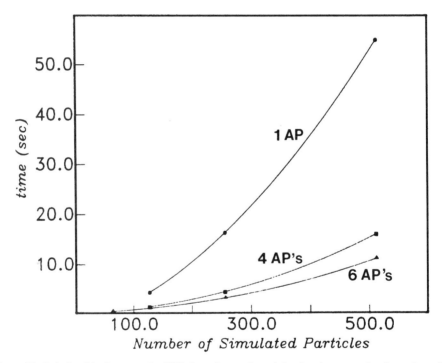

Figure 11. Relationships between the CPU time, the number of simulated water molecules and number of AP's.

VIII. Liquid Water by Molecular Dynamics

In this section we present results, obtained in a Molecular Dynamics simulation of liquid water using a potential model where the three-body interactions are included. Two different ab initio pair potentials are used: The MCY potential (38) and a recent potential by V. Carravetta and E. Clementi (42) (denoted hereafter as the VCC potential), while the three-body potential is the CC potential (27) discussed also above in the Monte Carlo study.

The computations consist of two main runs, denoted as I and II. As two-body potential, the MCY and the VCC potentials have been used for run I and II, respectively. For both runs the number of water molecules considered is 512. At the density of 0.997 g/cm^3, corresponding to a molar volume of 18.07 cm^3/mol this gives edge length of 24.86 A for the cubic box simulation cell. The dynamics of water molecules is generated using integrators based on the so called leap-frog (43) schemes, both for the centre of mass motion and for the rotational motion (44). Using a time step as high as 2.0×10^{-15} seconds the energy is still stable. This is generally one of the advantages in using leap-frog integrators. The spatial arrangement of the fractional charges gives a dipole moment of 2.19 D for the single MCY water molecule and 2.01 D in the VCC water. The experimental value for the dipole moment in water molecule is measured to 1.86 D (45). For a comparison, it can be mentioned that the corresponding values for the water dipole moments in some other potential models are (in Debyes), Rowlinson (46): 1.84, BNS (47): 2.17, ST2 (48): 2.353 and CF (49): 1.86. The long-range coulombic interactions due to the fractional charges are taken into account by the Ewald summation technique. The number of terms in the Ewald Fourier series (calculated in the reciprocal space) is 309.

The two simulations have been performed at 300 K and at the normal liquid density at 1 atm. Cubic periodic boundary conditions are used with the cut-off radius at one half of the box length, together with the minimum image convention. In all the runs the equilibrium phase was chosen to be 2000 steps, followed by a production phase of another 2000 steps.

The computed internal energies are -38.6 Kj/mole and -38.5 Kj/mole for the two-and the three-body using either MCY or VCC two-body potentials. The three body contribution is -4.2 Kj/mole and -4.0 Kj/mole for the MCY or VCC case, respectively.

The inclusion of the three-body potential in the simulations considerably narrows the gap between the experimental value at 300 K (50), -41.0 kJ/mol and the calculated value when the pairwise approximation is used. The calculated values here for the internal energy are in agreement with the corresponding values, reported by Clementi and Corongiu (27) using the MCY and CC potentials in their MC study of liquid water. As pointed out in Section VI, quantum mechanical corrections to thermodynamical quantities for liquid water have been investigated in a recent paper of Berens et al. (31), who report 4.2 kJ/mol for the total intermolecular energy and 2.2 kJ/mol for the free energy. Our estimate to the free energy, calculated from the mean square forces and torques is 3.0 kJ/mol with the two-body potential and 2.9 kJ/mol after the three-body potential is included.

The x, y, and z components of the velocity autocorrelation functions are calculated for the centre of mass motion. The motion of the centre of mass is correlated to the molecular axis system at t=0. In Figure 12 (top left inset), we report the components of the velocity auto-correlation function (VACF), calculated from the trajectories generated using the MCY-potential. The molecule lies on the yz-plane with the

Centre of mass VACF Components

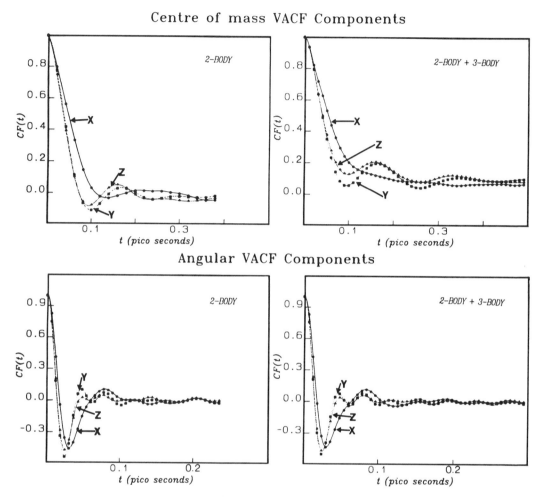

Angular VACF Components

Figure 12. Cartesian x- y- and z-components of the centre of mass velocity auto correlation function, calculated with MCY potential (top left) and with MCY+CC (top right). The components of the angular velocity autocorrelation function about the principal x- y- and z-axes of the moment of inertia, calculated with the MCY potential (bottom left) and with MCY+CC (bottom right).

z-axis as a C_{2v} axis. It is clear from Figure 12, that the simulated water molecules are more mobile in the x-direction (perpendicular to the molecular plane) than in y- or z direction within the actual time scale. This is in agreement with the observation of Impey et al. (51). In Fig. 12 (right top inset) we report the corresponding VACF for the centre of mass motion *after including the three-body potential.* Now, the displacement in the x-direction increases significantly. At the same time, the whole molecule has become more mobile. The VACF components do not exhibit negative values after the first drop-off as observed in the two-body VACF (Fig. 12, left side). Also a new feature is that the y- and z-components are more resolved, indicating a more diffusive motion in the y-direction than in the z-direction, in the time-scale around 0.1 picoseconds.

The components of the *reorientational motion* have been also analyzed. In Fig. 12 (bottom left inset) we report the angular velocity autocorrelation functions about the three principal axes x, y, and z of the moment of inertia. The molecular axis system is the same as used in the calculation of the centre of mass VACFs. Although the moment of inertia values are 2.94, 1.02 and 1.92 (in units of 10^{-47} kgm^2) for the xx, yy and zz (diagonalized) tensor components for the water geometry used in this study, the y- and z- components are only slightly resolved. This behaviour is observed by Impey et al. (51) and Tanaka et al. (52). As a whole, the inclusion of the three-body potential does not seem to effect the reorientational motion in the actual time scale.

From this simulation we can conclude, once more, that many-body corrections should be included in M.C. or M.D. simulations. Likely the time for pairwise simulation is slowly coming to an end especially for liquids and solutions with hydrogen bonded connectivity. For additional details on this M.D. simulation, we refer elsewhere (41). We are now planning to introduce four-body terms into the M.D. program. Clearly the complexity of the computation will increase substantially, but it will remain a reasonable task for the LCAP system.

Concluding Remarks

For some time we have been experimenting in our laboratory with our LCAP. We find the system notably adaptable and flexible, both to produce new applications (coding and debugging) and to carry out extremely demanding simulations such as the quantum mechanical and the statistical mechanical simulations reported in this work. There are a number of drawbacks which can be grouped into two classes. First, the FPS-164, even if most attractive in its price-performance and compiling facilities, has size, electric power and noise problems, which are only compounded by collecting ten FPS-164's. Second, the lack of a compiler (or even a precompiler) *for parallelism* limits the present use of LCAP to scientists who are familiar with the details of their theory and computer programs. Of course, once compiled, the application programs can be used as easily as any sequential program.

At present we bear with the first problem (quite easily indeed); concerning the second we are planning to write a compiler, rudimentary at first. In addition, at the hardware level, we plan to improve on the $E = Tr/Tt$ ratio by connecting the AP's with a fast bus and by establishing a second communication avenue which makes use of shared memories between AP's. The increase in communication will couple more and more the AP's; eventually we plan to evolve from LCAP to CAP. Finally, the *"visiting scientists"* experiment should be paramount in establishing how broadly this "brand" of parallelism can be succesfully used in scientific/engineering applications. To computational chemistry, our long-lasting arena, our brand of parallelism seems to fit nearly optimally.

Acknowledgement

It is our pleasure to thank Dr. Arthur G. Anderson and Mr. Earl F. Wheeler for their full support in our experimentation on parallel systems and applications.

References and Footnotes

1. See for example the papers in *Revs. Mod. Phys., 32,* (1960).
2. See for example E. Clementi, *J. Chem. Phys., 36,* 33 (1962); R.K. Nesbet, *J. Chem. Phys. 36,* 1518 (1962); R. G. Parr, *"The Quantum Theory of Molecular Electronic Structure,"* W. E. Benjamin, Inc., New York, 1963.
3. R. S. Mulliken, Nobel Prize Acceptance Speech, Stockholm, 1967.
4. a) *"International Conference on Parallel Processing"* August 25-28 (1981), Bellaire, MI., R. H. Kuhn, D. A. Padue, Eds., Los Angeles, CA. b) Hockney, R. W., *"Parallel Computers: Architecture, Programming and Algorithms",* R. W. Hockney, C. R. Jesshope, Eds., Bristol, Adam Hilger, Ltd. (1981). c) *"Parallel Processing Systems",* J. Evans, Ed., Cambridge, NY, Cambridge Univ. Press (1982). d) Y. Wallach, *"Alternating Sequential/Parallel Processing,"* Lecture Note in *Computer Science, 124,* Springer Verlag, Berlin, NY (1982). e) K. R. Wilson, in *Computer Networking and Chemistry,* P. Lykos, Ed., ACS Symposium Series, *19,* American Chemical Society (1975). f) J. D. Kuck, Y. Muraoka, and S-C. Chen, *IEEE Trans. Comp., Vol. C-12, No. 12,* (Dec. 1972). g) D. Kuck, P. Budnik, S-C Chen, E. Davis Jr., J. Han, P. Kraska, D. Lawrie, Y. Muraoka, R. Strebendt, and R. Towle, *IEEE Trans. Comp., Vol. C-23, No. 1,* (Jan. 1974). h) Arvind, *Proc. IEEE Conf. on Parallel Processing, 1980,* p. 7. i) *Parallel Architecture Workshop Visuals,* Univ. of Colorado, Boulder, (Jan. 1983). j) P M. Kogge, *"The Architecture of Pipelined Computers",* McGraw-Hill, New York. k) H. Fromm, U. Hercksen, U. Herzog, K-H. John, R. Klar, and W. Kleinoder, *IEEE Trans. Comp. Vol. C-32, No. 1,* (Jan. 1983).
5. E. Clementi, G. Corongiu and J. H. Detrich (to be published).
6. J. H. Detrich, G. Corongiu and E. Clementi, *Int. J. Quantum Chem.* (1984 Sanibel Symposium, in press).
7. H. L. Nguyen, H. Kahnmohammadbaigi and E. Clementi (to be published).
8. E. Clementi, J. M. Andre, M. C. Andre, D. Klint and D. Hahn, *Acta Physica Hungaricae, 27,* 493 (1969).
9. E. Clementi, J.W. Mehl and W. von Niessen, *J. Chem. Phys., 54,* 508 (1971)
10. E. Clementi and G. Corongiu, *Int. J. Quantum Chem., 9,* 213 (1982).
11. D. R. Hartree, *"The Calculation of Atomic Structures",* John Wiley & Sons, Inc., New York (1957).
12. R. S. Mulliken, and C. C. J. Roothaan, *Proc. Natl. Acad. Sci., U.S.A., 45,* 394 (1949); C. C. J. Roothaan, *Rev. Mod. Phys. 23,* 69 (1951); C. C. J. Roothaan, *Rev. Mod. Phys. 32,* 179 (1960); G. G. Hall, *Proc. R. Soc. Sect. A., 208,* 328 (1951).
13. R. McWeeny, *Proc. Roy. Soc., A196,* 215 (1946); S. F. Boys, *Proc. Roy. Soc. (London), A200,* 542 (1950); *ibid. A201,* 125 (1950).
14. D. R. Davis and E. Clementi, *"IBMOL: Computation of wave function for molecules of general geometry. An IBM 7094 program using the LCGO-MO-SCF Method"*—IBM Technical Report— (December 1965); E. Clementi and D. R. Davis, *J. Chem. Phys. 45,* 2593 (1966).
15. E. Clementi and D. R. Davis, J. of Comp. Phys. 1 , 223 (1966).
16. a) A. Veillard and E. Clementi, *"IBMOL: Computation of Wave Functions for Molecules of General Geometry — Version 2",* IBM Technical Report, (August 1966). b) J. W. Mehl and E. Clementi, *"IBM System 360 IBMOL-5 Program Quantum Mechanical Concepts and Algorithms",* IBM Technical Report RJ #883, (June 22, 1971). c) R. Pavani and L. Gianolio, *"IBMOL-6 Program — User's Guide",* Ist. Ricerche G. Donegani, Novara-Italy, Technical Report DDC-771 (January 1977). d) E. Ortoleva G. Castiglione and E. Clementi, *Comp. Phys. Comm., 19,* 337 (1980). e) C. Lupo and E. Clementi, *"Integrals Computation for Molecular Wave Functions: A Program With The Introduction of Local Symmetry"*—IBM Technical Report POK-18, (December 18, 1982). f) P. Habitz and E. Clementi.

Comp. Phys. Comm., 29, 301 (1983). g) E. Clementi and G. Corongiu, *Chem. Phys. Letters, 90,* 359 (1982). h) E. Clementi and J. W. Mehl, *"IBM System 360 IBMOL-5 Program General Information Manual",* IBM Technical Report RJ #853, (May 3, 1971). See also Ref. 16b).

17. E. Clementi, *Phys. of Elec. and Atomic Coll., VII ICPEAC,* North-Holland, 399 (1971). See also ref. 16h).

18. E. Clementi, *IBM Journal Res. and Dev. Supplement 9,* 2 (1965); E. Clementi, *Proc. Natl. Acad. Sci., U.S.A., 69,* 2942 (2972); G. C. Lie and E. Clementi, *J. Chem. Phys. 60,* 1275 (1974); *ibid. 60,* 1288 (1974).

19. a) M. Dupuis, J. Rys and H. F. King, *J. Chem. Phy. 65,* 111 (1976); H. F. King and M. Dupuis, *J. Comp. Phy. 21,* 144 (1976). b) V. R. Saunders and M. F. Guest, *Comp. Phys. Comm. 26,* 389 (1982). c) J. Almlof, K. Faegri, Jr. and K. Korsell, *J. Comp. Chem. 3,* 385 (1982). d) W. J. Hehre, W. A. Lathan, R. Ditchfield, M. D. Newton and J. A. Pople, *Quantum Chemistry Program Exchange, Program No. 368,* Indiana University, Bloomington, IN. e) L. E. McMurchie and E. R. Davidson, *J. Comp. Phys. 26,* 218 (1978). f) P. S. Bagus and A. R. Williams, *IBM J. of Research and Develop. 25,* 793 (1981). g) K. Ohno, in *Horizons of Quantum Chemistry,* K. Fukui and B. Pullman, Eds., pp. 245-266, (1979), Reidel, Dordrecht.

20. P. O. Löwdin, *Reviews of Modern Physics 35,* 724 (1963).

21. L. Gianolio, R. Pavani and E. Clementi, *Gazz. Chim. Ital., 108,* 181, (1978).

22. T. J. Zielinski, M. Shibata and R. Rein, *Int. J. Quantum Chem. 19,* 171 (1981).

23. A. Mondragon and I. Ortega Blanke, *Int. J. Quantum Chem. 22,* 89 (1982).

24. H. Chojmeck, J. Lipinski and W. A. Sokalski, *Int. J. Quantum Chem. 19,* 339 (1981).

25. S. Scheiner and C. W. Kern, *J. Am. Chem. Soc. 101,* 4081 (1979); *Chem. Phys. Letters. 57,* 331 (1978).

26. See, for example, the monographic work by E. Clementi *"Computational Aspects for Large Chemical Systems",* Lecture Notes in Chemistry, Vol. 19, Springer-Verlag, Berlin, New York, (1980).

27. E. Clementi and G. Corongiu, *Int. J. Quantum Chem. Symp. 10,* 31 (1983).

28. J. H. Detrich, G. Corongiu and E. Clementi, *Int. J. Quantum Chem. Symp.* (1984) (in press).

29. S. F. Boys and F. Bernardi, *Mol. Phys. 19,* 553 (1970)

30. *FPS-164 Operating System Manual, Vol.2* (Publication No. 860-7491-000B), Floating Point Systems, Inc. (January, 1983).

31. P. H. Berens, D. H. J. Mackay, G. M. White, and K. R. Wilson, *J. Chem. Phys., 79,* 2375 (1983).

32. B. J. Alder and T. E. Wainwright, *J. Chem. Phys. 31,* 459 (1959).

33. A. Rahman, *Phys. Rev. 136,* A405 (1964).

34. D. Levesque and L. Verlet, *Phys. Rev. A2,* 2514 (1970).

35. N. Anatasiou and D. Fincham, *Comp. Phys. Comm., 25,* 159 (1982).

36. D. Fincham, *"Information Quaterly-CCP5",* 2, 6 (1981).

37. H. L. Nguyen, H. Khanmohammadbaigi and E. Clementi, to be published.

38. O. Matsuoka, M. Yoshmine, and E. Clementi, *J. Chem. Phys., 64,* 2314 (1976).

39. D. Ceperly, *Nat. Resource Comput. Chem. Software Catalogue, Vol.1.* Prog. No. 5501(CLAMPS), 1980.

40. H. L. Nguyen, Ph.D. Thesis, Purdue University, 1983.

41. A. Laaksonen, E. Clementi (submitted for publication).

42. V. Carravetta and E. Clementi, *J. Chem. Phys.,* In Press.

43. L. Verlet, *Phys. Rev., 165,* 21 (1967)

44. D. Fincham, *"Information Quarterly — CCP5",* 2, 6 (1981)

45. T. R. Dyke, J. S. Muenter, *J. Chem. Phys., 59,* 3125 (1973)

46. J. S. Rowlinson, *Trans. Far. Soc., 47,* 120 (1951)

47. A. Ben-Naim and F. H. Stillinger, " Aspects of the Statistical Mechanical Theory of Water" in *Structure and Transport Processes in Water and Aqueous Solutions,* Ed., R. A. Horne, Wiley-Interscience, New York (1972)

48. F. H. Stillinger and A. Rahman, *J. Chem. Phys., 60,* 1545 (1974)

49. H. L. Lemberg and F. H. Stillinger, *J. Chem. Phys., 62,* 1677 (1975)

50. V. G. Dashevsky and G. N. Sarkisov, *Mol. Phys., 27,* 1271 (1974)

51. R.W. Impey, P. A. Madden and I.R. McDonald, *Mol. Phys., 46,* 513 (1982).

52. H. Tanaka, K. Nakanishi and N. Watanabe, *J. Chem. Phys., 78,* 2626 (1983).

Structure & Motion: Membranes, Nucleic Acids & Proteins,
Eds., E. Clementi, G. Corongiu, M. H. Sarma & R. H. Sarma,
ISBN 0-940030-12-8, Adenine Press,   Copyright Adenine Press, 1985.

Proteins, Cells and Recognition Sites

Ivar Giaever and Charles R. Keese

General Electric Corporation Research and Development
Schenectady, New York 12301 USA

Abstract

The concept of macromolecular binding by close stereochemical fit (the lock and key mechanism) is discussed as it relates to solid state antigen-antibody reactions and cell-substrate interactions. Experimentation in a variety of related areas is briefly described including the use of indium-coated slides to detect protein adsorption, the effect of intense UV irradiation on the properties of adsorbed protein layers, the ability of protein layers at oil-water interfaces to serve as substrates for culturing mammalian cells and the application of weak electric fields to detect cell motion.

Introduction

The lock and key concept has been very useful in biology, conjuring up a mental picture of the sterochemical fit between different large molecules. The concept has been successfully used to describe interactions between substrates and enzymes, hormones and receptors, and antigens and antibodies. In this brief discussion of current work in our laboratory, we shall be concerned with the immunology reaction when either the antigen or the antibody is preadsorbed onto a solid surface. We shall then compare this interaction with the attachment of mammalian cells to artificial substrates when the cells are cultured *in vitro* and show how these interactions, in our opinion, differ from the ordinary lock and key concept.

Solid Phase Immunology

In general, globular proteins will adsorb on any solid surface to form a monomolecular layer. To study such adsorption we employ a specially prepared surface that renders protein monolayers visible (1,2,3). Briefly speaking the surface consists of a glass slide onto which indium has been vacuum evaporated. As the indium vapors condense on the slide they bead into small metallic islands, and under proper conditions the dimensions of these islands are of the order of the wavelength of visible light. If a protein monolayer is adsorbed to this surface, the light scattering properties of the indium particles are altered, and the layer becomes visible to the naked eye.

It is possible to use the indium slide to detect the immunological reaction (Fig. 1). First, the antigen is adsorbed to the slide; next the slide is exposed to a solution that may or may not contain specific antibody. If specific antibody is present, the antibody attaches to the adsorbed antigen by a variety of interactions that require a close molecular fit, and a double molecular layer of protein is formed. Because of the unique properties of the indium slide this double protein layer is easily distinguished by visual inspection from a single layer of protein. If specific antibody is not present, the indium slide will not have changed in appearance.

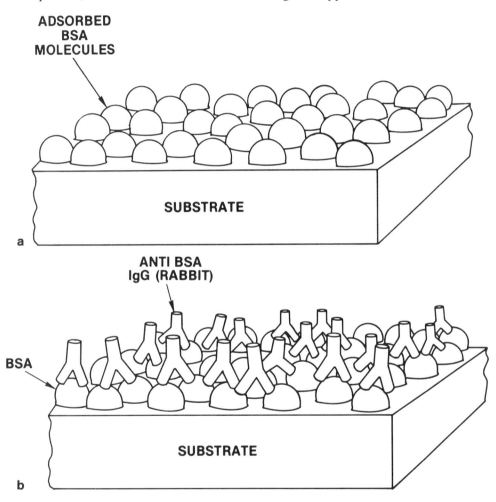

Figure 1. *SOLID PHASE IMMUNOLOGY* (a) In a solid phase immunological assay, such as that using the indium slide, antigen (e.g. bovine serum albumin) is first adsorbed to a substrate. (b) The adsorbed monolayer of protein is then placed in a solution to be tested. If specific antibody is present, it binds to the adsorbed layer forming a double molecular layer. (c) In this photograph of an indium slide viewed in transmitted light, one edge was dipped in a BSA solution (1 mg/ml); the opposite edge was dipped in an egg albumin solution (1 mg/ml). The slide was then rinsed and dried, and the bottom edge was allowed to react 10 minutes with a 1/10 dilution of anti-BSA rabbit serum. The darkest square has a double layer of protein consisting of BSA and antibody.

It is important to recognize that solid phase immunology is not symmetric with regards to antigen and antibody. In general an average size protein molecule will contain about 10 to 20 antigenic sites, each unique from the others. Thus, when the molecule adsorbs onto a surface, the likelihood of all sites being blocked in the adsorption process is very small and, hence, some of the antigenic sites will usually face the solution and be available for complexing with the antibody. If the antibody, on the other hand, is adsorbed on the surface (as long as we are dealing with the IgG molecules) very often the two binding sites will be facing the substrate or otherwise be blocked by steric hindrance. The lock and key concept is still valid, but the lock has been obstructed by the surface adsorption. We have determined that only 1 to 10 percent of the total adsorbed antibodies can react with antigen when the antibodies are adsorbed onto a surface.

UV Radiation and Loss of Antigenicity

In other studies, we have found that the antigenic sites on proteins are very sensitive to ultraviolet light when irradiated in air and can be easily altered. The mechanism is presumably that the UV light breaks chemical bonds in the adsorbed protein layer and simultaneously produces ozone. The broken bonds are then

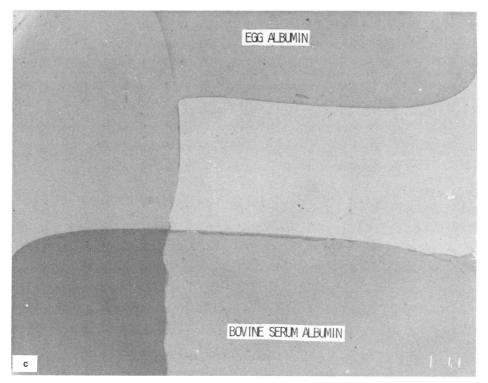

Figure 1c; for legend see previous page.

targets for the highly reactive ozone, resulting in a progressive oxidation of the protein layer (4). The protein layer can be totally removed from the surface if irradiated for long periods of time (10 min). Following such a treatment it can no longer be seen if the substrate is an indium slide. If the protein layer adsorbed on an indium slide is exposed for a relatively short time (30 sec), the optical density does not change, but the antigenic sites are altered, and antibodies will no longer bind to the adsorbed antigen (Fig. 2). Here we have slightly changed the shape or charge of the antigenic site, such that the close molecular fit between the antigen and antibody can no longer be achieved.

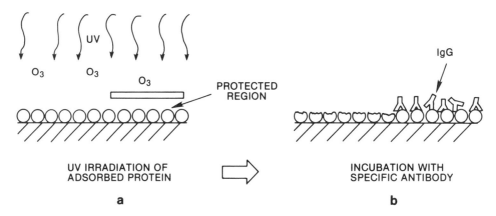

Figure 2. *PARTIAL OXIDATION OF ADSORBED PROTEINS AND LOSS OF ANTIGENICITY* (a) A preadsorbed layer of antigen is briefly exposed, in air, to an intense ultraviolet source resulting in the production of ozone (O_3). The combination of UV damage and ozone results in a partial oxidation of the adsorbed proteins. The area of protein that is shielded from the irradiation is not oxidized. (b) Following such treatment the protein-coated substrate is incubated with antiserum specific for the adsorbed antigen. The partially oxidized protein layer is no longer antigenic, and IgG molecules are able to bind only to the portion of the antigen layer that was previously shielded.

Cultured Cells on Solid Substrates

At this point let us consider the attachment of anchorage-dependent mammalian fibroblast cells to tissue culture dishes. In standard tissue culture protocol, cells of this type require attachment and spreading on a solid support before cell division can occur. Since serum is normally used in the culture medium that bathes the cells, the plastic surfaces that are generally used as substrates are covered with a layer of adsorbed protein before the cells attach (Fig. 3). If cells are seeded into a dish in a protein-free medium, there is no preadsorbed layer, and the cells adsorb directly onto the naked surface. This is a physical phenomenon and does not depend on any active processes in the living cell, neither does it depend on a specific sterochemical fit. If cells are seeded in a protein-free medium onto a dish precoated with an arbitrary protein layer, the cells will attach but will not spread as spreading is an active process, requiring certain components of the serum. Upon addition of serum, however, the cells will spread normally.

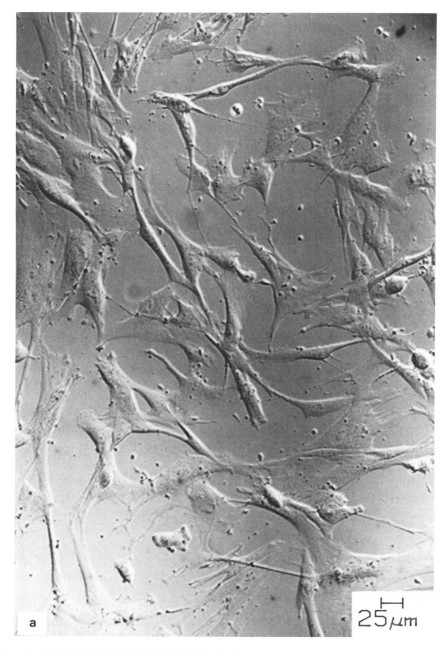

Figure 3. *CELL SPREADING ON SOLID SUBSTRATES* (a) Typical mammalian fibroblasts (3T3 murine cell line) that have spread on a protein-coated solid substrate are shown. (b) The drawing depicts the interaction of a spread mammalian fibroblast with a protein-coated solid substrate such as glass or plastic as viewed from the side. Such cells are thought to normally make contact with the protein layer at a limited number of small areas termed adhesion plaques or focal contacts. The arrows at the left-hand side of the cells represent the action-reaction pair of forces of the cell on the substrate and of the substrate on the cell.

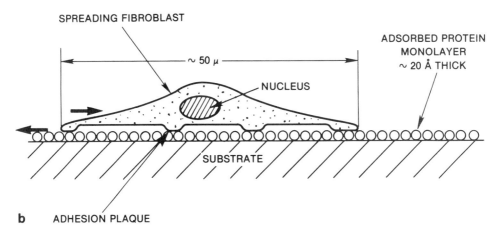

Figure 3b; for legend see previous page.

We have experimented with various precoatings of the plastic dishes to find out if the cells have a preference for certain proteins and to find out whether the lock and key concept can be used to model the results. So far most all adsorbed proteins that have been examined serve as competent surfaces for cell growth, and essentially only one protein, the immunoglobulin molecule, is very different from the others (5). Cells do not attach and spread well on preadsorbed IgG molecules, and if these molecules are specifically attached to a preadsorbed antigen, the cells are unable to attach and consequently to spread and grow on this surface (Fig. 4). Very interestingly, if this surface is irradiated for a short period (30 sec) with UV irradiation and then seeded with cells, the cells attach and grow. This illustrates that the cell's aversion to the adsorbed IgG is most likely a specific, biological effect that can be overcome by slight alterations to the bound IgG's surface properties. It is as if the mammalian cells have a "master key", i.e., one universal attachment molecule, that attaches to most protein molecules with IgG as an exception. An excellent candidate for such a multibinding protein is fibronectin that has been demonstrated to exhibit attachment to a wide variety of molecules.

The situation regarding cell attachment to adsorbed protein layers is the exact opposite of the immune reaction. Here an antibody molecule is specific for a single antigenic site. If the site is slightly altered, the antibody will no longer bind. The cells, on the other hand, will attach to almost any molecule except the IgG molecule. If the IgG molecule is slightly altered via UV-ozone exposure, the cells will attach to this molecule as well.

Cultured Cells on Liquid Substrates

Many problems exist in interpreting results when cells are grown on adsorbed protein layers in tissue culture. One important problem is the difficulty in knowing whether the cells remove the adsorbed protein molecules and form attachments

Figure 4. *HAPTOTACTIC BEHAVIOR OF CELLS ON A SPECIALLY PREPARED SUBSTRATE*
The gray background of this photograph is the result of numerous stained cells on a glass slide, whereas the letters "GE 100" are regions almost entirely devoid of cells. This unusual distribution of cells was obtained by applying two observations mentioned in this paper. A layer of BSA was first adsorbed to the glass slide. The slide was then partially covered with a shield that blocked UV from irradiating a region having the pattern "GE 100". Following incubation with anti-BSA serum, the slide was used as a substrate for the attachment and spreading of normal human fibroblasts (WI-38). The slide illustrates both the effect of partial oxidation on protein antigenicity and the inability of cells to attach to substrates coated with specifically adsorbed IgG molecules.

directly to the substrate. In our laboratory we have addressed this question by growing cells at the interface between an oil phase and the cell culture medium (6,7,8). Under proper conditions we have observed attachment, spreading, and growth of a number of different cell lines on the oil-water interfaces. In this novel arrangement, the cells can spread and grow only if they interact with the protein layer that is adsorbed to the interface, as the liquid oil cannot withstand any shearing forces. The oils we most commonly employ are fluorocarbon fluids. The strength of the protein layer that forms at the interface between the tissue culture medium and the more dense fluorocarbon oil can be regulated by the addition of surface active molecules to the oil phase such as pentafluorobenzoyl chloride (Fig. 5). If no surface active compound is present, the protein layer is not strong enough to support forces exerted by spreading cells. With an intermediate amount of additive, the cells will attach and spread, but often they fracture the protein layer. If large amounts of surface active compound are added, the cells appear to spread on the liquid surfaces in the same manner as on a solid support. In principle it is a great advantage to grow cells on protein layers at liquid interfaces, as, in general, these surfaces are much more reproducible than those on a solid support. We are presently

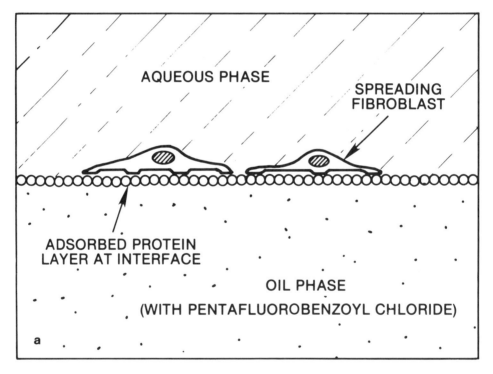

Figure 5. *CELL GROWTH AT OIL-WATER INTERFACES* (a) The attachment and spreading of mammalian fibroblasts on the adsorbed protein layer between two immiscible liquids is illustrated in the figure. When dealing with purified fluorocarbon oils, the addition of a surface active compound, such as pentafluorobenzoyl chloride, is required to produce an elastic protein layer than can withstand the forces exerted on it by the spreading cells. (b and c) The micrographs show 3T3 cells at an oil-water interface (perfluorohexane and tissue culture medium). Without pentafluorobenzoyl chloride added to the oil phase, the cells are unable to spread and ultimately stick to each other to form clusters of rounded cells (b). When a small quantity of the acid chloride is present, cell attachment, spreading, and growth occur to form a confluent layer of cells (c).

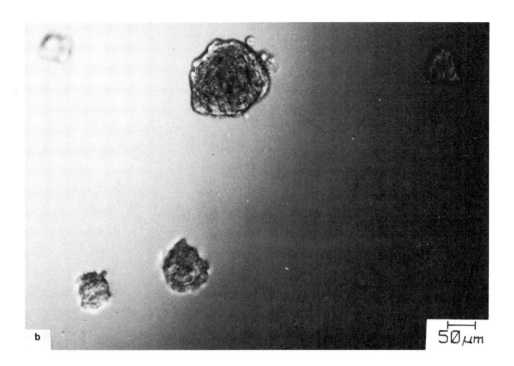

Figure 5b and 5c; for legend see previous page.

in the process of measuring the strength of these protein layers using a modified surface viscometer. The behavior of the protein at the interface varies from a fluid to an elastic membrane, depending on the type and amount of surface active compound added to the fluorocarbon fluid.

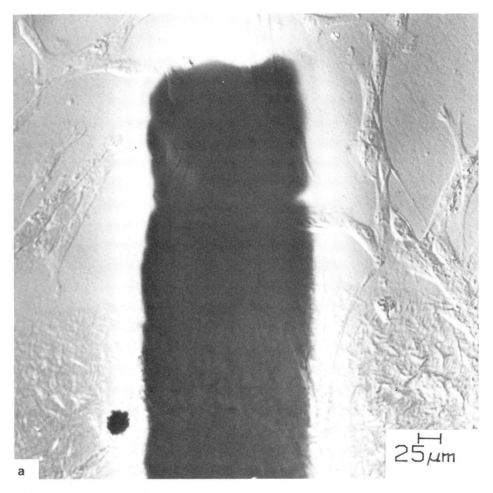

Figure 6. *MONITORING CELL BEHAVIOR WITH ELECTRIC FIELDS* (a) A micrograph of the small (10^{-4}cm¹) electrode used to detect cell motion in culture is shown. Upon careful observation, one can discern a few fibroblasts (WI-38) which are making contact with the gold electrode (the dark rectangular surface at the center of the figure). (b) The voltage across the electrodes and in phase with the applied 4000 Hz signal is shown as a function of time. Following the introduction of cells, the voltage increased slightly and then began to fluctuate with time. Fixing the cells with 10% formalin caused the fluctuations to cease demonstrating that their origin was unquestionably due to living cells. (c) The "noise-like" fluctuations observed in (b) have been recorded and transferred to a computer where they are processed and presented as a power spectrum. In the figure, the power spectrum obtained from normal (WI-38) and transformed (WI-38/VA13) cells are compared. As can be seen, the power spectrum of the transformed (cancerous) cells, although similar in shape, is approximately ten times larger than that obtained from normal cells. Lower frequency bands than those shown have not yet been studied.

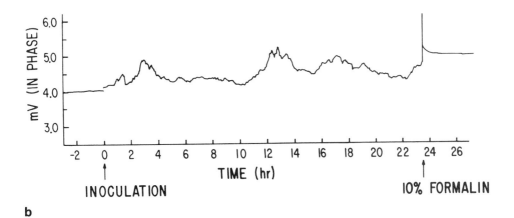

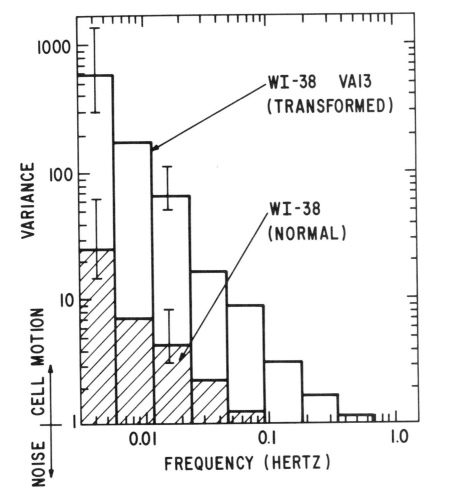

b

c

Figure 6b and 6c; for legend see previous page.

Dynamics of Cell Substrate Interaction

Attachment and spreading of cells to adsorbed protein layers in tissue culture is really a dynamic process, and to fully understand these phenomena it is necessary to follow events in time. We are presently investigating a new technique using weak electric fields to measure this dynamic behavior (9). Cells are culture on gold electrodes that have been evaporated onto the bottom of polystyrene tissue culture dishes. In the specific arrangement we use, we have both a large electrode ($2cm^2$) and a very small electrode ($10^{-4}cm^2$) (Fig. 6). The impedance of the system is such that the impedance of the small electrode to solution dominates the system. By plating out cells on the electrodes, the cells act as insulating particles because of their membranes, and a confluent layer of cells on the small electrode changes the impedance of the system by a factor of maybe 4. We use an applied AC signal of 4000 Hz and monitor both the in phase and the out of phase signal. After an initial rise in the impedance, the impedance will fluctuate as a function of time because the cells are in constant motion, making and breaking contacts with the surface. We have preliminary results on the power spectrum of the fluctuations caused by cancer cells and normal cells. The conclusions are that cancer cells change their contacts with the surface much more often than normal cells and that the magnitude of the noise is also much larger. We now plan to utilize this new method to study the interaction of cells with electrodes that have been precoated with different protein layers. These studies will allow us to determine if the cells recognize different adsorbed proteins as revealed by their dynamic behavior.

Conclusions

The detailed interaction mechanism of cells with their substrates when cells are cultured *in vitro* is still shrouded in mystery. It is clear to us that the specificity of the interaction is very different from the ordinary lock and key mechanisms used in biology. The attachment and spreading of cells on protein surfaces occurs as if the cells had a universal attachment molecule or "master key" that allows the cells to grow on a variety of protein molecules. It is almost opposite to the immune reaction; an antibody can only attach to a specific antigenic site, while a cell can interact with a great variety of sites except for a few specific ones. It should be remembered that the study of cell-substrate interaction has an important bearing on cancer, as it is the lack of specificity that allows cancer cells to metastasize and wander about in an organism. Ultimately, the hope is that when the attachment of cells to substrates *in vitro* has been clarified, we will also have learned something about the *in vivo* situation.

Acknowledgments

These studies were conducted in part pursuant to a contract with the National Foundation for Cancer Research.

References and Footnotes

1. Giaever, I., *J. of Immunology 110,* 1424 (1973).
2. Giaever, I., *J. of Immunology 116,* 776 (1976).
3. Rej, R., C.R. Keese and I. Giaever, *Clin. Chem. 27,* 1597 (1981).
4. Panitz, J. A. and I. Giaever, *Surface Science 97,* 25 (1980).
5. Giaever, I. and E. Ward, *Proc. Natl. Acad. Sci., USA 75,* 1366 (1978).
6. Giaever, I. and C.R. Keese, *Proc. Natl. Acad. Sci. USA 80,* 219 (1983).
7. Keese, C.R. and I. Giaever, *Science 219,* 1448 (1983).
8. Keese, C.R. and I. Giaever, *Proc. Natl. Acad. Sci. USA 80,* 5622 (1983).
9. Giaever, I. and C.R. Keese, *Proc. Natl. Acad. Sci. USA 81,* (1984) (in press).

Structure & Motion: Membranes, Nucleic Acids & Proteins,
Eds., E. Clementi, G. Corongiu, M. H. Sarma & R. H. Sarma,
ISBN 0-940030-12-8, Adenine Press, Copyright Adenine Press, 1985.

Protobiological Selforganization

Sidney W. Fox
Institute for Molecular and Cellular Evolution
University of Miami
Coral Cables, Florida 33123 U.S.A.

Abstract

Pasteur's question on spontaneous generation: "Can matter organize itself?" was answered affirmatively by Wald in 1954 in a theoretical treatment involving primitive protein. Our laboratory demonstrated before 1960 a benchtop retracement of such an event, in which selforganizing phenomena were prominent.

The initial step is one in which mixtures of amino acids are combined geothermally. This selfordering of amino acids has been studied especially in molten mixtures. Since the resultant thermal copolyamino acids (proteinoids, thermal proteins) are highly nonrandom, they are able to carry initial biomacromolecular information. The mechanism of selfordering is only partly understood.

Proteinoids react easily with water to yield populations of highly ordered laboratory protocells. These possess in some degree or form the roots of numerous functional properties of evolved cells: uniform microstructure, ability to synthesize peptide and internucleotide bonds, propensity to proliferate (in an act of selforganization), selective permeability in membrane, and electrical excitability. The events from amino acids through thermal proteins to polyfunctional cellular structures occur with no DNA/RNA in their history, but basic proteinoids do catalyze the formation of at least small polynucleotides in either aqueous solution or suspension.

Principles and phenomena that have been highlighted by the experiments include: (a) selforganization of thermal copolyamino acids to protocells, (b) selfordering of mixed amino acids as the *sine qua non* for selforganizing polymers, (c) the manifestation of (a) and (b) in steps, (d) proteins emerged as the first informational biomacromolecules, and (e) evolvable cells emerged early. The steps of (a)-(e) are all derived experimentally.

The demonstrated selfordering of α-amino acids has yielded a revised analysis of Darwinian evolution (S. W. Fox in Eds., M. W. Ho and P. T. Saunders, Beyond Neo-Darwinism, Academic Press, London, 1984).

Introduction

The studies reported here relate an experimental retracement of early steps in the emergence of cellular life. The experiments were constructed at the outset for the more limited goal of understanding the emergence of prebiotic protein. We assumed that original protein was necessarily ordered (1,2), in contrast to Oparin, who assumed that the first proteins were disordered (3, p.290). We did not think of ordering of protein by prior nucleic acids; the Watson-Crick formulation of the double helix (4) had not yet been announced, and the translation of nucleic acid into protein was not modelled until years later (5). We had, however, the experimental clue that amino acids order themselves (1,2) as they combine to form peptides.

Proteins have been of interest for understanding life because they are the carriers of numerous biological functions (6). What was learned early in the retracement experiments is that unsubstituted amino acids indeed order themselves under the influence of geological heat, that prebiotic (thermal) proteins have numerous specific, weak enzymelike activities (7-9), and that the thermal proteins aggregate to cellular structures quickly, richly, and inexorably in water. Since the unit of life is the cell, and a cellular unit emerged from the experiments, the problem had itself evolved from the emergence of prebiotic protein to the emergence of (proto)life. The problem of the relationship of the protocell to the modern cell is necessarily that of evolvability of the protocell to the modern cell.

Because of the fundamental importance of nucleic acid in all of modern cellular life, it is crucial to recognize that the known catalytic activities of thermal proteins include the ability to synthesize phosphodiester internucleotide bonds, as well as peptide bonds. This they can do in suspensions of cellular structures of proteinoid.

The indications of protein(cells)-first, and that protein might then be encoded in nucleic acids made by such protein, provide a necessary continuity in molecular evolution theory. At the same time these results help to explain the early emergence of biological information from amino acid mixtures. The early biological information was carried by the thermal proteins. This order of events is in keeping with the second law of thermodynamics operating in an ordered universe (10).

The studies have had to contend with numerous preconceptions; these have been reviewed quite comprehensively (11-13).

Selforganization

The role of selforganization in concepts of the origin of life has a long history. Since the possibility was first mentioned by Pasteur in 1864 (14), it was positively proposed by Haeckel (15), and then more explicitly based on selforganization of protein into a cell by Wald in 1954 (16,17). The selforganization of primitive protein to a protocell was experimentally modelled in 1959-1960 in this laboratory (18), while selforganization of nucleic acid, another kind of macromolecule, was proposed by Eigen in 1971 (19). Beginning in the early 1950s, selforganization has received increasing attention as a universal process both in protobiology and biology (e.g. 20).

Oparin (3) discussed the coacervate droplets (Fig. 1) introduced by Bungenberg de Jong (21) for their similarities to cells. He recognized the "very interesting possibil-

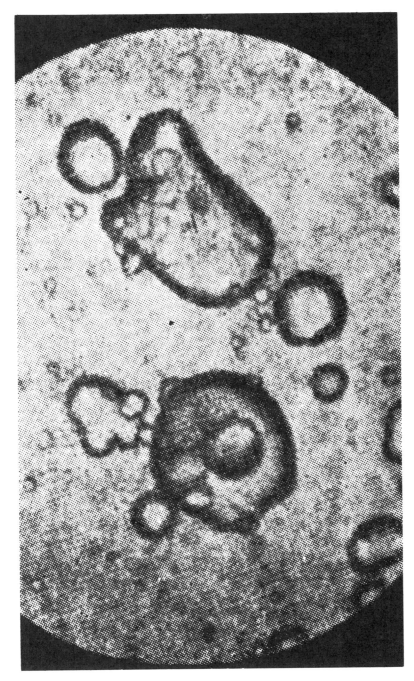

Figure 1. Coacervate droplets of Oparin, after Evereinova. Droplets composed of gum arabic, gelatin, and ribonucleic acid.

ity of forming protein-like multimolecular systems by the action of heat on mixtures of amino acids under natural conditions" (22). He felt, however, that it was unjustifiable to discuss the possibility that specific proteins or nucleic acids might have been formed in primaeval waters (22). Oparin performed interesting experiments by including evolved enzymes in coacervate droplets (23). An adequate model of the first cell, however, must of course arise from precursor polymers not obtained from evolved cells. Because of Oparin's use of polymers prepared by biological entities rather than under geological conditions, he missed crucial *ab initio* contributions of such geologically derived polymers to initial cellular structure and dynamic function.

These qualities were provided by microspheres formed by the selforganization of proteinoid (8). As in Fig. 2, the microspheres are, as can be seen, ordered; this may be partly a consequence of the nonrandomness of the molecules from which they are aggregated. Fig. 3 displays transmission electron micrographs.

Approaches to Protobiological Selforganization

Table I lists the approaches that have been exercised in inferring protobiological selforganization.

Table I

Approaches to Protobiological Selforganization

Reassembly
Materials, modern
Processes, Modern
Ab Initio Assembly
Materials, inferred primitive

Oparin's production of, and interpretation of, coacervate droplets, in the context of the origin of life, represents the use of modern materials for this purpose. Much can be inferred by this approach. If primordial materials, however obtained, had special properties that were significant in subsequent stages of evolution, key features would be missed by such investigations.

Instead of modern materials; modern processes, phenomena, and principles can be melded theoretically to provide a picture of how the first organism was assembled from its precursors. This mode is undoubtedly the one that has been most practiced.

The third mode of investigation involves allowing geological action on materials that are inferred to be primitive. This is a forward process of retracement of evolution. A well-known example is the Urey-Miller experiment, in which the primitive atmosphere inferred by Urey, following Oparin (3), was sparked by Miller (24). The results cannot, however, be more correct than is the inference of the identity of the materials; in this case the reducing atmosphere has been extensively

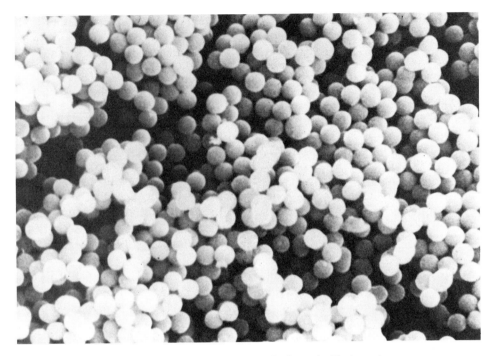

Figure 2. Scanning electron micrograph of proteinoid microspheres.

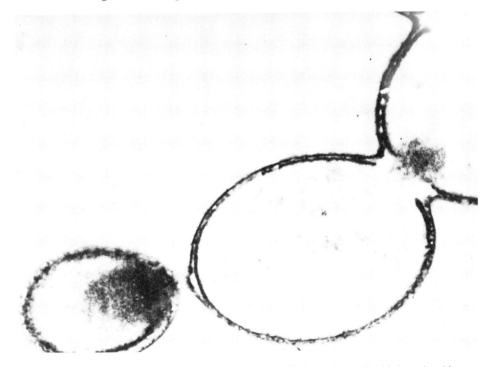

Figure 3. Transmission electron micrograph of proteinoid microspheres. Double layer is evident.

doubted (25). The retracement of forward evolution is nevertheless a relevant approach. The heating of amino acid mixtures is in the same mode. The approaches based on analysis of modern systems have an Aristotelean quality. To quote T. O. Fox (26), they are "not only not the right way to answer the question, they are not the right way to ask them."

Flowsheet

Fig. 4 presents the flowsheet of *ab initio* events in molecular evolution that had been identified by 1975 (27), and in which the microspheres emerged by self-organization, in an evolutionarily forward direction (28). The quality of nonrandomness of arrays of proteinoid molecules has been found to be crucial to the total evolution. Deterministic nonrandomness in the arrays of proteinoid molecules has been shown in a number of ways, quite strikingly in a discgel electropherogram of hemoproteinoid (29) made by heating heme with 20 kinds of amino acid (Fig. 5).

Numerous data on the proteinoids have been collected (8). These are mostly examples of specific catalytic activities that play the roles of protoenzymes. This is analogous to isolation of enzymes from evolved cells, which culminated a long controversy about vitalism. Enzymelike proteinoids are now indicated as having arisen separately from, but in advance of, the protocells which can subsequently arise from them by selforganization.

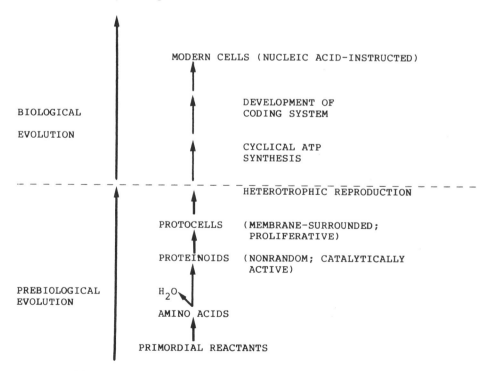

Figure 4. Flowsheet of molecular evolution, modified from Lehninger (27).

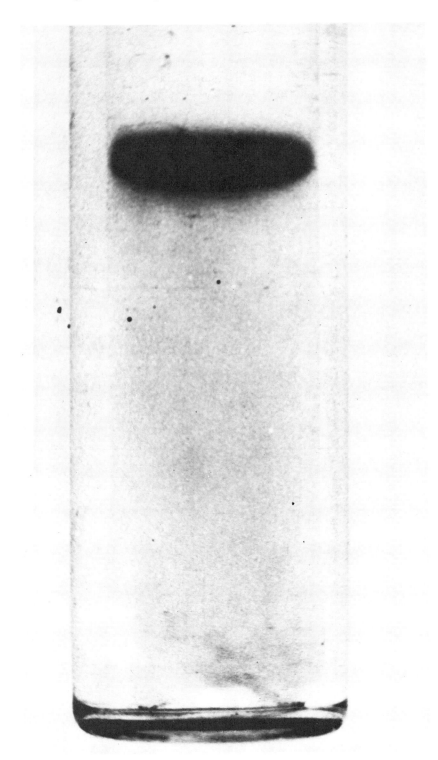

Figure 5. Discgel electropherogram of hemoproteinoid made by heating heme with amino acids (29).

Self-Ordering (Self-Sequencing)

The deterministic nature of molecular evolution from the amino acids is evident in the remarkable degree of self-ordering that amino acids undergo when they react to form copolyamino acids under the influence of geomimetic temperatures (30).

"Theoretically" possible combinations of three amino acids

Glu	-	Gly	-	Tyr
Glu	-	Tyr	-	Gly
Gly	-	Glu	-	Tyr
Gly	-	Tyr	-	Glu
Tyr	-	Gly	-	Glu
Tyr	-	Glu	-	Gly

All arrangements are equally probable with playing cards which are arranged randomly. Amino acids are <u>not</u> playing cards. Each of the twenty kinds has its own shape and chemical reactivity.

The <u>experimental</u> <u>result</u> is that amino acids react selectively due to different shapes and chemical reactivities. (Shapes above and below are all **exaggerations**; only one of each amino acid type is included).

Actual combinations of three amino acids by heating

Glu	-	Gly	-	Tyr
Glu	-	Tyr	-	Gly

Figure 6. Amino acids symbolized as playing cards and as molecules.

This special phenomenon has been reviewed a number of times (31,32,10), and has been observed in a number of laboratories. The essential distinction between the nonrandom reactions displayed by molecules and the random behavior of playing cards is graphically presented in Fig. 6. The ordering effect is observed on a residue-by-residue basis (10), much as Fig. 6 shows it to be.

The self-ordering, and the self-sequencing, of amino acids is an explanation for the origin of protobiological information *ab initio*. It has come to be viewed as the heart of molecular evolution preceding organisms. It has, however, had to contend with a pervasive view of randomness in several subject matter areas (33). Unquestionably, the extensive use of the playing card analogy has provided the concept of randomness as a useful boundary condition in many theoretical treatments. It fails, however, to take into account the nature of the units of evolution, which are expressed at the molecular level, in which shapes of molecules differ in contrast to playing cards.

Selforganization of Proteinoid: The Protocell

As stated, the concept of selforganization of thermal protein to form a protocell has led to investigational activity that proceeds in a direction opposite to that employed in most biological studies. In biological studies the mode is predominantly

Table II
Salient Properties of Proteinoid Microspheres

Protobiochemical

Esterolytic
Phosphatatic
Decarboxylatic
Peroxidatic
Synthetic, with P-O-P or ATP
    For peptides
    For polynucleotides
Photodecarboxylatic

Protophysiological

Electrotactic
Protometabolic (Catalytic)
Aggregative
Protomotile
Osmotic
Permselective
Fissive
Protoreproductive
Conjugative
Protocommunicative
Excitable

deductive in abstract, and analytical in practice. Inductive reasoning has been much less used in science than deductive reasoning. The investigative activity that corresponds to inductive reasoning is synthetic, or constructionistic (34). The word "constructionistic" embraces both molecular synthesis and selforganization. The special contribution of constructionism to the problem of life's emergence is beginning to be recognized for its unique relevance (35-38). It is a special route to identifying emergent properties.

The identification of those properties represents a major fraction of the research that has been conducted in this laboratory (Table II). The remarkable nature of these properties was recognized rather early (39). One should recall how many of the functions and the specificity of modern cells is rooted in protein (40,6, p. 122). It was necessary, however, to recognize in addition that a first cell, a protocell, would not be a fully evolved cell; if it had been such the process would have been a form of instant creation. Instead, the modern cell could have evolved in a stepwise fashion from a photocell, in which not all functions were fully developed (41). Natural selection was, in this view, already operative—as was natural variation by virtue of the fact that the matrix was not random (30).

Instead of reviewing the numerous remarkable properties of proteinoid microspheres, we shall discuss here only the most recent studies of origins of synthesis of macromolecules by proteinoid microspheres.

Anabolic Activities in Proteinoid Microspheres

The catabolic activities of proteinoids and of proteinoid microspheres have been studied in several laboratories, and an extensive catalog has accumulated (10).

The proteinoid microspheres, sometimes referred to as laboratory protocells, are both structurally stable and dynamic in a biochemical and cytophysical sense. Those containing basic proteinoid in mixture with acidic proteinoid (42) are more dynamic and less stable than those made from acidic proteinoid alone. In other words, a primary contribution to the primitive evolving cell was structural stability, while the main contribution of the basic proteinoids was protometabolic activities.

The two anabolic activities of primary concern are the synthesis of peptide bonds and internucleotide bonds. These syntheses are catalyzed by the same agent, basic proteinoid (42). Since such catalysis occurs also within suspensions of microspheres, microspheres in suspension can simultaneously catalyze the synthesis of the two types of macromolecular bond: peptide and internucleotide phosphate ester.

The early microspheres, then, provided evolution with a suitable microlocale for the origin and evolution of genetic coding (43). Numerous studies have indicated that tactic proximities of polyamino acids and polynucleotides display selectivities that could have served as the basis for codonicity or anticodonicity (8,44).

Principles in the Selforganization of Protobiological Units (Protocells)

Many principles and phenomena in the selforganization of protocells can now be reviewed relative to one another (33). Principal ones of these are briefly stated here:

1. Selforganization of Polyamino Acids

Pasteur's question of 1864, Can matter organize itself? has been answered affirmatively in theoretical treatments by Haeckel and especially by Wald (15,16). Pasteur's question was asked in a context of the first living beings, which is the context in which Wald answered it. Wald visualized primitive protein as the material that organized itself.

Experiments of the 1950s were already reported in this context in 1960 under the title Self-organizing Phenomena and the First Life (18).

2. Self-Sequencing of Amino Acids

Many viewers of the scene have been unable to recognize how life could have emerged from a random matrix. We now recognize that the matrically closest precursor to the first cell was very nonrandom, as established by experiments. The meaningful selforganization was thus selforganization of selfordered polymers. The self-sequencing emerges as the *sine qua non* of self-organization. It results in nonrandomness within and between molecules.

3. Stepwiseness

The experiments indicate that evolution at the molecular level is not gradual. Nor is it instantaneous. In the approach to a protocell, and in subsequent development to a cell, evolution is stepwise. The steps consists of a sequence of individually likely changes, whereas an overall single change has not been visualized.

Molecular evolution can thus be characterized as stepwise, self-ordered self-organization.

4. Protein-First

Since DNA/RNA are incapable of replicating themselves or of making anything else biochemical, proteins had to be on the scene first, along with an energy source such as ATP or other pyrophosphates. Primitive, or thermal, proteins have the same properties qualitatively as modern proteins. These enzymic activities are a litany of catabolic and anabolic activities (9).

The flowsheet (Fig. 4) is compatible with the protein-first concept. The DNA-first concept has not moved past the armchair stage into the laboratory.

5. Cells-Early

By the pure logic of the analytic approach, one can argue that cells should have arisen last. The whole cell represents more than the sum of its components, and for this reason appears to be an appropriate "wrapup" entity. The experiments, however, imply otherwise, as follows.

A first cell would have arisen with the greatest of ease. The production of proteinoid microspheres is simple, rapid, and inexorable. The experiments show further that the first cell was not a modern cell. It became a modern cell in a stepwise sequence. Natural selection would of course apply to adaptive modifications in the individual cell as well as to those in the total multicellular organism or populations.

The requirements for the experimental model of molecular evolution have been similar to the above. Those stipulations are not identical, however, and a summary of requirements is in Table III.

Table III
Requirements for Experimental Retracement of Molecular Evolution

1. Use of geological conditions *ab initio*
2. Emergence of proteins as first informational molecules
3. Emergence of cells-early (self-organization)
4. Identification of emergent properties (Constructionism; emergentism)

Requirements must be met *simultaneously.*

As we have learned, these requirements had to be met simultaneously.

Relationships to Other Aspects of Science

A number of investigators have concerned themselves with interdigitating the proteinoid results with the hypercycle of Eigen and Schuster (45). These include, for example, the concept of polypeptide assignment catalysts (46) and the suggestion of a bridging protohypercycle (47). In addition, we have for some time been probing reverse translation (8) experimentally; these experiments may illuminate such a possibility.

The proteinoid theory of origins has been brought into simultaneous discussion with "scientific creationism" (13), Darwinian theory (33), determinism (30), psychobehavior (48), the mind (41), and cosmogenesis (10). The first four of the above six are chapters that were sought by experts in other disciplines; the evidence thus indicates recognition of the relevance of origin of life studies to a wide interface of bioscience.

State of the Art on Broad Questions

Thanks fundamentally to the principle of selfsequencing of amino acids, major questions of origins have been opened to disciplinary investigation, and answers have been advanced. Such questions are listed in Table IV.

Table IV
Beginnings

Origin of order in proteins
Origin of metabolism
Origin of cells
Origin of membranes
Origin of cellular reproduction
Origin of coded genetics
Origin of excitability
Source of variants for natural selection
Origin of consciousness

The emphasis is on origin, not on origin and full-blown evolutionary development. When the latter is assumed, as is sometimes true, the tacit conceptual objective may be a kind of instant creation. As has been mentioned, however, evolution is a stepwise sequence. We may not assume that the sequence will ever be complete.

Acknowledgements

This work was supported by grants from the National Foundation for Cancer Research and National Aeronautics and Space Administration (NGR 10-007-008). Contribution No. 371 of the Institute for Molecular and Cellular Evolution.

References and Footnotes

1. S. W. Fox, M. Winitz, and C. W. Pettinga, *Journal of the American Chemical Society 75*, 5539 (1953).
2. S. W. Fox, *American Scientist 44*, 347 (1956).
3. A. I. Oparin, *The Origin of Life on the Earth*, 3rd ed., Academic Press, New York, (1957).
4. J. D. Watson and F. H. C. Crick, *Nature 171*, 737 (1953)
5. M. W. Nirenberg and J. H. Matthaei, *Proceedings National Academy Sciences USA 47*, 1588 (1961).
6. A. L. Lehninger, *Principles of Biochemistry*, Worth, New York (1982).
7. D. L. Rohlfing and S. W. Fox, *Advances in Catalysis 20*, 373 (1969).
8. S. W. Fox and K. Dose, *Molecular Evolution and the Origin of Life*, revised ed., Marcel Dekker, New York (1977).
9. S. W. Fox, *Naturwissenschaften 67*, 378 (1980).
10. S. W. Fox, *Naturwissenschaften 67*, 576 (1980).
11. S. W. Fox, *Naturwissenschaften 60*, 359 (1973).
12. D. L. Rohlfing in *Molecular Evolution and Protobiology*, Eds., K. Matsuno, K. Dose, K. Harada, and D. L. Rohlfing, Plenum Press, New York, p. 29 (1984).

13. S. W. Fox in *Science and Creationism,* Ed., A. Montagu, Oxford University Press, p. 194 (1984).

14. L. Pasteur cited in P. Vallery-Radot, *Oeuvres de Pasteur II,* Masson, Paris, p. 328 (1922).

15. E. Haeckel, *Naturliche Schopfungsgeschicte,* G. Riemer, Berlin (1868).

16. G. Wald, *Scientific American 191* (2), 44 (1954).

17. A misreading of Wald's pioneering contribution is by Folsome, who claimed that selforganization to a first cell was recognized after Wald's paper, cf. S. W. Fox, *Origins of Life 10,* 67 (1980). This assertion of Folsome's followed earlier overreaching claims by him for his own experiments in self-organization, cf. S. W. Fox, *Naturwissenschaften 64,* 380 (1977).

18. S. W. Fox, *Yearbook Society General Systems Research 5,* 57 (1960).

19. M. Eigen, *Naturwissenschaften 58,* 465 (1971).

20. S. W. Fox (Ed.), *BioSystems 12,* 131-330 (1980).

21. H. G. Bungenberg de Jong in *Colloid Science II,* Ed., H. R. Kruyt, Elsevier Publishing Co., New York, p. 232 (1949).

22. A. I. Oparin, *Life, Its Nature, Origin, and Development,* Oliver and Boyd, Ltd., Edinburgh, p. 67f (1960).

23. A. I. Oparin, *Subcellular Biochemistry 1,* 75 (1971).

24. S. L. Miller, *Science 117,* 528 (1953).

25. R. A. Kerr, *Science 210,* 42 (1980).

26. T. O. Fox, personal communication.

27. A. L. Lehninger, *Biochemistry,* 2nd edition, Worth and Co., New York (1975).

28. S. W. Fox and K. Matsuno, *Journal of Theoretical Biology 101,* 321 (1983).

29. K. Dose and L. Zaki, *Zeitschrift fur Naturforschung 26b,* 144 (1971).

30. S. W. Fox and T. Nakashima in *Individuality and Determinism,* Plenum Press, New York p. 185 (1984).

31. K. Dose and H. Rauchfuss in *Molecular Evolution,* Eds., D. L. Rohlfing, A. I. Oparin, Plenum Press, New York, p. 199 (1972).

32. P. Melius and V. Nicolau in *Molecular Evolution and Protobiology,* Plenum Press, New York, p. 125 (1984).

33. S. W. Fox in *Beyond Neo-Darwinism,* Eds., M. W. Ho and P. T. Saunders, Academic Press, London, p. 15 (1984).

34. S. W. Fox in *Bioorganic Chemistry Vol III,* Ed., E. E. van Tamelen, Academic Press, New York, p. 21 (1977).

35. K. Dose in *Molecular Evolution and Protobiology,* Eds., K. Matsuno, K. Dose, K. Harada and D. L. Rohlfing, Plenum Press, New York, p. 1 (1984).

36. D. L. Rohlfing in *Molecular Evolution and Protobiology,* Eds., K. Matsuno, K. Dose, K. Harada, and D. L. Rohlfing, Plenum Press, New York, p. 29 (1984).

37. R. S. Young in *Molecular Evolution and Protobiology,* Eds., K. Matsuno, K. Dose, K. Harada, and D. L. Rohlfing, Plenum Press, New York, p. 45 (1984).

38. A. Cherkin in *Molecular Evolution and Protobiology,* Eds., K. Matsuno, K. Dose, K. Harada, and D. L. Rohlfing, Plenum Press, New York, p. 49 (1984).

39. A. L. Lehninger, *Biochemistry,* Worth and Co., New York (1970).

40. D. Nachmansohn, *Science 168,* 1059 (1970).

41. S. W. Fox, *International Journal of Quantum Chemistry Quantum Biology Symposium,* in press.

42. S. W. Fox and T. Nakashima, *BioSystems 12,* 155 (1980).

43. S. W. Fox in *Science and Scientists,* Eds., M. Kageyama, K. Nakamura, T. Oshima, and T. Uchida, Japan Sc. Soc. Press, Tokyo (1981).

44. J.C. Lacey Jr., D. P. Stephens, and S. W. Fox, *BioSystems 11,* 9 (1979).

45. M. Eigen and P. Schuster, *The Hypercycle,* Springer-Verlag, Berlin (1979).

46. V. Bedian, *Origins of Life 12,* 181 (1982).

47. K. Matsuno, *Journal of Theoretical Biology 105,* 185 (1983).

48. S. W. Fox in *Levels of Integration and Evolution of Behavior,* Eds., G. Greenberg and E. Tobach, Lawrence Erlbaum Associates, Hillsdale, NJ, in press (1984).

Structure & Motion: Membranes, Nucleic Acids & Proteins,
Eds., E. Clementi, G. Corongiu, M. H. Sarma & R. H. Sarma,
ISBN 0-940030-12-8, Adenine Press,   Copyright Adenine Press, 1985.

Electrical Aspects of Cellular Reproduction

Herbert A. Pohl
Director, Pohl Cancer Research Laboratory
& Department of Physics
Oklahoma State University
Stillwater, Oklahoma 74078

Abstract

All species of living cells examined to date and that are capable of reproduction, appear to produce natural ac electric fields. The fields appear to be maximal at or near mitosis, at least in the unicellular yeast, *Saccharomyces cervisiae.* Such conclusions follow from three independent experimental lines: direct electrical sensing, cellular spin resonance, and dielectrophoresis studies. The evidence for these natural ac oscillations and their implications for cellular reproduction are discussed. If cellular reproduction is found to involve electrical ac oscillations, then an exciting new avenue, an electrical one, will open up into the understanding of processes governing the four phases of mammalian life, viz.: that during *embryonic* gowth, during natural *somatic* cell *replacement,* during *wound-healing,* and in *tumors.*

Introduction

Experimental evidence from several lines in the last three years leads us to know that individual cells emit oscillating electric fields. Not all cells do this. To date, only those cells which are capable of reproducing appear to do this. Cells of wide ranging types, including bacterial, fungal, algal, piscine, avian, and mammalian cells exhibit the alternating current (ac) electrical oscillations. Cellular electrical oscillations thus appear to be "universal". Moreover, the oscillations appear to be maximal at or near mitosis, at least in yeast, where one now can readily identify the phase (1) in the cellular life cycle by the cellular shape. These two basic observations, the "universality", and occurrence during the reproductive phase lead one to a new theory, or model, for the control of cellular growth, namely; that cellular reproduction requires electrical oscillations (CRREO); and that without them, reproduction cannot occur.

The Experimental Results

The lines of experimental evidence are three. In the one, using the phenomenon of dielectrophoresis, the experimenter compares the way in which cells appear to attract highly polarizable (high permittivity) particles as compared to the way in which poorly polarizable one gather on individual cells (2-9). As highly polarizable particles, numerous materials can serve, such as tiny crystals of $BaTiO_3$, $SrTiO_3$, $LiNbO_3$, or certain ekaconjugated polymers such as the PAQR polymer (10-11) obtained by the condensation polymerization of anthraquinone and pyromellitic dianhydride (10-12). As controls, the low permittivity ones such as SiO_2, Al_2O_3, or $BaSO_4$ can serve. The experiment requires very little in the way of apparatus . . . little more than a good microscope and a conductance meter, since no external fields other than the ones supplied by the cells themselves are required. In brief, one puts a mixture of the cells and selected tiny (about 2 micrometer diameter) particles under the microscope and counts the fraction of the particles associated with the cells. The average number, n, of particles per cell, and concentration of particles, p, is taken. The ratio, n/p, relfects the degree of association of the particles with the cell. The n/p ratios for particles of high permittivity is then compared with that for those of low permittivity. Table I below gives typical results obtained.

Table I
The preference of various living cells for highly polarizable particles.
(As determined with micro-dielectrophoresis)

Cell Type	Ratio, Polarizable to Non-Polarizable Particles: $(n/p)_{polar}$ to $(n/p)_{non-polar}$
Bacteria, B. cereus,synchronized	4.0
Yeast, Saccharomyces cerevisiae	2.0
Alga, Chlorella pyrenoidosa	2.2
Avian, Chicken red blood cell	2.0
Mammalian, Mouse "L" fibroblasts	2.0
Mammalian, Mouse ascites tumor fibroblasts	2.2
Mammalian, Mouse, fetal fibroblasts	2.0

The preferential gathering of the electrically sensitive highly polarizable particles as by dielectrophoresis (13,14) is direct evidence of ac fields about the cells.

In the second type of experiment, *lone cells* are observed by researchers here (13,15-20) and by Lamprecht and Mischel (21-22) in Germany to spin in a sharply resonant way (cellular spin resonance or "CSR") as an ac field is applied. Just about the only way this can be understood is in terms of having the external oscillating field interacting with an oscillating field of the cell.

In the third type of experiment, researchers here (23-24) and with Cyril Smith in England (25) have observed directly the ac output of individual cells. That for *Netrium digitus* was reported to occur at about 7 and 33 kHz (23-24). That for baker's yeast was reported by both groups to occur at about 7-8 MHz. The fields emitted by cells are very weak indeed, involving in the order of magnitude only about 10^{-14} watts per cell. This makes the detection difficult. Action potentials of macrophages derived from human monocytes were observed to oscillate at about 20 to 80 Hz (26). The power spectra of mormyrid fish *(Pollimyrus isidori)* ranges from about one to fifty KHz, and is involved with the electric organs (27). The output is used by such "weakly electric" fish as the South American gymnotoid fish and the mormyriform fish of Africa for "electro-location". Yano et al. (28) have already observed much slower oscillations (at a few Herz) of the human placental syncytiotrophoblast. Given the existence of such natural cellular ac oscillations, however, it now becomes necessary to try to picture just how and why they might arise.

Discussion

There are several plausible mechanisms for the natural electrical oscillations. Fröhlich suggests the cooperative oscillations of cellular dipoles occur, driven in a laser-like way by cellular metabolism (29-32). There is much experimental support for his view, especially for oscillations occurring (33-51) in the 100 GHz range (10^{11} Hz). But the three experiments above indicate oscillations in the much lower range of only 100 to 10,000,000 Hz. As a consequence, a second mechanism, that involving oscillating chemical reactions (44,45,52-56,28,57-59) which produce pulsating charge waves in the cell, is preferred. Further experiments are needed to help clarify the picture.

Given the above experimental facts, it is interesting to ask what natural ac oscillations might mean in the life of a cell? Are they a necessity or a frill? The "universality", from bacterium to mammal, implies that the natural ac oscillations have been carried on throughout the three or so billion year evolutionary process, and hence must reflect some needed aspect of cell life. The increased strength of the natural ac oscillations at or near mitosis, implies that they are connected somehow with the reproductive process. One can speculate as to what that might be, and for a start, assume that cellular reproduction cannot occur unless the oscillations are present. To this picture we add the knowledge that the needed oscillations could be damped out or prevented by the presence of energy absorbing material, and that we also know that such electrical energy absorption can typically be strong only in a fairly narrow frequency range (13,60-65) and finally that such absorption ranges vary considerably from material (60-62). These ideas, when put together, present us with a means to explain cellular growth control.

Picture a cell in a cell-free body fluid, and which normally has a chemistry producing electrical oscillations unabsorbed by such a body fluid which is relatively free of cells. The cell will then be free to oscillate, free to reproduce, and it might then do so. But as it becomes more tightly surrounded by other cells, picture them as now absorbing the electrical energies emitted by the cell and now preventing it from oscillating or reproducing. The situation is something like that of a mechanical alarm clock. Wind it, put it up on the dresser, and it runs. But plunge it into water, or oil, and it stops. The cell free to oscillate at a frequency fixed by its *normal* chemistry is free to reproduce when other cells are not too plentiful nearby. But it cannot reproduce by oscillating at that same (now strongly damped) frequency when many other cells are nearby. Another frequency quite different and well away from the absorption range is then required. This could be provided, we suggest, by the unlimbering of another oscillating reaction mechanism, under control of a special gene. Ahuja and Anders in 1977 (66) showed that a gene present in most cells, the "Tu gene", could evoke reproduction. It is normally inactive, repressed by an "R gene" on a separate chromosome. We suggest that the natural ac oscillations may be under such control. The operation of a gene such as the Tu gene (66) could alter the frequency of the needed electrical oscillation, and free the cell to reproduce even though it were surrounded by other cells.

The theory, or CRREO model, can be tested in many ways. One way is to ask if it can explain what is known as "contact" or density inhibition of cell growth. Here, when a culture dish containing a good culture medium at the proper temperature, etc., is inoculated with a normal mammalian cell line, the cells grow until they cover the medium in a thin monolayer, then stop reproducing. Several plausible explanations have been advanced for this behavior, including (a) the lack of food supply and (b) the possible emission of poisons and inhibitors by cells. But if such were the case, then actively flushing fresh medium across the monolayer of these "confluent" cells should initiate regrowth. It does not. This shows that the food/inhibitor explanations are probably not the complete answer and that there must be some other aspect, some other cause, possibly via gap junctions (67). The electrical mechanism suggested above, however, also fills the bill. Cells in their normal state (Tu gene repressed) and in diluted circumstance are free to oscillate, free to reproduce, but when in crowded circumstance are not free to do so. How do cells in crowded circumstance actively reproduce? According to the CRREO model, only by shifting (Tu gene active) their oscillation frequency away from the absorption frequency range characteristic of cells *en masse*. These cancer cells do, perhaps, and as "transformed" cells grow not only to form a monolayer, but in the up and down directions too, as if they were free to oscillate, free to reproduce no matter the presence of neighboring cells.

In summary, we have seen that there are three lines of experimental evidence pointing to the fact that there are natural cellular ac oscillations. From this evidence we can derive a theory of growth control, one involving electrical oscillations. The theory is testable, and up to the present time has passed several tests. It may be helpful in understanding cell growth and its control during embryonic growth, natural somatic replacement, wound-healing, and oncogenic growth. Much more research needs to be done, for the new theoretical model, CELLULAR REPRODUCTION REQUIRES ELECTRICAL OSCILLATION, though suggestive, is certainly much too simplistic.

Acknowledgement

This research has been supported in part by the Pohl Cancer Research Laboratory, Inc., to whom thanks are due.

References and Footnotes

1. Hartwell, L. H., Culotti, J., Pringle, J. R., and Reing, B. J., (1974) *Science 183*, 46.
2. Pohl, H. A. (1980) "Micro-dielectrophoresis of Dividing Cells", in *Bioelectrochemistry*, pp. 273-295, edited by H. Keyzer and F. Gutmann, Plenum Press.
3. Pohl, H. A. (1980), *Int. J. Quantum Chem. 7*, 411
4. Pohl, H. A., (1980) *J. Biol. Phys. 7*, 1.
5. Pohl, H. A., "Do cells in the reproductive state exhibit Fermi-Pasta-Ulam-Fröhlich resonance and emit electromagnetic radiation?", *Collective Phenomena 3*, (1981), pp. 221-244.
6. Pohl, H. A., (1981), *J. Theor. Biol. 93*, 207.
7. Pohl, H. A. (1981), *J. Bioenerg. Biomembranes 13*, 149.
8. Pohl, H. A., Braden, T., Robinson, S., Piclardi, J., Pohl, D. G., (1981), *J. Biol. Phys. 1*, 133.
9. Pohl, H. A. and Braden, T. (1982), *J. Biol. Physics 10*, 17.
10. Pohl, H. A. and Wyhof, (1972), *J. Non-Cryst. Solids, 11*, 137.
11. Pohl, H. A., "Quasi one-dimensional electronic conduction and nomadic polarization in polymers", *J. Biol. Phys. 2*, 113-172, (1974)
12. Pohl, H. A. and Pollak, M., "Nomadic Polarization in Quasi One-dimensional Solids", *J. Chem. Phys. 66*, 4031-4040 (1977)
13. Pohl, H. A. (1978), *"Dielectrophoresis, The Behavior of Matter in Nonuniform Electric Fields"*, Cambridge University Press.
14. Pohl, H. A., (1951), *J. Appl. Phys. 22*, 869.
15. Pohl, H. A. and Crane, J. S. (1971), *Biophys. J. 11*, 711.
16. Chen, C. S. (1973), *On the Nature and Origins of Biological Dielectrophoresis"*, Ph.D. Thesis, Oklahoma State University, Stillwater, Oklahoma 74078
17. Pohl, H. A.(1982), *Int. J. Quantum Chem. 9*, 399
18. Pohl, H. A. (1983), "Cellular Spinning in Pulsed Rotating Electric Fields", *J. Biol. Physics, 11*, 59; 66.
19. Mischel, M., Voss, A., and Pohl, H.A., (1982) *J. Biol. Physics 10*, 233.
20. Mischel, M., and Pohl, H. A. (1983) "Cellular Spin Resonance in Rotating Electric Fields", *J. Biol. Physics, 11*, 98
21. Mischel, M. and Lamprecht, I. (1980) *Z. Naturforsch, 35c*, 1111.
22. Mischel, M. and Lamprecht, I. (1983), "Rotation of Cells in Nonuniform Rotating Alternating Fields", *J. Biol. Physics, 11*, 43.
23. Rivera, H. (1983) Private Communication.
24. Rivera, H., Biscar, J. P., and Pohl, H. A., (1983) (private communication).
25. Smith, Cyril W. (1983) (private communication).
26. McCann, F. V., Cole, J. J., Greyre, P. M., and Russell, J. A. G. (1983) *Science 219*, 991.
27. Westby, M., and Kirschbaum, F. (1982) *J. Compar. Physiol., 145*, 399.
28. Yano, J., Okada, Y., Tsuchiya, W., Kinoshita, M., Tominaga, T., and Nishimura, M.(1981), *Acta Obst. Gyn.* Japan, *33*, 137.
29. Fröhlich, H., (1968) *Int. J. Quantum Chem. 2*, 641.
30. Fröhlich, H. (1977) *Neurosci. Res. Program Bull. 15*, 641.
31. Fröhlich, H. (1980) *Adv. Electron. Electron Physics 53*, 85.
32. Fröhlich, H., *Coherent Excitations in Biological Systems*, Ed. by H. Frohlich and F. Kremer (Springer-Verlag, 1983) pp. 1-6.
33. Adey, W. R. (1983) *Biological Effects of Radiation*, Edited by M. Grandolfo, S. M. Michaelson, and Rindi, A., p. 359, 561.
34. Adey, W. R. (1981) *Physiological Reviews 61*, 435.
35. Adey, W. R. (1981) *Biological Effects of Nonionizing Radiation*, Ed. by K. Illinger, p. 271, (1982).
36. Albanese, R. A. (1983) "Radiofrequency Radiation and Chemical Rations Dynamics", *International Conference on "Nonuniform Electrodynamics in Biological Systems"*, Loma Linda, CA., Edited by W. R. Adey, June 1983.
37. Blackman, C. F., Elder, J. A., Weil, C. M., Benane, S. G., Eichinger, D. C. and House, D. E. (1979), *Radio Science 14*, 93.
38. Blackman, C. F., Benane, S. G., Elder, J. A., House, D. E., Lampe, J. A., and Faulk, J. N., (1980), *Bioelectromagnetics 1*,

39. Blackman, C. F., Benane, S. G., Joines, W. T., Hollis, M. A., and House, D. A., (1980) *Bioelectromagnetics 1,* 277.
40. Blackman, C. F., Benane, S. G., Kinney, L. S. Joines, W. T., and House, D. E., (1982) *Radiation Research 92,* 510.
41. Del Giudice, E., Doglia, S., and Milani, M. (1982) *Physica Scripta 26,* 232.
42. Joines, W. T. and Blackman, C. F., (1980) *Bioelectromagnetics 1,* 271.
43. Joines, W. T. and Blackman, C. F. (1981) *Bioelectromagnetics 2,* 411.
44. Kaiser, F. (1983) International Conf. on *"Nonuniform Electrodynamics in Biological Systems",* Loma Linda, CA., Edited by W. R. Adey, June 1983.
45. Kaiser, F., *Coherent Excitations in Biological Systems,* Ed. by H. Fröhlich and F. Kremer (Springer-Verlag, 1983) pp. 128-133.
46. Lawrence, A. F. and Adey, W. R., (1982) *Neurological Research 4,* 115.
47. Rowlands, S., Sewchand, L. S., Lovlin, R. E., Beck, J. S., and Enns, E. G., (1981) *Phys. Lett. 82A,* 426.
48. Rowlands, S., Sewchand, L. S., Enns, E. G. (1982) *Can. J. Physiol. Pharm. 60,* 52.
49. Rowlands, S. (1982) *J. Biol. Phys. 10,* 199
50. Rowlands, S. Eisenberg, C. P., and Sewchand, L. S. (1983) "Contractils: Quantum Mechanical Fibrils", *J. Biol. Phys. 11,* 1
51. Webb, S. J. (1980) *Phys. Rep. 60,* 201.
52. Noyes, R. M. and Field, R. J. (1974) *Ann. Rev. Phys. Chem. 25,* 95.
53. Noyes, R. M. (1977) *Acc. Chem. Res. 10,* 214, and 273.
54. Okada, Y., Tsuchiya, W., and Inouye, A. (1979), *J. Membrane Biol., 47,* 357.
55. Rapp, P. E. (1979) *J. Exp. Biol. 81,* 281.
56. Treherne, J. E., Foster, W. A., and Schofield, P. K. (1979) "Cellular Oscillators" *J. Exp. Biol. 81* (review volume).
57. Zeuthren, E. (1958) "Artificial and Induced Periodicity in Cells", *Adv. Biol. Med. Physics 6,* pp. 37-73.
58. Schmidt, S., and Ortoleva, P. (1979) *J. Chem. Phys. 71,* 1010.
59. Epstein, I. R., Kustin, K., de Pepper, and Orban, M. (1981) *J. Amer. Chem. Soc. 103,* 2133.
60. Von Hippel, A. H., *Dielectrics and Waves* (1954), J. Wiley & Sons, N.Y.
61. Bottcher, C. J. F., Van Belle, O. C., Bordewijk, P., and Rip. P., *Theory of Polarization* (1973) *Vol. I & II,* Elseviere, N.Y.
62. Hill, N. E., Vaughn, W. E., Price, A. H., and Davies, M., *Dielectric Properties and Molecular Behavior,* (1969), Van Nostrand Reinhold Co., N.Y.
63. Coelho, R., *Physics of Dielectrics,* (1979), Elsevier, N.Y.
64. Gabler, R., *Electrical Interactions in Molecular Biophysics,* (1978) Academic Press, N.Y.
65. Pethig, R., *Dielectric and Electronic Properties of Biological Materials,* (1979) J. Wiley & Sons, N.Y.
66. Ahuja, M. R., and Anders, F., (1977) in *Recent Advances in Cancer Research, Vol. I,* Edited by R. C. Gallo; CRC Press pp. 103-117.
67. Manjanath, C. K., and Page, E., (1982) *Biochem. J. 205,* 189.

Structure & Motion: Membranes, Nucleic Acids & Proteins,
Eds., E. Clementi, G. Corongiu, M. H. Sarma & R. H. Sarma,
ISBN 0-940030-12-8, Adenine Press,   Copyright Adenine Press, 1985.

Structural and Kinetic Properties of Water
Near a Charged Membrane Model:
A Computer Simulation

Michele Marchesi and Giulio Barabino
Istituto Circuiti Elettronici, C.N.R.
Genova, Italy

Abstract

A molecular dynamics simulation has been performed with 200 water molecules between two walls. The intermolecular interaction potential is the CI pair potential. One wall models a charged membrane, interacting with water molecules through an electric potential which simulates the effect of a negative charge distribution on the wall and of a ionic double layer. The whole simulation lasted about 18 ps, at an average system temperature of 305.4 K. Static and dynamics properties have been computed and discussed.

Introduction

The study of the structure of water at solid interfaces is important for understanding a large variety of physical phenomena, such as electrode processes, lubrication, formation of lamellar liquid cristals, and also wetting and the interaction of macroscopic particles.

From a biological point of view, understanding the solvent molecular rearrangement (hydrophobic hydration) that occurs at the interface between a membrane and water is an important step on the path to explaining many physical phenomena that take place at membrane level, such as ionic and molecular transport, ion adsorption, protein binding, the behavior of receptor sites and so on, up to the formation of biological membranes themselves. Hence the necessity to study the physical properties of interfacial water, and to determine whether they are similar to those of liquid water, or not.

Computer simulation techniques are a well established tool for providing important insights into the molecular structure and dynamics of pure liquid water (1-3). From the extent of success in reproducing measured properties of real water by means of these simulations, it follows that it is advisable to perform simulations of water at solid interfaces. However, the goodness of a computer simulation, and of the physical properties derived by it, depends on the goodness of the models used. In the case of pure liquid water, the molecular structure is quite simple, and deriving a suitable water model, with a good water-water interaction potential is feasible. Many different "good" water models have been proposed in literature (1,2,4,5). The situation is quite different if we deal with the interactions of water with a membrane, or with another kind of physical interface. In this case, the modeling is much more complicated for two different reasons. The first one is the difficulty of performing energetic computations with quantum mechanical techniques for a system which is composed by several tens of atoms, and whose geometry is changing with time. The second one is that the dimensions of the components of the membrane are much bigger than water dimensions. Furthermore, the ions present in the system, adsorbed on the membrane surface or soluted in water, are very few compared with the number of water molecules. At a normality $N = 0.1$, which is a value typically found in biological systems, we have about one ion in 500 water molecules. Consequently, a simulation with physical significance, involving, say 20 ions, should consider at least 10,000 water molecules, which is a number not feasible on currently available computers.

What we can do at present is to perform simulations of water near simplified interface models, while taking into account average properties of the modeled interfaces. Recently, a number of computer simulations of this kind have been performed (6-9), dealing with the interaction of liquid water with very simplified wall models. These simulations were able to show that the presence of an interface leads to structural modifications in the water close to it, even if the wall-water interaction potential is very simple, usually purely repulsive.

In this work, a new wall-water interaction potential is introduced, in order to simulate, in a very simplified way, a charged membrane and its counterion atmosphere. The potential is to be considered as a further step toward the simulation of a "physical" membrane model, with its ionic environment.

The model consists of a "hard wall", to which is added an electric potential, in form of a decreasing exponential function. The membrane charges and the counterions are not individually present, for the reasons discussed above, but their average effect is taken into account through this electric potential. The potential parameters are chosen in order to give a membrane potential of about -1.4 V, with a length constant of 10 A. The value of the membrane potential is more than tenfold that of an usual biomembrane, and it has been chosen in order to enhance its effects on water molecules.

The model system is a "box", surrounded by two rigid walls. The first wall models the charged membrane, while the wall at the opposite end of the cell is not charged. The uncharged wall acts as a "boundary condition" to prevent water molecules from flowing out of the cell, and is a good reference for discriminating the effects on water structure due to the electric potential from the effects simply due to the presence of an interface.

The whole simulation, involving 200 water molecules, lasted about 18 ps. Many quantities of physical interest are presented and discussed.

Model and Method of Computation

1. Model

The water-water interaction potential used in the MD simulation is the CI pair potential of Matsuoka, Clementi and Yoshimine (4), based on extensive ab initio quantum mechanical interaction calculations on the water dimer. This potential is one of the most widely used in computer simulations. The cut-off radius used in the simulation is r = 7.44 A.

The interaction between the water molecules and the membrane model is described by a potential φ, which is divided into two parts:

$$\varphi = \varphi_w + \varphi_e \tag{1}$$

The first part φ_w represents the repulsive interaction of the membrane with the water molecule. It acts both on the oxygen and hydrogen atoms of water, and has the form:

$$\varphi_w = \alpha \exp(\beta z) \tag{2}$$

where z is the distance of the atom from the membrane. The parameter values are $\alpha = 1.0521 \times 10^{-20}$ J and $\beta = -0.32258 \; 10^{12}$ m^{-1} (8). φ_w is neglected for z > 0.31 A. This potential increases very steeply as z decreases, and consequently the membrane may be considered as a repulsive "hard wall".

The second part of the membrane-water potential is the electric potential φ_e which models the presence of negative charges on the membrane surface and of a counter-ion atmosphere near the membrane itself.

The ionic double layer which is found in the medium surrounding a charged membrane has been studied both theoretically, starting from a classical theory developed seventy years ago by Gouy (10) and Chapman (11), and with computer simulations (12). In all cases, the resulting average electric potential in the medium, though quite complicated in its functional expression, may be well approximated by a decreasing exponential function. Considering the many simplifications of the whole model for our purposes, the use of this exponential function should give a good indication of the average behavior near the wall. Also it greatly simplifies the potential and forces computation during the simulation. The electric potential φ_e used in the simulation has the form:

$$\varphi_e(z) = A \exp(Bz) \tag{3}$$

and acts on the three charges of the water model used in the simulation. The electric field due to this potential is not constant, so its effect on water dipoles is not only an induced dipole rotation, but also a translation, directed toward the membrane or in the opposite sense depending on dipole orientation.

The parameters A and B are directly given by the hypotheses reported in the introduction. Their values are: $A = -1.4$ V, $B = -10^9$ m^{-1}. No cut-off is applied to the electric potential, whose "tail" is therefore present also near the wall opposite to the membrane model, but the electric field strength is about 14 times greater near the membrane than near the uncharged wall. The wall-water interaction energy near the membrane for a hydrogen atom is shown in Fig. 1a, for a distance up to 0.3 A from the membrane. The difference between the very steep wall potential (dotted line) and the much softer electric potential energy (dashed line) appears clear. The full membrane-hydrogen potential (solid line) reaches a flat minimum at about 0.05 A, and for distances greater than this minimum it is practically equivalent to the electric energy.

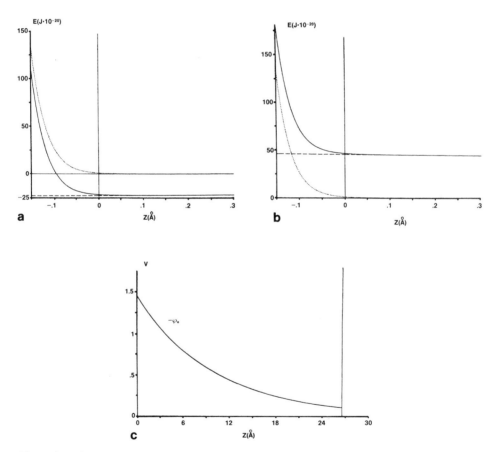

Figure 1. a) Electric interaction energy (dashed line), "hard wall" repulsive potential (dotted line) and total interaction energy (solid line), of water hydrogen atom with the membrane model. b) Electric interaction energy (dashed line), "hard wall" repulsive potential (dotted line) and total interaction energy (solid line), of water oxygen atom with the membrane model. c) The electric potential φ_e, modeling the charged membrane and the counterions double layer, throughout the box. The potential is shown in absolute value.

The membrane-oxygen interaction potential presents a different behavior, being the electric interaction energy always positive, with an absolute value doubled with respect to hydrogen's one. The resulting energy is shown in Fig. 1b. The force acting on oxygen atoms is always repulsive.

Fig 1c shows the electric potential due to the membrane charge, throughout the box. The electric field associated to it has the same decay constant as z increases, and the figure gives a graphical view of the different intensities of the electric force near the membrane, and in the other parts of the box.

It is worth noting that these electric forces are about two order of magnitude lower than the electric and quantum mechanical forces acting between two neighboring molecules. Therefore, the effects of the membrane charge on molecular motion and orientation can be appreciated only after very long simulation runs. During short simulations, these electric forces are only "noise" superimposed to the actual intermolecular forces.

To conclude this section, let us discuss some of the effects which are not taken into account in the proposed wall-water potential. Besides the already mentioned interactions with actual membrane atoms and charges, and with single counterions, the interaction potential does not take into account two important phenomena.

The first one is the reaction field due to the water molecules whose distance from the molecule considered is greater than the cut-off radius. The reaction field has effects mainly on dipolar orientations, and on the Kirkwood g factor. While in the case of liquid water it is quite straightforward to take into account the reaction field (13, 14), this is not the case for a water layer between two walls. In this situation, the reaction field depends on the z coordinate and on the orientation of the water molecule considered and of all the other molecules within the cut-off radius. Its computation would require an enormous amount of additional computer time.

The second effect not considered is the interaction of water electric charges with the dielectric discontinuity due to the presence of the walls. A MD simulation has been reported in literature with and without the presence of image charges (9), and the results support the conclusion that the effect due to this kind of interaction can be neglected, at least in an approximate wall-water interaction model.

2. Method

The model system is shown in Fig. 2. It consists of a box bounded by two walls: the membrane and the opposite wall. The projection of the box on the wall plane (the XY plane) is a square of a side a = 15 A, while the distance between the walls is b = 26.56 A. The membrane surface is the XY plane, situated at z = 0, while the opposite wall is a plane situated at z = 26.56 A.

Inside the box, 200 water molecules were simulated, integrating the equation of

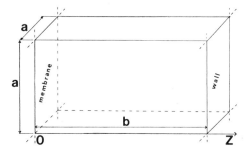

Figure 2. Box used in Molecular Dynamics simulations.

motion according to Molecular Dynamics (MD) technique, with a water density of about 1 g/cm³. The numerical integration of Newton's equations of motion was made using a technique due to Memon, Hockney and Mitra (15), which is an improvement of the better known "SHAKE" algorithm (16). The integration is made only in cartesian coordinates, because the motion of each single atom is considered separately. The electric forces acting on the negative charge of the CI model, due to the charges of the surrounding molecules and to the membrane potential, are transformed in an equivalent force system, acting on the three water atoms, and giving the same resultant and the same momentum. In the integration algorithm, besides the external forces, the Lagrangian forces due to the geometrical constraints for each molecule are also taken into account. The motion of every single atom is then computed, using Verlet's leapfrog 2nd order algorithm (17), with a subsequent correction which exactly preserves the molecular geometry, but maintains the order of convergence of the algorithm.

The time step of the integration was 2.4198×10^{-4} ps, while its duration was 75,000 steps, that is about 18 ps. The positions and velocities of the molecules were recorded, for future analysis, every 60 time steps. The whole system temperature was periodically adjusted, every 600 time steps, in order to maintain the constancy of the total energy. The target temperature was 300 K, while the computed average system temperature during the whole simulation was 305.4 K, with a standard deviation of 10.1 K. It is worth noting that, extensive simulations with a time step three times longer than that used here, showed about the same temperature standard deviation. This fact leads to the conclusion that the total energy drift during the simulation was not due to numerical "noise" generated by the integration method, but to other factors, probably to the cut-off radius applied to the water-water interaction potential.

Periodic boundary conditions were used along X and Y axes. The starting configurations was an ice lattice, with the 200 water molecules in a $5 \times 5 \times 8$ regular lattice, with random orientations. The "aging" phase was itself a long computer simulation, lasting 90,000 time steps of $7.2567 \ 10^{-4}$ ps, and so covering a time interval of about 65 ps. Such a long time is necessary in order to reach an equilibrated configuration of water dipoles under the influence of an external electric potential. The water dielectric relaxation time, in fact, is about 7.5 ps at these temperatures, and equilibrium is reached after many times the relaxation time.

The results of the simulation have been analyzed discriminating between the water close to the membrane model (membrane water, MW) the "bulk" water (BW) in the middle of the box, and the water close to the wall opposite to the membrane (wall water, WW). A water molecule is considered close to a wall if the difference between the z coordinates of its oxygen atom and of the wall is less than 2.2 A, in absolute value, while it is considered in the bulk if the z coordinate of its oxygen atom lies between 10.28 A and 16.28 A, that is in a water layer 6 A thick and centered in the middle of the box.

Results and Discussion

1. Density Profiles

The oxygen and hydrogen density profiles along z axis are presented in Fig. 3. They show the usual damped oscillatory behavior found in this kind of simulations of a fluid near a wall. These results do not differ substantially from those reported by Jönsson (8) and by Barabino et al. (9). They present an hydrogen density peak very close to the wall, followed by an oxygen peak and by subsequent density oscillations for both atoms. The extent of these density oscillations reaches about 10 A from the walls. The average water density in the box center is slightly lower than one, owing to a tendency of water molecules to stay closer to the walls. No significant difference is observed in the atom densities close to the two walls, although the density peaks near the membrane (z = 0) are slightly less pronounced (about 5% less) than the peaks close to the opposite wall.

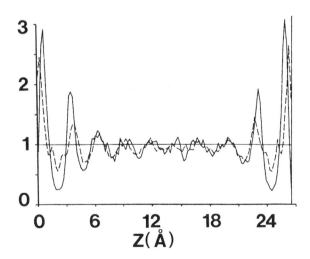

Figure 3. The oxygen (solid line) and hydrogen (dashed line) density profiles along the Z axis.

The radial distribution functions g_{OO} (r), g_{OH} (r) and g_{HH} (r) were computed for the bulk water, and are reported in Fig. 4a. They compare well with analogous func-

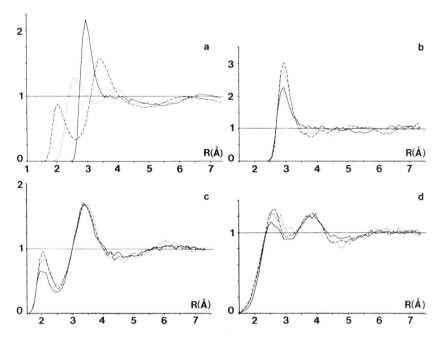

Figure 4. a) The radial distribution functions g_{OO} (solid line), g_{OH} (dashed line) and g_{HH} (dotted line) for BW. See text for the definition of "bulk water" BW. b) The oxygen-oxygen correlation function $g_{OO}^{(XY)}$ for BW (solid line), MW (dashed line) and WW (dotted line). All the functions are normalized to give the same limiting value as r tends to infinity. See text for the definition of $g_{OO}^{(XY)}$ and of BW, MW and WW water layers. c) The oxygen-hydrogen correlation function $g_{OH}^{(XY)}$. d) The hydrogen-hydrogen correlation function $g_{HH}^{(XY)}$.

tions reported in literature for simulations of liquid water, using the CI water-water potential (3,14), supporting the conclusion that, from this point of view, the water at a distance greater than about 10 A from the walls behaves like ordinary liquid water.

An analysis has also been undertaken of the projections of the radial distribution functions along XY plane. These functions, as defined in (8), are obtained taking into account only the molecules belonging to layers 1 A thick, parallel to the walls. The atoms coordinates are projected on XY plane, and here the two-dimensional radial distribution functions between the various kinds of atoms are computed. They are of the form:

$$g_{BA}^{(XY)}(r) = \frac{1}{2\pi \, \rho_B \, rd} \, \frac{dN_{BA}(r)}{dr} \tag{4}$$

where ρ_B is the atomic density, d is the layer thickness (1 A in our case), and $N_{BA}(r)$ is the average number of B atoms within a cylinder of radius r around atom A. These functions give a quantitative description of the molecular order along directions parallel to the walls, depending on the molecular position with respect to the walls. After their calculation, these functions have been rescaled of a constant factor, in order to make them tend to one for increasing r. In fact, the average density in the considered layers is much different, being close to one in the bulk, and the much higher (almost two) in the layers close to the walls, and rescaling is necessary for making comparisons. The layers near the walls are at a distance between 0.5 A and 1.5 A from the walls, while the results for the bulk refer to the average of six layers 1 A thick, covering the interval between 10.28 and 16.28 A.

Fig. 4b shows the results for the oxygen-oxygen correlation function. The function referring to BW (solid line), is very similar to the radial correlation function reported in Fig. 3, while the functions referring to MW (dashed line) and to WW (dotted line), present a peak significantly higher, confirming previous results about the more ordered structure of water at solid interfaces (6,8,9). Moreover, near the membrane, the peak of this function is slightly higher than near the opposite wall, showing that the presence of a greater electric field tend to further increase the order of water molecules along XY plane.

The oxygen-hydrogen and hydrogen-hydrogen spatial correlation functions along the XY plane are shown in Figs. 4c and 4d, respectively. Unlike the oxygen-oxygen correlation function, the functions referring to BW are not very similar to the corresponding radial spatial correlation functions showed in Fig. 4a, especially for short distance, and particularly for $g_{HH}^{(XY)}$. A previous simulation, made in absence of electric field, and not yet published, had shown no significant differences between the radial spatial correlations functions in the bulk, and their projections on XY plane. Owing to this fact, the observed differences are probably due to the electric field influence. Also in these functions, a more pronounced behavior is observed near the walls, and more near the membrane than near the other wall.

Finally, it is worth noting that these functions are much more irregular than the radial ones, owing to the lower number of cases averaged to obtain them.

2. Angular Averages and Dipolar Correlations

The first reported quantity of physical interest dealing with the orientational behavior of water molecules is the average of the cosine of the angle which water dipoles form with Z axis, as a function of z. This quantity, $g_\mu(z)$, is given by the equation:

$$g_\mu(z) = \langle \hat{\mu}_i \cdot \hat{z} \rangle_{z, z+dz} \qquad (5)$$

where $\hat{\mu}_i$ is the versor of the water electric dipole, and the average is taken over the water molecules whose oxygen z coordinate lies between z and dz. $g_\mu(z)$, which is shown in Fig. 5a, gives information about the water molecules orientation in the box. The tendency of water molecules to orient with the hydrogen atoms toward the walls, which is suggested by the curves in Fig. 4a, is clearly shown by the behavior of g_μ near the walls. The $g_\mu(z)$ curve also shows a clear tendency of water molecules throughout the box to orient according to the external electric field,

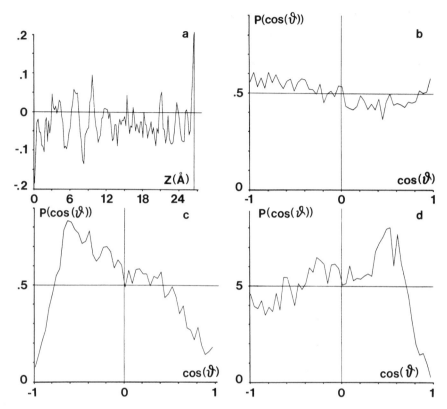

Figure 5. The average of the cosine of the angle which water dipoles form with Z axis, along Z axis (inset a). The angular probability distribution function P(cos θ) for BW (inset b); for MW (inset c) and for WW (inset d).

being its average clearly negative. This should be expected owing to the long aging phase which was carried out before the present simulation. Another characteristic of this function is an oscillatory behavior, which is particularly evident near the membrane model, and which corresponds, at least in the box half near the membrane, with oxygen and hydrogen density oscillations. The physical meaning of this behavior is unclear.

The angular probability distribution function P(cos θ)(6), where θ is the angle formed by the water dipole with z-axis, has been recorded for BW, MW and WW, and is shown in Figs. 5b, 5c and 5d, respectively. These figures illustrate again the water dipole orientation due to the impressed electric field, except for WW. Here there appears an induced dipole counter-orientation, which is unexpected because it is not observed near hard walls in absence of an electric field. A comparison of Figs. 5 with the corresponding figures of Refs. (6) and (9) stresses very impressive differences between those simulations and the present one.

3. Energetics

The average potential and kinetic energies for BW, MW and WW have been calculated for each molecule. In this comparison, given every molecule belonging to the examined layer, its interaction with all the other molecules within the cut-off radius is calculated. The potential energy is split, for each molecule, in the energy of interaction with the other molecules, and with the walls. The former is then split again in energy due to the electric interaction of the water molecule charges, and in "quantum mechanical" energy due to the exponential interaction between the atoms, accordingly to CI water model. The interaction with the walls is divided in electric interaction with the membrane charges and ionic double layer model, and in the interaction with the steep repulsive "hard wall" potential.

The energy values are reported in Table I, together with their standard deviation. As regards intermolecular interactions, BW shows energy values higher in their absolute value, and a total interaction energy lower than MW and WW. This is to be expected, owing to the lower number of neighboring molecules that are present near the walls.

Table I

Average temperatures and energies of water molecules in the different layers, and their standard deviations. The energy values are given in kJ/mol.

	BW	MW	WW
Temperature (K)	307.9	296.6	301.9
σT	25.9	36.5	36.7
Kinetic energy E	7.68	7.40	7.53
σE_c	.65	.91	.92
Intermolecular electric binding energy E	−111.4	−102.1	−100.3
σE_e	3.8	5.2	4.5
Quantum mechanics binding energy E_q	43.1	37.7	36.9
σE_q	3.2	4.2	3.7
Interaction energy with φ_e potential E	−.16	−.27	−0.06
σE_m	.08	.26	.02
Interaction energy with φ_w wall potential E_w	0	.008	.006
σE_w	0	.04	.04

The interaction energy with the electric potential due to the membrane reflects the average orientation of water molecules, and is lower in the layer close to the membrane. This energy is about two order of magnitude smaller than the intermolecular binding energy. The interactions with the repulsive wall potentials is negligible, as it should be, with respect to other forms of interactions.

The standard deviation values are quite high, but this fact is probably due more to the natural fluctuations of the system during its time, rather than to actual statistical fluctuations due to numerical noise.

Figs. 6a, 6b and 6c show the normalized distribution of the total intermolecular binding energy of each molecule in the solution, with the exclusion of the interactions with the walls. These figures refer to BW, MW and WW, respectively. The computed distributions are seen to be continuous, unimodal curves that are approximately symmetric, with the same width and all having qualitatively the same shape. The BW distribution function is centered on a more negative value than the others, reflecting the lower average intermolecular energy.

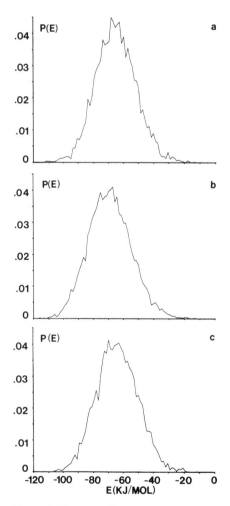

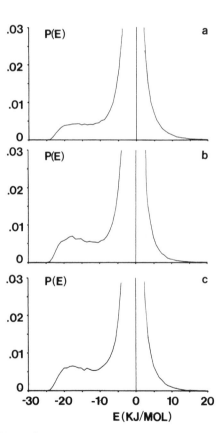

Figure 6. The normalized binding energy distribution function for BW (inset a); for MW (inset b) and for WW (inset c).

Figure 7. The normalized pair interaction energy distribution function for BW (inset a); for MW (inset b) and for WW (inset c).

The normalized distribution of water-water pair interaction energies is shown in Figs. 7a, 7b and 7c, referring, as usual, to BW, MW and WW, respectively. Each curve gives the probability of observing a pair of molecules with interaction energy E. The peak at $E = 0$ includes the relatively large number of molecular pairs which are well separated in space and therefore have very small molecule-molecule binding energies. Each figure refers to a single water layer. It reports the distribution function of the molecule-molecule pair interactions, for which at least one molecule is contained in the layer.

As one can see, close to the walls there is a significantly higher number of molecule pairs with interaction energies lower than -15 kJ/mol, with respect to the bulk. The shape of the curve below -15 kJ/mol is different from, and more "peaked" for MW and WW than for BW probably owing to the higher molecular density near the walls (see Fig. 3) (which increases the probability of finding lower interaction energies). It seems not to depend on the membrane electric potential, as there are no significant differences between MW and WW. A corresponding more "peaked" behavior in pair interaction energies for water molecules near a solute methane molecule and near a dipeptide model has been noted in other computer simulations recently reported (19,20).

The number of hydrogen bonds formed by water molecules in each of the three groups is presented in Table II. The energy chosen to discriminate between intermolecular bonds is $E_{HB} = -10$ kJ/mol, which has been deduced from Fig. 7. As in the previous energetic analysis, the bonds of each molecule belonging to the given layer with all the other molecules within the cut-off radius are considered. The number of hydrogen bonds is slightly higher in the bulk, rather than in wall layers. The two wall layers, again, do not show any remarkable difference.

Table II

Fraction of water molecules with a given number of hydrogen bonds in the BW, MW and WW layers. The hydrogen bond energy is $E_{HB} = -10$ kJ/mol.

Number of hydrogen bonds	BW	MW	WW
0	.002	.003	.002
1	.035	.034	.046
2	.181	.202	.221
3	.381	.417	.414
4	.319	.307	.282
5	.076	.036	.033
6	.006	.001	.001
7	0	0	0

4. Kinetic Properties

The water self-diffusion coefficient can be determined by the long-time limiting slope of the molecular center-of-mass mean-square displacement (Fig. 8a):

$$D = \frac{1}{6t}\langle[R(o) - R(t)]^2\rangle \qquad (6)$$

where R(t) is the center of mass coordinate vector of the generic j-th molecule at time t, and the average is taken over all the molecules considered. The average is calculated using only those water molecules that are within the considered layer at all times. The resulting values of the self-diffusion coefficient in the three layers can be obtained from the slope of the presented curves. Although these curves still exhibit appreciable statistical fluctuations, the center-of-mass mean-square displacement of BW tends to increase more steeply, thus yielding a higher value of D, which is estimated to be about 20% larger than for MW and WW. This result confirms those of previous simulations (8,9,21).

The normalized center-of-mass velocity autocorrelation function $C_v(t)$:

$$C_v(t) = \frac{\langle v(o) \cdot v(t)\rangle}{\langle \underline{v}(o) \cdot \underline{v}(o)\rangle} \qquad (7)$$

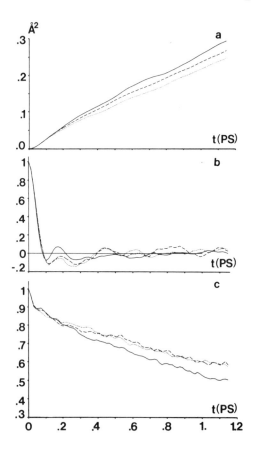

Figure 8. a) The molecular centre of mas mean-square displacement for BW (solid line), MW (dashed line) and WW (dotted line). b) The velocity autocorrelation function C_v for BW (solid line), MW (dashed line) and WW (dotted line). c) The dipole-dipole autocorrelation function Γ_1 for BW (solid line), MW (dashed line) and WW (dotted line).

is presented in Fig. 8b, following the usual convention of presenting BW, MW and WW layers with different line types. The initial decay is the same for the three water layers. Then, bulk water shows an oscillatory behavior, while C_v for water close to both walls remains negative for a longer time. The oscillations in MW and WW functions for times longer than some tenths of picoseconds are mainly due to statistical noise, owing to the limited number of molecules considered in these layers.

The oscillatory behavior of bulk water does not agree with a previously reported result (18), where C_v for liquid water shows, after the initial decay and the first minimum, a gradual and monotonic approach to zero value. The different, more "physical" way of rescaling temperatures used in that work, however, could explain the difference.

The dipole-dipole autocorrelation function $\Gamma_1(t)$:

$$\Gamma_1(t) = \langle \underline{\mu}_i(o) \cdot \underline{\mu}_i(t) \rangle \qquad (8)$$

has been computed, and is shown in Fig. 8c, for the three water layers. The initial rapid decay of Γ_1 during the first 0.03 ps corresponds to the molecular libration, and is the same for the three layers. This similarity holds for all the simulations on the same subject reported in literature (20,21). For longer times, the difference in decay rates of Γ_1 is significant: BW molecules tend to reorient more quickly than water molecules near the walls. No significant differences can be appreciated between MW and WW. Fitting a simple exponential function to these Γ's results in the decay times reported in Table III, together with their standard deviation. Considering the similar result reported in a previous simulation of ST2 water molecules near a hard wall (8), and the very similar behavior of both MW and WW, it can be argued that the difference in decay times is due to the presence of the hard wall, while the electric field due to the membrane has no effect on this kind of reorientational motion.

Table III
Dipolar relaxation times, in ps, with their standard deviations for BW, MW and WW layers.

Water layer	τ	σ_t
BW	1.9	.4
MW	2.8	.6
WW	2.6	.6

Conclusions

This work is part of a research project which started in 1982 in our Institute, aiming to study through molecular dynamics simulations the structural and dielectric behavior of water near large biological structures, like membranes or macromolecules. This "bound water" is reported to have a different dielectric permittivity than free liquid water, both in its static permittivity value and in its dielectric relaxation frequency, which should be more than tenfold lower than in liquid water (22,23).

The first molecular dynamics simulation performed by our group on this subject was the simulation of ST2 water molecules between two "hard walls" (8). In that work, differences were noted in the structural and dynamics properties of water near the walls, with respect to free water, which could yield differences both in the static dielectric permittivity value and in the dielectric relaxation frequency. These differences were very far from the measured values, particularly for bound water relaxation frequency, but the roughness of the used wall-water interaction potential could explain the discrepancy. The use of a more physical interaction potential was expected to yield values more similar to the measured values for bound water.

After almost two years of studies and simulations, however, the situation is quite different. Improving the wall-water interaction potential with a "softer" one, making it possible to take into account, at least qualitatively, the Lennard-Jones interactions of the walls with water molecules, did not lead to any significant improvement in the agreement between simulated and measured bound water dielectric properties (9). On the contrary, the dipolar relaxation times did not show any significant difference between bound and free water, and gave worse agreement with experimental data than the previous simulation, which was made with a rougher interaction potential.

In the presented simulation the wall-water potential is further improved, with water molecules subject to an average electric field having a very high value near the "membrane" (more than 1 MV/m), which should be close to its actual physical value. Moreover, before performing the presented simulation, the system has been simulated for a time of more than 60 ps, thus eliminating transient effects which could distort the results. Even with these improvements, however, the simulation results are the same: near an interface modeling a charged membrane, water structural properties are changed with respect to liquid water, yielding a more ordered water structure and confirming previous results about simulations of water near walls or near nonpolar solutes, but the properties related to dielectric permittivity do not improve their agreement with experimental data. The increase in dipolar relaxation time, for example, is the same near the membrane model and near the uncharged wall, and thus it is clearly due to the presence of the hard wall, and not to the electric field of the membrane model.

It is not possible to draw conclusion, because only after simulations of water near membrane models fully taking into account the atom-atom interactions between water and the lipidic polar head and also the interactions of the single counterions with the membrane and the solvent, it will be possible to state whether hydration water dielectric properties are similar to the experimentally measured values, or not.

Moreover, the reported simulation has been performed at a pressure of about 500 atm, which must be greatly reduced to obtain full physical significance.

We can only conclude that improving the wall-water potential, taking into account, "on the average", the electric phenomena at biomembrane level, does not improve the fitting between data derived from the simulation, and experimental data, as regards bound water dielectric properties. This is a challenge to perform future, more accurate, simulations of this kind, and perhaps also to study theoretically more in detail the dielectric "δ" relaxation, which is thought to be associated with bound water.

References and Footnotes

1. Rahman, A. and Stillinger, F. H., *J. Chem. Phys. 55,* 3336 (1971).
2. Stillinger, F. H. and Rahman, A., *J. Chem. Phys. 60,* 1545 (1974).
3. Lie, G. C., Clementi, E., and Yoshimine, M., *J. Chem. Phys. 64,* 2314 (1976),
4. Matsuoka, O., Clementi, E. and Yoshimine, M., *J. Chem. Phys. 64,* 1351 (1976).
5. Jorgensen, W. L., *J. Chem. Phys. 77,* 5757 (1982).
6. Jönsson, B., *Chem. Phys. Lett. 82,* 520 (1981).
7. Christou, N. I., Whitehouse, J. S., Nicholson, D. and Parsonage, N. G., *Far. Symp. Chem. Soc. 16,* 139 (1981).
8. Marchesi, M., *Chem. Phys. Lett. 97,* 224 (1983).
9. Barabino, G., Gavotti, C. and Marchesi, M., *Chem. Phys. Lett. 104,* 478 (1984).
10. Gouy, C., *J. Phys. 9,* 457 (1910).
11. Chapman, D. L., *Phyl. Mag. 25,* 475 (1913).
12. Tortie, G. M., Valleau, J. P. and Patey, G. N., *J. Chem. Phys. 76,* 4615 (1982).
13. van Gunsteren, W. F., Berendsen, H. J. C. and Rullmann, J. A. C., *Faraday Discuss. 66,* 58 (1978).
14. Steinhauser, O., *Mol. Phys. 45,* 335 (1982).
15. Memon, M. K., Hockney, R. W. and Mitra, S. K., *J. Comput. Phys. 43,* 345 (1981).
16. Ryckaert, J. P., Ciccotti, G. and Berendsen, H. J. C., *J. Comput. Phys. 23,* 327 (1977).
17. Verlet, L., *Phys. Rev. 59,* 8 (1967).
18. Tanaka, H., Nakanishi, K. and Watanabe, N., *J. Chem. Phys. 78,* 2626 (1983).
19. Owicki, J. C. and Scheraga, H. A., *J. Am. Chem. Soc. 99,* 7403 (1977).
20. Rossky, P. J. and Karplus, M., *J. Am. Chem. Soc. 101,* 1913 (1979).
21. Geiger, A., Rahman, A. and Stillinger, F. H., *J. Chem. Phys. 70,* 263 (1979).
22. Grant, E. H., Keefe, S. E. and Takashima, S., *J. Phys. Chem. 72,* 4373 (1968).
23. Pennock, B. E. and Schwan, H. P., *J. Phys. Chem. 73,* 2600 (1969).

Structure & Motion: Membranes, Nucleic Acids & Proteins,
Eds., E. Clementi, G. Corongiu, M. H. Sarma & R. H. Sarma,
ISBN 0-940030-12-8, Adenine Press,   Copyright Adenine Press, 1985.

Selective Carrier-Induced Transport Processes in Artificial Membranes

W. E. Morf, E. Pretsch and W. Simon

Department of Organic Chemistry
Swiss Federal Institute of Technology,
CH-8092 Zurich/Switzerland

Abstract

For the neutral carrier mediated ion transport across bilayer lipid membranes and solvent polymeric bulk membranes a unified formal description is presented. The model is applied for rationalizing current-voltage characteristics, electric conductances, ion-transport versus potentiometric selectivities, and rectification phenomena on asymmetric membranes. Methods for calculating ion-ionophore interaction energies are discussed.

Introduction

Ion carriers (ionophores) have found wide application in thin model membranes (< 10 nm), e.g. bilayer lipid membranes (BLM), to mimick certain ion transport processes in biological systems (1,2). On the other hand, various electrically neutral ionophores became relevant as components of ion-selective electrodes (3-5). In these analytical devices, the ionophores are usually incorporated into solvent polymeric membranes (SPM) of a thickness of around 0.1 mm.

Here we try to give a unified formal description of the carrier-mediated ion transport properties of BLM- and SPM-systems. Emphasis will be placed on the current-voltage characteristics, the membrane potential, and the ion selectivity of these two types of membranes.

A unified approach to the theory of neutral carrier membranes

Ion transport in symmetric membranes

Comprehensive experimental investigations and theoretical treatments of the carrier-mediated ion transport across lipid bilayer membranes or solvent polymeric bulk membranes were offered by the groups of Eisenman (6-10), Läuger (11-13), and others (14-20). It was shown that the presence of ionophores in such membrane systems may lead to widely different electric phenomena. This was mainly ascribed to the fact that bilayer membranes represent space-charge regions whereas bulk membranes are nearly electroneutral phases. Another difference exists for the electric field within the membranes which assumes extreme values of $\sim 10^5$ V/cm in bilayers, but only $\sim 10^3$ V/cm in solvent polymeric membranes. Nevertheless, the ion transport behavior of bilayer and bulk membranes can be formally described on the basis of a unified approach, as is shown below.

Valuable information on the ion transport across carrier membranes was obtained from the study of current-voltage characteristics (for a review, see (1, 2, 4, 17, 19)). In such experiments the membrane is usually interposed between two aqueous solutions of identical composition that contain only one sort of permeating ions I^{z_i}. The measured electric current density j is then directly related to the mass flux of the transported ions, $J_i = j/z_i F$, where F is the Faraday constant.

In the normal case ("equilibrium domain" of ion transport (6)), the flux of permeating ions is controlled by the rate of translocation of carrier complexes within the membrane. If the interior of the membrane is treated as a series of N activation energy barriers, the migration of ions can be described by the following relation (4,19):

$$(J_i)_{eq} = 2N \, \bar{k}_i c_i \, \sinh(z_i FV/2NRT) \qquad (1)$$

where $\bar{k}_i = D_i/d$ is the rate constant for the translocation of the complexed ions across the membrane interior (thickness d, diffusion coefficient D_i), c_i is the boundary concentration of complexes in the membrane under equilibrium conditions, V is the transmembrane potential, i.e. the voltage applied to the membrane, and RT/F is the Nernst factor. As shown in Figure 1, Eq. 1 with N=2 can be used to fit the hyperbolic current-voltage curves that were reported for the transport of alkali cations or Ca^{2+} ions across lipid bilayers in the presence of macrotetrolide antibiotics (12) or synthetic neutral ionophores (4,19). On the other hand, the current-voltage behavior observed for bilayers in the presence of lipid-soluble anions (without ionophores (13,21)) conforms to Eq. 1 with N=1 (Figure 1). In contrast, the carrier-mediated ion transport across permselective bulk membranes (4,17,19) is characterized by N$\Rightarrow\infty$ for which case Eq. 1 predicts ohmic behavior (see Figure 1).

From conductance measurements on lipid bilayers the decisive concentration c_i of charge carriers in the membrane was found to be linearly related to the ion activity a_i in the aqueous solutions (6,7,12,13). The concentration of mobile cationic complexes in macroscopic liquid membranes, however, turned out to be at best proportional to $a_i^{1/(z_i+1)}$ (see Figure 2). From the thermodynamic standpoint, the following general result is expected:

$$c_i = K_i a_i \exp(-z_i F \Delta\phi_{eq}/RT) \qquad (2)$$

where K_i is the overall distribution coefficient of the species I^{z_i} between external solution and membrane, taking into account the solubility-enhancing effect and the ion-binding affinity of the carriers involved. Evidently, the interfacial potential difference $\Delta\phi_{eq}$ corresponds to the given surface potential in the case of bilayer

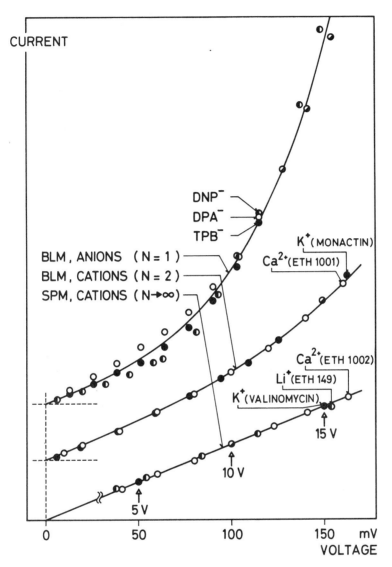

Figure 1. Current-voltage characteristics for bilayer lipid membranes (BLM) in the presence of lipophilic anions (top) and of cation-selective carriers (middle) as well as for solvent polymeric membranes (SPM) containing cation carriers (bottom). The experimental points were fitted by theoretical curves; details are given in the text. DNP^-: 2,4-dinitrophenolate; DPA^-: dipicrylamide; TPB^-: tetraphenylborate; monactin, valinomycin: carrier antibiotics; ETH 1001, ETH 1002: Ca^{2+}-selective synthetic ionophores; ETH 149: Li^+-selective synthetic ionophore.

membranes, but is regulated by the Donnan distribution equilibria of all participating ions in the case of electroneutral membranes.

Deviations from the normal current-voltage behavior can be caused by slow

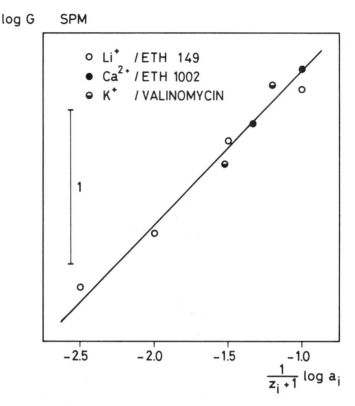

Figure 2. Dependence of the conductance G [S/cm²] of carrier-based solvent polymeric membranes (17,22,25) on the activity a_i of the external aqueous phases (chloride solutions of the cations I^{z_i} preferred by the ionophores).

interfacial reactions ("kinetic domain" of ion transport (6)). If the ion flux is completely limited by the interfacial kinetics, the following relation applies:

$$(J_i)_{kin} = \overleftarrow{k}_i c_i \frac{\sinh(z_i FV/2RT)}{\cosh((1+\beta)z_i FV/4RT)} \tag{3}$$

where $\overleftarrow{k}_i$ is the overall rate constant of the process by which the ions cross the interface from the membrane into the aqueous phase (i.e. the decomplexation reaction), including the contribution of the surface potential $\Delta\phi_{eq}$. The parameter β denotes the portion of the applied voltage that operates in the "kinetic domain" along the N energy barriers of the membrane interior. The results obtained on lipid bilayers indicate that nearly the whole potential drop occurs within such membranes (4,8-13), i.e. $\beta \cong 1$. Hence the voltage-dependent function in Eq. 3 reduces to $\tanh(z_i FV/2RT)$ and predicts saturation of current (see Figure 3). In contrast, it was suggested that $\beta \cong 0$ should hold for bulk membranes in the "kinetic domain" of ion transport (17). Accordingly, the current would increase with $\sinh(z_i FV/4RT)$, which is in analogy to the symmetric Butler-Volmer equation derived for metal electrodes

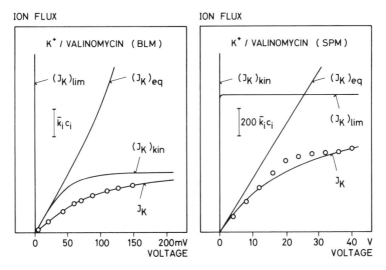

Figure 3. Partial fluxes $(J_K)_{eq}$, $(J_K)_{kin}$ and $(J_K)_{lim}$, and total flux J_K of K^+ ions across valinomycin-based membranes (left: bilayer lipid membrane; right: solvent polymeric membrane). The experimental values (12,18) were approximated by curves calculated from Eqs. 1 and 3-5 using basically the same parameters, e.g. $\bar{k}_i/\bar{k} = 0.46$ (for d = 7 nm) and $\bar{k}_i c_i/\bar{k}_s c_s = 0.002$.

(23,24). However, no kinetic limitations of this type were actually observed for 0.1 mm thick carrier membranes (see Figures 1 and 3). The reason is that the rate constant of ion translocation across such solvent polymeric membranes is only $\bar{k}_i \approx 10^{-6}$ cm/s (for $D_i \approx 10^{-8}$ cm²/s (18)), as compared to the value of $\bar{k}_i \approx 10^{-2}$ cm/s expected for lipid bilayers.

Another limitation of the carrier-mediated ion transport across organic membranes results from the fact that the ionophores are virtually trapped within the membrane because of their high lipophilicity. This implies that, at steady-state, the translocation of 1 mol of ion/carrier complexes (stoichiometry 1:n_i) always requires the back-diffusion of n_i mol of free carrier molecules. If this becomes the dominating rate-determining process, the following flux equation holds (11,17):

$$(J_i)_{lim} = \frac{2}{n_i}\bar{k}_s c_s \tanh(z_i FV/2n_i RT) \tag{4}$$

where $\bar{k}_s = D_s/d$ is the diffusional rate constant and c_s the mean concentration of free carriers S in the membrane. Evidently, Eq. 4 predicts a saturation of the ion flux, respectively the electric current, at high voltages. The reason for this phenomenon is that the driving force for the back-diffusion of ionophores, namely their concentration gradient, cannot exceed the value $2c_s/d$. Saturation behavior of this type was observed for bulk membranes based on different neutral ionophores (see Figure 3 and (17)). In agreement with Eq. 4, the limiting currents at high voltages were found to decrease with decreasing carrier concentration (17) (i.e. with increasing ion activity a_i) and with increasing membrane thickness (25).

A general treatment of the ion transport behavior of symmetric membrane systems should encompass all the limiting cases given in Eqs. 1, 3, and 4. Summarizing all the theoretical descriptions of ion fluxes presented so far, we can derive the following universal approach:

$$\frac{1}{J_i} = \frac{1}{(J_i)_{eq}} + \frac{1}{(J_i)_{kin}} + \frac{1}{(J_i)_{lim}} \tag{5}$$

Equation 5 is applicable to both macroscopic membranes (as an approximation) and bilayer membranes based on neutral carriers ($n_i > 0$) as well as to membrane systems without ionophores ($n_i = 0$). The result clearly shows that the processes of membrane-internal ion translocation, of interfacial reactions, and of back-diffusion of free carriers represent a series of resistances to ion migration. The process related to the highest resistance in the end determines the shape of the current-voltage curve (see Figure 3).

Ion transport in asymmetric membranes

The formal description given in Eqs. 1-5 is clearly restricted to symmetric membrane systems for which a strictly symmetric current-voltage curve is predicted, that is $J_i(-V) = -J_i(V)$. Asymmetric ion-transport characteristics, respectively rectification of the current, may arise from different factors:

(a) asymmetries in the electrolyte compositions on the two sides of the membrane ($a_i' \neq a_i''$),

(b) asymmetries in the surface potentials and interfacial kinetics ($\Delta\phi_{eq}' \neq \Delta\phi_{eq}''$, $K_i' \neq K_i''$, $\overleftarrow{k_i'} \neq \overleftarrow{k_i''}$), and

(c) asymmetries in the activation energy barriers for ion translocation within the membrane.

The extension of the model to asymmetric membrane systems leads to the following key result which is valid for $J_i \ll (J_i)_{lim}$ (26):

$$J_i = \overline{K_i}\overline{k_i} \frac{a_i' \exp(z_i FV/2RT) - a_i'' \exp(-z_i FV/2RT)}{F_i(V) + w_i' \exp((1+\beta)z_i FV/4RT) + w_i'' \exp(-(1+\beta)z_i FV/4RT)} \tag{6}$$

where $\overline{K_i} = \sqrt{c_i' c_i''/a_i' a_i''}$ is the mean distribution coefficient of the ions at the two interfaces including the electric potential terms (see Eq. 2), and $w_i' = \overline{k_i}/\overleftarrow{k_i'}$ and $w_i'' = \overline{k_i}/\overleftarrow{k_i''}$ are kinetic parameters. The voltage-dependent function $F_i(V)$ takes into account the number N, the location x, and the relative height $\Delta\omega_i$ (in units of RT) of the membrane-internal energy barriers:

$$F_i(V) = \frac{1}{N} \sum_{n=1}^{N} \exp\left[\frac{z_i F}{RT} (V - V_A) \frac{d/2 - x(n)}{d} + \Delta\omega_i(n)\right] \tag{7}$$

The asymmetry potential V_A corresponds to the difference between the two surface potentials at equilibrium. Equation 6 (see also Figure 4) clearly demonstrates that a net flux of ions cannot be generated by asymmetries of the membrane per se. A driving force is generally required, given in Eq. 6 by the difference in the electrochemical activities of the external solutions. Due to the rectification documented in Figure 4 for asymmetric membranes, however, an enrichment of permeating ions on one side of the membrane can be simply achieved by fluctuations of the membrane potential or of other driving forces. The principle of asymmetric membranes may be selected by nature in order to maintain or even establish asymmetries between intra- and extracellular fluids.

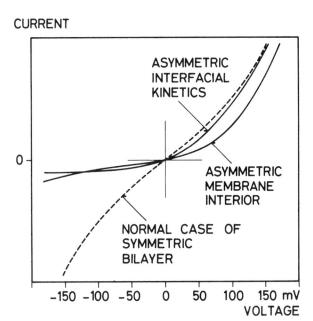

CURRENT

VOLTAGE

Figure 4. Current-voltage curves for symmetric and asymmetric bilayer membranes in the presence of ionophores. Symmetric curve (see also Figure 1, middle): calculated from Eqs. 6 and 7 using $a_i' = a_i''$, $V_A = 0$, $N = 2$, $\Delta\omega_i(1) = \Delta\omega_i(2)$, and $w_i' = w_i'' \cong 0$. Asymmetric curves: calculated as before but with $w_i'' \exp(-\Delta\omega_i) = 1$ (asymmetric interfacial kinetics), respectively with $\Delta\omega_i(2) = \Delta\omega_i(1) + 2$ (asymmetric membrane interior).

Selectivity of ion transport

The selectivity of carrier-mediated ion transport was extensively studied on symmetric membrane systems (4,6-10,16-20,27-29). From conductances G measured on bilayer membranes as well as from ionic transference numbers t determined on bulk membranes, the same selectivity between ions of the same charge was established at relatively low voltages (for $J_i \ll (J_i)_{lim}$, see Eqs. 1-5):

$$\left(\frac{G_j}{G_i}\right)_V = \left(\frac{t_j}{t_i}\right)_V = \frac{J_j(V)}{J_i(V)} = \frac{K_{ij}^{Tr} a_j}{a_i} \qquad (8)$$

where the ion-transport selectivity factor is defined as

$$K_{ij}^{Tr} = \frac{K_j \bar{k}_j}{K_i \bar{k}_i} \frac{1 + 2w_i \cosh((1+\beta)z_iFV/4RT)\dfrac{N \sinh(z_iFV/2NRT)}{\sinh(z_iFV/2RT)}}{1 + 2w_j \cosh((1+\beta)z_iFV/4RT)\dfrac{N \sinh(z_iFV/2NRT)}{\sinh(z_iFV/2RT)}} \tag{9}$$

The following limiting cases are discerned:

$$(K_{ij}^{Tr})_{eq} = \frac{K_j \bar{k}_j}{K_i \bar{k}_i} \tag{9a}$$

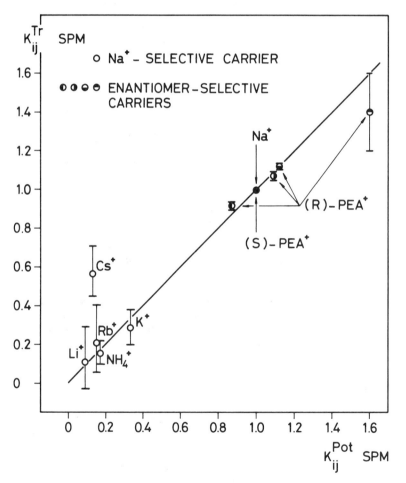

Figure 5. Correlation between potentiometric selectivities K_{ij}^{Pot} and transport selectivities K_{ij}^{Tr} (ratio of electric transference numbers) for solvent polymeric membranes containing different synthetic ionophores (27-29). For the Na⁺-selective carrier: $I^{z_i} = Na^+$, $J^{z_j} = $ alkali cations; for the enantiomer-selective carriers: $I^{z_i} = $ (S)-α-phenylethylammonium ion, $J^{z_j} = $ (R)-α-phenylethylammonium ion.

$$(K_{ij}^{Tr})_{kin} = \frac{K_j \overline{k}_j / w_j}{K_i \overline{k}_i / w_i} \tag{9b}$$

Since the ratio of kinetic parameters, w_j / w_i, was shown to roughly correlate with $(K_{ij}^{Tr})_{eq}$ (4,19), a clear loss of ion selectivity is obtained in the "kinetic domain". In contrast, the selectivity exhibited in the "equilibrium domain" mainly reflects the ion-binding selectivity K_j / K_i of the ionophores involved.

Membrane potential and potentiometric selectivity

The selectivity behavior of carrier membranes was also investigated potentiometrically (3-10,16-20,27-29). For systems containing two ions of the same charge, the zero-current membrane potentials or voltages V_o are given by the Nicolsky-Eisenman equation:

$$V_o = \frac{RT}{z_i F} \ln \frac{a_i'' + K_{ij}^{Pot} a_j''}{a_i' + K_{ij}^{Pot} a_j'} \tag{10}$$

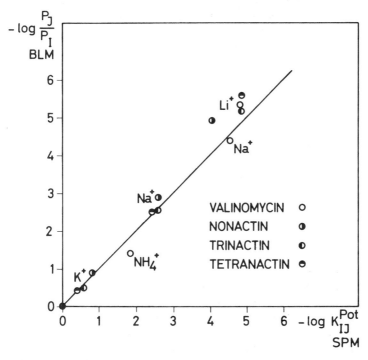

Figure 6. Comparison of the selectivities of neutral-carrier-modified solvent polymeric membranes (SPM) and bilayer lipid membranes (BLM) (27). The permeability ratios P_J / P_I (at "equilibrium" as far as available) fulfilled for the glyceryl dioleate BLM's are taken from figs. 10 and 11 in (8). Values on the SPM's were obtained using 0.1M solutions of the aqueous chlorides and membranes of the composition: 33.1 wt.-% polyvinyl chloride, 66.2 wt.-% dioctyl adipate, 0.7 wt.-% carrier. For the macrotetrolides: $I^{z_i} = NH_4^+$; for valinomycin: $I^{z_i} = K^+$.

The potentiometric selectivity factor agrees with the corresponding ion-transport selectivity (Figure 5) and with the so-called permeability ratio P_j/P_i (Figure 6):

$$K_{ij}^{Pot} = K_{ij}^{Tr}(V_o) \qquad (11)$$

Similar relations were found for the monovalent/divalent ion selectivity (4,17). Accordingly, the remarkable ion selectivities exhibited by natural or synthetic ionophores can be fully exploited in sensor membranes for specific potentiometric determinations of ion activities (for a review, see (3-5)). The basic requirement is that such ion-selective electrodes operate reversibly which means that the "equilibrium domain" of ion transport is realized in these membrane systems.

Calculation of the ion-ionophore interaction energies
in view of designing ion selectivity of membranes

In the "equilibrium domain" the transport selectivity is controlled by the overall distribution coefficients K_i and K_j (Eq. 9a). These depend on the distribution coefficients of the free cations, on the stability constants of the cation/carrier complexes in the membrane phase, and on the concentration of free carriers. The most obvious selectivity-determining parameters are the stability constants of the ion/carrier complexes. Currently research efforts are focused on the calculation of the corresponding interaction energies. The most important contributions are related to:

1) Ion-ligand interactions

2) Structural changes (torsion angles, bond lengths, bond angles) of the ligand through complexation

3) Interaction of ligand, ion/carrier complexes and free ions with the environment.

Unfortunately, calculations of high precision are needed since the ion selectivity ranges of practical carrier ligands correspond to free energy changes of ≤ 30 kJ/mol. The individual terms 1) and 3) are by at least one order of magnitude larger.

In the past decade much effort has been invested into the calculation of ion-ligand interaction energies (30-33). Only ab initio calculations with carefully chosen basis sets (34) gave satisfactory values, whereas less sophisticated methods often led to unrealistic results (31). Therefore, interaction energies of larger molecules have been approximated by using pair potentials which were derived from ab initio calculations on small systems (33). Unfortunately the different methods available for the computation of conformational energy changes (see 2) above) are still too uncertain (35). The optimization of bond lengths and bond angles has been neglected in almost all calculations of ion-ligand interaction energies. Recent results document that such effects are significant even in the case of small ligand molecules (36).

For a realistic ligand design a prediction of the structure of the ion/carrier complex is a fundamental step. For example by fixing the ligand in a given conformation (e.g. as derived from X-ray analysis), the position of the complexed ions can be predicted very reliably (37,38). This is nicely demonstrated in Figures 7-10.

The structure of the complex of NaSCN with 1,4,7,10,13-pentaoxa-16-thiacyclooctadecane (monothia-18-crown-6) was determined by X-ray crystallography (39) (see Figure 7). Using the coordinates of the ligand atoms, the potential surface for a Na^+ ion was calculated with pair potentials derived by ab initio calculations (40) (see Figure 8). The most stable calculated position of Na^+ deviates by only 6 pm from the experimentally found location.

18,18′-Spirobi(19-crown-6) forms a 2:1 (Li^+:ligand) complex with LiI which contains 4 H_2O molecules per ligand (41) (see Figure 9). The potential surfaces were calculated with pair potentials (42) in absence (Figure 10, top) and in presence of the four water molecules (Figure 10, bottom). The computed absolute energy minima are located within 51 and 49 pm, respectively, to the experimentally observed positions of Li^+.

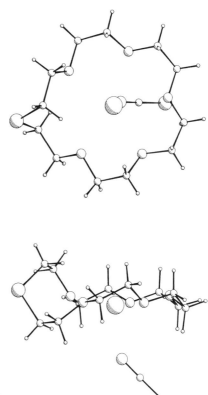

Figure 7. Structure of the NaSCN complex of monothia-18-crown-6 (see(39)).

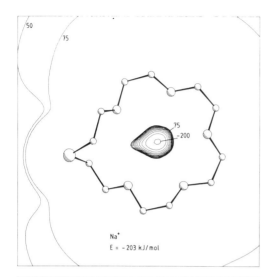

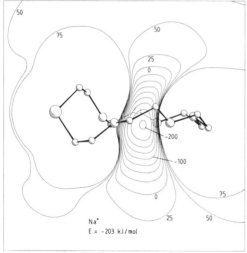

Figure 8. Isoenergy contour maps for the interaction of Na$^+$ with monothia-18-crown-6. The planes of projection correspond to Fig. 7 and contain the calculated optimal position of Na$^+$.

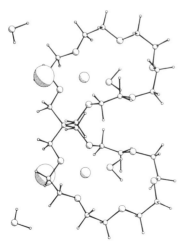

Figure 9. Structure of the 2:1 (Li$^+$:Ligand) complex of LiI with 18, 18′-spirobi-(19-crown-6) ($C_{25}H_{48}O_{12}$·2LiI·4H$_2$O, see (41)).

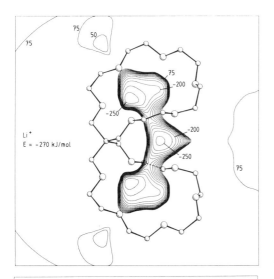

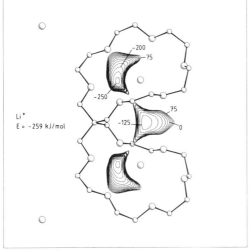

Figure 10. Isoenergy contour maps for the interaction of Li$^+$ with 18, 18'-spirobi-(19-crown-6). Top without, bottom with the four water molecules.

Acknowledgement

This work was partly supported by the *Swiss National Science Foundation*. We thank M. Welti for the computations for the figures 8 and 10.

References and Footnotes

1. G. Eisenman, Ed., *"Membranes"*, Vol. 2, *"Lipid bilayers and antibiotics"*, Dekker, New York, 1973.
2. A. E. Shamoo, Ed., *"Carriers and Channels in Biological Systems"*, Ann. New York Acad. Sci. U.S.A., *Vol. 264,* (1975).
3. D. Ammann, W. E. Morf, P. Anker, P. C. Meier, E. Pretsch, and W. Simon, *"Neutral carrier based ion-selective electrodes"*, in *Ion-Selective Electrode Rev.* 5, 3-92 (1983).

4. W. E. Morf, *The Principles of Ion-Selective Electrodes and of Membrane Transport,* Akadémiai Kiadó, Budapest, 1981/Elsevier, Amsterdam, New York, 1981.

5. W. E. Morf and W. Simon, *"Ion-selective electrodes based on neutral carriers",* in *"Ion-Selective Electrodes in Analytical Chemistry",* Ed., H. Freiser, Plenum Press, New York, 1978.

6. S. M. Ciani, G. Eisenman, R. Laprade, and G. Szabo, *"Theoretical analysis of carrier-mediated electrical properties of bilayer membranes",* chapter 2 of ref. (1).

7. G. Szabo, G. Eisenman, S. M. Ciani, R. Laprade, and S. Krasne, *"Experimentally observed effects of carriers on the electrical properties of membranes",* chapter 3 of ref. (1).

8. G. Eisenman, S. Krasne and S. Ciani, in ref. (2), p. 34.

9. S. Krasne and G. Eisenman, *J. Membrane Biol. 30,* 1 (1976).

10. S. Ciani, *J. Membrane Biol. 30,* 45 (1976).

11. P. Läuger and G. Stark, *Biochim. Biophys. Acta 211,* 458 (1970).

12. P. Läuger, *Science 178,* 24 (1972).

13. P. Läuger and B. Neumcke, *"Theoretical analysis of ion conductance in lipid bilayer membranes",* chapter 1 of ref. (1).

14. D. A. Haydon and S. B. Hladky, *Q. Rev. Biophys. 5,* 187 (1972).

15. S. B. Hladky, *Biochim. Biophys. Acta 352,* 71 (1974).

16. W. Simon, W. E. Morf, E. Pretsch, and P. Wuhrmann, in *"Calcium Transport in Contraction and Secretion",* Ed., E. Carafoli, North-Holland Publishing Company, Amsterdam, 1975.

17. W. E. Morf, P. Wuhrmann, and W. Simon, *Anal. Chem, 48,* 1031 (1976).

18. A. P. Thoma, A. Viviani-Nauer, S. Arvanitis, W. E. Morf, and W. Simon, *Anal. Chem. 49,* 1567 (1977).

19. W. E. Morf, D. Ammann, R. Bissig, E. Pretsch, and W. Simon, *"Cation selectivity of neutral macrocyclic and nonmacrocyclic complexing agents in membranes",* in *"Progress in Macrocyclic Chemistry",* Vol. 1, Eds., R. M. Izatt and J. J. Christensen, Wiley-Interscience, New York, 1979.

20. A. A. Lev, V. V. Malev, and V. V. Osipov, *"Electrochemical properties of thick membranes with macrocyclic antibiotics",* chapter 7 of ref. (1).

21. P. Läuger, "Ion transport across lipid bilayer membranes", in *"Physical Principles of Biological Membranes",* Ed., F. Snell, Gordon & Breach, New York, 1970.

22. S. Arvanitis, Dissertation ETH Nr. 6261, Zurich, 1978.

23. J. A. V. Butler, *Trans. Faraday Soc. 19,* 729 (1924).

24. T. Erdey-Gruz and M. Volmer, *Z. Phys. Chem. (Leipzig) 150,* 203 (1930).

25. A. P. Thoma, Dissertation ETH Nr. 6062, Zurich, 1977.

26. W. E. Morf, H. Ruprecht, P. Oggenfuss, and W. Simon, *Proceedings of the Meeting on Theory and Application of Ion-Selective Electrodes in Physiology and Medicine,* held at Burg Rabenstein, BRD, Sept. 1983, in press.

27. W. Simon, *"Selective Transport Processes in Artificial Membranes",* in *"Molecular Movements and Chemical Reactivity as Conditioned by Membranes, Enzymes and Other Macromolecules",* Eds., R. Lefever and A. Goldbeter, John Wiley & Sons, Inc., New York, 1978, p. 287.

28. W. E. Morf and W. Simon, *"Transport properties of neutral carrier membranes",* in *"Ion-Selective Electrodes",* Ed., E. Pungor, Akadémiai Kiadó, Budapest, 1977, p. 25.

29. A. P. Thoma, A. Viviani-Nauer, K. H. Schellenberg, D. Bedeković, E. Pretsch, V. Prelog, and W. Simon, *Helv. Chim. Acta 62,* 2303 (1979).

30. E. Clementi, *Lecture Notes in Chemistry 2,* 1 (1976).

31. P. Schuster, W. Jakubetz, and W. Marius, *Topics in Current Chemistry 60,* 1 (1975).

32. A. Pullman, H. Berthod, and N. Gresh, *Int. J. of Quant. Chem. Symp. 10,* 59 (1976).

33. E. Clementi, *Lecture Notes in Chemistry 19,* 1 (1980).

34. W. Kolos, *Theor. Chim. Acta 54,* 187 (1980).

35. J. Bendl and E. Pretsch, *J. Comp. Chem. 3,* 580 (1982).

36. P. Portmann, M. Welti, and E. Pretsch, in preparation.

37. E. Pretsch, J. Bendl, P. Portmann, and M. Welti, in *"Proceedings of the Symposium on Steric Effects in Biomolecules, Eger, Hungary, 1981",* Ed., G. Naray-Szabo, Elsevier-Akadémiai Kiadó, 1982, p. 85.

38. M. Welti, E. Pretsch, E. Clementi, and W. Simon, *Helv. Chim. Acta 65,* 1996 (1982).

39. M. L. Campbell, N. K. Dalley, R. M. Izatt, and J. D. Lamb, *Acta Cryst. B37,* 1664 (1981).

40. G. Corongiu, E. Clementi, E. Pretsch, and W. Simon, *J. Chem. Phys. 70,* 1266 (1979).
41. M. Czugler and E. Weber, *JCS. Chem. Commun. 472* (1981).
42. G. Corongiu, E. Clementi, E. Pretsch, and W. Simon, *J. Chem. Phys. 72,* 3096 (1980).

Structure & Motion: Membranes, Nucleic Acids & Proteins,
Eds., E. Clementi, G. Corongiu, M. H. Sarma & R. H. Sarma,
ISBN 0-940030-12-8, Adenine Press, Copyright Adenine Press, 1985.

Ultrafast Dynamics in Proteins and Membranes by Time Resolved Fluorescence Spectroscopy

Enrico Gratton
Department of Physics
University of Illinois at Urbana-Champaign
Urbana Il 61801

and

Joseph R. Lakowicz
Department of Biological Chemistry
University of Maryland
Baltimore, Md 21201

Abstract

The recent development of multifrequency phase and modulation fluorometry allowed us to attain picosecond resolution in fluorescence decay mesurements using a variety of light sources. Continuous-wave lasers, the pulsed synchrotron radiation produced in storage rings and conventional arc lamps have been used in conjunction with the cross-correlation detection method. The measurement over a wide frequency range of phase and modulation of the emission with respect to the excitation permitted us to analyze complex emission processes in important biological systems. In particular we have measured time resolved emission spectra from few hundred picoseconds to several hundred nanoseconds in proteins and in membranes. The spectral drift during the excited state lifetime originates from reorientations and displacements of dipoles around the excited fluorophore. The phenomenological models currently used to describe the relaxation process are inadequate. Non-exponential spectral center of gravity shifts have been observed for TNS in apomyoglobin. Also for PATMAN-labeled vesicles the spectral width changed as a function of time.

Introduction

Fluorescence spectroscopy is a widely used technique to investigate dynamical properties in proteins, nucleic acids and membranes. The appeal of fluorescence methodologies lies in their intrinsic sensitivity, in the localization of the excitation, and also in the characteristic time scale of the emission process. The excited state lifetime is typically of the order of few nanoseconds. This period corresponds to the time scale of many important biological processes, including diffusion or transport of small molecules over a distance of several nanometers, rotational motions of proteins and internal residues in proteins, rotational motions of probes in membranes,

155

segmental motions in proteins and nucleic acids, proton transfer reactions, dipolar reorientations around the excited state dipole and others.

The characteristic fluorescence parameters, steady-state excitation and emission spectra, emission anisotropy and the time course of the fluorescence emission and anisotropy can reveal structural and dynamical characteristics of the biological system under investigation. The fluorescence emission process in biological system is generally complex, and even in simple biological systems, it is characterized by a time course wich deviate from the single exponential behavior expected for the decay of the excited state of a simple system. The non-single exponential behavior may arise as a consequence of the number and types of the emitting species present, from the complexity of the fluorophore environment or from processes which take place during the lifetime of the excited state. The characterization of the heterogeneity of the emission and the assignment to the different physical mechanisms involved is the more chellanging problem in fluorescence spectroscopy (1-3).

In the past few years the development of multifrequency phase and modulation fluorometry allowed us to attain picosecond resolution in fluorescence decay measurements and to analyze complex emission processes in biological systems. In the following we present the principle of multifrequency phase and modulation fluorometry and some results obtained on important biological systems.

Multifrequency phase and modulation fluorometry

Conventionally, the decay of the fluorescence emission is recorded in the time domain by the popular technique of single photon counting. In this case the sample under investigation is illuminated by a short light pulse and the time delay between the exciting pulse and the detection of a photon by the collecting system is measured. The number of events measured as a function of the time delay gives the pulse response of the system if the width of the excitation pulse can be neglected. Otherwise complex deconvolution methods must be applied. An alternative technique consists of measuring the harmonic response of the system. The fluorescence sample is illuminated by a light beam whose intensity is sinusoidally modulated and the phase delay and the modulation ratio of the emitted radiation with respect to the excitation is measured (Figure 1). For an exponentially decaying system, the relation between the phase shift ϕ and the modulation ratio M and the characteristic time τ is given by

$$\tan \phi = \omega \, \tau^P$$

$$M = 1/(1 + (\omega \, \tau^M)^2)^{-\frac{1}{2}}$$

where τ^P and τ^M represent two independent measurements of τ, and ω is the angular frequency of light modulation. In the past, one of the major criticisms to the harmonic response method was that complex decays cannot be analyzed. However, if the harmonic response of the system is measured over a large frequency range and the pulse response over a large time range, then the harmonic

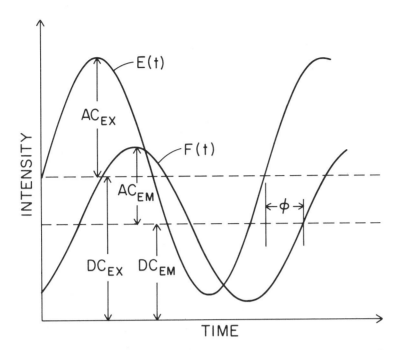

Figure 1. Schematic representation of the excitation E(t) and fluorescence F(t) waveforms in the harmonic response method. Fluorescence is delayed by an angle τ and demodulated with respect to the excitation.

response and the pulse response are equivalent, being the Fourier transform one of the other. In practice the collection of data in the time domain is performed over a restricted time interval and also, the acquisition of data in the frequency domain in multifrequency phase fluorometry, is performed over a limited frequency range. This limitation originates important practical differences between the two methods. A comparative discussion of the merits of the two methods was presented else-where (4). A further comparison between the two methods can be performed in such a cases in which the pulsed light source can be used to perform measurements directly in the frequency domain taking advantage of the high harmonic content of a high repetition frequency pulsed light source. Recently a multifrequency phase and modulation fluorometer was built at the PULS laboratory in Frascati using the pulsed synchrotron radiation produced at the ADONE storage ring. The details of the instrument and some applications have been reported elsewhere (4). For this instrument the results obtained using the harmonic method compared favorably with the pulse response method.

Among the biological systems investigated using multifrequency phase and modula-tion fluorometry are the diffusion of oxygen in iron free myoglobin and hemoglobin (5,6), the rotational behavior of probe molecules in synthetic and natural mem-branes (7-9), the energy transfer between tyrosine and tryptophan in BSA, and the dipolar reorientation after excitation in isotropic solvents, in proteins and in membranes (10). In the following we discuss in some detail the measurement of dipolar reorientations in proteins and in membranes.

Dipolar relaxations in proteins and membranes

The effect of the dielectric constant upon the absorption and emission spectra have been the subject of a number of studies (11-14). The effect of dipolar relaxation upon the electronic levels can be rationalized with relation to figure 2. In the ground state the solvent molecules are statistically organized around a fluorophore to minimize the energy corresponding to the interaction of the chromophore with the surrounding solvent. At the time of excitation, due to the different electronic configuration of the excited state, a Franck-Condon state is obtained in which the solvent molecules can be in an unfavorable orientation with respect to the excited state dipole moment. Large variations in orientation and magnitude of the excited state dipole with respect to the ground state can be obtained in certain molecules giving origin to an unrelaxed Franck-Condon state which may differ by as much as 5000 cm^{-1} from the relaxed equilibrium configuration. This effect gives a large spectral shift of the emission with respect to the excitation and, under favorable circumstances, to a time dependent effect that involves the relaxation of the initial Franck-Condon state to the final relaxed state. A characteristic feature of dipolar relaxations is the shift of the emission spectrum with time. At short times after excitation the emission comes primarily from the unrelaxed or 'blue' state. As time evolves, the emission from the relaxed or 'red' state becomes predominant. The measurement of the spectral characteristics (spectral center of gravity and spectral

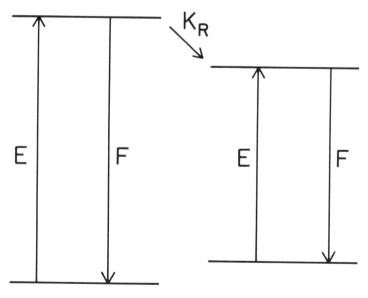

Figure 2. Schematic representation of the shift of the excited state level due to dipolar reorientations occuring during the excited state lifetime.

width) as a function of time is used to investigate the modalities of the relaxation process. Measurements of time resolved spectra in a protein and synthetic membranes using multifrequency phase and modulation fluorometry have recently appeared (10).

In figure 3 some typical fluorophores used to investigate dipolar relaxations are shown. These fluorophores have in the excited state a large charge separation which amount approximatly to one electronic charge over a distance of several angstroms. In the case of PRODAN the excited state dipole has a magnitude of about 8 Debyes.

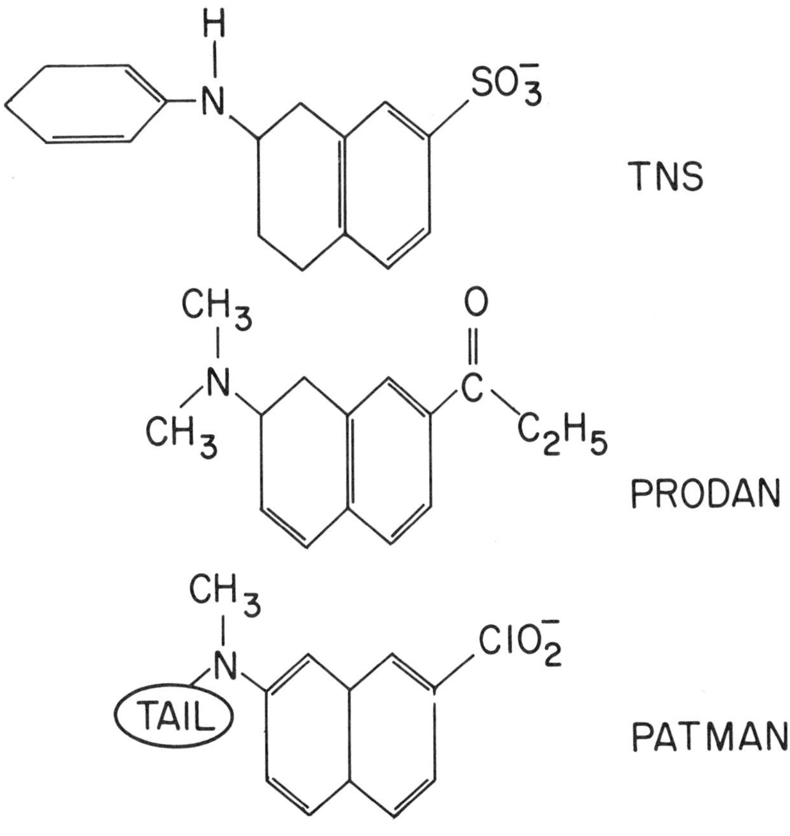

Figure 3. Naphtalene derivatives used to study dipolar reorientations. TNS: 2-p-toluidinyl-6-naphtalene sulfonic acid. PRODAN: 2-dimethylamino-6-propionyl-naphtalene. PATMAN: 6-palmitoyl-2-[(2-trimethyl-ammonium)ethyl]-methylamino-naphtalene chloride.

Models for dipolar relaxations

The models currently used to describe the influence of the dielectric properties of the solvent upon the emission spectra consider the solvent as a continuum and the dielectric constant of the medium determines the levels of the electronic states. Generally these models are used to calculate the difference in energy between the unrelaxed and the relaxed state. A complete dynamical theory describing the time course of the relaxation process is not available. Two different approaches have been proposed, one called the continuous model and the other the two-step model. The continuous model, based on a phenomenological description of the relaxation process, was introduced by Bakshiev et al. (15). These investigators propose to describe the spectral shift using a continuous model in which the decay of the intensity is described by an exponential shift of the maximum convoluted with the exponential decay of the excited state population. The spectral characteristics are described by expressions of the form

$$I(t,\lambda) = [f(\bar{\nu} - \Delta\xi e^{-t/\tau^r}]e^{-t/\tau^f}$$

where τ^f is the fluorescence decay time, $\Delta\xi$ the total shift of the spectral center of gravity, $f(\nu)$ describes the average spectral shape and τ^r is the charactersitsc relaxation time of the spectral center of gravity. This empirical approach was used to compare the values of the relaxation time in different solvents, and at several temperatures and viscosities. In this approach the spectral shape is assumed to be constant. However, it has been observed that this latter condition is seldom satisfied. In several experiments a variation of the spectral width, as the system evolves from the unrelaxed to the relaxed state, has been reported.

The second approach assumes that the relaxation process can be described by a two-state system as shown in figure 4. In this case the rate of transfer from the unrelaxed to the relaxed state is k_r. The spectral characteristics are described by a bimodal distribution of a 'blue' and a 'red' spectrum. The time behavior of such a system is characterized, from the point of view of the relaxation process, by an exponential decay of the 'blue' spectrum and by an exponential growth of the 'red'

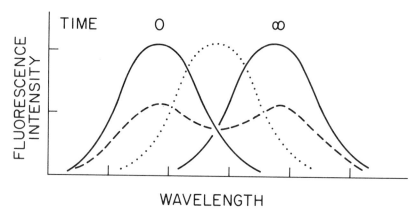

Figure 4. Schematic representation of the continuous and two-state models of solvent relaxation. The expected instantaneous spectra at intermediate times are shown for the two-state (———) and the continuous (······) relaxation models. Also shown are the expected spectra at time t=0 and time t=∞.

spectrum. Superimposed, is a convolution of the total emitted fluorescence by the decay of the excited state population. The extent of back reaction is neglected. The equation describing the spectral characteristics is

$$I(t,\lambda) = i_1(t) \, f_1(\lambda) + i_2(t) \, f_2(\lambda)$$

In the two-state model the spectral shape is a function of time (figure 4), and two exponentials describe the decay process at each wavelength.

An alternative approach to the continuous model and to the two state model has recently been proposed by G. Weber (16). Weber observed that the separation of charge in the excited fluorophore is over a distance which is large compared with the average dimension of a solvent molecule (typically 3-4 A). The excited fluorophore appears as two point charges to the solvent molecules. Assuming a given distance of the solvent molecules with respect to the charge, it was possible to calculate the average extra energy of the excited state. Weber assumed a Langevin distribution of the solvent dipoles in the field of a monopole. Using this simple approach the spectral center of gravity and the spectral width of the emission spectrum were obtained. Using the same model Weber attempted a calculation of the dynamics of the relaxation process. The results indicated that the spectral width varied with time. At excitation a relatively narrow distribution exists around the fluorophore due to the ground state interactions. Also a narrow distribution is present in the final relaxed state. At intermediate time a broader distribution is present due to the assumed Langevin distribution of the dipoles.

Experimetal observation of dipolar relaxation in proteins

We have measured dipolar relaxation in pure solvents, in proteins and in membranes. A measurement at a single emission wavelength consisted of a set of phase shift and modulation ratio values recorded at several different modulation frequencies. Then the emission wavelength was changed and a new set of phase and modulation values was acquired. This procedure was repeated for several values of the emission wavelength. For each wavelength value the set of phase and modulation data was fitted to a sum of exponentials using a non linear-least squares routine described elsewhere (17,18). The number of exponential components used varied depending of the wavelength. Generally two or three exponentials were used. But there was no case in which two exponentials described the relaxation process at all wavelengths as required by the two state model. Notice that we attributed no physical significance to the rates and preexponential factors obtained by the fitting procedure. The fit was used to generate an analytical function that accurately described the time evolution of the emission at each wavelength. After proper normalization, we calculated the intensity of the emission at different wavelengths for selected values of time. The time resolved spectra obtained using this procedure for TNS in glycerol and TNS in apomyoglobin are shown in figure 5. The value of the spectral center of gravity (and spectral shift) as a function of time was calculated and is reported in figure 6. The relevant properties of figure 5 and 6 are

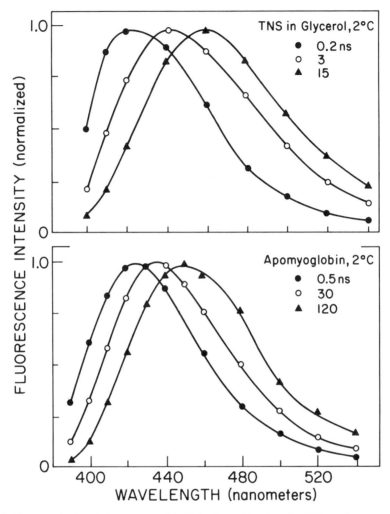

Figure 5. Time-resolved emission spectra of TNS in glycerol (top) and TNS bound to apomyoglobin (bottom).

related to the differences of the relaxation process of TNS in glycerol with respect to apomyglobin. The relaxation process in glycerol was reasonably well described by an exponential shift of the spectral center of gravity. Also the relaxation process in glycerol was strongly temperature dependent (not shown). The total spectral shift was about 2000 cm^{-1}. Furthermore, the broadening of the spectrum at intermediate time is clearly visible in figure 5.

In apomyoglobin instead, the total spectral shift was about 500 cm^{-1} after 30 nanoseconds, but the spectral center of gravity continued to shift almost linearly with time (see spectrum at 120 nanoseconds in figure 5). Also, in apomyoglobin, the relaxation process was largely insensitive to temperature changes in the range 2°C

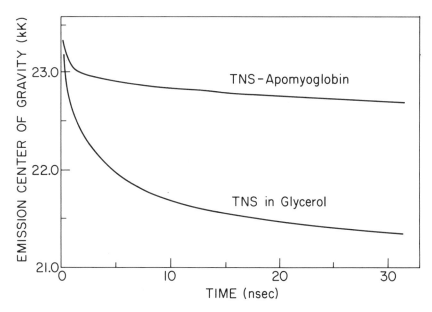

Figure 6. Displacement of spectral center of gravity of TNS in glycerol and bound to apomyoglobin.

to 30°C. The spectral width was narrower in apomyoglobin and remained constant during the relaxation process. It is also interesting to note the time scale of the relaxation process. In glycerol, the relaxation rate extimated by the exponential shift of the spectral center of gravity corresponded well to the rotational rate of glycerol molecules (at a given temperature). In apomyoglobin, because of the non-exponential shift of the spectral center of gravity, it was difficult to define a characteristic relaxation time. However if we arbitrarly assume that the total energy released is of the order of 1000 cm^{-1}, then the average time to complete the relaxation process is of the order of 30 nsec. Of course this is a very crude extimate until a more detailed analysis is performed.

It is of interest to discuss the possible origin of the relaxation process observed in apomyoglobin. Fluorescence polarization experiments have shown that TNS is rigidly bound in the myoglobin pocket. Furthermore, due to the nature of dipolar interactions, only dipoles vhich are in the myoglobin pocket or close to the TNS molecule would partecipate to the relaxation process. In a protein there are dipoles in the peptide backbone, in the bound water, and dipoles and charges on specific residues. A reorientation of the peptide dipoles that can give origin to the energy changes observed between unrelaxed and relaxed states would involve large conformational changes of the protein structure. It is unlikely that these changes can occur in a time scale of the order of nanoseconds. Instead, we believe that we are observing reorientation effects of bound water dipoles and/or movements of specific protein residues. From figure 6 it appears that there is a fast relaxation that is completed in about 1 nanosecond followed by a much slower process. Because of the hydrophobic nature of the myoglobin pocket, it is possible that the fast process corresponds to a reorientation of a protein residue. Changes in specific interactions of the fluorophore upon excitation can also partecipate.

We have also measured time resolved spectra for TNS and PATMAN in phopholipid vesicles. The steady state spectra of TNS in DPPC below and above the gel to liquid-crystal phase transition showed a small variation indicating that this probe is not very sensitive to the transition. Instead, a dramatic spectral shift between the gel and the liquid-crystalline phase was observed using PATMAN (Figure 7). Time resolved spectra at 0.2 nsec, 2 nsec and 20 nsec are shown in figure 8 for PATMAN in DPPC at 8°C and 46°C. A large relaxation is also observed for PATMAN in

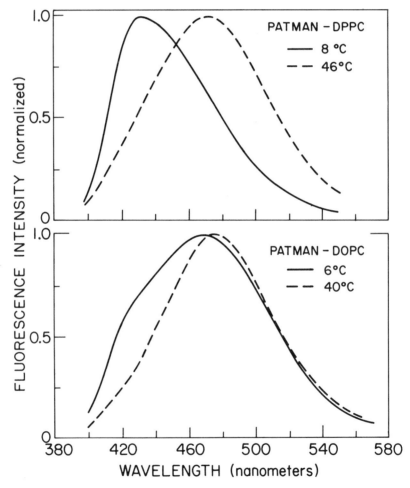

Figure 7. Steady-state emission spectra of PATMAN-labeled phospholipid vesicles. Top: DPPC vesicles at 8°C (——) and 46°C (-------), bottom: DOPC vesicles at 6°C (——) and at 40°C (-------).

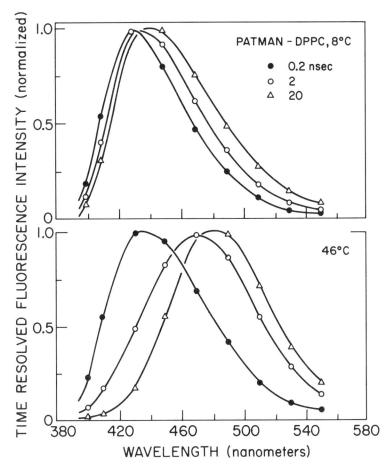

Figure 8. Time-resolved emission spectra of PATMAN-labeled DPPC vesicles. Top: 8°C; bottom: 46°C. Time resolved spectra are shown at 0.2 nsec, 2 nsec and 20 nsec.

DOPC (figure 9). DOPC is in the liquid-crystalline phase at all temperatures investigated. In fact for PATMAN in DOPC the relaxation process is not strongly temperature sensitive indicating that is the physical state of the phopholipid vesicle that determines the relaxation process rather than the temperature change. In Figure 10 the position of the spectral center of gravity is shown as a function of time. For PATMAN the total energy released during the relaxation is about 2000 cm^{-1} for the fluid phase and only about 500 cm^{-1} for the gel phase. Also the spectral center of gravity drift seems to be exponential. The brodening of the spectrum at intermediate time is very evident in figure 8 and 9. The spectral width as a function of time for PATMAN in DOPC and DPPC increased during the first nanosecond and then decreased to give a relaxed narrower spectrum (figure 11).

The physical origin of the relaxation process in membranes is unknown. The probe used in this work, PATMAN, has a polar fluorescent head, which presumably lies

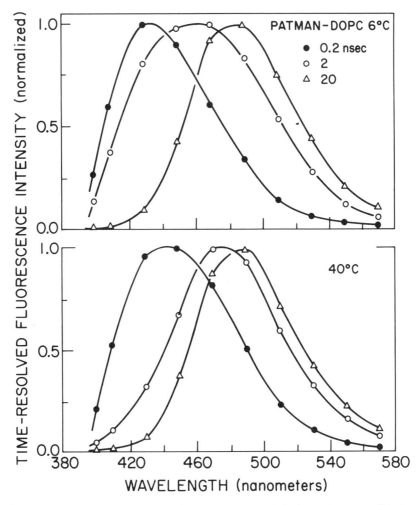

Figure 9. Time-resolved emission spectra of PATMAN-labeled DOPC vesicles. Top: 6°C, bottom: 40°C. Time-resolved emission spectra are shown at 0.2 nsec, 2 nsec and 20 nsec.

near the membrane surface, and a long hydrophobic tail which anchor the molecule to the membrane. Solvent dipoles are likely to play a minor role in the observed relaxation. The large sensitivity to the physical state of the membrane indicates that the phopholipid moiety is directly involved. Most likely a rotation or displacement of the polar head is responsible for the observed relaxation.

In conclusion, we have measured fast processes in proteins and membranes which originate from dipolar reorientation and/or displacements of charged groups. Time resolved spectra were obtained from few hundred picoseconds after excitation to hundreds of nanoseconds. The continuous model and the two state model are inadequate to describe the spectral characteristics of the observed relaxation. Weber's approach qualitatively reproduces some of the features observed. However, we

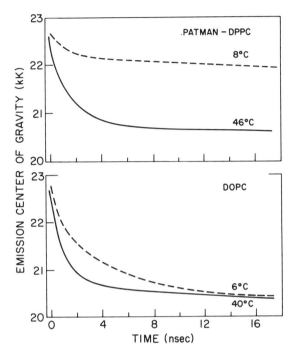

Figure 10. Displacement of the spectral center of gravity of PATMAN labeled vesicles. Top: DPPC at 8°C (——) and at 46°C (-------). Bottom: DOPC at 6°C (——) and at 40°C (-------).

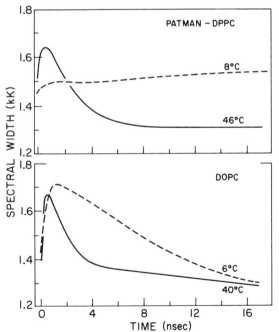

Figure 11. Variations of the spectral width of PATMAN-labeled vesicles. Top: DPPC at 8°C (——) and at 46°C (-------). Bottom: DOPC at 6°C (——) and at 40°C (-------).

are still very far from an understanding at a microscopic level of the relaxation processes observed in proteins and membranes.

References and Footnotes

1. D.M. Jameson, E. Gratton and R.D. Hall, *Ap. Spect. Rev., 20,* 55 (1984).
2. E. Gratton, D.M. Jameson and R.D. Hall, *Ann. Rev. Biophys. Bioeng., 13,* 105 (1984).
3. W.R. Ware, in *Creation and detection of the excited state, Vol 1A* (A.A. Lamola ed.), Dekker, N.Y., p. 123 (1971).
4. E. Gratton, D.M. Jameson, N. Rosato and G. Weber, *Rev. Sci. Instrum., 55,* 486 (1984).
5. E. Gratton, B. Alpert, D.M. Jameson and G. Weber, *Biophys. J., 45,* 789 (1984).
6. D.M. Jameson, E. Gratton, B. Alpert and G. Weber, *Biophys. J., 45,* 795 (1984).
7. T. Parasassi, F. Conti and E. Gratton, *Biochemistry 23,* 5660-5664 (1984).
8. F. Conti, T. Parasassi, A. Sapora and E. Gratton, *B.B.A. 805,* 117-122 (1984).
9. J.R. Lakowicz, H. Cherek, B.P. Maliwal and E. Gratton, *Biochemistry,* in press (1984).
10. J.R. Lakowicz, H. Cherek, B.P. Maliwal, G. Laczko and E. Gratton, *J. Biol. Chem. 259,* 10967-10977 (1984).
11. E. Lippert, Z. *Elektrochem., 61,* 962 (1957).
12. N.S. Bayliss and E.G. McRae, *J. Phys. Chem., 58,* 1002, (1954).
13. Y. Ooshika, *J. Phys. Soc. Japan, 9,* 594 (1954).
14. N. Mataga, Y. Kaifu and M. Koizumi, *Bull. Chem. Soc. Japan, 29,* 465 (1956).
15. N.G. Bakshiev, *Opt. Spectros., 16,* 566 (1964),
16. R. B. McGregor, Jr. and G. Weber, *Ann. N. Y. Acad. Sci., 366,* 140, (1981).
17. D.M. Jameson and E. Gratton, in *New Directions in Molecular Luminescence* (D. Fastwood, ed.) ASTM STP 822, American Society for Testing and Materials, Philadelphia, p. 67, (1983).
18. J.R. Lakowicz, E. Gratton, G. Laczko, H. Cherek and M. Limkeman, *Biophysical J. 46,* 463-478 (1984).

Structure & Motion: Membranes, Nucleic Acids & Proteins,
Eds., E. Clementi, G. Corongiu, M. H. Sarma & R. H. Sarma,
ISBN 0-940030-12-8, Adenine Press, Copyright Adenine Press, 1985.

Dynamic Properties of Ion Channels and Ion Pumps

P. Läuger
Department of Biology, University of Konstanz,
D—7750 Konstanz, F. R. G.

Abstract

Recent studies of protein dymanics suggest that ionic channels can assume many conformational substates. Long-lived substates have been directly observed in single-channel current records. In many cases, however, the lifetimes of conformational states will be far below the theoretical limit of time resolution of single-channel experiments. The existence of such "hidden" substates may strongly influence the observable (time-averaged) properties of a channel, such as the concentration dependence of conductance. In special situations the rate of ion translocation becomes limited by the rate of conformational transitions; in this case the channel may approach the kinetic behaviour of a carrier. As a result of the strong coulombic interaction between an ion in a binding site and polar groups of the protein, rate constants of conformational transitions may depend on the occupancy of the binding site. Under this condition a nonequilibrium distribution of conformational states is created when ions are driven through the channel by an external force. The concept of channels with multiple conformational states may also be applied to ion pumps. An ion channel functions as a pump when the energy profile of the channel is transiently modified in an appropriate way by an energy-supplying reaction. Absorption of a light-quantum or phosphorylation of the channel protein may alter the binding constant of an ion-binding site in the channel and, at the same time, change the height of adjacent barriers. In this way an ion may be preferentially released to one side of the membrane, while during the transition back to the original state of the channel another ion is taken up from the opposite side.

Introduction

Ions permeate through cellular membranes by means of special mechanisms different from simple diffusion through the lipid bilayer. In the discussion of possible passive transport pathways, two alternatives are usually considered: carrier and channel mechanisms. A carrier (in its simplest form) may be defined as a transport system with a binding side that is exposed alternately to the left and to the right side (but not to both sides simultaneously). A channel, on the other hand, consists of one or several binding sites arranged in a transmembrane sequence and is accessible from both sides at the same time.

Clear-cut examples of carrier and channel mechanisms in ion transport have been obtained from the study of certain small or medium-sized peptides and depsipeptides produced by microorganisms. Cyclodepsipeptides, such as valinomycin, have been shown to act by a translatory carrier mechanism which involves a movement of the whole carrier molecule with respect to the lipid matrix of the membrane. A well-characterized ion channel is the channel formed by the linear pentadecapeptide gramicidine A. In these cases the distinction between a channel which is more or less fixed within the membrane and a carrier moving within the lipid matrix is unambiguous.

The discrimination between carrier and channel mechanisms becomes less obvious in the case of large transport proteins spanning the cell membrane. Such a protein is unlikely to move as a whole within the membrane. It still can act as carrier (according to the definition given above), however, if a conformational change within the protein switches the binding site from a left-exposed to a right-exposed state. A channel, on the other hand, does not necessarily have a fixed, time-independent structure. Proteins may assume many conformational substates and move from one state to the other. Accordingly, a channel may carry out conformational transitions between states differing in the height of the energy barriers that restrict the movement of the ion. It can be shown that such a channel with multiple conformational states may approach the kinetic behaviour of a carrier. Channel and carrier models should therefore not be regarded as mutually exclusive possiblities, but rather as limiting cases of a more general mechanism.

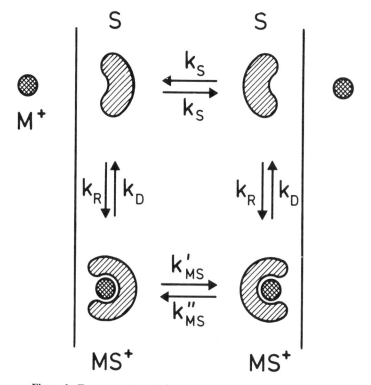

Figure 1. Transport of ion M^+ mediated by a translatory carrier S.

Kinetics of Translatory Ion Carriers

Incorporation of certain macrocyclic antibiotics, such as valinomycin, enniatin B, and the macrotetrolides into artificial lipid bilayer membranes result in a strong increase of potassium permeability of the membrane (1,2). In solution these compounds form complexes with alkali ions in which the primary hydration shell of the ion is replaced by ligand groups from the macrocycle. All experiments carried out so far are consistent with the notion that valinomycin and related compounds act by a translatory carrier mechanism. Both the stationary (2,3) as well as the nonstationary (4) behaviour of macrocyclic ion carriers in planar bilayer membranes can be approximately described on the basis of the assumption that the overall transport occurs in four distinct steps (Fig. 1): (i) association of ion M^+ and carrier S in the interface, (ii) translocation of the complex MS^+, (iii) dissociation of MS^+, (iv) back transport of free carriers S. In the analysis of the experiments it is further assumed that the individual reaction steps can be described in the same way as chemical processes, introducing reaction rate constants.

The kinetics of ion carriers may be analyzed by studying the electrical conductance of artificial planar bilayer membranes. In particular, electrical relaxation techniques, such as the voltage-jump (4), or charge-pulse method (5) may be used to evaluate the individual rate constants. As a specific example, we consider the results obtained for valinomycin/Rb^+ in a monoolein/n-decane membrane (5).

$$k_R \simeq 3 \times 10^5 \, M^{-1} \, s^{-1} \qquad (25°C, \, 1M \, RbCl)$$

$$k_D \simeq 2 \times 10^5 \, s^{-1}$$

$$k_{MS} \simeq 3 \times 10^5 \, s^{-1}$$

$$k_S \simeq 4 \times 10^4 \, s^{-1}.$$

At one-molar concentration of the transported ion ($c_M = 1$ M), the rate constants of association ($c_M k_R$), dissociation (k_D) and translocation of the loaded carrier (k_{MS}) are approximately equal ($2 - 3 \times 10^5 \, s^{-1}$). The rate-determining step in this system is the back transport of the free carrier ($k_S \simeq 4 \times 10^4 \, s^{-1}$). $k_{MS} \simeq 3 \times 10^5 \, s^{-1}$ is the frequency by which the ion-carrier complex crosses the central barrier; the reciprocal value $1/k_{MS} \simeq 3 \, \mu$sec is the average time required for translocation. This time may be compared with the diffusion time $\tau = d^2/2D$ of a spherical particle of the size of the carrier across the same distance (membrane thickness $d \simeq 5$ nm) in water (diffusion coefficient $D \simeq 3 \times 10^{-6} \, cm^2 \, s^{-1}$), which is about $0.04 \, \mu$sec.

An important quantity of the kinetic description of carriers (and other transport systems) is the maximum turnover rate f. f is defined as the limiting transport rate which is approached under short-circuit conditions for infinite ion concentration on the cis-side and zero ion concentration of the trans-side. In the above example, f is about $3 \times 10^4 \, s^{-1}$. The high efficiency of valinomycin as an ion carrier mainly results from this high turnover rate, whereas the binding constant for the ion, $k_R/k_D \simeq 1.5 \, M^{-1}$, is rather low.

The Gramicidin Channel as a Model Channel

The finding that gramicidin A, a hydrophobic peptide with known primary structure, forms alkali-ion permeable channels in lipid bilayer membranes (6) opened up the possibility of studying ion permeation through channels in a simple model system. Gramicidin A is a linear pentadecapeptide with the sequence HCO-L-Val-Gly-L-Ala-D-Leu-L-Ala-D-Val-L-Val-D-Val-L-Trp-D-Leu-L-Trp-D-Leu-L-Trp-D-Leu-L-Trp-NHCH$_2$CH$_2$OH. Evidence that gramicidin A forms channels (and does not act as a mobile carrier) has been obtained in experiments in which very small amount of the peptide were added to a planar bilayer membrane (6). Under this condition the membrane current under a constant applied voltage fluctuates in a step-like manner (Fig. 2). The size of the single conductance step is about 90 pS in 1 M Cs$^+$, corresponding to a transfer of 6 x 10^7 Cs$^+$ ions per sec. This transport rate is larger by a factor of about one thousand than the turnover number of a mobile carrier of the valinomycin type (see above), making a translatory carrier mechanism highly unlikely.

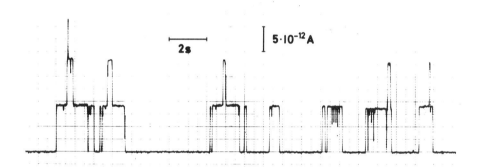

Figure 2. Conductance fluctuations of a planar bilayer membrane (glycerolmonooleate/n-decane) in the presence of very small amounts of gramicidin A. A voltage of 100 mV was applied between the aqueous solutions (1 M CsCl, 25°C). The fluctuations of membrane current arise from the formation and disappearance of single cation-permeable channels. The observed single-channel current corresponds to a flow of 6 x 10^7 Cs$^+$ ions per second (10).

A structural model of the gramicidin channel has been proposed by Urry (7,8). According to this model which is now supported by many experimental findings (9), the channel consists of a helical dimer that is formed by head-to-head (formyl end to formyl end) association of two gramicidin monomers and is stabilized by intra- and inter-molecular hydrogen bonds (Fig. 3). The central hole along the axis of the π^6 (L, D)-helix has a diameter of about 0.4 nm and is lined with oxygen atoms of the peptide carbonyls, whereas the hydrophobic amino-acid residues lie on the exterior surface of the helix. The total length of the dimer is about 2.5-3.0 nm, the lower limit of the hydrophobic thickness of the lipid bilayer.

The entry of the ion into the channel is made energetically favorable by interaction with the peptide carbonyls. In the 0.4-nm wide channel, water molecules can pre-

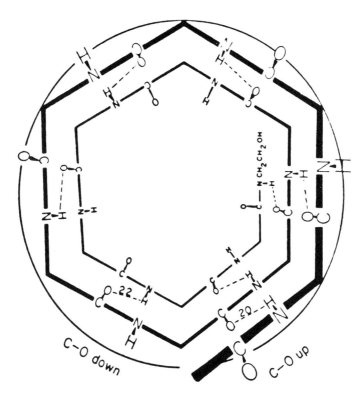

Figure 3. Structure of the π^6 (L, D)-helix of gramicidin A (8). The hole along the helix axis has a diameter of 0.4 nm and is lined with oxygen atoms of the peptide carbonyls. Hydrophobic amino-acid residues located at the periphery of the helix are not shown. The transmembrane channel consists of two helices joined at the formyl ends.

cede and follow the ion through the channel so that probably only part of the primary hydration shell is striped off. The interaction with the ligand groups creates a series of potential energy minima along the pathway of the ion. Superimposed onto this potential is the dielectric interaction of the ion with the water phase and the membrane lipid which gives rise to a broad energy barrier with the peak in the middle of the membrane. This picture predicts the existence of an energetically favorable "binding" site for the ion at either end of the channel.

Channels with Multiple Conformational States

Ion transport in a channel may be described as a series of thermally activated processes in which the ion moves from a binding site across an energy barrier to an adjacent site. In this treatment the energy levels of wells and barriers are usually considered to be fixed, i.e., independent of time and not influenced by the movement of the ion. This description, which corresponds to an essentially static picture of protein structure, represents a useful approximation in certain cases. Recent findings on the dynamics of protein molecules, however, suggest the use of a more general concept of barrier structure.

A protein molecule in thermal equilibrium may exist in a large number of conformational states and may rapidly move from one state to the other (11,12). Evidence for fluctuations of protein structure comes from X-ray diffraction and Mössbauer studies (13,14), fluorescence depolarisation experiments (15) and NMR measurements (16). These and other studies have shown that internal motions in proteins occur in a wide time range, from picoseconds to seconds.

Direct evidence that ionic channels may assume different conformational states comes from single-channel current records obtained by the patch-clamp technique (17). Intermediate conductance levels between the fully open and the fully closed state have been observed, for instance, with acetylcholine-activated endplate channels (18,19,20). In these systems the lifetimes of the substates were sufficiently long so that transitions could be observed directly in the current records. The detection of fast transitions between substates is limited, however, by the finite bandwidth of the measurement. This means that in many cases the observed single-channel current represents merely an average over unresolved conductance states. As will be discussed below, the existence of such "hidden" substates may strongly influence the observable properties of the channel, such as the current-voltage characteristic or the concentration dependence of conductance.

Of particular interest is the possibility that ion translocation in the channel becomes coupled to conformational transitions (21,22). In this case the conductance of the channel explicitly depends on the rate constants of conformational transitions. Such coupling may occur when the average lifetimes of conformational states are of the same order or lower than the dwelling times of ions in the binding sites. The channel may then exhibit unusual flux-coupling behaviour in experiments with more than one permeable ion species.

As a result of the strong coulombic interaction between an ion in a binding site and polar groups of the protein, transition rate constants may depend on the occupancy of the binding site. Under this condition a nonequilibrium distribution of conformational states is created when ions are driven through the channel by an external force (a difference of electrochemical potential). As will be shown below, this may lead to an apparent violation of microscopic reversibility, i.e., to a situation in which the frequency of transitions from state A to state B is no longer equal to the transition frequency from B to A.

Two-State Channel with Single Binding-Site

We consider a channel that fluctuates between two conducting states A and B and assume that the rate of ion flow through the channel is limited by two (main) barriers on either side of a single (main) binding site (Fig. 4). In series with the rate-limiting barriers, smaller barriers may be present along the pathway of the ion. This model corresponds to a channel consisting of a wide, water-filled pore and a narrow part acting as a selectivity filter (23). Depending on the occupancy of the binding site, the channel may exist in four substates (Fig. 5): A°: conformation A,

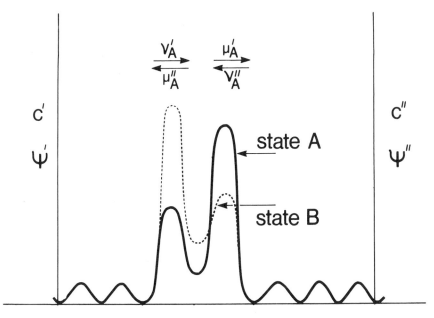

Figure 4. Energy profile of a channel with two conformational states A and B. ν_A' and ν_A'' are the frequencies of jumps from the solutions into the empty site (state A); μ_A' and μ_A'' are the frequencies of jumps from the occupied site into solutions; c', c'' and ψ', ψ'' are the ion concentrations and the electrical potentials in the left and right aqueous solutions.

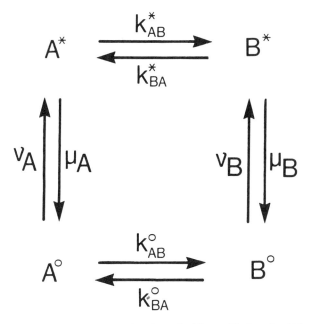

Figure 5. Transitions among four substates of a channel with one binding site (A°: conformation A, empty; A^*: conformation A, occupied; B°: conformation B, empty; B^*: conformation B, occupied).

empty; A*: conformation A, occupied; B°: conformation B, empty; B*: conformation B, occupied. Since an ion in the binding site interacts electrostatically with neighbouring polar groups of the protein, the rate constants for transitions between A and B depend, in general, on whether the binding site is empty or occupied (i.e., $k_{AB}^{o} \neq k_{AB}^{*}$ and $k_{BA}^{o} \neq k_{BA}^{*}$).

Transitions between empty and occupied states occur by exchange of an ion between the binding site and the left or right aqueous phase (Fig. 4):

$$\nu_A = \nu_A' + \nu_A'' = c'\rho_A' + c''\rho_A'' \tag{1}$$

$$\mu_A = \mu_A' + \mu_A'' \tag{2}$$

Similar equations hold for state B. In Eq. 1 it is assumed that ions in the energy wells outside the rate-limiting barriers are always in equilibrium with the corresponding aqueous phase. The jumping frequencies ν_A' and ν_A'' into the empty site are then proportional to the aqueous ion concentrations c' and c'', respectively, whereas the rate constants μ_A' and μ_A'' for leaving the site are independent of c' and c''.

The principle of microscopic reversibility requires that the rate constants obey the following relationship (22):

$$\frac{\nu_A' \mu_A'}{\nu_A'' \mu_A''} = \frac{\nu_B' \mu_B'}{\nu_B'' \mu_B''} = \frac{\nu_A' \mu_B'}{\nu_B'' \mu_A''} \cdot \frac{k_{AB}^{*} k_{BA}^{o}}{k_{AB}^{o} k_{BA}^{*}} = \exp\left[z(u - u_o)\right] \tag{3}$$

z is the valence of the permeable ion species, u the voltage across the channel, and u_o the equilibrium voltage of the ion, both expressed in units of kT/e_o (k, Boltzmann's constant; T, absolute temperature; e_o, elementary charge):

$$u = \frac{\psi' - \psi''}{kT/e_o} \tag{4}$$

$$zu_o = \ln(c''/c') \tag{5}$$

If p(X) is the probability that a given channel molecule is in state X, the time-averaged single-channel ion flux ϕ from solution' to solution'' (Fig. 4) is given by

$$\phi = \nu_A' \, p(A^o) - \mu_A'' \, p(A^*) + \nu_B' \, p(B^o) - \mu_B'' \, p(B^*) \tag{6}$$

The probabilities p(X) may be obtained from the steady-state conditions dp(X)/dt = 0, introducing the equilibrium constants H° and H* of the conformational transitions (22):

$$H^o = k_{AB}^{o}/k_{BA}^{o} \quad ; \quad H^* = k_{AB}^{*}/k_{BA}^{*} \tag{7}$$

The result reads:

$$\phi = (1/\sigma)\,[1-\exp(zu_o-zu)]\,[\nu_A'\,\mu_A'\,(1 + \nu_B/k_{BA}^o + \mu_B/k_{BA}^*) +$$
$$+ \nu_B'\,\mu_B'\,(H^oH^* + H^*\nu_A/k_{BA}^o + H^o\mu_A/k_{BA}^*) + H^*\nu_A'\,\mu_B' + H^o\nu_B'\,\mu_A' \qquad (8)$$

$$\sigma \equiv (1 + H^o)\,(\mu_A + H^*\mu_B + \mu_A\,\mu_B/k_{BA}^*) + (1 + H^*)\,(\nu_A + H^o\nu_B + \nu_A\nu_B/k_{BA}^o) +$$
$$+ \nu_A\,\mu_B\,(H^*/k_{BA}^o + 1/k_{BA}^*) + \nu_B\,\mu_A\,(H^o/k_{BA}^* + 1/k_{BA}^o) \qquad (9)$$

It is seen from Eqs. 8 and 9 that the ion flux ϕ explicitly depends on the rate constants k_{AB}^o, k_{BA}^o, k_{AB}^* and k_{BA}^*. This is an expression of the phenomenon of coupling between ion translocation and conformational transitions. Similarly, the steady-state probabilities $p(X)$ not only contain the equilibrium constants H^o, H^*, ν_A/μ_A and ν_B/μ_B but depend explicitly on the translocation rate constants μ_A', μ_A'', μ_B' and μ_B''.

An essential condition for the occurrence of coupling is the assumption that transitions between the two conformations can take place both in the empty and in the occupied state of the binding site. If transitions start only from one of the states (k_{AB}^o, $k_{BA}^o = 0$ or $k_{AB}^* $, $k_{BA}^* = 0$), then the dependence of ϕ on the rate constants of conformational transitions is lost.

Introducing the condition k_{AB}^o, k_{AB}^*, k_{BA}^o, $k_{BA}^* \ll \nu_A$, ν_B, μ_A, μ_B (frequency of conformational transitions much smaller than frequency of ion jumps) the ohmic single-channel conductance $\Lambda(c)$ is obtained from Eq. 8 in the form:

$$\Lambda(c) = p_A\,\Lambda_A + (1 - p_A)\,\Lambda_B \qquad (c' = c'' = c) \qquad (11)$$

$p_A = p(A^o) + p(A^*)$ is the probability of finding the channel in state A (A^o or A^*) which, in the vicinity of equilibrium, is given by:

$$p_A = \frac{1+cK_A}{1+H^o+(1+H^*)\,cK_A} \qquad (12)$$

$K_A = \rho_A/\mu_A$ and $K_B = \rho_B/\mu_B$ are the equilibrium constants of ion binding in states A and B, which are connected by the relation $H^*K_A = H^oK_B$. Λ_A is the conductance of the channel in state A (z is the valency of the ion):

$$\Lambda_A = \frac{z^2e_o^2}{kT} \cdot \frac{cK_A}{1+cK_A} \cdot \frac{\mu_A'\,\mu_A''}{\mu_A} \qquad (13)$$

A similar equation holds for Λ_B. Since p_A is a function of ion concentration c, the concentration dependence of the observed average conductance Λ is different from the simple saturation characteristic of a channel with time-independent potential profile (Eq. 13).

Deviations from a saturation behaviour of conductance are usually explained on the basis of ion-ion interactions in the channel (24,25) or the presence of regulatory binding sites. In the channel mechanism discussed here, the single-channel conductance Λ is influenced by the concentration-dependent distribution of conformational states (Eq. 12). The concentration dependence of the probability p_A results from the fact that the equilibrium constants H° and H^* of conformational transitions in the empty and the occupied state of the binding site are, in general, different. Only in the limiting case $H^\circ \approx H^*$ does p_A assume a concentration-independent value $1/(1+H^\circ)$.

Nonequilibrium Distribution of Long-lived Channel States

In the following we consider a channel which fluctuates between three conformational states A, B and C (18,19,26); A may be the fully open state, C the closed state and B a conductive substate. We assume that the lifetimes of these states are so long that the transitions can be directly observed in a single-channel current record (Fig. 6). A macroscopically observable transition, say, from A to B can result, at the microscopic level, from a transition $A^\circ \Rightarrow B^\circ$ (binding site empty) or from a transition $A^* \Rightarrow B^*$ (binding site occupied). (The distinction between these two elementary processes is always meaningful as long as the actual duration of a conformational transition is shorter than the mean lifetimes of the empty and occupied states of the binding site.) Accordingly, the microscopic description of the channel may be based on the scheme shown in Fig. 7.

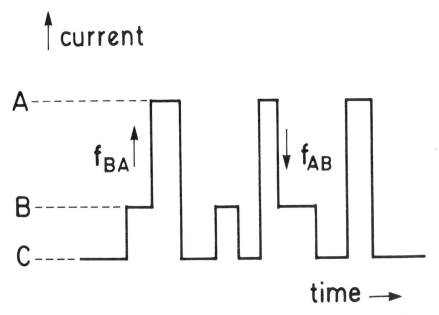

Figure 6. Single-channel current record showing transitions among three different conformational states A, B and C of the channel. f_{AB} and f_{BA} are the frequencies of transitions $A \Rightarrow B$ and $B \Rightarrow A$, respectively.

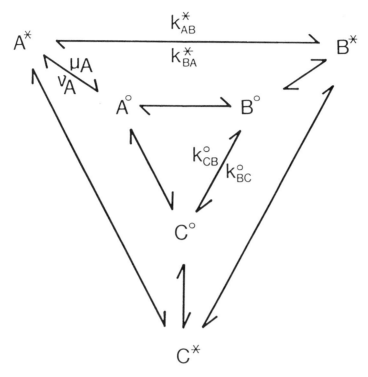

Figure 7. Transitions among three conformational states A, B and C of a channel. States with empty binding site are denoted by A°, B°, and C°, states with occupied binding site by A^*, B^* and C^*.

From single-channel records such as shown in Fig. 6 the frequency f_{XY} of transitions from state X to state Y may be obtained (X, Y = A, B, C). For a cyclic interconversion of three states:

the principle of microscopic reversibility requires that under equilibrium conditions the transition frequencies in both directions are the same ($f_{XY} = f_{YX}$). If, however, transitions between conformational states depend on the occupancy of ion binding sites, one may expect that the symmetry relationship $f_{XY} = f_{YX}$ no longer holds when ions are driven through the channel by an external force.

The expectation that f_{XY} and f_{YX} may become unequal in the presence of a driving force for ion flow is born out by an analysis of the reaction scheme of Fig. 7. Since A, B and C are assumed to be observable conductance states, their lifetimes must be much larger than the average dwell time of ions in the binding site. (If state C is nonconducting, the binding site may be assumed to be in fast exchange with one of the aqueous solutions.) The probabilities $p(X^{\circ})$ and $p(X^*)$ for empty and occupied binding site are then given by the steady-state condition of a channel which is in a (quasi) permanent state X (X = A, B, C):

$$\frac{p(X^*)}{p(X^{\circ})} = \frac{\nu_X}{\mu_X} \tag{14}$$

Microscopic reversibility requires that the following relations hold (22):

$$k^{\circ}_{AB} k^{\circ}_{BC} k^{\circ}_{CA} = k^{\circ}_{BA} k^{\circ}_{AC} k^{\circ}_{CB} \quad ; \quad k^*_{AB} k^*_{BC} k^*_{CA} = k^*_{BA} k^*_{AC} k^*_{CB} \tag{15}$$

$$\frac{\nu'_X \mu'_X}{\nu''_X \mu''_X} = \frac{\nu'_X \mu'_Y}{\nu''_Y \mu''_X} \cdot \frac{k^*_{XY} k^{\circ}_{YX}}{k^{\circ}_{XY} k^*_{YX}} = \exp\left[z(u - u_o)\right] \tag{16}$$

(X, Y = A, B, C). Since transitions X $\Rightarrow$ Y may occur in the empty and in the occupied state of the binding site, the total transition frequency f_{XY} is given by

$$f_{XY} = p(X^{\circ}) k^{\circ}_{XY} + p(X^*) k^*_{XY} \tag{17}$$

The asymmetry in the transition frequencies may be expressed by the quantity ρ_{XY}:

$$\rho_{XY} \equiv \frac{f_{XY} - f_{YX}}{f_{XY} + f_{YX}} \tag{18}$$

ρ_{XY} is obtained by calculating the probabilities $p(A^{\circ})$, $p(A^*)$, etc., in the steady state. The result reads:

$$\rho_{AB} = \frac{1}{D_{AB}} [1 - \exp(zu - zu_o)] (q_{ABC} + q_{BCA} + q_{CAB}) \tag{19}$$

with the abbreviations:

$$q_{XYZ} \equiv (\nu'_X \nu''_Y - \nu''_X \nu'_Y) \left(\frac{\mu''_X}{\nu'_X} \nu_Z k^{\circ}_{XY} k^*_{YZ} k^*_{ZX} - \frac{\mu''_Y}{\nu'_Y} \mu_Z k^*_{XY} k^{\circ}_{YZ} k^{\circ}_{ZX} \right) \tag{20}$$

$$D_{AB} \equiv 2r_{AB} r_{BA}(r_{CA} + r_{CB}) + r_{AB} r_{BC} r_{CA} + r_{AC} r_{CB} r_{BA} \tag{21}$$

$$r_{XY} \equiv \mu_X k^{\circ}_{XY} + \nu_X k^*_{XY} \tag{22}$$

The quantities ρ_{BC} and ρ_{CA} are obtained from Eq. 19 by cyclic permutation of the subscripts A, B and C.

The result contained in Eqs. 19-22 may be summarized in the following way. Transition frequencies are asymmetric ($\rho_{XY} \neq 0$), as long as a driving force for ion flow is present (u $\neq$ u$_o$). At equilibrium where $\exp(zu - zu_o)$ becomes equal to unity, the asymmetry disappears ($\rho_{XY} = 0$). This has to be expected, since the asymmetry of transition frequencies results from a nonequilibrium distribution of conformational states which is created by an ion flow through the channel. Furthermore, by using Eq. 15 it may be shown that ρ_{XY} vanishes when the transition rate constants for empty and occupied binding site are identical ($k_{XY}^{o} = k_{XY}^{*}$). In general, however, k_{XY}^{o} and k_{XY}^{*} are different, since the presence of a charge in the binding site changes the energy of the different conformational states.

The problem of cyclic interconversion of channel states discussed here has a well-known counterpart in ordinary chemical kinetics (27,28). In a cyclic reaction among three chemical species A, B and C an equilibrium state may be maintained, in priniciple, by a fast clockwise reaction A $\Rightarrow$ B $\Rightarrow$ C $\Rightarrow$ A and a simultaneous slow counterclockwise reaction A $\Rightarrow$ C $\Rightarrow$ B $\Rightarrow$ A. This possibility is excluded by the principle of microscopic reversibility or detailed balance (29) which requires that in the equilibrium state the rates of each elementary reaction step X $\rightleftarrows$ Y are the same in both directions. If, however, a nonequilibrium steady state is maintained by continuous supply or withdrawal of reactants, the reaction rates become unequal in forward and backward direction. In the case of the ionic channel the transition frequencies become asymmetric when interconversion among the conformational states is coupled to a dissipative process, the transfer of ion from high to low electrochemical potential. In the presence of a difference of electrochemical potential the cycle is driven preferentially in one (clockwise or counterclockwise) direction, meaning that $f_{AB} - f_{BA} = f_{BC} - f_{CB} = f_{CA} - f_{AC} \neq 0$.

Ion pumps

The concept of channels with multiple conformational states may also be applied to ion pumps. An ion channel functions as a pump when the energy profile of the channel is transiently modified in an appropriate way by an energy-supplying reaction (30-32). Absorption of a light-quantum or phosphorylation of the channel protein may alter the binding constant of an ion-binding site in the channel and, at the same time, change the height of adjacent barriers. In this way an ion may be preferentially released to one side of the membrane, while during the transition back to the original state of the channel another ion is taken up from the opposite side.

As a specific example we consider the light-driven proton pump in the purple membrane of halobacteria (33). A minimum model (32,34) of the pumping cycle is depicted in Fig. 8. It is assumed that in the ground state (HP) of the pump a proton is located in a binding site which is accessible from the left (cytoplasmic) phase but separated from the right (external) phase by a high barrier. Absorption of a proton created an activated state HP* (via short-lived intermediates) in which the energy level of the binding site is shifted upward (corresponding to a decrease of binding strength). In the activated state the barrier heights are changed in such a way that the proton is released preferentially to the right (external) solution. After dissociation of the proton, the channel molecule relaxes back to a conformation with a low barrier on the left (cytoplasmic) side (P* $\Rightarrow$ P). The original state is restored by uptake of H$^+$ from the cytoplasmic side (P $\Rightarrow$ HP). The whole cycle is equivalent to a net transfer of a proton from the cytoplasm (phase$'$) to the external medium (phase$''$).

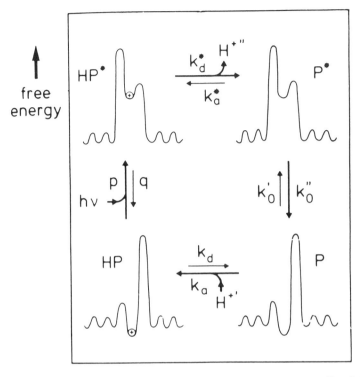

Figure 8. Channel mechanism for a light-driven proton pump. The energy profile of the channel is transiently modified by absorption of a photon. In the ground state HP the proton binding site is exposed to the left-hand aqueous phase (phase') and in excited state HP* to the right-hand phase (phase"). During the decay of the excited state back to the ground state (via P* and P) a proton is released to phase" and another proton is taken up from phase'.

The rate constant p for transitions from the ground state to the excited state depends on light intensity J:

$$p = p_o + \gamma J. \qquad (23)$$

p_o is the rate constant of spontaneous transitions HP $\Rightarrow$ HP* in the dark, and γ is the absorption cross section. If the pump starts to work at zero initial voltage ($V \equiv \psi' - \psi'' = 0$) and equal proton activities in both aqueous phases (a' = a''), a difference in the electrochemical potential $\tilde{\mu}_H$ of H$^+$ builds up, which consists partly of a voltage V and partly of a pH difference. With increasing $\Delta\tilde{\mu}_H$ the rates of the reverse processes (P* + H^{+}" $\Rightarrow$ HP, HP $\Rightarrow$ P + H^{+}') are enhanced so that eventually a limiting value of $\Delta\tilde{\mu}_H$ is reached at which the net rate of proton transfer vanishes. In the absence of leakage pathways this limiting value is given by (32):

$$(\Delta\tilde{\mu}_H)_{max} = -(RT\ln\frac{a'}{a''} + FV)_{max} = RT\ln\frac{p_o + \gamma J}{p_o + \gamma J \exp(-h\nu/kT)} \qquad (24)$$

$(\Delta\tilde{\mu}_H)_{max}/F$ is the so-called proton-motive force of the pump. As the photon energy $h\nu$ is much higher than the thermal energy kT, the term $\gamma J \exp(-h\nu/kT)$ in the denominator can often be neglected. On the other hand, it is seen from Eq. 24 that $(\Delta\tilde{\mu}_H)_{max}$ can never exceed the value $Lh\nu$ ($L = R/k$ is Avogadro's constant).

It is interesting to note that the expression for $(\Delta\tilde{\mu}_M)_{max}$ does not contain the proton dissociation constants K and K^*. This means that a pK difference between ground state and excited state is not a necessary condition for a large proton-motive force. The essential feature of the pumping mechanism is a light-induced change of barrier structure which switches the proton binding site from a left-exposed to a right-exposed state. On the other hand it may be shown that a low pK of the excited state ($K^* \gg a$) and a high pK of the ground state ($K \ll a$) is favourable for a high turnover rate of the pump.

References and Footnotes

1. Burgermeister, W. and Winkler-Oswatitsch, R., *Top. Curr. Chem. 69*, 91 (1977).
2. Szabo, G., Eisenman, G., Laprade, R., Ciani, S. M., and Krasne S., in *Membranes—A Series of Advances,* Ed. Eisenman, G., Vol. II, pp. 179-328. Dekker, M., New York.
3. Stark, G. and Benz, R., *J. Membrane Biol. 5,* 133 (1971).
4. Stark, G., Ketterer, B., Benz, R., and Läuger, P., *Biophys. J. 11,* 981 (1971).
5. Benz, R. and Läuger, P., *J. Membrane Biol. 27,* 171 (1976).
6. Hladky, S. B., and Haydon, D. A., *Biochim. Biophys. Acta 274,* 294 (1972).
7. Urry, D. W., *Proc. Nat. Acad. Sci. USA 68,* 672 (1971).
8. Urry, D. W., Goodall, M. C., Glickson, J. D., and Mayers, D. F., *Proc. Nat. Acad. Sci. USA 68,* 1907 (1971).
9. Finkelstein, A. and Andersen, O. S., *J. Membrane Biol. 59, 155* (1981).
10. Bamberg, E., Alpes, H., Apell, H.-J., Benz, R., Janko, K., Kolb, H.-A., Läuger, P., and Gross, E., in *Biochemistry of Membrane Transport.* FEBS-Symposium No. 12, Semenza, G. and Carafoli, E., editors, pp. 179-201. Springer, Berlin (1977).
11. Frauenfelder, H., Petsko, G. A., and Tsernoglu, D., *Nature* (Lond.) *280,* 558 (1979).
12. Karplus, M. and McCammon, J. A., *Ann. Rev. Biochem. 52,* 263 (1983).
13. Huber, R., Deisenhofer, J., Colman, P. M., Matshushima, M., and Palm, W., *Nature* (Lond.) 264, *415* (1976).
14. Parak, F., Frolov, E. N., Mössbauer, R. L., and Goldanskii. V. I., *J. Mol. Biol. 145,* 825 (1981).
15. Lakowicz, J. R., Maliwal, B. P., Cherek, H., and Balter, A., *Biochem. 22,* 1741 (1983).
16. Wagner, G., *Quart. Rev. Biophys. 16,* 1 (1983).
17. Sakmann, B. and Neher, E., (editors), *Single-Channel Recording,* Plenum Press, New York (1983).
18. Hamill, O. P. and Sakmann, B., *Nature* (Lond.) *294,* 462 (1981).
19. Trautmann, A., *Nature* (Lond.) *298,* 272 (1982).
20. Auerbach, A. and Sachs, F., *Biophys. J. 42,* 1 (1983).
21. Frehland, E., *Biophys. Struct. Mechanism 5,* 91 (1979).
22. Läuger, P., Stephan, W., and Frehland, E., *Biochim. Biophys. Acta 602,* 167 (1980).
23. Sakmann, B., Noma, A., and Trautwein, W., *Nature* (Lond.) *303,* 250 (1983).
24. Hille, B. and Schwarz, W., *J. Gen. Physiol. 72,* 409 (1978).
25. Sandbloom, J., Eisenman, G., and Hägglund, J., *J. Membrane Biol. 71,* 61 (1983).
26. Hamill, O. P., Bormann, J., and Sakmann, B., *Nature* (Lond.) *305,* 805 (1983).
27. Skrabal, A., *Z. Phys. Chem. Abt. B, 6,* 382 (1930).
28. Hearon, J. Z., *Bull. Math. Biophys. 15,* 121 (1953).
29. Onsager, L., *Phys. Rev. 37,* 405 (1931).
30. Patlak, C. S., *Bull. Math. Biophys. 19,* 209 (1957).
31. Jardetzky, O., *Nature* (Lond.) *211,* 969 (1966).

32. Läuger, P., *Biochim. Biophys. Acta 553,* 143 (1979).
33. Stoeckenius, W., Lozier, R. M. and Bogomolni, R. A., *Biochim. Biophys. Acta 505,* 215 (1979).
34. Stoeckenius, W., in *Membrane Transduction Mechanisms.* Cone and Dowling, editors, Raven Press, New York (1979).

Structure & Motion: Membranes, Nucleic Acids & Proteins,
Eds., E. Clementi, G. Corongiu, M. H. Sarma & R. H. Sarma,
ISBN 0-940030-12-8, Adenine Press, Copyright Adenine Press, 1985.

Molecular Structures and Librational Processes in Sequential Polypeptides: From Ion Channel Mechanism to Bioelastomers

D. W. Urry, C. M. Venkatachalam, S. A. Wood and K. U. Prasad

Laboratory of Molecular Biophysics
University of Alabama in Birmingham
School of Medicine
Birmingham, Alabama 35294

Abstract

Polypeptide dynamics of ion channel mechanisms and bioelastomers are here considered within a common conceptual basis. The structure of the ion selective Gramicidin A transmembrane channel is briefly described and the differential behavior of the L—D and D—L peptide moieties is experimentally demonstrated and then demonstrated in terms of calculations of helical librational states of the single stranded β-helix. The results of temperature studies in solution and in crystalline states are noted wherein it is seen that heating can convert poly(L,D)dipeptide from α-helix to β-helix; and with the numbers of states observed in peptide moiety librational (lambda) plots within a given energy cutoff, a qualitative estimate of entropy change is obtained using the Boltzmann relation. This estimate is compared with experimental results.

With regard to the bioelastomer problem, the properties of the polypentapeptide of elastin are briefly considered; data is noted which shows that this molecular system self-assembles into anisotropic fibers, fibrils and filaments. Thermoelasticity studies are reported on bands of γ-irradiation cross-linked polypentapeptide which show the cross-linked matrix to be an entropic elastomer. Molecular structure studies are briefly reviewed which give rise to a proposed family of closely related β-spiral structures for the polypentapeptide. This family of structures is considered in relaxed and extended states and lambda plots are reported for a 1 kcal/mole-residue energy cutoff for both states. Again the Boltzmann relation is used to obtain a qualitative estimate of entropy change to see if entropy of the proposed structure decreases on extension and whether such a decrease would be sufficient to provide an explanation for an entropic elastomer.

I. Introduction

In two major areas of research in the Laboratory of Molecular Biophysics, elements of polypeptide dynamics appear to be obvious and integral aspects of function. Both areas involve sequential polypeptides and, in both, the polypeptide mobility appears to be describable in terms of rocking motions of segments of polypeptide backbone resulting from compensating changes of torsion angles. The simplest example would be a peptide libration, i.e., the rocking of a peptide moiety resulting from the coordinated rotations of $\pm\psi_i$ with $\mp\phi_{i+1}$. Because in the trans peptide unit the C_i^a—C_i' and $N_{\overline{i+1}}C_{i+1}^a$ bonds are nearly colinear, peptide libration results in minimal perturbation to the remainder of the polypeptide chain. In the single stranded β-helical conformation of the Gramicidin A transmembrane channel, peptide librations are the dynamic process whereby a monovalent ion is able to pass through the channel (2) and the energetics of libration are considered relevant to ion selectivity (3). In the study of the elasticity of a polypentapeptide, such motions appear to be an important source of the entropy giving rise to an entropic restoring force on extension (4-6). In each of the two applications, the same polypeptide chain is experimentally obtained in two different structural states with different degrees of librational freedom; the extents of librational freedoms are compared by so-called lambda plots (plots of ψ_i vs ϕ_{i+1}) for a given energy cutoff which provide a visual picture of the extent of motion; and the change in entropy between the two conformational classes is qualitatively compared in terms of the Boltzmann relation by means of a simple ratio of numbers of states counted on the same grid and within the same low energy cutoff.

II. Structure and Dynamics of Poly(L,D)dipeptides: β-helical Channels

A. The Gramicidin A Channel Structure

Gramicidin A, HCO−L·Val^1−Gly^2−L·Ala^3−D·Leu^4−L·Ala^5−D·Val^6−L·Val^7−D·Val^8−L·Trp^9−D·Leu^{10}−L·Trp^{11}−D·Leu^{12}−L·Trp^{13}−D·Leu^{14}−L·Trp^{15}−NHCH$_2$-CH$_2$OH(7), forms monovalent cation selective channels across lipid bilayer membranes. Interestingly these channels exhibit phenomenology strikingly similar to that of the "on" states of physiological channels. This phenomenology includes magnitude of single channel currents, ideality of cation selectivity, ion saturation, competition and block, extra noise in the current while open, fast interruption, sublevels, and single filing of ion and water (8-11). The molecular structure is given in Figure 1 (2, 12). The structure is an amino end to amino end dimerized, single stranded, left-handed β-helix with just over 6 residues/turn and with peptide carbonyls pointing alternately parallel and antiparallel to the helix axis. This places three peptide carbonyls pointing outward into solution like sentinels guarding a channel with a maximal radius of near 2 A. The peptide moieties are considered as capable of libration such that the carbonyl oxygens can move into the channel to achieve direct lateral coordination of ions with radius less than 2 A.

A direct means of deriving the structure of Figure 1 and, at the same time, of locating the ion binding sites and providing information on the relative movements of the L·residue and D·residue C−O components of the L−D and D−L peptide moieties, respectively, is achieved by mapping the ion induced carbonyl carbon chemical shifts (13, 14). As shown in Figure 2, the thallium ion induced carbonyl carbon chemical shifts identify two localized binding sites. Attempts to plot the experimental chemical shifts in terms of other structures lead to contradictions

A.   B.

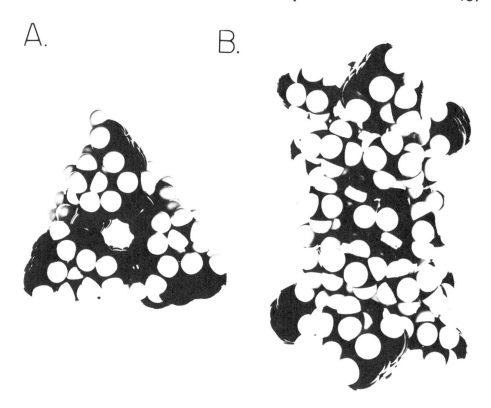

Figure 1. Molecular structure of the Gramicidin A transmembrane channel. A. Channel view of space filling model showing 3 peptide oxygens of residues 15, 13 and 11 directed outward. B. Space filling model showing two molecules associated head to head (amino end to amino end) with the formyl hydrogens, one from each chain, in contact at center. Reproduced with permission from Urry, et. al, *Ann. NY Acad. Sci., 264,* 203 (1975).

allowing the other structures to be eliminated (13). A comparison of all of the L·residue C—O chemical shifts with those exhibited by the D·Val[8] and D·Leu[14] carbonyl carbons allows the conclusion that the correct structure has a left-handed helical sense. The structure including helical sense are as proposed in 1971 (2). Interestingly, for the concerns of this manuscript, the data also show that the L·residue carbonyl carbon chemical shifts are downfield on ion binding as occurs when carbonyl oxygens of analogs of Gramicidin A directly coordinate monovalent cations (15), whereas the ion induced chemical shift of the D·residue carbonyl carbon is upfield as would occur if the D—L peptide moiety librated in a manner moving the C—O away from the polar channel and toward the lipid side of the channel. The picture is one of the oxygen end of the L·residue C—O moving into the channel to coordinate the ion while the D·residue C—O moves outward toward the less polar lipid. As seen below, this is the type of associated or cooperative motion that occurs with the peptide moieties of the dipeptide repeat.

WIRE MODEL AND ION BINDING SITES OF THE

GRAMICIDIN A TRANSMEMBRANE CHANNEL

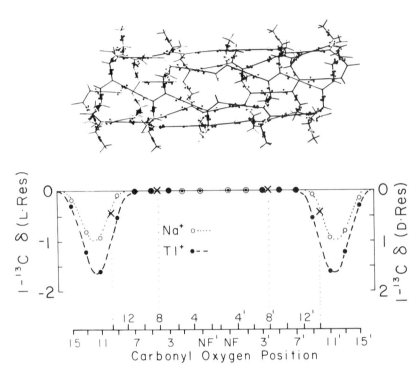

Figure 2. *Above:* Wire model of the Gramicidin A transmembrane channel displayed exactly as in Figure 1B. *Below:* Ion induced carbonyl carbon chemical shifts, plotted with respect to carbonyls of the wire model above, showing two binding sites. The downfield chemical shifts exhibited by the L·C—O residues are plotted with respect to the left-hand ordinate. The chemical shifts exhibited by the D·Leu[14] and D·Val[8] carbonyls are plotted with respect to the right-hand ordinate and indicated by X. The L-residue carbonyl resonces shift downfield as occurs when carbonyls coordinate an electron withdrawing cation. The D·Leu[14] carbonyl resonance shifts upfield as would occur if this carbonyl librated outward in concert with the L-carbonyl librating inward. This is qualitatively seen in Figure 3. Reproduced with permission from D. W. Urry in *The Enzymes of Biological Membranes,* E., A. N. Martonosi, Plenum Publishing Corporation, New York, (in press).

B. Helical Librational States of the Channel

While maintaining the helical parameters, the motion of the backbone of the $\beta^{6 \cdot 2}$-helix can be examined in terms of four torsion angles. In taking the α-carbon to α-carbon dipeptide repeat, the set of angles would, for example, be ψ_i^L, ϕ_{i+1}^D for the first peptide moiety and ψ_{i+1}^D, ϕ_{i+2}^L for the second peptide moiety. The potential functions of Scheraga and coworkers were used (16,17) and the cyclization method of Go and Scheraga (18) as adapted to helical constraints by Go and Okuyama (19) was the approach to determine allowed torsion angles. Given the helical parameters of n = −3.1 dipeptides/turn and h = 1.53 A/dipeptide (20), the calculations provide the set of allowed torsion angles at 5° intervals that can occur within a chosen kcal/mole-residue cutoff energy above the lowest energy conformation (21). Figure 3 contains a stereo pair plot of conformational states with a 2 kcal/mole-residue cutoff and the motional characteristics are seen to be describable in terms of peptide librations with the L-residue C−O moving into the channel while the D-residue C−O moves outward with a smaller amplitude of oscillation. Figure 4 contains the lambda plots at 5° torsion angle intervals of the conformational states within a 1 kcal/mole-residue cutoff energy. These figures provide a visualization of the character of allowed motions and of the number of states. This visual estimation of numbers of states using a selected grid and energy cutoff is particularly useful in comparing the same structure in entirely different conformations, for example α-helix and β-helix (see below).

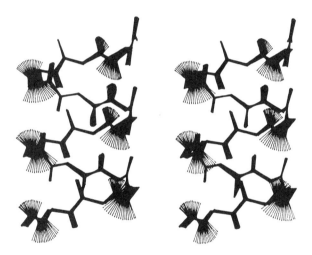

Figure 3. A stereo plot of the skeletal atoms of a β-helix with an L, D repeat dipeptide. A set of conformations with energies within 2 kcal/mole-residue of the minimum energy for poly(L,D)alanine and all with the same helical parameters, is depicted by superposition. The figure illustrates one mode of helical libration where the L−D peptide unit is tilted with minimal displacement of the D−L moiety. That the L−D unit can be librated to a large extent without al' ring the helical parameters of the chain is an interesting property of the β-helical conformation.

C. Temperature Studies Estimating Entropy Changes

Temperature studies have previously been reported on hydrogenated Gramicidin A′ in trifluoroethanol solution using circular dichroism to follow the conformational change (12). Gramicidin A′ was hydrogenated to remove side chain chromophores and, thereby, to be assured that changes in backbone conformation alone are being characterized. A broad transition is observed; it is centered near 10°C, starts below −30°C ($[\theta]^{223} \simeq 1 \times 10^4$) and ends about 70°C ($[\theta]^{223} \simeq -3.2 \times 10^4$). The large

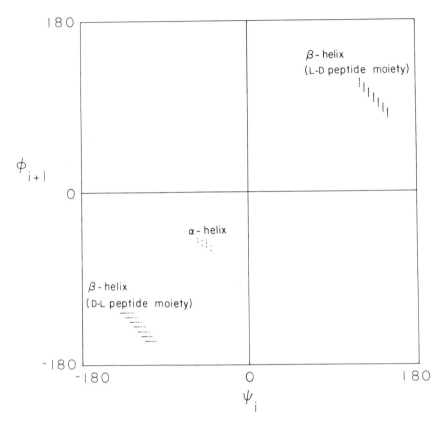

Figure 4. A Lambda plot delineating the extent of helical librations possible in poly(L,D)alanine in the α-helical conformation (n = 1.8 dipeptides per turn and h = 3 A/dipeptide) and in the β-helical conformation (n = −3.1 dipeptides/turn and h = 1.53 A/dipeptide). The plot shows the set of conformations with the chosen helical parameters within 1 kcal/mole-residue from the minimum energy conformation.

negative ellipticity at 223 nm indicates the presence of an ordered state at elevated temperature. IR data, obtained on films cast from trifluoroethanol at room temperature, give an amide I band at 1650 cm^{-1} and an amide II band at 1547 cm^{-1}. This indicates the presence of α-helix. Treating the broad transition as a unimolecular conformational change allows calculation of an equilibrium constant as a function of temperature and an estimate of entropy change of 23.6 eU/mole Gramicidin. On a per residue basis for the pentadecapeptide, this would give ~1.6 eU/residue.

Interestingly, Lotz et. al (22) have examined the temperature dependence of poly (γ−benzyl−D,L−glutamate) cast as a film from chloroform. Following their report, at room temperature the polypeptide is found in the α-helical conformation; heating the film at 130°C and cooling converts the sample to single stranded β-helix (π_{DL}-helix following our earlier nomenclature); and further heating at 220 to 230°C and cooling converts the sample to a double stranded β-helical conformation ($\pi\pi_{DL}$ in their nomenclature). X-ray diffraction patterns were obtained for each of these states. Even though these are ordered conformations, the β-helices have much higher entropy than the α-helix. It is of particular interest with respect to the bioelastomer problem to note that large changes in entropy between two conformations do not require that the higher entropy conformation be random or disordered.

D. Comparison of Mobilities of α-helical and Single-stranded β-helical Poly(L,D)alanines

Also plotted in Figure 4 are the states within a 1 kcal/mole-residue cutoff energy for the L·Ala—D·Ala dipeptide of poly(L,D)alanine in the α-helical conformation. It is possible to see directly from the lambda plots that the α-helical conformation has far fewer states accessible within a 1 kcal/mole-residue than does the β-helical conformation. Another way of expressing this difference is by means of the Boltzmann relation

$$S = R \ln W \qquad (1)$$

where S is the entropy of the conformation, R is the gas constant (1.987 cal/mole-degree) and W is the number of *a priori* equally probable states accessible to the dipeptide within the conformation. The change in entropy on going to the β-helical conformation from the α-helical conformation could be written

$$\Delta S = S^\beta - S^\alpha = R \ln W^\beta / W^\alpha \qquad (2)$$

Taking all of the states within 1 kcal/mole-residue of the lowest energy state to be equally probable, a qualitative estimate of the entropy change would be given by the ratio of the number of states contained in the lambda plot of Figure 4. The number of states per dipeptide repeat in the β-helical conformation is 79 and the number per dipeptide in the α-helical conformation is only 9. This gives for ΔS a value of about 2 eU per residue.

Somewhat surprisingly this simple estimate derived from a single counting of states gives a value which is close to that obtained from the conformational transformation in trifluoroethanol exhibited by hydrogenated Gramicidin A' on going from α-helix to β-helix (i.e. 1.6 eU/reside). It may also be useful to note that the estimate for entropy is well-behaved as the energy cutoff is raised or lowered. For example, for a 2 kcal/mole residue cutoff the ratio W^β / W^α is 190/23 and for a 0.6 kcal/mole residue cutoff the ratio is 37/4. In all three cases, the calculated value for ΔS is just over 4 eU/dipeptide or about 2 eU per residue with a drift toward slightly higher values as the cutoff is lowered. Within the uncertainties of the values for the potential functions, it is not unreasonable to consider all of the states within a 1 kcal/mole-residue energy band to be essentially at the same energy and thus to be equally accessible. If it were more satisfying, one could, of course, take the summation over states, $\sum_i \exp(-\epsilon_i / kT)$ for each conformation and then take the ratio of these values to estimate an entropy change. One could go a step further and include the difference in internal energy of the two conformations divided by the absolute temperature (23). Perhaps one reason the calculation appears useful is that many factors, not explicitly considered, cancel out in the ratio for a unimolecular conformational change. In the next section, a similar approach is taken to consider two families of conformations of the polypentapeptide of elastin, one when relaxed and the other resulting from extension (stretching).

III. Structure and Dynamics of the Polypentapeptide of Elastin: β-Spirals and Bioelastomers

The precursor protein of the biological elastic fiber has been shown by Sandberg and coworkers (24-26) to contain repeating peptide sequences. The major repeats are a polyhexapeptide $(L \cdot Ala^1 - L \cdot Pro^2 - Gly^3 - L \cdot Val^4 - Gly^5 - L \cdot Val^6)_n$ and a polypentapeptide $(L \cdot Val^1 - L \cdot Pro^2 - Gly^3 - L \cdot Val^4 - Gly^5)_n$. The polypentapeptide is the most striking primary structural feature with n = 11 in pig and n = 13 without a single variation in chick (27). The polyhexapeptide has an n of 5 and a fraction in pig (26). High polymers of the hexapeptide can be formed into a cellophane-like sheet which is non-elastomeric (28). It is thought to have a structural (29) and a chemotactic role (30). High molecular weight polypentapeptide, on the other hand, when cross-linked is elastomeric and can, depending on number of cross-links and water content, exhibit the same elastic modulus as the native elastic fiber (31).

A. Properties of the Polypentapeptide (PPP) of Elastin

The polypentapeptide (PPP) is soluble in all concentrations in water below 20°C. When n is as low as 6 or as great as 200 or more, raising the temperature causes reversible, concentration and chain length dependent aggregation. On standing at elevated temperature, the aggregates settle to form a viscoelastic phase called the coacervate. The coacervate is about 50% water by volume. The aggregation as a function of temperature can be followed by light scattering at 300 nm. This is shown in Figure 5 for a range of concentrations of PPP with n > 200. The process of coacervation may be thought of as a process of water extrusion and, indeed, as the temperature of the coacervate phase is raised, it continues to extrude water up to 60°C. Interestingly if a droplet of the cloudy solution is placed on a carbon coated grid and negatively stained with uranyl acetate, the aggregates appear as elongated structures, comprised of filaments aligned in parallel. Optical diffraction of the electron micrograph gives a periodicity of about 50 Å between filaments (32). While such observations on organic polymers using negative staining have been dismissed as artifacts, additional microscopic characterization provides compelling evidence for anisotropy.

B. Self-alignment into Fibers Comprised of Fibrils During Cross-Linking

When an occasional lysine residue replaces Val^4 in some chains and a glutamic acid residue occasionally replaces Val^4 in other chains, these Glu and Lys side chains can be chemically coupled during coacervation in a water bath shaker with a water soluble carbodiimide (31). Observation of the product in water using a light micro-

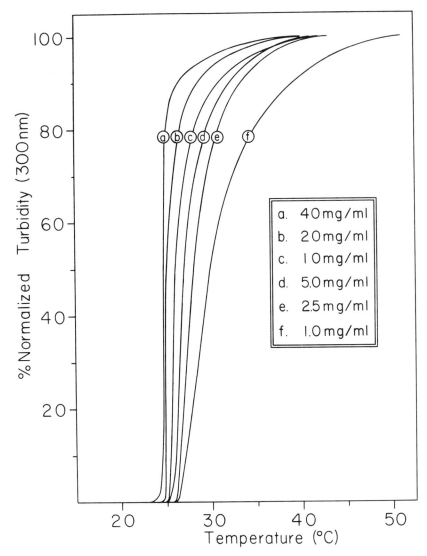

Figure 5. Temperature profile of coacervation for different concentrations of the polypentapeptide of elastin having n ≥ 200. This aggregation, followed by light scattering at 300 nm, is considered to be an inverse temperature transition as it gives rise to filaments observed in the electron microscope on negative staining of droplets of the cloudy solution. The process is one of water extrusion and hydrophobic association. The 40 mg/ml curve is plotted also in Figure 8A to show the similarity of temperature elicited transitions.

scope shows the presence of fibers (see Figure 6A). Observation of the product in a scanning electron microscope after drying and coating with aluminum shows fibers that splay-out into many fine fibrils which recoalesce back into the same diameter fiber (see Figures 6B and C). And negative staining of the product and obsevation

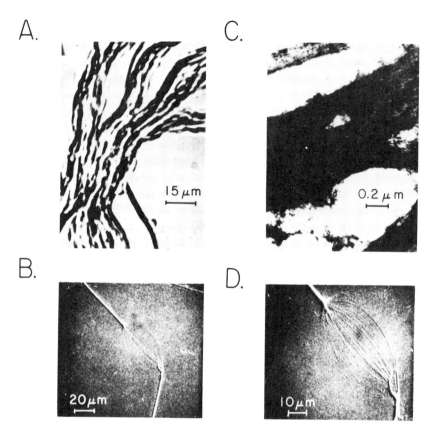

Figure 6. Micrographs of polypentapeptide which had been chemically cross-linked during coacervation. A. Light micrograph of sample in water showing fibrillar nature of product. (Reproduced with permission from Urry, *Ultrastruct. Pathol., 4,* 227, (1983). B. Scanning electron micrograph showing a fiber to splay-out into many fibrils and to recoalesce back into the same sized fiber. C. Enlargement of B showing fibrils in greater detail (B and C reproduced with permission from Urry, et. al, *Biochemistry, 15,* 4083 (1976)) D. Transmission electron micrograph showing the fine fibrils to be comprised of parallel aligned filaments (Reproduced with permission from Urry and Long, *Adv. Exp. Med. Biol., 79,* 685 (1977)).

at high magnification in a transmission electron microscope shows the fibrils to be comprised of parallel aligned filaments (see Figure 6D). At each stage of magnification the anisotropy is apparent and is seen to derive from parallel aligned filaments. The conclusion seems inescapable that the polypentapeptide forms anisotropic, non-random fibers under coacervation conditions.

C. *Thermoelasticity of γ-irradiation Cross-Linked Polypentapeptide*

Elastomeric bands suitable for careful stress-strain and thermoelasticity studies can be prepared by means of γ-irradiation cross-linking of coacervate formed from very high molecular weight PPP. The coacervate is placed in the bottom of a Cryotube. A pestle, with a channel or a groove turned in it, is inserted into the Cryotube; and

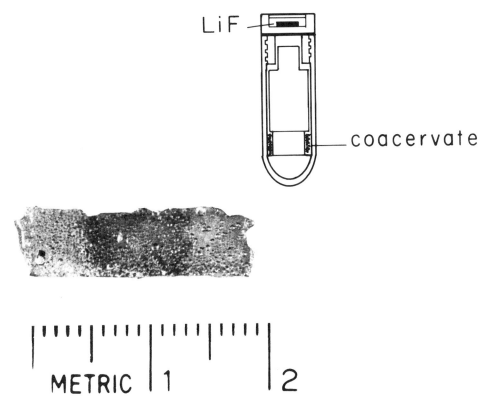

Figure 7. Elastomeric band formed on γ-irradiation cross-linking of coacervate when shaped as shown in Cryotube.

the coacervate slowly flows into and fills the groove, as shown in Figure 7. After γ-irradiation cross-linking, the band of cross-linked polypentapeptide is removed from the pestle. A typical elastomeric band is also shown in Figure 7. The band, immersed in water and equilibrated at low temperature, is stretched to about 50% elongation; the temperature is then raised to 75°C and after suitable equilibration time the force, f, is measured. The temperature is decreased; re-equilibration is awaited and a new value for the force is measured. The data, plotted as ln(f/T) versus temperature for a 15 MRAD cross-linked sample, is shown in Figure 8A. This plot follows the approach of Flory and coworkers and utilizes the thermodynamic equation of state for elasticity (33-36). The total elastomeric force, f, is the sum of an internal energy component, f_e, and an entropy component, f_s, i.e.

$$f = f_e + f_s \tag{3}$$

At constant volume and length, the ratio, f_e/f, can be expressed (34,35) as

$$f_e/f = -\overline{T}\left(\frac{\ln(f/T)}{T}\right)_{V,L} \tag{4}$$

or at constant pressure and length it becomes

$$f_e/f = -\overline{T}\left(\frac{\ln(f/T)}{T}\right)_{P,L} - \frac{\beta\,\overline{T}}{\alpha^3 - 1} \tag{5}$$

where $\alpha = L/L_i$ with L_i being the initial length and L the extended length, and β is the thermal expansion coefficient. Constant composition is assumed. Also included in Figure 8A is the temperature profile for coacervation. The correspondence of the transition temperatures is striking and it can be argued therefore that the force is being affected by solvent (water) extrusion as the temperature is raised through the transition. The decreasing positive slope up until 60°C is likely due in part to extrusion of solvent. Above 60°C the slope is nearly zero and taking $\beta=0$, the f_e/f ratio calculates to be 0.12. With the assumption that the composition is constant above 60°C, the γ-irradiation cross-linked polypentapeptide coacervate appears to be an essentially ideal elastomer with the force being almost entirely due to entropy. This assumption and an accurate estimate of β can readily be checked by examining a coacervate-filled graduated tube and determining the volume of the coacervate and of an overlying quantity of solvent (water) as a function of temperature. As shown by data obtained under identical conditions on elastic fiber purified from bovine ligamentum nuchae (See Figure 8B), the behavior of the native elastic fiber is quite analogous. Under carefully controlled conditions, it has also been concluded that the biological elastic fiber is dominantly an entropic elastomer (37,38). Thus the polypentapeptide must be an anisotropic, entropic elastomer. Below is considered the molecular conformation of the polypentapeptide and the possible source of entropy change on extension. It was demonstrated above with the α-helix to β-helix conversion that there can be large entropy changes with both conformations having regular describable conformations.

D. Conformational Studies on the Polpentapeptide of Elastin

Early conformational studies on the polypentapeptide addressed the issue of describing the most ordered conformation attainable. The conformation, so described, contained a repeating pentamer with the dominant secondary structural feature being a Pro^2-Gly^3 Type II β-turn and the pentamer repeated on a helix axis with no hydrogen bonding between repeats. The conformation was called a β-spiral, β to

---

Figure 8. Thermoelasticity studies in water plotting ln (f/T) on the left-hand ordinates for both the γ-irradiation cross-linked polypentapeptide (A.) and the elastic fiber purified from ligamentum nuchae (B.). Also included are the temperature profiles for coacervation in A. of the same polypentapeptide as used after cross-linking for the thermoelasticity studies and in B. of α-elastin, a 70,000 molecular weight oxalic acid fragmentation product, derived from the purified elastic fiber of ligamentum nuchae. The relationship between the temperature profiles for aggregation and the transition in the force curves is striking suggesting that similar solvent processes are occurring; an the similarity between the synthetic polypentapeptide data and the natural fiber data is equally striking. This suggests that the processes giving rise to the elastomeric force are the same. Both sets of data give an essentially zero slope above 60°C which reflects the dominant contribution of entropy to the elastomeric force.

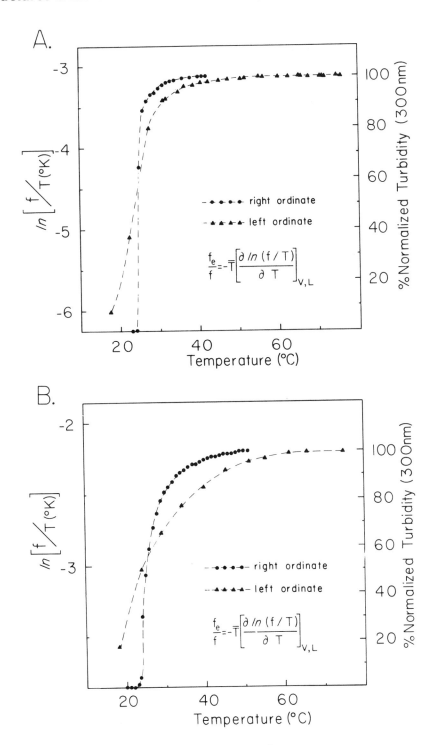

Caption for Figure 8 on previous page.

signify the dominant β-turn feature and spiral to draw a distinction between the commonly known relatively rigid helices which are characterized by hydrogen bonding between repeating units. The most ordered conformational state was described with 2.8 pentapeptide units per turn (n = 2.8) and a 3.2 A translation per repeat along the spiral axis (39).

With so many residues per turn of helix, going from a linear to a cyclic structure could be achieved with only small changes in torsion angles. Thus we can consider cyclic conformations and see if a cyclic structure has a conformation that is very similar to the linear polypentapeptide. The set of cyclic analogs, $^{\Gamma}$(L·Val^1—L·Pro^2—Gly^3—L·Val^4—Gly)$_n$$^{\lrcorner}$, were synthesized with n = 1, 2, 3, 4, 5 and 6 and their conformations compared to that of the linear PPP (40). It was found that the cyclopentadecapeptide (n = 3) exhibited a conformation *almost* identical to that of the linear PPP and that n = 1 and n = 2, though having the same primary structure, exhibited very different conformations. The conformation of the cyclic conformational correlate (the cyclopentadecapeptide) was derived in solution (41) and was determined in the crystal (42) and the two were very similar. The Pro^2—Gly^3 Type II β-turn described many years earlier (39) was found in the crystal. The solution conformation was used with the aid of molecular mechanics calculations to convert to a linear structure and the result was a dynamic β-spiral structure with 2.7 repeats per turn and a 3.5 A per pentamer translation along the helix axis (43). The β-turn, the derived β-spiral and schematic of the β-spiral are shown in Figure 9 as parts A, B and C, respectively. The β-spiral has water within it (as does the cyclopentadecapeptide in the crystal and solution structure); the β-turns function as spacers between turns of the β-spiral (much as the β-turns function as legs in the stacked cyclopentapeptides of the crystal), and the L·Val^4—Gly^5—L·Val^1 segment on slight extension becomes a suspended segment with much motional freedom due in part to being effectively surrounded by water and due to the presence of the Gly^5 residues. The force which causes the β-spiral to form in the relaxed state, e.g, during coacervation, is considered to be hydrophobic interturn interactions and hydrophobic intermolecular interactions give rise to the filament and fibril formation (4).

E. Librational Entropy Mechanism of Elasticity

Interest is now in the number of ways that the pentamer segment can span the gap in the β-spiral from one Val^1 α-carbon to the next and, in particular, in the number of ways that this can be achieved within, for example, 1 kcal/mole-residue of the lowest energy state. Using the same general approach as applied to the poly(L,D)alanine in the β-helix conformation and a 5° torsion angle step, it is found that there are 762 states. If the allowed ψ_4 angles are plotted versus ϕ_5 in one Lambda plot and ψ_5 versus ϕ_1 in a second, as shown in Figure 10, A and B, it is seen that the allowed values fall on or near a line at about 45° with each axis just as occurred for the librations of the β-helix. Thus the states can be described as being related by peptide librations. Now increasing the translation along the spiral axis from 3.5 to 8.0 A, as in stretching 130%, the same process can be followed. The

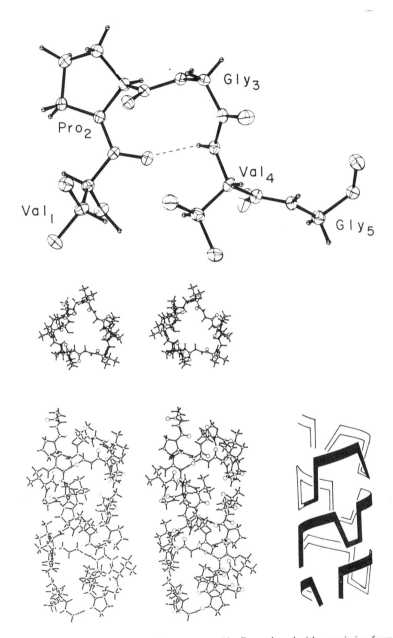

Figure 9. A. Pro[2]−Gly[3], Type II β-turn of the pentapeptide. Reproduced with permission from Cook et. al, *J. Amer. Chem. Soc., 102,* 5502 (1980). B. Stereo pair plots of the proposed β-spiral for the elastomeric polypentapeptide of elastin. *Above:* View along spiral axis showing space for water within the spiral. *Below:* Side view showing β-turns with hydrohobic interturn interactions functioning as spacers between turns for the relaxed state (Reproduced with permission from Venkatachalam and Urry, *Macromolecules, 14,* 1225 (1981)). On slight extension, the interturn contacts would cease. C. Schematic representation of β-spiral structure showing open spaces on the surface through which intraspiral water could exchange with interspiral water and showing the β-turn to be a spacer between turns of spiral in the relaxed state. (Reproduced with permission from Urry, *Ultrastruct. Pathol., 4,* 227 (1983).

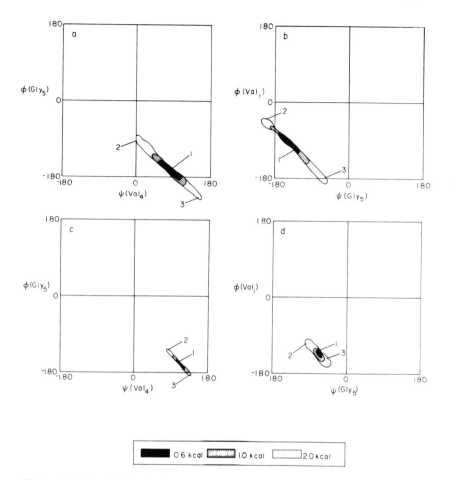

Figure 10. Lambda plots showing the librational nature of the local deformations that can occur within 0.6 kcal/mole-residue, 1 kcal/mole-residue and 2 kcal/mole-residue from the minimum energy conformation. The freedom of motion of the Val_4-Gly_5 and the Gly_5-Val_1 peptide moieties; (a) and (b) are the relaxed conformation of the polypentapeptide with h = 3.5 A while (c) and (d) are the corresponding plots for the extended state with h = 8 A. It is immediately clear that the effect of stretching is one of markedly reducing the degree of freedom of the $Val_4-Gly_5-Val_1$ segment.

number of states within 1 kcal/mole-residue using the same $5°$ torsion angle step is now reduced to 58. As seen in the Lambda plots (see Figure 10C and D), these too lie on the $45°$ angle line and can be thought of as being related by peptide librations. Stereo pairs of the relaxed and extended conformations are shown in Figure 11. In each there are plotted three pentamers with each pentamer in the lowest energy conformation and the central pentamer is also shown in librational states on both sides of the lowest energy conformation. This gives a qualitative picture of the extent of motional damping that results from extension.

As seen with the α-helix to β-helix transformation, a qualitative estimate of the entropy can be obtained from the ratio of the number of states in the relaxed

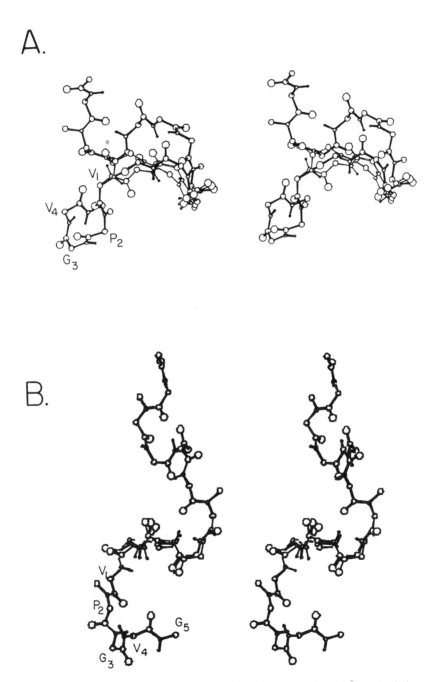

Figure 11. Stereo plots of a three-pentamer segment of a polypentapeptide chain in a β-spiral conformation delineating the librational excursions of the internal pentamer segment. The equilibrium conformation and the two librated exteme conformations are respectively those marked 1, 2 and 3 in Figure 10. A. relaxed state and B. extended state.

conformation, W^R, to the number of states in the extended conformation, W^E. i.e. $\Delta S = R \ln W^R/W^E = R \ln 762/58 = 5.12$ eU/pentapeptide or about 1 eU per residue. This quantity also is well-behaved as the cutoff energy is raised or lowered. For a 2 kcal/mole-residue energy cutoff $W^R = 1853$ and $W^E = 162$ giving a ΔS of 4.84 eU/pentapeptide and for a 0.6 kcal/mole-residue energy cutoff, $W^R = 342$ and $W^E = 24$ giving a ΔS of 5.28 eU/pentapeptide. Using the Boltzmann summation over states for all the states within two kcal/mole-residue gives a ΔS of 5.06 eU/pentapeptide.

The obvious choice of plotting the 4-5 and 5-1 peptide librations was due to their occurrence in the suspended segment, L·Val⁴—Gly⁵—L·Val¹, and their flanking the Gly⁵ residue where the absence of a side chain gave greater freedom to the libration. This also suggests a means of checking the importance of this motion to elasticity. The question becomes what would be the effect of placing a residue with a side chain in position 5, even of a modest replacement of an α-hydrogen with a methyl moiety. Indeed both the L·Ala⁵ polypentapeptide and the D·Ala⁵-polypentapeptide have been synthesized and characterized. Each substitution destroys elasticity. The L·Ala⁵ substitution, while having a temperature profile for aggregation which is identical to that of PPP, results in a granular precipitate on heating. This simply fragments on handling rather than appear as a viscoelastic coacervate (5). The D·Ala⁵ substitution, while having a dense phase which reluctantly settles and associates as a viscoelastic coacervate. Once cross-linked, on stretching it simply fragments into a cloudy solution (44). Replacing the Gly³ of the β-turn on the other hand with a D·Ala³ residue which stabilizes the β-turn, results in an increase in the elastic modulus (45). Thus evidence is accumulating for a new mechanism of elasticity exhibited by certain sequential polypeptides derived from the sequence of the natural elastic fiber.

Acknowledgement

This work was supported in part by the National Institutes of Health, grant nos. GM-26898 and HL-29578. The authors wish to thank Dr. Marianna M. Long for preparing the α-elastin utilized in Figure 8B.

References and Footnotes

1. In the use of the term "libration", the root word may be taken as libra (the scales or balances) and it reflects the oscillatory process of a balance before it comes to rest at equilibrium. In a polypeptide, it is the limited rotation or rocking of a moiety that has insufficient energy for a full rotation.
2. Urry, D. W., Goodall, M. C., Glickson, J. D. and Mayers, D. F., *Proc. Natl. Acad. Sci. USA, 68,* 1907 (1971).
3. Urry, D. W. in *Conformation of Biological Molecules and Polymers—The Jerusalem Symposia on Quantum Chemistry and Biochemistry, V.,* Eds. Bergmann, E. D. and Pulmann, B., Israel Academy of Sciences, Jerusalem, p. 723 (1973).
4. Urry, D. W., Venkatachalam, C. M., Long, M. M. and Prasad, K. U. in *Conformation in Biology,* Eds. Srinivasan, R. and Sarma, R. H., G. N. Ramachandran Festschrift Volume, Adenine Press, USA, p. 11 (1982).

5. Urry, D. W., Trapane, T. L., Long, M. M. and Prasad, K. U., *J. Chem Soc., Faraday Trans. I., 79,* 853 (1983).
6. Urry, D. W. and Venkatachalam, C. M., *Int. J. Quantum Chem.: Quantum Biology Symp. No. 10,* 81 (1983).
7. Sarges, R. and Witkop, B., *Biochemistry, 4,* 2491 (1965).
8. Finkelstein, A. and Andersen, O. S., *J. Membr. Biol., 59,* 155 (1981).
9. Eisenman, G., Sandblom, J. and Neher, E. in *Metal-Ligand Interactions in Organic Chemistry and Biochemistry,* Eds., Pullman, B. and Goldblum, N., Part 2, D. Reidel, Dordrecht-Holland, p.1 (1977).
10. Neher, E., *Techniques in Cellular Physiology, 121,* 1 (1982).
11. Szabo, G. and Urry, D. W., *Science, 203,* 55 (1979).
12. Urry, D. W., Long, M. M., Jacobs, M. and Harris, R. D., *Ann. NY Acad. Sci., 264,* 203 (1975).
13. Urry, D. W., Trapane, T. L. and Prasad, K. U., *Science, 221,* 1064 (1983).
14. Urry, D. W. in *The Enzymes of Biological Membranes,* Ed., Martonosi, A. N., Plenum Publishing Corporation, New York, (in press).
15. Urry, D. W. in *Enzymes of Bilogical Membranes,* Ed., Martonosi, A. N., Plenum Publishing Corporation, New York, Vol. 1, p. 31 (1976).
16. Momany, F. A., Carruthers, L. M., McGuire, R. F. and Scheraga, H. A., *J. Phys. Chem., 78,* 1595 (1974).
17. Momany, F. A., McGuire, R. F., Burgess, A. W. and Scheraga, H. A., *J. Phys. Chem., 79,* 2361 (1975).
18. Go, N. and Scheraga, H. A., *Macromolecules, 6,* 273 (1973).
19. Go, N. and Okuyama, K., *Macromolecules, 9,* 867 (1976).
20. Venkatachalam, C. M. and Urry, D. W., *J. Comput. Ch., 4,* 461 (1983).
21. Venkatachalam, C. M. and Urry, D. W., *J. Comput. Ch. 5,* 64 (1984).
22. Lotz, B., Colonna-Cesari, F., Heitz, F. and Spach, G., *J. Mol. Biol., 106,* 15 (1976).
23. Eyring, H., Henderson, D., Stover, B. J. and Eyring, E. M. in *Statistical Mechanics and Dynamics,* John Wiley and Sons, Inc., New York, 1964.
24. Sandberg, L. B., Gray, W. R., Foster, J. A., Torres, A. R., Alvarez, V. L. and Janata, J., *Adv. Exp. Med. Biol., 79,* 277 (1977).
25. Sandberg, L. B., Soskel, N. T., and Leslie, J. B., N. Engl. *J. Med., 304,* 566 (1981).
26. Gray, W. R., Sandberg, L. B. and Foster, J. A., *Nature, 246,* 461 (1973).
27. Sandberg, L. B. (private communication).
28. Rapaka, R. S., Okamoto, K. and Urry, D. W., *Int. J. Pept. Protein Res., 11,* 109 (1978).
29. Urry, D. W., *Perspect. Biol. Med., 21,* 265 (1978).
30. Senior, R. M., Griffin, G. L., Mecham, R. P., Wrenn, D. S. Prasad, K. U. and Urry, D. W., *J. Cell Biol. 99,* 870 (1984).
31. Urry, D. W., Okamoto, K., Harris, R. D., Hendrix, C. F. and Long, M. M., *Biochemistry, 15,* 4083 (1976).
32. Volpin, D., Urry, D. W., Pasquali-Ronchetti, I. and Gotte, L., *Micron, 7,* 193 (1976).
33. Flory, P. J., *Rubber Chemistry and Technology, 41,* G41-G48 (1968).
34. Flory, P. J., Cifferri, A., Hoeve, C. A. J., *J. Polymer Sci., XLV,* 235 (1960).
35. Mark, J. E., *Rubber Chem. Technol., 46,* 593 (1973).
36. Mark, J. E., *J. Polymer Sci., Macromolecular Reviews, 11,* 135 (1976).
37. Hoeve, C. A. J. and Flory, P. J., *Biopolymers, 13,* 677 (1974).
38. Andrady, A. L. and Mark, J. E., *Biopolymers, 19,* 849 (1980).
39. Urry, D. W. and Long, M. M., *CRC Crit. Rev. Biochem., 4,* 1 (1976).
40. Urry, D. W., Trapane, T. L., Sugano, H. and Prasad, K. U., *J. Am. Chem. Soc., 103,* 2080 (1981).
41. Venkatachalam, C. M., Khaled, M. A., Sugano, H. and Urry, D. W., *J. Am. Chem. Soc., 103,* 2372 (1981).
42. Cook, W. J., Einspahr, H. M., Trapane, T. L., Urry, D. W. and Bugg, C. E., *J. Am. Chem. Soc., 102,* 5502 (1980).
43. Venkatachalam, C. M. and Urry, D. W., *Macromolecules, 14,* 1225 (1981).
44. Urry, D. W., Trapane, T. L., Wood, S. A., Walker, J. T., Harris, R. D. and Prasad, K. U., *Int. J. Pept. Protein Res., 22,* 164 (1983).
45. Urry, D. W., Trapane, T. L., Wood, S. A., Harris, R. D., Walker, J. T. and Prasad, K. U., *Int. J. Peptide Protein Res. 23,* 425 (1984).

Structure & Motion: Membranes, Nucleic Acids & Proteins,
Eds., E. Clementi, G. Corongiu, M. H. Sarma & R. H. Sarma,
ISBN 0-940030-12-8, Adenine Press, Copyright Adenine Press, 1985.

Ligand Binding and Protein Dynamics

Hans Frauenfelder
Department of Physics
University of Illinois at Urbana-Champaign
1110 West Green Street
Urbana, IL 61801

Abstract

Protein dynamics is explored by using the binding of carbon monoxide to myoglobin as a probe. The concept of functionally important motions ("fims") is introduced. The experiments reveal the existence of different classes of fims. The fastest motion—fim 3—occurs at and near the iron. Fim 2 probably involves the heme group and some parts of the protein; it can be studied between 60 and 160 K by observing the behavior of the near-infrared line at 760 nm. The investigation demonstrates that the activation energy for fim 2 is not sharp, but distributed. Fim 2 can explain Mössbauer effect data and may also be connected to the motion of the ligand into the protein matrix. Fim 1 is responsible for overall relaxation of the protein and may be related to protein unfolding. A comparison of the experimental results with theory implies that only fim 3 can be explained in a straightforward way through molecular dynamics calculations.

1. Introduction

It is by now well established that proteins and nucleic acids are not rigid, but are breathing and wiggling and that the motion is important for their function (1-4). Much remains to be done, however, before the dynamic features of biomolecules are fully understood. It is likely that fluctuations occur in a vast time range, say from picoseconds to seconds, but not all of these fluctuations may be functionally relevant. We can consequently introduce the concept of *functionally important motions (fims),* motions that can be shown to be related to a particular function of a biomolecule, and ask the following questions:

(i) Can we find and investigate fims for a particular protein function?

(ii) If yes, how can we characterize these motions? Are they uniformly distributed in time (or activation energies) or do they bunch in clusters:

(iii) Can fims be predicted by theory, for instance by molecular dynamics calculations? If not, how can theory account for the observed characteristics of fims?

Even resting biomolecules will move and fluctuate, but it is difficult to assess which of these motions are functionally relevant. In order to find and study fims, we must look at proteins in action, observe and classify the motions that occur during a given process, and try to determine if a particular motion is important for the function under consideration. In the present contribution, we will discuss one of the simplest protein functions, the association and dissociation of carbon monoxide (CO) to myoglobin (Mb), and show that three candidates for fims can be recognized. Future studies will have to decide if these motions are indeed functionally important and whether additional fims exist.

2. *The Binding of CO to Myoglobin*

The binding of CO to Mb has been studied by many groups with various tools (5). The most detailed information has come from flash photolysis experiments over wide ranges in time and temperature (6,7). The method is, at least in principle, simple: MbCO is placed in a cryostat. The bond between the heme iron and the CO molecule is broken by a laser flash. The photodissociated CO moves away from the binding site, but later rebinds. Dissociation and association are followed, for instance by monitoring the Soret line. The main experimental results are summarized in Figs. 1 and 2, where N(t) is the fraction of Mb molecules that have not rebound a CO at the time t after photodissociation. In both figures, log N(t) is plotted versus log (t). The data are interpreted with the help of Fig. 3 which gives a schematic cross section through Mb.

At temperatures below about 180 K, a photodissociated CO molecule remains within the heme pocket and rebinds from there. We denote this geminate process by I and it is shown in Fig. 1. At higher temperatures, a CO molecule can either rebind directly or move into the protein matrix and return from there, giving rise to the matrix process denoted by M. Above about 220 K, the CO molecules have the further choice of moving through the protein matrix into the solvent. Binding from

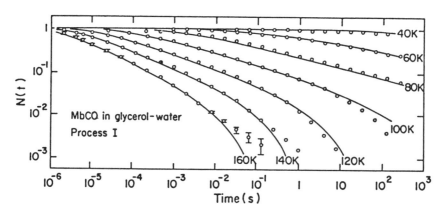

Figure 1. Binding of CO to Mb after photodissociation between 40 K and 160 K. N(t) is the fraction of Mb molecules that have not rebound a CO molecule at the time t after photodissociation. (After ref. 6.)

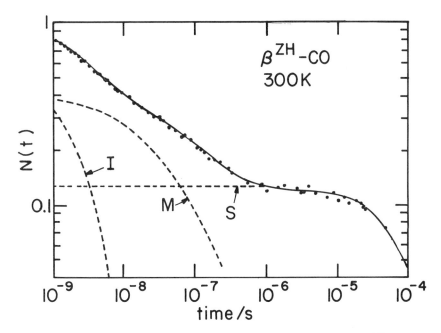

Figure 2. Binding of CO to the separated beta chain of human hemoglobin Zurich, β^{ZH}. (After ref. 7.) Because the separation into the different processes discussed in the text is more clearly apparent in β^{ZH} that in Mb, β^{ZH}CO is shown here. The essential features are similar to MbCO.

the solvent gives rise to the solvent process S. All three processes are shown schematically in Fig. 3.

Dynamic features are involved in all the processes shown in Figs. 1-3. Of particular interest for the following discussion are:

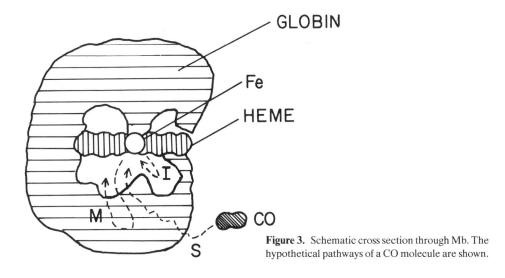

Figure 3. Schematic cross section through Mb. The hypothetical pathways of a CO molecule are shown.

(1) Association to, and dissociation from, the heme iron involve motion of the heme group, the iron, and possibly parts of the F helix.

(2) Motion of the CO through the protein matrix cannot occur if the protein is rigid (8) and it consequently must involve protein relaxation (9).

(3) As Fig. 1 demonstrates, binding below 180 K is not exponential in time. The protein cannot be in a unique state but must be able to exist in a number of conformational substates (10) with different barrier heights. The fraction $N(t)$ is given by

$$N(t) = \int g(H) \exp[-k(H)t]dH, \tag{1}$$

Here $g(H)dH$ is the probability of having a barrier with activation enthalpy between H and $H + dH$ and $k(H)$ is the rate coefficient for a barrier of height H (11,6). Assuming that $k(H)$ is given by an Arrhenius relation, $k(H) = A \exp(-H/RT)$, $g(H)$ can be found from the experimentally measured $N(t)$ by inversion of Eq. (1). The distribution function $g(H)$ for the binding of CO to Mb is shown in Fig. 4. Above about 250 K, the observed binding from the solvent, process S in Fig. 2, is visible and exponential in time. At these temperatures, the protein must be able to fluctuate from one substate to another, thus averaging out the distribution.

In the following sections we will discuss the dynamic information that can be extracted from the features (1)-(3).

3. *Fim 3: Motion of Heme and Iron*

In the bound state (MbCO) the heme group is essentially planar, the iron atom lies close to the mean heme plane and has spin 0. In the deoxy state (Mb) the heme is

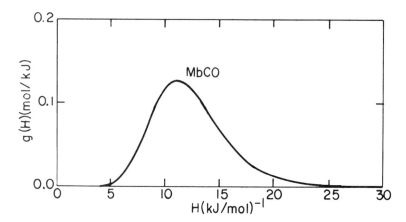

Figure 4. Activation energy distribution for binding of CO to Mb; $g(H)dH$ is the probability of finding a barrier with activation enthalpy between H and $H + dH$.

domed, the iron is about 0.5 Å away from the mean heme plane and has spin 2 (12). Upon photodissociation of MbCO, both the heme and the iron atom move and the iron atom changes spin. Since in Mb the iron atom is linked to the proximal histidine (His F 93), the motion of the iron and the heme group is communicated also to the protein matrix. Indeed, the F helix also shifts in going from MbCO to Mb (12). We can ask: Does the entire motion happen at once? How fast is the motion? A tentative answer to these questions comes from optical and Raman studies (13-17) which suggest that photodissociation occurs faster than ps at room temperature and faster than 20 ns even at 20 K. Experiments at temperatures below about 30 K indicate, however, that the protein does not immediately relax to the deoxy state, but reaches a state Mb* (18-22). The Soret bands in Mb and Mb* are essentially identical. The porphyrin core sizes and the magnetic properties, however, are close but not identical. We tentatively identify Mb* as the state in which the iron has partially (or completely) moved out of the MbCO position, but the protein has not yet relaxed and we call the transition MbCO $\Rightarrow$ Mb* fim 3.

Fim 3 is largely an electronic process, but most likely involves some nuclear motion. If we characterize fim 3 by an effective activation energy E_3^r, the limit $\tau < 20$ ns at 20 K implies $E_3^r < 4$ kJ/mol.

The functional importance of fim 3 is not clearly established. We can speculate, however, that the rapid motion of the iron out of the MbCO position prevents the CO molecule from immediate rebinding and makes thermal dissociation faster. This speculation is supported by the observation that a fast rebinding process is observed in cm cyt c (23) and in HRP (24) which do not serve as natural storage proteins.

4. Fim 2: Matrix Motion and Mössbauer Effect

The results sketched in the previous section raise the question of when and how the protein relaxes to the deoxy state after photodissociation. Studies of this question require suitable markers, for instance Raman or optical lines, whose positions differ in the states Mb* and Mb. We have investigated fim 2 by observing a near-infrared line at 758 nm. Yonetani and coworkers (25,18) first noticed that this line is red-shifted by about 10 nm after photodissociation at very low temperatures. It consequently can serve as a marker for the transition Mb* $\Rightarrow$ Mb.

Recent measurements (26,27) give peak wavelengths of 758 nm for Mb and 766 nm for Mb*. Eaton and coworkers call the 758 nm line (in deoxyhemoglobin) band III and interpret it as a charge-transfer transition which involves porphyrin and iron states, $a_{2u} \Rightarrow d_{yz}$ (28-30). This assignment makes it plausible that the position of the iron with respect to the heme plane affects the wavelength of band III and explains why band III can be used to monitor the relaxation Mb* $\Rightarrow$ Mb. Experiments with band III are, however, somewhat cumbersome because of the small extinction coefficient (150 M^{-1} cm^{-1}), nearly a thousand times smaller than that of the main Soret band at 420 nm.

Using infrared diodes as light sources, we have monitored band III after photodissociation at temperatures between 20 and 240 K and times from about 5 μs to 10 s. The band behaves as expected: At low temperatures and short times, the peak appears at 766 nm, the wavelength of Mb*. With increasing temperature and also with increasing time, the peak shifts towards 758 nm, characteristic of Mb. At 160 K and about 10 ms, the peak position is close to 758 nm; relaxation is nearly complete.

The detailed data evaluation is complicated by the fact that two other lines contribute background (28) and that relaxation and nonexponential rebinding (Fig. 1) occur simultaneously. The essential features of fim 2, however, are clear: The relaxation is nonexponential in time and consequently must be described by a distribution (31), $g(E^r)$, where $g(E^r)dE^r$ denotes the probability of finding a relaxation energy between E^r and E^r+dE^r. Note that $g(H)$ and $g(E^r)$ refer to two different processes: $g(H)$ characterizes the distribution of barrier heights for CO binding to the heme iron, $g(E^r)$ the relaxation Mb* $\Rightarrow$ Mb. For the preliminary analysis here we assume the square distribution given in Fig. 5. A crude fit of this distribution to the data, after correcting for binding (Mb* $\Rightarrow$ MbCO) and assuming an Arrhenius relation, $k(E^r) = A^r \exp(-E^r/RT)$, gives

$$A^r = 10^{13\pm2}\ s^{-1},\ E^r_{min} = 12\pm3\ kJ/mol,\ E^r_{max} = 36\pm10\ kJ/mol. \qquad (2)$$

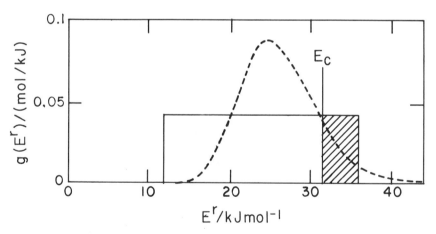

Figure 5. Distribution of the activation energy for fim 2, the relaxation Mb* $\Rightarrow$ Mb, as observed by the shift of the line near 758 nm after photodissociation. The solid line gives the crudest approximation, the dashed line is the distribution that fits the Mössbauer data.

This distribution is remarkably wide. An idea of the range of rates covered is obtained if we plot the rate coefficients corresponding to the values $E^r_{min} = 12$ kJ/mol and $E^r_{max} = 36$ kJ/mol as function of 1/T in Fig. 6. In the same figure we also give corresponding curves for the distribution Fig. 4 that characterizes the low-temperature binding of CO. Fig. 6 shows that fim 2 stretches over more than 10 orders of

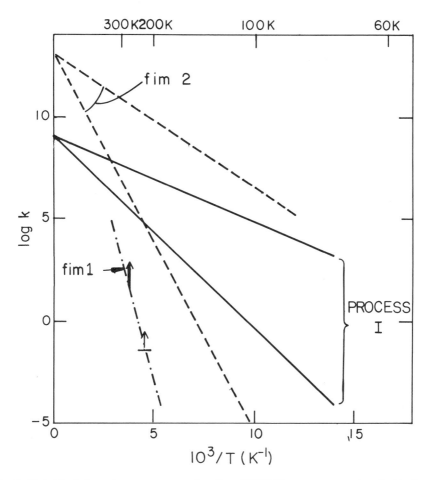

Figure 6. Plot of log k for various processes as a function of $10^3/T$. The processes are described in the text.

magnitude in time at 100 K. Fig. 6 also exposes another suggestive property of fim 2: At 300 K, the time range covered by the extrapolation goes from 200 ns to 10 ps, times that may be important for protein function. To follow this suggestion, we consider two aspects, the Mössbauer effect and motion into the matrix.

The Mössbauer effect in myoglobin has been studied by a number of groups and we consider here the data of Parak et al and of Bauminger et al (32,33) on deoxyMb. In these experiments, the fraction f(T) of gamma rays that are emitted with essentially the natural line width is determined as a function of temperature. Up to about 200 K, f behaves as in a solid; above 200 K f decreases much more rapidly. Fim 2 can explain the striking feature of f(T) in a natural way. The characteristic rate for the ^{57}Fe Mössbauer lines is given by $k_{Mö} \cong 10^7 \, s^{-1}$. We assume that fim 2 moves the iron with a rate k_2 which depends on the relevant activation energy E^r. As simplest "model" we further assume that Fe nuclei moving with a rate $k_2 < k_{Mö}$ contribute fully to the natural line, but that nuclei with $k_2 > k_{Mö}$ do not contribute at all. In terms of Fig. 5, this assumption means that only nuclei in the shaded area above E_c contribute to the sharp line where

$$E_c = RT \ \ln(A^r/k_{Mö}). \tag{3}$$

The fraction f is given by

$$f = \frac{E_{max} - E_c}{E_{max} - E_{min}}. \tag{4}$$

Fig. 7 shows f(T), calculated from Eq. (4) with the values of Eq. (2), together with the data of Parak et al, corrected for the low-temperature slope. Eq. (4) predicts the general trend of the Mössbauer data well. Note that Eq. (4) is not a fit, but a prediction, based entirely on IR data obtained at lower temperatures. The fact that the details are not reproduced well can be blamed on the oversimplified form of $g(E^r)$. Indeed, we can fit the data within experimental errors by using the distribution shown dashed in Fig. 5. The corresponding fit is shown as dashed line in Fig. 7.

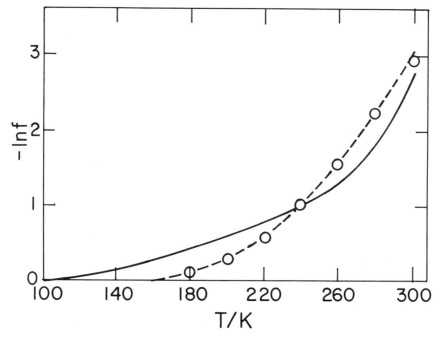

Figure 7. Mössbauer effect data, prediction, and fit. The circles denote the experimental values of $-\ln f$ for ^{57}Fe in deoxy Mb after subtraction of the extrapolated low-temperature contribution (after Parak et al., ref. 32). The solid curve is the prediction based on the square distribution of Fig. 5, the dashed line gives the fit corresponding to the dashed distribution in Fig. 5.

The motion of the ligand into the protein matrix after flash photolysis (M in Figs. 2 and 3) may also be connected to fim 2. It is experimentally difficult to accurately determine the rates for this motion in the case of MbCO, but the rate as found (6) fits into the region for fim 2 given in Fig. 6.

Resonance Raman experiments on hemoglobin by Friedman et al. (34) near 300K show the presence of a relaxation mode that is nonexponential in time and may be related to fim 2.

A detailed description of the motion involved in fim 2 will require many more experiments, but we can speculate that it may involve a sliding of the F helix into the final deoxy position. The coincidence of the time scale with that required for motion of the CO into the protein matrix implies that some fluctuations occur also on the distal side. The functional importance of fim 2 may be connected with these fluctuations; fim 2 may permit the migration of the ligand through the matrix.

5. *Fim 1: Large-Scale Motions and Unfolding*

Figures 1, 4, and 6 together pose a puzzle: The data in the first two figures imply that the distribution g(H), characteristic of CO binding, does not depend on temperature up to about 200 K. Fig. 6, however, demonstrates that much of fim 2 is already faster than rebinding at 150 K. These two observations suggest that fim 2 is not responsible for the relaxation of g(H) and for the exponential time dependence of CO binding above 250 K. Another motion, presumably governed by larger activation energies, must be involved. To search for this fim, a marker is needed again. The only one that we have found so far is the distribution g(H) itself which we have studied in preliminary experiments by using "pressure titration" (35,36). These experiments provide definitive proof of the existence of another motion, fim 1.

The idea of *pressure titration* is simple: Under pressure, the binding of CO to Mb speeds up. The effect is particularly noticeable at low temperatures. At 100 K, the time at which the N(t) curve (Fig. 1) cuts the line N(t) = 0.1 can shift by about a factor 50 at 2 kbar. The shift depends, however, on the path by which the experimental conditions are reached (Fig. 8). If we first cool and then pressurize (a $\Rightarrow$ b $\Rightarrow$ c) the speedup is smaller than if we first pressurize and then cool (a $\Rightarrow$ b$'$ $\Rightarrow$ c$'$). The observation finds a natural explanation in the concept of substates (6,10). At c, the protein is in conformational substates characteristic of 1 bar while at c$'$ the substates correspond to 2 kbar. Since CO binding at 100 K is much faster in c$'$ than in c, the two states can be distinguished by measuring N(t). We now denote transitions between c and c$'$ as fim 1. Fim 1 can be studied as follows. The sample is first brought to c along the path a $\Rightarrow$ b $\Rightarrow$ c and N(t) is measured at 100 K and 2 kbar. At 2 kbar, the temperature is then increased to a value T_m, left there for 1 min, cooled again to 100 K and N(t) is remeasured. This procedure is repeated with increasing T_m until N(t) follows the curve corresponding to state c$'$. The result is dramatic: Up to T_m = 200 K, N(t) is unchanged but at T_m = 220 K it follows the curve for state c$'$. The transition c $\Rightarrow$ c$'$ consequently occurs between 200 and 220 K, and the experi-

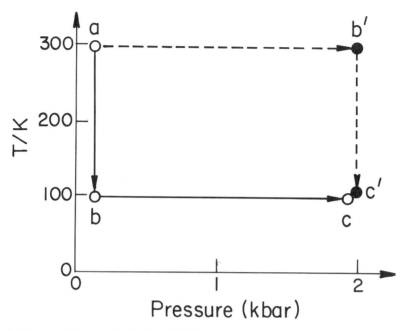

Figure 8. Pressure effect on the binding of CO to Mb: The MbCO system can be brought to a low-temperature high-pressure state along the two pathways shown.

ment yields limits on the rate k_1 of fim 1, $k_1(200K) < 10^{-3}$ s^{-1} and $k_1(220K) > 0.05$ s^{-1}. These limits are shown in Fig. 6 and they lead to $E_1^r > 70$ kJ/mol and $A_1^r > 10^{16}$ s^{-1} for the activation energy and preexponential of fim 1. These values suggest that fim 1 may not be an Arrhenius transition, but may be a phase (glass) transition. More experiments are required to explore the properties of fim 1, to study the role of surface water, and to investigate the connection between fim 1 and unfolding (6).

6. Summary and Comparison with Theories.

We can now give some answers to the questions posed in the introduction. Even if some of these answers are still tentative, the experiments described in the last sections will, in time, provide more information and either reinforce the speculations given here or destroy them.

The data presented here suggest that there are at least three distinct classes of functionally important motions in Mb (37). Fim 3 is the motion of the iron and the heme group immediately after photodissociation into a metastable state Mb* with a conformation intermediate between that of MbCO and Mb. This motion is very fast even at 2 K, and probably does not involve the globin. Fim 2 is the relaxtion Mb* $\Rightarrow$ Mb, with motion of the globin. Fim 1 is an overall relaxation of either a large part or possibly the entire protein which may be connected to partial unfolding (38). As

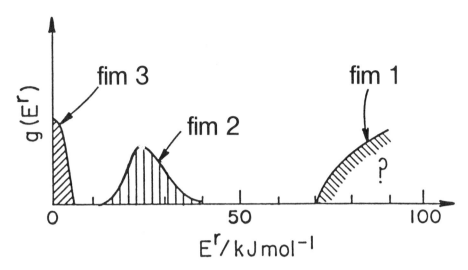

Figure 9. Schematic representation of the activation energies for the three classes of functionally important motions discussed in the text.

sketched in Fig. 9, the three classes possess different activation energies. It is likely that also the preexponential factors and hence the activation entropies differ.

In the one case, fim 2, where we have been able to study the motion over a wide range of temperature, the activation energy is not sharp but given by a broad distribution as shown in Fig. 5. It will be interesting to see if fims 1 and 3 are also distributed.

The dissililarity of fim 2 and fim 1 is remarkable. Originally we had hoped to explain the motion of the ligand through the protein matrix, the relaxation of the heme iron as seen in the Mössbauer effect, and the relaxation of g(H) by one class of motion. The data presented here demonstrate that such a unification is not possible. Fim 2 and fim 1 represent disparate phenomena; their activation energies and entropies differ and they may involve different parts of the protein. Indeed, even the early measurements (6) indicate that when Mb molecules are held rigidly in poly (vinyl alcohol), the photodissociated ligands can still migrate into the protein matrix, but the distribution g(H) does not relax; rebinding remains nonexponential in time up to at least 320 K.

The connection of the three classes of fims with theory also remains to be investigated in detail. It is likely that molecular dynamics calculations may lead to an understanding of fim 3. It appears, however, that molecular dynamics cannot reach fim 1 and fim 2. The time scales are so long that present computers are inadequate. There may even be a more fundamental obstacle: The large barriers observed in fim 3 make substates a realistic feature. It is not clear whether dynamics calculations, starting from a given protein conformation, may not remain confined within one substate, unable to reach many of the other substates that play important roles. Fim 2 and fim 1 consequently call for a dynamical theory of activated processes, as pioneered by Karplus and collaborators (39). Even with such a theory, an understanding of the process under consideration and a knowledge of the energy surfaces is required before it can be applied. The theory explains a posteriori and cannot predict what motions will occur while a protein performs a particular function.

Theories of a more general type, presently applied to glasses and spin glasses, may offer new insights (40-46). Concepts such as frustration and percolation may turn out to explain some of the features of protein dynamics. Indeed, the separation of fims into classes, if confirmed by more experiments, may imply the existence of general phenomena. We are only at a beginning and a close collaboration between theorists and experimentalists will be required for progress.

Acknowledgements

I thank A. Ansari, J. Berendzen, S. F. Bowne, I. E. T. Iben, T. B. Sauke, and E. Shyamsunder for their collaboration and for many stimulating discussions. I am grateful to J. O. Alben, P. G. Debrunner, W. A. Eaton, F. Fiamingo, V. I. Goldanskii, E. Gratton, M. W. Makinen, H. Roder, D. L. Rousseau, and P. G. Wolynes for useful data and comments. The work was supported by Grant PHS GM 18051 from the Department of Health and Human Services and by Grant PCM 82-09616 from the National Science Foundation.

References and Footnotes

1. Protein dynamics has been described in a number of reviews and has been the subject of many conferences. The conference proceedings (2)-(4) provide excellent surveys and contain references to the reviews and most seminal papers.
2. *Structure and Dynamics: Nucleic Acids and Proteins,* Eds. Clementi, E. and Sarma, R. H., Adenine Press, New York (1983).
3. *Mobility and Function in Proteins and Nucleic Acids,* Eds. Porter, R., O'Connor, M., and Whelan, J., *Ciba Foundation Symposium 93,* Pitman, London (1983).
4. *Structure, Dynamics, Interactions and Evolution of Biological Macromolecules,* Eds., Hélène, C., Reidel, Dordrecht (1983).
5. Antonini, E. and Brunori, M., *Hemoglobin and Myoglobin in Their Reactions with Ligands,* North-Holland, Amsterdam (1971).
6. Austin, R. H., Beeson, K. W., Eisenstein, L., Frauenfelder, H., and Gunsalus, I. C., *Biochemistry 14,* 5355 (1975).
7. Dlott, D. D., Frauenfelder, H., Langer, P., Roder, H., and DiIorio, E. E., *Proc. Natl. Acad. Sci. USA 80,* 6239 (1983).
8. Perutz, M. F. and Mathews, F. S., *J. Mol. Biol. 21,* 199 (1966).
9. Case, D. A. and Karplus, M., *J. Mol. Biol. 132,* 343 (1979).
10. Frauenfelder, H., Petsko, G. A., and Tsernoglou, D., *Nature 280,* 558 (1979).
11. Austin, R. H., Beeson, K., Eisenstein, L., Frauenfelder, H., Gunsalus, I. C., and Marshall, V. P., *Phys. Rev. Letters 32,* 403 (1974).
12. Dickerson, R. E. and Geiss, I., *Hemoglobin: Structure, Function, Evolution and Pathology.* Benjamin-Cummings, Menlo Park (1983).
13. Shank, C. V., Ippen, E. P., and Bersohn, R., *Science 193,* 50 (1976).
14. Reynolds, A. H., Rand, S. D., and Rentzepis, P. M., *Proc. Natl. Acad. Sci. USA 78,* 2292 (1981).

15. Terner, J. Strong, J. P., Spiro, T. G., Nagumo, M., Nicol, M. F., and El-Sayed, M. A., *Proc. Natl. Acad. Sci. USA 78*, 1313 (1981).

16. Rousseau, D. L. private communication.

17. Alberding, N., Austin, R. H., Beeson, K. W., Chan, S. S., Eisenstein, L., Frauenfelder, H., and Nordlund, T. M., *Science 192*, 1002 (1976).

18. Iikzuka, T., Yamamoto, H., Kotani, M., and Yonetani, T., *Biochem. Biophys. Acta 371*, 126 (1974).

19. Spartalian, K., Lang, G., and Yonetani, T., *Biochem. Biophys. Acta 428*, 281 (1976).

20. Friedman, J. M., Rousseau, D. L., and Ondrias, M. R., *Ann. Rev. Phys. Chem. 33*, 471 (1982).

21. Ondrias, M. R., Rousseau, D. L., and Simon, S. R., *J. Biol. Chem. 258*, 5638 (1983).

22. Roder, H., Berendzen, J., Bowne, S. F., Frauenfelder, H., Sauke, T. B., Shyamsunder, E., and Weissman, M. B., *Proc. Natl. Acad. Sci. USA 81*, (1984).

23. Alberding, N., Austin, R. H., Chan, S. S., Eisenstein, L., Frauenfelder, H., Good, D., Kaufmann, K., Marden, M., Nordlund, T. M., Reinisch, L., Reynolds, A. H., Sorensen, L. B., Wagner, G. C., and Yue, K. T., *Biophysical J. 24*, 319 (1978).

24. Doster, W., et al., to be published.

25. Yonetani, T., Iizuka, T., Yamamoto, H., and Chance, B., in *Oxidases and Related Redox Systems*, Eds., King, T. E., Mason, H. S., and Morrison, M., University Park Press, Baltimore, p. 401 (1973).

26. Sharanov, Y. A., Sharaonva, N. A., Figlovsky, V. A., and Grigorjev, V. A., *Biochim. Biophys. Acta 709*, 332 (1982).

27. Alben, J. O. and Fiamingo, F. personal communication

28. Eaton, W. A., Hanson, L. K., Stephens, P. J., Sutherland, J. C., and Dunn, J. B. R., *J. Am. Chem. Soc. 100*, 4991 (1978).

29. Eaton, W. A. and Hofrichter, J., in *Methods in Enzymology, Vol. 76*, Academic Press, New York, p. 175 (1981). (See also ref. 30)

30. Makinen, M. W. and Churg, A. K., in *Iron Porphyrins Part I*. Eds. Lever, A. B. P. and Gray, H. B., Addison-Wesley Reading, MA, p. 191 (1983).

31. A distribution of activation energies is not the only way to explain the nonexponential time dependence; it is also possible that diffusion in a conformation space takes place. This problem has been treated by Agmon and Hopfield: Agmon, N. and Hopfield, J. J., *J. Chem. Phys. 78*, 6947 (1983), *79*, 2042 (1983).

32. Parak, F., Knapp, E. W., and Kucheida, D., *J. Mol. Biol. 161*, 177 (1982).

33. Bauminger, E. R., Cohen, S. G., Nowik, I., Ofer, S., and Yariv, J., *Proc. Natl. Acad. Sci. USA 80*, 736 (1983).

34. Friedman, J. M., Scott, T. W., Stepnoski, R. A., Ikeda-Saito, M., and Yonetani, T., *J. Biol. Chem. 258*, 10564 (1983).

35. Eisenstein, L. and Frauenfelder, H., in *Frontiers of Biological Energetics, Vol. 1*, Academic Press, New York, p. 680 (1978).

36. Alberding, N., Ph.D. Thesis, University of Illinois (1978).

37. Tony Crofts has suggested that one should not only introduce fims, but also bums, biologically unimportant motions.

38. Englander, S. W. and Kallenbach, N. R., *Quarterly Rev. Biophys., 16*, 521 (1984).

39. Northrup, S. H., Pear, M. P., Lee, C. Y., McCammon, J. A., and Karplus, M., *Proc. Natl. Acad. Sci. USA 79*, 4035 (1982).

40. Edwards, S. F. and Anderson, P. W., *J. Phys. F5*, 65 (1975).

41. Kirkpatrick, S. and Sherrington, D., *Phys. Rev. B17*, 4384 (1978).

42. Druger, S. D., Nitzan, A., and Ratner, M. A., *J. Chem. Phys. 79*, 3133 (1983).

43. Goldanskii, V. I., Krupyanskii, Yu. F., and Fleurov, V. N., *Doklady AN SSSR 272*, 978 (1983).

44. Mezard, M., Parisi, G., Sourlas, N., Toulouse, G., and Virasoro, M., *Phys. Rev. Lett. 52*, 1156 (1984).

45. Singh, G. P., Schink, H. J., Lohneysen, H. V., Parak, F., and Hunklinger, S., *Z. Phys. B 55*, 23 (1984).

46. Palmer, R. G., Stein, D. L., Abrahams, E., and Anderson, P. W., *Phys. Rev. Lett. 53*, 958 (1984).

Structure & Motion: Membranes, Nucleic Acids & Proteins,
Eds., E. Clementi, G. Corongiu, M. H. Sarma & R. H. Sarma,
ISBN 0-940030-12-8, Adenine Press,   Copyright Adenine Press, 1985.

Neutron Spectroscopy and Protein Dynamics

H. D. Middendorf
Department of Biophysics
King's College London
London WC2B 5RL, U.K.

J. T. Randall[‡]
Department of Zoology
University of Edinburgh
Edinburgh EH9 3JT, U.K.

Abstract

Thermal and cold neutrons with de Broglie wavelengths between 1 and 20 Å are versatile probes of the structure and dynamics of organic molecules. By performing an energy analysis at each scattering angle in addition to measuring the angular dependence of the scattered intensity, familiar diffraction methods are extended to give dynamic structure factors $S(\mathbf{Q},\omega)$ which describe both the time-averaged and the dynamical aspects of atomic and molecular interactions. During the past 10 years, as the result of increased flux levels and advances in instrumentation, the scope of neutron spectroscopy has broadened considerably. Biomolecular applications of quasi-elastic and inelastic neutron scattering are now being pursued more actively, and are beginning to provide valuable spatiotemporal information on intermolecular and intramolecular processes with scale lengths in the 0.1 to 100 Å region and characteristic times between 10^{-13} and 10^{-6} sec. This contribution surveys the present experimental situation, outlines the principles underlying advanced neutron techniques, and discusses current work on proteins.

Introduction

Our knowledge of the detailed three-dimensional structure of biological molecules and their assemblies derives largely from X-ray, neutron, and electron diffraction experiments. The structural information obtained is in the form of scattering density maps which, if resolved to better than a few Å, may be refined to give more or less complete lists of the time-averaged atomic coordinates. It has become possible also, in favorable circumstances, to extract a certain amount of dynamical information from diffraction data by analyzing individually the temperature factors of all heavy atoms or groups, and to determine the dimensions of their "thermal clouds" (1,2).

[‡]Deceased June 16th, 1984.

The basic conservation equations governing the interaction between a quantum of incident radiation and a scattering center are given by

$$\hbar\mathbf{Q} = \hbar(\mathbf{k}-\mathbf{k_o}) \qquad \text{(momentum)} \qquad (1a)$$

$$\hbar\omega = E - E_o \qquad \text{(energy)} \qquad (1b)$$

where $\hbar\mathbf{Q}$ and $\hbar\omega$ are the momentum and energy transfers, respectively ($\hbar$=Planck's constant/2π). The initial and final wavevectors and energies are denoted by $\mathbf{k_o},\mathbf{k}$ and E_o,E; if particles of rest mass m are employed the corresponding velocities are $\mathbf{v_o},\mathbf{v}$. It is the coherent elastic scattering characterized by $k \simeq k_o$, or $v \simeq v_o$, which produces $\mathbf{Q}$-dependent diffraction patterns related by Fourier transformation to the real-space structure. Some fraction of the radiation will however be scattered inelastically, i.e. exchange energy with thermal vibrations and other excitations in the sample, and this component carries information on atomic and molecular motions beyond that already contained in the temperature factors modifying the elastic scattering. In order to fully exploit the information content of the scattered radiation field, one must perform an energy analysis of the intensity at each scattering angle in addition to measuring the angular dependence, and aim to derive dynamic structure factors $S(\mathbf{Q},\omega)$ which describe both the time-averaged and the dynamical aspects of the scattering process. In practice, with the notable exception of X-ray spectroscopy using the Mössbauer effect (3), the demands on resolution are such that this is feasible only when the incident beam energy E_o is of the order of, or smaller than, the thermal energy k_BT of the sample.

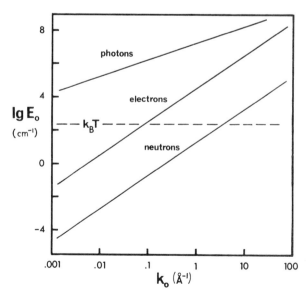

Figure 1. Energy-momentum relations $E_o=f(k_o)$ for electromagnetic quanta, electrons, and neutrons.

To appreciate the potential of different radiation scattering techniques for probing Q,ω-regions relevant to biomolecular studies, one may compare the free-space dispersion relation $E_o = c\hbar k_o$ for electromagnetic waves with that for electrons and neutrons, $E_o = \frac{1}{2}mv_o^2 = \hbar^2 k_o^2/2m$ (Figure 1). It is apparent that X-ray quanta are much too energetic to be analyzed spectrally with a resolution well below $k_B T$, whereas radiation of non-zero rest mass offers considerably more scope. The basic energy-momentum relation $E_o = f(k_o)$ is particularly favorable for thermal and cold neutrons which have de Broglie wavelengths $\lambda_o = 2\pi\hbar/mv_o = 2\pi/k_o$ between 1 and 20 Å. Neutron scattering techniques are in fact unique in that they allow small energy changes characteristic of weakly interacting molecular systems to be resolved along with spatial information in the 0.1 to 100 Å range.

The simplicity of nuclear scattering and the closeness of measured spectra to fundamental correlation functions makes neutron scattering a valuable experimental complement to numerical simulation studies. It will be of interest, therefore, in the context of this Symposium, to survey the present experimental situation, to outline the principles underlying advanced neutron techniques, and to discuss a few representative results.

Scattering Regimes, Interaction Parameters, Correlation Functions

The quasi-elastic and inelastic scattering of cold neutrons was first used to investigate the dynamics of fibrous biopolymers about 15 years ago (4,5). Following this early work and a number of preliminary studies at Harwell (6,7) between 1972 and 1975, biomolecular applications of neutron spectroscopy have developed only very slowly (8). This is mainly due to the fact that the basic limitations of neutron scattering techniques, i.e. low flux and relative inaccessibility of instruments, are particularly severe in experiments requiring an energy analysis of the scattered radiation, but it also reflects the limited Q,ω-coverage of the spectrometers available until the mid-1970s. Since then, instruments employing novel principles of energy analysis have been built and perfected mainly at the Institut Laue-Langevin (ILL) in Grenoble (9), and these have in particular opened up the inner quasi-elastic scattering regime down to energy transfers $\hbar\omega$ of a few neV, or 10^{-5} cm^{-1}. A reasonably uniform coverage of large parts of the Q,ω-plane by new spectrometers has been achieved gradually in recent years. Although there are certain low-Q regions which are not yet accessible but of great interest for macromolecular work, it must be recognized that the instruments now in routine use at the ILL and a few other research centers provide excellent tools for developing the application of neutron scattering to the molecular dynamics of biomolecules. The flux levels at which these instruments operate, however, are still rather low compared with the sources used for photon or electron spectroscopy, and the fact that spatiotemporal information is obtained in the form of sets of Q-dependent spectra can be a mixed blessing as it is necessary in a single run to measure 6 to 12 spectra with equally good counting statistics. Fairly large samples (50-500 mg) and long run times (6-24 hr) are required to accomplish this under typical experimental conditions, and therefore the total time available on a particular instrument is always the overriding factor determining experimental strategy and the extent to which it is possible to vary sample parameters.

Next to the momentum-energy domain covered, the second important aspect of a radiation scattering technique is given by the physical nature of the elementary interaction. This determines the strength of the scattering, its directional properties, and the possibilities for modifying the scattering from sample constituents of special interest. In the absence of magnetic interactions, which is normally the case for biological materials, neutrons are scattered only from the atomic nuclei. Each neutron-nucleus interaction contributes a scattered wavelet, or probability distribution, that is spherical and characterized by a single empirical parameter, the scattering length b. There are no form factors as in X-ray or electron scattering, but b depends on the spin state I of the target nucleus in such a way that it can assume two values, b^+ and b^-. These correspond to the two possible compound states $I \pm \frac{1}{2}$ formed during a scattering event, the neutron spin being $\frac{1}{2}$.

In an assembly of N isotopically pure nuclei located at r_i (i=1,2,...,N), the spin orientations will normally be random, i.e., not correlated with r_i, and it is appropriate then to distinguish between two components in the scattered radiation field: *coherent scattering* $\sim \langle b_i \rangle^2$, due to the superposition of wavelets bearing a definite phase relationship to the structure and therefore capable of producing an interference pattern; and *incoherent scattering* due to randomly distributed spin disorder, this component being proportional to the mean-square deviation from $\langle b_i \rangle$. The incoherent cross-section is thus given by $\sigma_{inc} = 4\pi (\langle b_i^2 \rangle - \langle b_i \rangle^2)$, and the corresponding coherent cross-section is $\sigma_{coh} = 4\pi b_{coh}^2$ where $b_{coh} = \langle b_i \rangle$ denotes the coherent scattering length. Isotopic impurities are a second source of incoherent scattering; these are negligible for all elements encountered in biophysical work with the exception of deuterium which, however, is the principal isotope used for systematic substitution and is therefore treated as a separate element. The neutron scattering lengths and cross-sections (10) for H, D, C, O, and N are listed in Table I together with the mean abundance of these elements in eight globular proteins; the total

Table I

Scattering lengths b_{coh}, cross-sections σ_{coh} and σ_{inc}, and mean abundances f_{ch} for the five most common elements in natural and deuterated proteins (abundances for 8 proteins from Ref. 11).* Fractional abundance-weighted cross-sections $\bar{\sigma}_{coh}$ and $\bar{\sigma}_{inc}$ calculated for fully hydrogenous and fully deuterated molecules; the total cross-section ratios are $(\bar{\sigma}_{inc}/\bar{\sigma}_{coh})_t = 10.1$ and 0.18, respectively.

	spin	b_{coh} (Fermi)	σ_{coh} (barn)	σ_{inc} (barn)	f_{ch} %	$\bar{\sigma}_{coh}$ %	$\bar{\sigma}_{inc}$ %	$\bar{\sigma}_{coh}$ %	$\bar{\sigma}_{inc}$ %
H	$\frac{1}{2}$	−3.740	1.758	79.7	50.2	22.3	99.9		
D	1	6.674	5.597	2.0	50.2			47.7	95.9
C	0	6.648	5.554	<0.02	30.9	43.3	0.0	29.1	0.3
O	0	5.83	4.27	<0.02	10.6	11.4	0.0	7.7	0.1
N	1	9.36	11.01	0.46	8.3	23.0	0.1	15.5	3.6

*1 Fermi = 10^{-13} cm, 1 barn = 10^{-24} cm².

relative cross-sections resulting for these abundances are also given and compared for the two limiting cases of fully hydrogenous and fully deuterated molecules.

Because of $|b_i| \ll \lambda_o$, a sum over the Fermi pseudo-potential $(2\pi\hbar^2/m)b_i \; \delta(\mathbf{r}-\mathbf{r}_i)$ may be used to describe the scattering density of a sample comprising N nuclei at $\mathbf{r}_i$. For a rigid assembly of nuclei, the scattering is entirely elastic and the differential cross-section given by (12)

$$\frac{d\sigma}{d\Omega} = \sum_{i,j} b_i b_j \exp(-i\mathbf{Q}\cdot\mathbf{r}_i) \exp(i\mathbf{Q}\cdot\mathbf{r}_j) = N[b_{inc}^2 \, S_{inc}(\mathbf{Q}) + b_{coh}^2 \, S_{coh}(\mathbf{Q})] \qquad (2a)$$

where

$$S_{coh}(\mathbf{Q}) = \frac{1}{N} \sum_{i,j} \exp(-i\mathbf{Q}\cdot\mathbf{r}_i) \exp(i\mathbf{Q}\cdot\mathbf{r}_j) \qquad (2b)$$

represents the static structure factor familiar from diffraction work, and $S_{inc}(\mathbf{Q}) \equiv 1$ is a unit incoherent background. To generalize Eq. 2 for inelastic scattering, it is necessary to introduce time-dependent position vectors $\mathbf{r}_i(t)$ for the N nuclei considered, and to construct time correlation functions by averaging over a thermal ensemble of nuclear trajectories as follows:

$$F_{coh}(\mathbf{Q},t) = \frac{1}{N} \langle \sum_{i,j} \exp(-i\mathbf{Q}\cdot\mathbf{r}_i(0)) \exp(i\mathbf{Q}\cdot\mathbf{r}_j(t)) \rangle_{thermal} \qquad (3)$$

For arbitrary (i,j) this function describes all pair correlations including N self-correlation terms; it therefore relates to coherent scattering in the widest sense, i.e. the interference effects produced by stationary as well as moving nuclei. The associated time correlation function for incoherent scattering, denoted $F_{inc}(\mathbf{Q},t)$, is obtained from Eq. 3 by summing only the diagonal terms i=j. Although $F_{inc}(\mathbf{Q},t)$ reduces to a flat background in the static limit t=0, its spectral structure turns out to be a valuable source of information on atomic and molecular motions. Equation 3 is often referred to as the "intermediate" scattering function because its spectral representation, obtained on Fourier transformation with respect to time, is nothing but the dynamic structure factor $S_{coh}(\mathbf{Q},\omega)$, whereas Fourier transformation with respect to $\mathbf{Q}$ gives the van Hove space-time correlation function G(r,t) (14). The corresponding incoherent scattering functions are similarly related to the van Hove self-correlation function $G_s(\mathbf{r},t)$. These transform relationships may be summarized symbolically as

$$S_{coh}(\mathbf{Q},\omega) \Leftrightarrow F_{coh}(\mathbf{Q},t) \Leftrightarrow G(\mathbf{r},t) \qquad (4a)$$

$$S_{inc}(\mathbf{Q},\omega) \Leftrightarrow F_{inc}(\mathbf{Q},t) \Leftrightarrow G_s(\mathbf{r},t) \qquad (4b)$$

The space-time correlation functions appearing here are of fundamental interest; they connect the experimentally accessible quantities (mainly S, but also F) with a substantial body of theoretical work on the statistical mechanics of systems of interacting particles (15).

Molecular Systems, Quasi-Elastic and Inelastic Scattering

The correlation function formalism outlined above provides a powerful and conceptually attractive framework for describing the scattering of thermal and cold neutrons from an assembly of moving point-like nuclei, but the actual experiment
always consists of collecting angle and time dependent intensity data recorded by
an array of detectors around the sample. Most spectrometers in current use work
with unpolarized neutrons and produce data in the form of a double differential
cross-section, $d^2\sigma/d\Omega dE$, which gives the intensity of neutrons with energies between E and E+dE scattered into a solid angle element $d\Omega$. The appropriate
generalization of Eq. 2a is

$$\frac{d^2\sigma}{d\Omega d\omega} = N \frac{k}{k_o} [b_{inc}^2 S_{inc}(\mathbf{Q},\omega) + b_{coh}^2 S_{coh}(\mathbf{Q},\omega)] \tag{5}$$

where the factor $k/k_o = v/v_o$ results from converting neutron density to flux. Measurements relating directly to $F(\mathbf{Q},t)$ can only be carried out by means of the spin-
echo technique discussed below. Equation 5 is strictly valid for monatomic scatterers
but will be a good approximation for polyatomic systems when the scattering from
a single species of nuclei predominates greatly, the prime example of this being
hydrogen in organic molecules. Although $d^2\sigma/d\Omega dE$ can always be written as the
sum of an incoherent and a coherent part, it is not in general possible to factorize
each of these into a cross-section $\sim b^2$ and a dynamic structure factor $S(\mathbf{Q},\omega)$. For a
molecular sample, the incoherent part of Eq. 5 must be replaced by appropriate
sums over atomic species or groups and their particular scattering laws, but the
formulation of this is still relatively straightforward. The coherent part, however,
because it describes interference effects that reflect the relationship between structure and dynamics much more intimately, will give rather complicated expressions.
The superposition and interaction of a large number of vibrational, rotational, and
diffusive modes in a complex biomolecular system makes it very difficult or impossible to evaluate $d^2\sigma/d\Omega dE$ explicitly, and a full interpretation of inelastic scattering
data usually requires extensive model calculations and numerical simulations.

From an experimental point of view, it is important to realize that the dynamic
structure factors are intrinsic sample properties which depend only on the difference quantities $\mathbf{Q}$ and ω according to Eq. 1. Within certain limits, different combinations of $\mathbf{k}_o,\mathbf{k}$ and E_o,E can be chosen to give the same $\mathbf{Q},\omega$-values. To introduce
the scattering angle 2θ defined by $\cos2\theta = \mathbf{k}\cdot\mathbf{k}_o/kk_o$, Eq. 1 may be solved for Q and
the result is

$$Q^2 = k_o^2 [2+\overline{\omega} - 2(1+\overline{\omega})^{1/2} \cos2\theta] , \quad \overline{\omega} = \hbar\omega/E_o \tag{6}$$

At finite 2θ, an expansion in powers of $\overline{\omega}$ leads to

$$Q^2 = 4k_o^2(\sin^2\theta + \tfrac{1}{2}\overline{\omega} \sin^2\theta + \tfrac{1}{16}\overline{\omega}^2 \cos2\theta + ...) \tag{7}$$

and for $\overline{\omega}=0$ this reduces to the Q-value for elastic scattering, $Q_o=(4\pi/\lambda_o)\sin\theta$. In *quasi-elastic scattering* (16), the energy change of a neutron is small compared with E_o so that the Q-dependent Doppler-like broadening of the incident "line" due to diffusive processes is the principal effect observed; here the term linear in $\overline{\omega}$ suffices to describe the scattering kinematics. In the study of biomolecular dynamics, where energy transfers $\lesssim k_BT$ are of greatest interest, quasi-elastic scattering generally implies $|\hbar\omega| \ll E_o < k_BT$.

The domain of *inelastic scattering* proper (17) is characterized by $|\overline{\omega}| \gtrsim 1$ and may be reached either by "upscattering" (neutrons gaining energy from the sample nuclei) or by "downscattering" (neutrons losing energy). Vibrational and rotational motions with distinct energy levels $\hbar\omega_j$ give rise to the scattering here; these may be localized modes showing little or no dispersion, or they may represent phonons $\omega_j(\mathbf{q})$ propagating with wave vector $\mathbf{q}$. The differential intensity due to a quantized mode is proportional to (a) the cross-section σ_{coh} or σ_{inc}, (b) a Debye-Waller factor $\exp(-2W_p)$, (c) a "polarization" term $(\mathbf{Q}\cdot\mathbf{C}_{pj})^2$ where $\mathbf{C}_{pj}$ is the normalized amplitude vector of nucleus p vibrating in mode j, (d) a Bose-Einstein factor $\overline{n}_j = [\exp(\hbar\omega_j/k_BT)-1]^{-1}$ for thermally populated energy levels, and (e) a product of two δ-functions ensuring conservation of energy and momentum. In the simplest case of incoherent scattering there are no interference terms, i.e. no vectorial conditions on $\mathbf{Q}$, and the essential effects are displayed by the single-phonon approximation (13)

$$\frac{d^2\sigma}{d\Omega dE} \sim \frac{k}{k_o} b_{inc,p}^2 \sum_p \exp(-2W_p) \sum_{j,q} (\overline{n}_j + \tfrac{1}{2} \mp \tfrac{1}{2})(\mathbf{Q}\cdot\mathbf{C}_{pj})^2(2m\omega_j)^{-1}\delta(\omega \mp \omega_j) \tag{8}$$

where $p=1,2,...,N_p$ labels the protons in a unit cell or molecule ($\mathbf{C}_{pj}$, $\overline{n}_j$, and ω_j depend on $\mathbf{q}$, and the double sum extends over all modes). This equation describes the dominant scattering process in which single quanta $\hbar\omega_j$ are gained (upper signs) or lost (lower signs) by the neutrons. In general there will also be scattering events in which quanta of energy equal to integral multiples of $\hbar\omega_j$ are exchanged; these multi-phonon processes may contribute appreciably to $d^2\sigma/d\Omega dE$ at energies $\lesssim k_BT$ when several or many quantum levels are populated and the resulting nuclear amplitudes are larger than those associated with high-frequency modes. The thermal population factor $\overline{n}_j$ plays an important role in that it governs the character and observability of neutron spectra: at high frequencies we have $\overline{n}_j \Rightarrow \exp(-\hbar\omega_j/k_BT)$ for energy gain and $(\overline{n}_j+1) \Rightarrow 1$ for loss; in cold neutron experiments (upscattering) the spectra therefore become unobservable beyond $\hbar\omega \approx 5k_BT \approx 1000$ cm^{-1}. Apart from $\overline{n}_j$ and the Debye-Waller factor (which is rather insensitive to spectral detail), the differential cross-section reflects mainly the density of vibrational states weighted by the $\mathbf{Q}$-component of the proton amplitudes. Depending on sample geometry, the number of modes, and the dynamical properties of protonated groups in the molecule or unit cell, the j,$\mathbf{q}$-sum in Eq. 8 may be averaged and/or approximated in various ways. In solution scattering experiments, interest focuses on the spherically averaged hydrogen-weighted frequency distribution

$$\overline{Z}_p(\omega) = \frac{1}{3N_p} \sum_{j,q} C_{pj}^2 \, \delta(\omega - \omega_j) \tag{9}$$

which can be extracted from a set of measured spectra by $Q \Rightarrow O$ extrapolation. It is important to remember that the molecular dynamics contained in $\omega_j(\mathbf{q})$ is largely determined by the heavy atoms (mainly C,O,N), whereas the incoherent scattering comes almost exclusively from hydrogen atoms. Insofar as these can be regarded as rigidly attached to the former, and this is certainly a reasonable approximation at low frequencies, they serve as natural probes of the dynamics.

Biophysical Applications

What, now, can we learn about the dynamics of biomolecular systems from neutron scattering experiments, using the full range of spectroscopic techniques presently available? These span almost 7 decades in energy centered on 0.1-1 cm^{-1}, and 2-3 decades in momentum transfer. Given this large Q,ω-domain and the many possibilities for dynamic difference experiments, it is clear that a comprehensive set of neutron data on a protein or nucleic acid must contain a wealth of information. At this stage in the application of neutron spectroscopy to biophysical problems, in order to make good use of a valuable but limited resource, it is important to fully understand the principles involved and to be aware of the instrumental characteristics. The quantitative relation between the data obtained in a particular experimental setup and the molecular dynamics of the sample will depend very much on the method used for energy analysis, on a number of fundamental as well as practical constraints, and on the way in which the observables are affected by finite resolution.

In the sections to follow, beginning with the more conventional methods, we discuss the essential features of both quasi-elastic and inelastic neutron scattering techniques in the context of current work on proteins.

Crystal and Filter Analysis

The basic measuring process according to Eqs. 1 and 5 consists of two steps: preparation of a beam of neutrons with well-defined $\mathbf{k}_o$ and E_o, and determination of the fractional intensity of neutrons scattered with energy E as a function of the angular variables defining $d\Omega$ (i.e. scattering angle 2θ and possibly an azimuthal angle φ). The magnitude of the scattering vector $\mathbf{Q}$ then follows from Eq. 6. A straightforward but time-consuming way of doing this is to employ a single-crystal analyzer with which a point-by-point scan of $d^2\sigma/d\Omega dE$ along a chosen $\omega(Q)$ path may be performed. In an instrument of this kind, known as a triple-axis spectrometer (Figure 2a), incident-beam monochromatization is likewise carried out by Bragg reflection, and it is possible with suitable crystals to select a wide range of λ_o values

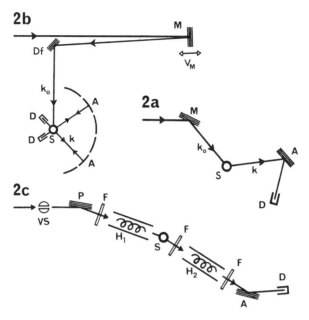

Figure 2. Diagrams illustrating the principles of (a) triple-axis measurements, (b) backscattering and (c) spin-echo spectroscopy. A=analyzer crystal(s), D=detector, Df=deflector crystal, F=spin flipper, H=magnetic guide field, M=monochromator crystal, P=polarizer, S=sample, VS=velocity selector. The two guide fields in (c) are of equal dimensions and strength.

from a "white" neutron source. The flexibility given by independent **Q** and ω variation through large regions of energy-momentum space makes this type of spectrometer ideal for the study of coherent inelastic scattering from lattice waves. Thus the low-frequency branches ($\hbar\omega \lesssim 100$ cm^{-1}) of the phonon dispersion diagrams have been mapped out for a number of molecular crystals of biophysical interest, e.g. for single crystals of deuterated amino acids and of hydrogenous DNA-type pyrimidines (18), and the data are being used in conjunction with model calculations to test and refine the basic atomic and molecular interaction parameters (force constants, elastic moduli). As an example, a set of dispersion curves for glycine measured at Chalk River is reproduced in Figure 3. Triple-axis spectrometers have also contributed valuable incoherent scattering data to studies of the hydration dynamics of oriented DNA fibers (5) and whole cells (19); here attention focuses on the quasi-elastic region which may be examined with a resolution (20) $\Gamma_o = 0.1$ to 0.5 cm^{-1}.

All methods of measuring differential cross-sections in the Q,ω-domain face the fundamental problem that, by Liouville's theorem, the phase-space density of neutrons extracted from a given source (moderator at temperature T_m) cannot be increased. Any reduction in the wavelength spread $\Delta\lambda$ of the incident beam, together with increased energy discrimination on the detector side and better collimation before and after scattering, carries a heavy intensity penalty. For fixed source and sample parameters, substantial improvements in energy resolution and/or

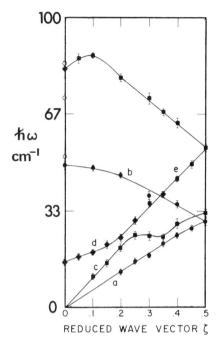

Figure 3. Dispersion curves for intermolecular modes propagating along b*-axis in perdeuterated α-glycine single crystal, measured on triple-axis spectrometers (reproduced from Ref. 18).

counting rate can only be achieved by means of multidetector arrays intercepting many more neutrons, and by relaxing the Q-resolution or constraining the range of Q. A drastic example of this is provided by a variant of the triple-axis spectrometer in which the analyzer is replaced by a polycrystalline beryllium, beryllium oxide, or graphite filter with a sharp cut-off at $\lambda_c=3.96$, 4.66, or 6.3 Å, respectively, so that only cold neutrons with $E<660 \lambda_c^{-2}$ cm^{-1} reach the detector. Energy-loss spectra in the $100<(-\hbar\omega)<5000$ cm^{-1} region are then measured by varying E_o; it follows from Eq. 1 that under these conditions

$$\hbar\omega = -\frac{\hbar^2}{2m} k_o^2 \qquad (10)$$

Filter detection enables large solid angles to be utilized and gives greatly increased counting rates, but this method of energy analysis is restricted essentially to a single $\omega(Q)$ scan along the energy-momentum curve shown in Figure 1. The resolution is quite bad below k_BT and improves with increasing energy transfer, but above a few 100 cm^{-1} we have $Q>4$ Å^{-1} and this means that filter-detector instruments are of interest primarily for incoherent scattering work. They have been used, e.g., in investigations of the librational modes of biological water (21) between 300 and 900 cm^{-1}.

Time-of-Flight Spectroscopy

Whereas triple-axis instruments and most of their variants make use of the wave properties of neutrons by employing crystal analysis in one way or another, a

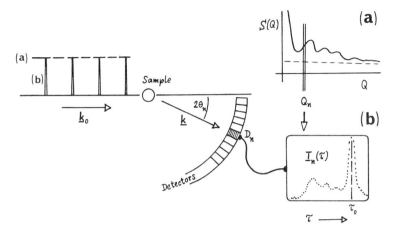

Figure 4. From diffraction to time-of-flight spectroscopy: (a) Incident monochromatic beam continuous, total intensity recorded by detector D_n proportional to $S(Q_n)dQ$; (b) incident beam chopped into train of equidistant pulses, scattered intensity $S(Q_n)dQ$ energy-analyzed into time-of-flight spectrum $I_n(\tau)$ (shown in insert as appearing on monitor screen).

different and powerful method of energy analysis is to measure directly the velocity changes of scattered neutrons. Because of their large rest mass, thermal and cold neutrons with $\lambda = 2$ to 10 Å have velocities from 2000 to 400 m/s, and it is relatively easy to detect energy differences $\Delta E = mv\Delta v$ corresponding to 0.1% of $k_B T$ at 300 K. To accomplish this, it is necessary to work with a pulsed beam which provides time windows for velocity analysis. The principle of time-of-flight spectroscopy is best understood with reference to a diffractometer equipped with a multidetector array (see Figure 4). Three essential changes are required to turn this instrument into a spectrometer: (a) the incident beam has to be chopped into a train of short pulses separated by $\Delta t = 5$ to 10 ms; (b) each detector needs to be connected to a multichannel analyzer capable of sorting detected neutrons into a histogram according to their time of arrival; (c) in order to observe incoherent scattering over a wide Q-range, the detectors have to be redistributed to cover almost the entire semi-circle around the sample. It is possible, then, to perform an energy analysis at each scattering angle as follows: Of the monoenergetic neutrons comprising a single pulse incident on the sample at time t_s, a certain fraction will be deflected by an angle $2\theta_n$. All neutrons in this bunch take off into the same direction, that is towards D_n, but with a distribution of velocities v spread around v_o according to the decelerations and accelerations received during their almost instantaneous interaction at $t = t_s$ with the nuclei in the sample. Over a flight path R of a few meters between sample and detector D_n, these velocity differences become observable as time-of-arrival differences at D_n, measured relative to t_s. For a neutron arriving at $t > t_s$, its "time of flight" is defined by $\tau = (t - t_s)/R$ and this is just the inverse velocity $1/v$; its energy therefore is $E = mv^2/2 = m\tau^{-2}/2$. In this way an elemental time-of-flight spectrum is built up in the electronic analyzer during the interval Δt between two successive pulses, and the process is repeated many times until the counting statistics in the accumulated spectrum is good enough. Each one of the other

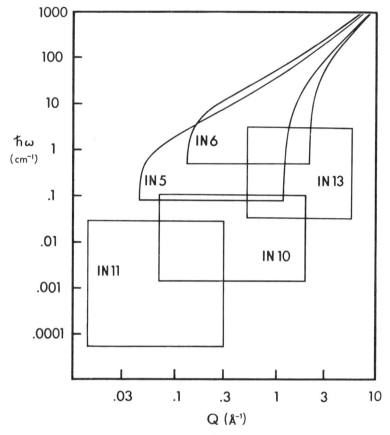

Figure 5. Regions in Q,ω-space covered by current quasi-elastic and inelastic neutron scattering experiments on biomolecular samples at the Institut Laue-Langevin, using spectrometers IN5 and IN6 (time-of-flight), IN10 and IN13 (backscattering), and IN11 (spin-echo). The domain boundaries shown are for incident wavelengths λ_o of 10, 5, 6.3, 2.5 and 8.3 Å, respectively.

detectors produces a similar τ-dependent spectrum simultaneously, "labelled" by its scattering angle 2θ or elastic momentum transfer Q_o.

It is obvious from Figure 5 that time-of-flight spectroscopy, because of the simultaneous acquisition of $d^2\sigma/d\Omega dE$ data over a large Q,ω-domain with emphasis on $\hbar\omega \lesssim k_B T$, is capable of providing much useful spatiotemporal information on the dynamics of biomolecules. In hydration studies, the aim is to characterize in detail the dynamics of water molecules closely associated with proteins. It is difficult to do this in solution, and the strategy adopted in most investigations is that of sorption experiments in which protein powders, films, or oriented fibers are hydrated in steps from the gently dried state up to levels close to the water content necessary for functional interactions (22). This approach has been used to obtain time-of-flight spectra for collagen fibers (23) and for biosynthetically deuterated C-phycocyanin (24). By measuring $d^2\sigma/d\Omega dE$ at several points along the sorption isotherm and carrying out separate runs for different H/D contrast combinations, it is possible to examine the structural and dynamical changes accompanying the sequence of hydration events at or near the protein surface. The cross-section ratio between sorbed H_2O molecules and the bulk of a protein is particularly favorable for covalently deuterated samples, and changes in the vibrational properties become measurable already at hydration levels of about 2% (g water/g protein), i.e. during the first hydration stage when interactions with ionizable side chains predominate. A set of low-hydration spectra from partially deuterated phycocyanin (Figure 6) shows the difference intensities obtained, and also illustrates the nonlinear relationship between time-of-flight τ and energy transfer $\hbar\omega$. Because of the large number of modes in a heterogeneous macromolecular system, the absence of selection rules in neutron scattering, and various intrinsic as well as instrumental broadening effects, such spectra always consist of a continuous intensity distribution carrying numerous more or less differentiated features. The complex pattern of intensity changes seen here reflects both the net increase and the repartition of the various hydration-activated degrees of freedom of the protein-water system. For any fixed covalent deuteration level of the protein, by increasing the D_2O fraction at constant total water content, one may "fade out" the proton signal from the water of hydration and those labile hydrogen positions that exchange on the time scale of the experiment. This reduces the incoherent scattering throughout the Q-range observed,

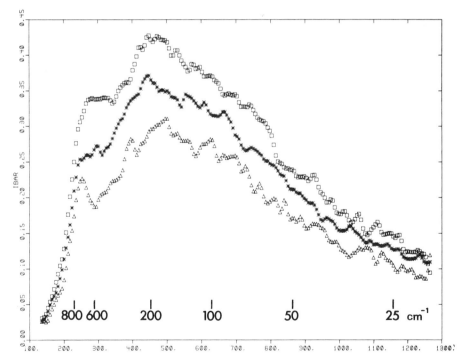

Figure 6. Time-of-flight spectra of H_2O-exchanged, 80% biosynthetically deuterated phycocyanin: nearly dry (Δ), H_2O-equilibrated at relative humidities of 43% (*) and 70%($\square$); T=25°C (from unpublished work at ILL in collaboration with H.L. Crespi). Time-of-flight scale from 100 to 1300 μs/m.

but it also shifts the emphasis in the wet-minus-dry difference spectra from single-particle motions to two-particle correlations, and gives characteristic Q-dependences which carry information on the collective aspects of the water dynamics and its coupling to the protein matrix. At higher hydration levels, equivalent to monolayer coverage and beyond, the fraction of bulk-like water molecules increases and we expect the vibrational coupling to the protein to become "softer" overall; this is borne out by the difference spectra which show a number of distinct intensity changes and peak shifts.

To observe intramolecular fluctuations or discrete modes, one must accentuate the scattering from the protein interior as a whole, from particular subunits or domains, or from elements of the secondary structure. The study of hydration and exchange processes is of course closely related to, and often inseparable from, studies of the intramolecular dynamics. Thus the hydration experiments discussed above yield information on the self-correlation of intramolecular motions in the limiting case of

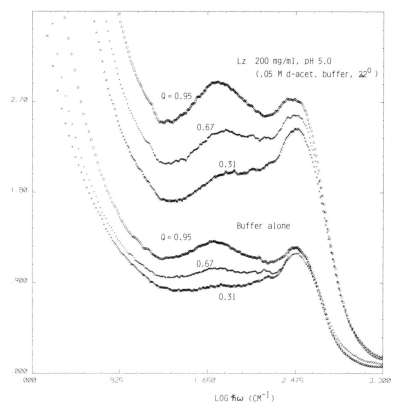

Figure 7. Time-of-flight spectra (converted to logarithmic $\hbar\omega$ scale) for a 200 mg/ml lysozyme solution compared with those for the perdeuterated buffer alone ('pH'=5.0, T=22°C). Spectra averaged over 8° to 15° wide sectors centered on $\langle 2\theta \rangle = 15°, 31°$, and 45°.

hydrogenous proteins hydrated with D_2O. While this is useful at low levels of water uptake, e.g. for determining the spectrum of internal fluctuations activated upon hydration from the dry or nearly dry state, greater interest attaches to experiments on fully hydrated molecules. Although the specific cross-section of a hydrogenous, D_2O-exchanged protein is an order of magnitude larger than that of heavy water (compare Table 1), much of the inelastic scattering will be due to the perdeuterated buffer even when the solution is very concentrated. The lysozyme spectra shown in Figure 7 illustrate this point. At concentrations in the 1 to 10% range, time-of-flight spectra with exceptionally good counting statistics are required to resolve intramolecular modes of proteins and to detect spectral changes resulting from ligand binding, subunit interaction, or complex formation.

Considerable impetus for solution scattering experiments has come from the new time-focusing spectrometer IN6 at Grenoble which combines moderately good resolution ($\Gamma_0 \approx 0.5$ cm^{-1}) with relatively high flux (9). This is achieved by means of an array of 3x7 monochromator crystals defining three beams with slightly different $\mathbf{k}_0$ focused geometrically onto the sample; after passing through a beryllium filter these beams are Fermi-chopped into pulses in such a way that the Δv_0 are compensated for by different times of chopping—hence "time-focusing". In current solution scattering work using IN6 and the multi-chopper spectrometer IN5, time-of-flight data are being collected for three systems: hexokinase (25) with and without glucose, lysozyme (26) with and without N-acetylglucosamine (GlcNAc) or its trimer, and bovine pancreatic trypsin inhibitor (27) (BPTI). These experiments address a number of important questions on the dynamics of proteins in relation to numerical simulation studies and to molecular interpretations of their thermodynamic properties (28). The first goal in enzyme experiments must be to spectrally resolve and quantify the well-known "stiffening" of an enzyme upon ligation, and to look for evidence in support of inter-domain motions such as the much-discusssed hinge-bending mode (29) in the 1 to 30 cm^{-1} region. Initial results obtained for lysozyme (Lz) at rather high concentrations (85 or 200 mg/ml) suggest that enhanced cooperative fluctuations of the tertiary structure of Lz relative to Lz·(GlcNAc)$_n$ occur at low Q_0 where the dynamics is sampled over distances between 10 and 15 Å; to substantiate this observation it will be essential to extend the data to lower concentrations. A second objective of current time-of-flight experiments is to make contact with *ab initio* simulations of nuclear motions in proteins with time scales in the 0.05 to 100 ps region (600 to 0.3 cm^{-1}). Extensive simulations of the molecular dynamics exist for BPTI (30), and this is obviously a good candidate for first comparisons. Spherically averaged single-phonon spectra of D_2O-exchanged BPTI have recently been calculated on the basis of a normal-mode analysis (27), and the results are being used to examine the relationship between hydrogen-weighted frequency distributions derived from IN6 spectra, the single-phonon scattering according to Eq. 8, and further calculations aimed at determining two-phonon corrections. For the small, compact protein investigated, this work is expected to shed light on the key question of how closely $\overline{Z}_p(\omega)$ approximates the "true" frequency distribution in different spectral regions.

Backscattering

Both triple-axis and time-of-flight spectrometers can give a coarse view of the quasi-elastic scattering regime $|\hbar\omega| \ll E_o$, with an optimal resolution between 0.05 and 0.5 cm^{-1} that depends strongly on λ_o. Time-of-flight instruments in fact provide sets of Q_o-dependent quasi-elastic lines simultaneously with the inelastic spectra since the data recorded extend from $\hbar\omega \approx -E_o$ through the elastic peak into the inelastic region proper. In the enzyme-inhibitor studies discussed above, for example, this allows changes in the translational and rotational Brownian motion properties of protein and/or inhibitor molecules as a whole to be detected. A distinction between quasi-elastic and inelastic scattering is easily made in small molecular systems when diffusive and vibrational modes of motion are well separated in energy; this is not possible in large, irregularly folded polymeric structures exhibiting a multiplicity of modes over a wide range of characteristic times. The outer quasi-elastic and the near-inelastic scattering regimes, bracketed roughly by $0.05 < |\hbar\omega| < 50$ cm^{-1}, cover the biophysically important transition region between purely diffusive and purely "discrete" processes; much of this is a *terra incognita* where experimental information is scarce, and where biomolecular systems are thought to have evolved particularly clever ways of exploiting the coupling or competition between cooperative and dissipative modes of motion (31).

To examine the Q,ω-dependence of quasi-elastically scattered neutrons in greater detail, i.e. below the 0.05 cm^{-1} resolution limit of time-of-flight spectrometers, one must employ backscattering techniques. By going to the limit of vertical incidence ($\theta_B = 80°$ to $90°$) in Bragg reflection from a given (*hkl*)-plane of a single crystal, it is possible to select a highly monochromatic line with a wavelength spread corresponding to a few 10^{-3} cm^{-1}. This method of energy analysis forms the basis of backscattering spectroscopy (16), and is implemented in a configuration resembling that of a triple-axis spectrometer (Figure 2b). Since both monochromator and analyzer crystals operate in backscattering or near-backscattering geometry with fixed θ_B, it is necessary to devise some way of scanning the energy within a window $E_o^{min} \leq \hbar\omega \leq E_o^{max}$. This is accomplished either by the Doppler effect, i.e. by imparting a periodic translational motion of a few Hz to the monochromator, or by temperature-cycling the lattice spacing of the monochromator crystal. The spectrometer IN10 at Grenoble uses the first method in exact backscattering and covers an energy window of typically ± 0.1 cm^{-1} with an optimal resolution of 1.4×10^{-3} cm^{-1} ($Q_o = 0.07$ to 2 Å^{-1}); its sister instrument IN13 employs the second method with $\theta_B \approx 80°$ which gives a resolution of 0.03 cm^{-1} in a 3 cm^{-1} wide window that can be shifted with respect to $\hbar\omega = 0$ ($Q_o = 0.5$ to 5.5 Å^{-1}) (9). The analyzers in backscattering spectrometers consist of large-area arrays of single crystals (up to 0.3 m^2) so that the Q-resolution is rather coarse; despite of this the flux penalty for the energy resolution is severe and the count rates are one to two orders of magnitude lower than those on time-of-flight instruments.

The backscattering technique has so far been used mainly in protein hydration studies (23,24,32). The quasi-elastic data obtained here consist of sets of Q=const.

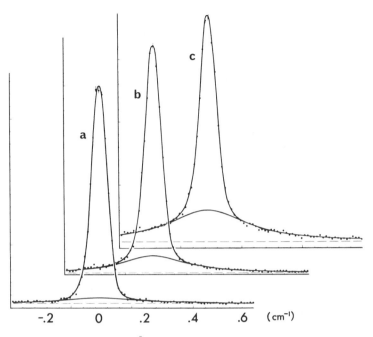

Figure 8. Quasi-elastic spectra at $Q_o=1.5 \text{ Å}^{-1}$ showing activation of sidechain mobility (broad component) during H_2O-hydration of 100% biosynthetically deuterated phycocyanin: (a) 0.06, (b) 0.19, (c) 0.32 g/g water uptake (from unpublished work at ILL in collaboration with I. Anderson and H. L. Crespi).

spectra which display the broadened elastic peaks at high resolution. These may possess shoulders, wings, or other subsidiary features essentially symmetrical with respect to $\hbar\omega=0$. It is often possible to distinguish two components in the scattering, and Figure 8 shows as an example three quasi-elastic spectra measured for fully deuterated phycocyanin (d-PC). The quantities of interest in an analysis of first-order effects are the total intensities S(Q) and the deconvoluted widths $\Delta E(Q)$ of the central and broad components, $S_c(Q,\omega)$ and $S_b(Q,\omega)$, considered separately as well as in relation to each other. The information contained in data of this kind is similar to that derived from quasi-elastic light scattering, e.g. from Rayleigh linewidth measurements, except that owing to their much lower wavelengths neutrons will "see" a great variety of diffusive motions at the atomic and molecular level over distances in the range $1<(2\pi/Q)<100$ Å.

At very low momentum transfers the scale lengths are such that line broadenings are determined essentially by the macroscopic transport coefficients (12); in the case of isotropic, infinite-domain diffusion described by Fick's law the time correlation is purely exponential, i.e. $F_{inc}(Q,t)=\exp(-D_TQ^2t)$, and the spectra therefore are Lorentzians of width $\Delta E_{inc}=2\hbar D_TQ^2$ where D_T is the translational diffusion coefficient. To observe the translational Brownian motion of H_2O molecules sorbed to d-PC, which gives rise to the very slight broadening of the elastic peaks in Figure 8, one must increase the resolution 20-fold and look at the inner quasi-elastic region

($|\hbar\omega|\lesssim 0.05$ cm^{-1}) on a low-Q backscattering spectrometer such as IN10. Using silicon (111) reflection, this instrument allows line shape changes of about 10^{-3} cm^{-1} to be detected, and hydration difference spectra then reveal an oscillatory dependence of ΔE_{inc} on Q between Q=0.3 and 1.5 Å^{-1}. With increasing hydration, the first maximum of $\Delta E_{inc}(Q)$ shifts from $Q_{max} \approx 0.4$ Å^{-1} to 0.8 Å^{-1} while gaining in intensity by a factor of two, and the following minimum between Q_{min}=0.7 and 1.1 Å^{-1} becomes progressively more shallow. This oscillatory Q-dependence, analyzed in conjunction with S(Q) measurements, is consistent with a model in which water molecules perform jump diffusion between a distribution of strong hydration sites developing into clusters or patches with increasing water uptake (32).

The broad component seen in Figure 8, i.e. $S_b(Q,\omega)$, appears to relate primarily to scattering from the rotational modes of relatively mobile, H$_2$O-hydrated and proton-exchanged side chains at or near the surface of d-PC. This component is barely detectable at low levels of hydration and becomes measurable when the water uptake has reached a level equivalent to about ½ monolayer coverage. The hydration difference broadenings derived from $S_b(Q,\omega)$ measurements range from 0.1 to 1 cm^{-1}, corresponding to rotational correlation times of the order of 10^{-10}s. The data shown in Figure 8 illustrate an important concept in neutron scattering, the "elastic incoherent structure factor" or EISF (33). This yields direct information on the geometry of proton motions whenever the self-correlation function $G_s(\mathbf{r},t)$ approaches a finite time-asymptotic distribution $G_s^R(\mathbf{r},\infty)$ within a bounded domain ΔV_R (such as the volume swept out by the protons during librational and rotational diffusion of a side chain). After Fourier transformation according to Eq. 4b the spectrum is described by

$$S_{inc}^R(\mathbf{Q},\omega) = A_o(\mathbf{Q})\,\delta(\omega) + S_*^R(\mathbf{Q},\omega) \tag{11}$$

and is seen to consist of a sharp line $\sim\delta(\omega)$ superimposed on a broader quasi-elastic feature. The EISF is given by the elastic contribution, i.e. $A_o(\mathbf{Q})$, and represents the square of the spatial Fourier transform of the stationary probability distribution, or probability of occupancy, set up by the ensemble of particle trajectories within ΔV_R as t$\Rightarrow\infty$.

Spin-Dependent Scattering

The experiments discussed so far were all performed with unpolarized neutrons, and the essential physics of the scattering process was a function of two more or less independent variables, i.e. momentum and energy transfer. During the past 5-10 years, great advances have been made in the production and handling of polarized neutron beams, and it has become possible to utilize the spin state of the neutron as a third variable in macromolecular scattering studies. The polarization $\mathbf{P}_o$ of a beam of neutrons with partially or fully aligned spins is defined as the average

$$\mathbf{P}_o = \frac{1}{N}\sum_n \mathbf{P}_{on} \tag{12}$$

where N denotes the number of neutrons and $0 \le P_o \le 1$. There are four possible ways for a neutron (spin 'up'$=+$, 'down'$=-$) to interact with a sample nucleus: $(++), (--), (+-),$ or $(-+)$. In the first two of these, the spin did not "flip" upon scattering, i.e. change direction by $180°$, whereas the last two cases imply spin-flip scattering. Detailed analysis (13) of these interactions shows that the final polarization $\mathbf{P}$ is unchanged (no spin-flip) for both coherent scattering and the scattering due to random isotope distributions. For spin-incoherent scattering, on the other hand, the resulting polarization is $\mathbf{P}=-\tfrac{1}{3}\mathbf{P}_o$ because in this case two cartesian components scatter with spin-flip and the third without. Polarization measurements thus provide a way to experimentally separate the coherent from the incoherent scattering, and are particularly valuable for organic substances because of the strong dependence of their spin properties on H/D composition.

Most of the spectrometer configurations discussed in the preceding sections could in principle be adapted for polarization analysis. The inevitable intensity losses and other limitations of existing polarization devices make this a difficult and costly proposition, however. At the ILL, two techniques employing polarized neutrons have been developed to the point of routine use. The first of these is spin-echo spectroscopy (34,35); it relies on a novel, extremely sensitive method of energy analysis (see Figure 2c): a beam of polarized neutrons aligned along $\mathbf{k}_o=k_o\hat{z}$ passes a $\pi/2$ spin-turn coil at the entrance of the guide field H_1 and starts precessing about H_1. Each neutron accumulates a total number N_1 of Larmor precessions along its flight path through H_1. Immediately after scattering, a π-coil reverses two of the spin components such that$(x,y,z) \Rightarrow (x,-y,-z)$, and the accumulated precessions of each neutron are "unwound" during passage through H_2, giving a total of N_2 precessions with the opposite sense. Any small energy change suffered by a neutron during interaction with a nucleus in the sample will result in a net difference $\Delta N=N_1-N_2$, and the probability of detection after polarization analysis is proportional to $\cos[2\pi(N_1-N_2)]$. Upon integration over the sample spectrum $S_{coh}(\mathbf{Q},\omega)$ for coherent scattering (i.e. without spin-flip) and the wavelength distribution $I(\lambda)$, the observed mean polarization is

$$\langle P_z \rangle = \int I(\lambda)\, d\lambda \int_{-\infty}^{\infty} S_{coh}(\mathbf{Q},\omega)\, \cos[t(\lambda)\omega]\, d\omega \qquad (13)$$

where $t(\lambda)=\overline{N}m\lambda^3/2\pi\,\hbar\overline{\lambda}$ defines a time scale which can be varied via the guide field strength $H_1=H_2$ ($\overline{\lambda}$ and $\overline{N}$ are averages over $I(\lambda)$). Important features of this method are, first, that the cosine Fourier transform in Eq. 13 is just $F_{coh}(\mathbf{Q},t)$ so that the Q-dependent time correlation function is measured directly. Second, there is no need to sharply monochromatize the incident beam because the quasi-elastic energy change is "encoded" in the form of a precession shift for each neutron individually, and all neutrons within a certain wavelength band ($\Delta\lambda/\overline{\lambda}\approx0.1$) are spin-focused at the analyzer position (as implied by its name, this technique may be regarded as the spatial analogue of NMR with a $\pi/2-\pi-\pi/2$ pulse sequence).

The spin-echo technique has already been used with considerable success in investigations of the dynamics of polymer (36) and macroionic (37) solutions, and it promises to give unique information in particular on coherent quasi-elastic scattering from oligomeric proteins and large biomolecular assemblies. Preliminary experiments on brome-grass virus and hemoglobin solutions (35) were undertaken between 1979 and 1981, and some initial results have also been reported on hydration-induced inter-subunit motions in phycocyanin (24). The first more detailed spin-echo study of low-frequency intramolecular modes (μs to ns range) in a rather flexible protein, immunoglobulin G, was performed recently (38). Because of the emphasis on coherent scattering, neutron spin-echo is primarily a low-Q technique ($Q_o = 0.015$ to 0.3 Å^{-1}) providing spatiotemporal data similar to quasi-elastic light scattering but over scale lengths down to 10 Å. It thus extends optical measurements into the important crossover region between local motions and hydrodynamic interactions, characterized by $0.1 < R_g Q_o < 10$ where R_g is the macromolecular radius of gyration. At low Q, the normalized intermediate scattering function for a system of coherent scatterers may be expanded according to

$$\overline{F}(Q,t) = \exp[-C_1(Q)t - \tfrac{1}{2}C_2(Q)t^2 - \ldots] \tag{14}$$

and the quantity of immediate interest is $C_1(Q)$, obtained from IN11 data as the slope of $\ln\overline{F}(Q,t)$ as $t \Rightarrow 0$. In the dilute limit, $C_1(Q)$ would be identified with $D_T Q^2$. For a more concentrated system, or significant intramolecular motions, $C_1(Q)$ contains information on the spatial correlation of the motions at times up to a few ns. In the frequency domain, this term transforms into a Lorentzian lineshape with Q-dependent width; $C_2(Q)$ describes deviations from exponential decay of the time correlation and gives a non-Lorentzian component. As an example, Figure 9 shows

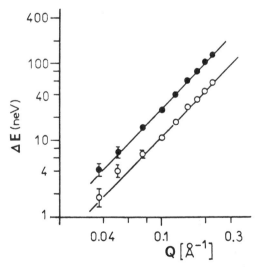

Figure 9. Spin-echo measurements of the quasi-elastic linewidths for hemoglobin in D$_2$O buffer at concentrations of 100 (●) and 260 (○) mg/ml (measured at ILL by Y. Alpert and F. Mezei; reproduced from Ref. 39).

the Q-dependence of linewidths for two hemoglobin solutions (39); these results demonstrate that at high concentrations the diffusion coefficient is reduced considerably but that this is not accompanied by changes in the dynamics: the Lorentzian scattering law, i.e. simple exponential decay of the intermediate scattering function, holds at least down to $2\pi/Q \approx 25$ Å.

It is also possible now to perform time-of-flight spectroscopy with polarized neutrons (40). A straightforward way of doing this is to polarize the incident beam (all spins 'up') before or after mechanical chopping, and to place an array of polarization analyzers in the flight paths between sample and D_n (compare Figure 4). The non-flip cross-section proportional to $[S_{coh}(\mathbf{Q},\omega) + \frac{1}{3}S_{inc}(\mathbf{Q},\omega)]$ is measured with this setup when only neutrons with 'up' spins reach D_n; from a second run, upon changing the analyzer configuration for the detection of neutrons with 'down' spins, one obtains $S_{inc}(\mathbf{Q},\omega)$ alone. A more sophisticated variant of this method employs a random sequence of polarized neutron bunches which are generated by an electronically driven pseudo-statistical flipper. A versatile instrument for the exploitation of such techniques has been commissioned recently, and is being used to study the molecular dynamics of plastic crystals (40) and selectively deuterated polymers (41). The $\hbar\omega$ range over which efficient polarization analysis can be performed extends from -15 cm^{-1} to about 100 cm^{-1} ($\Gamma_o \approx 0.5$ cm^{-1}, $Q_o = 0.1$ to 2.2 Å^{-1}), and it is obviously of great interest for future biomolecular work to be able to separate S_{coh} and S_{inc} experimentally in this Q,ω-region.

Concluding Remarks

Slow-neutron scattering was born in Rome exactly half a century ago with Fermi's discovery (42) that paraffin and other hydrogenous substances act as "moderators" in which fast neutrons rapidly lose their energy by thermalizing collisions with bound protons. It was only during the late 1960s, however, that detailed neutron scattering investigations of the structure and dynamics of large molecules became feasible. As the result mainly of instrumental advances, biomolecular applications of neutron spectrocopy are now being pursued more actively. Considering the experimental techniques and representative results discussed in this paper, there can be little doubt that neutron scattering will eventually rank among the three or four major techniques capable of contributing spatially and/or spectrally resolved data to the growing field of biomolecular dynamics. Whereas the Q,ω-range and resolving power of instruments available for developing these applications are excellent, the flux levels are still rather low. Large increases in thermal neutron fluxes are predicted for the spallation sources currently under construction or at the design stage (43,44). Higher rates of data collection will make it possible to study the effects of changes in temperature and other important sample parameters. These sources, by virtue of their pulsed nature, are likely to stimulate novel approaches to probing $S(\mathbf{Q},\omega)$ by intense polychromatic neutron bursts with duty cycles that are comparable with the characteristic relaxation times of a number of interesting biophysical processes.

Acknowledgements

We are indebted to the Institut Laue-Langevin for the use of facilities, and to several of its staff members for helpful discussions during the preparation of this survey. Thanks are due to Drs. P. Martel, B.M. Powell, and F. Mezei for permission to reproduce Figures 3 and 9.

References and Footnotes

1. *Mobility and Function in Proteins and Nuclei Acids.* Eds. Porter, R., O'Connor, M. and Whelan, J., CIBA Foundation Symposium 93, Pitman, London (1983).
2. Petsko, G.A. and Ringe, D., *Ann. Rev. Biophys. Bioeng. 13,* 331 (1984).
3. Albanese, G. and Deriu, A., *Riv. Nuovo Cimento 2,* 1 (1979).
4. Boutin, H. and Yip, S., *Molecular Spectroscopy with Neutrons.* MIT Press, Cambridge, MA (1968).
5. Dahlborg, U. and Rupprecht, A., *Biopolymers 10,* 849 (1971).
6. Middendorf, H.D. and Blaurock, A.E., *UK Neutron Beam Res. Committee, Ann. Rpt.,* p. 35 (1974). Randall, J. T. and Gilmour, S., unpubl. observations on C-phycocyanin (1975).
7. Hecht, A.M. and White, J.W., *J. Chem. Soc. Faraday Trans. II 72,* 439 (1976).
8. Middendorf, H.D., *Ann. Rev. Biophys. Bioeng. 13,* 425 (1984).
9. *Neutron Research Facilities at the ILL High Flux Reactor.* Eds. Maier, B. and Briggs, G.A., Institut Laue-Langevin, Grenoble (1983).
10. A useful up-to-date compilation may be found in Ref. 44.
11. Chothia, C., *Nature 254,* 304 (1975).
12. Egelstaff, P.A., *An Introduction to the Liquid State.* Academic, London (1967).
13. Marshall, W. and Lovesey, S.W., *Theory of Thermal Neutron Scattering.* Clarendon, Oxford (1971).
14. van Hove, L., *Phys. Rev. 95,* 249 (1954).
15. Martin, P.C., in *Problèmes à N Corps.* Eds. de Witt, C. and Balian, R., Les Houches Lectures, Gordon & Breach, New York (1967).
16. Springer, T., *Springer Tracts in Mod. Phys. 64,* 1 (1972).
17. "Inelastic scattering" is used here in its narrow sense; this term is also applied broadly to all conditions in which some of the scattered radiation differs in energy from E_0, by however small an amount.
18. Powell, B.M., and Martel, P., *Biophys. J. 34,* 311 (1981); *Chem. Phys. Letters 67,* 165 (1979).
19. Trantham, E.C., Rorschach, H.E., Clegg, J.S., Hazlewood, C.F., Nicklow, R.M. and Wakabayashi, N., *Biophys. J. 45,* 927 (1984).
20. We define Γ_0 as the half-width at half-maximum of the elastic peak, measured e.g. for a vanadium standard.
21. Martel, P., *J. Biol. Phys. 8,* 1 (1980); *Biochim. Biophys. Acta 714,* 65 (1982).
22. Rupley, J.A., Gratton, E. and Careri, G., *Trends Biochem. Sci. 8,* 18 (1983).
23. Jenkin, G.T., Miller, A., White, J.W. and White, S.W., *ILL Ann. Rep. Annex* pp. 320 & 373 (abstr.), Institut Laue-Langevin, Grenoble (1977); White, S.W., D. Phil. thesis, Oxford Univ. (1977).
24. Middendorf, H.D., Randall, J.T. and Crespi, H.L., in *Neutrons in Biology,* p. 381, Ed. Schoenborn, B.P., Plenum, New York (1984).
25. Jacrot, B., Cusack, S., Dianoux, A.J. and Engelman, D.M., *Nature 300,* 84 (1982).
26. Middendorf, H.D., et al. (unpublished results).
27. Cusack, S., *Comments Molec. Cell. Biophys.* (in press).
28. Cooper, A., *Prog. Biophys. Mol. Biol.* (in press).
29. McCammon, J.A. Gelin, B. R., Karplus, M. and Wolynes, P.G., *Nature 262,* 325 (1976).
30. McCammon, J.A., *Rep. Prog. Phys. 47,* 1 (1984).
31. Careri, G., Fasella, P. and Gratton, E., *Ann. Rev. Biophys. Bioeng. 8,* 69 (1979).
32. Middendorf, H. D. and Randall, J.T., *Phil. Trans. R. Soc. Lond. B290,* 639 (1980).
33. Leadbetter, A.J. and Lechner, R.E., in *The Plastically Crystalline State,* p. 285, Ed. Sherwood, J.N., Wiley, New York (1979).
34. Mezei, F., *Physica 120B,* 51 (1983).

35. Neutron Spin Echo, *Lecture Notes in Physics Vol 128*, Ed. Mezei, F., Springer, Berlin (1980).

36. Nicholson, L.K., Higgins, J.S. and Hayter, J.B., *Macromolecules 14*, 836 (1981).

37. Hayter, J.B. in *Scattering Techniques Applied to Supramolecular and Nonequilibrium Systems*, p. 49, Eds. Chen, S.-H, Chu, B., and Nossal, R., Plenum, New York (1981).

38. Alpert, Y., Cser, L., Farago, B., Franek, F., Mezei, F. and Ostanevich, Y.M., *Eur. J. Biochem.* (submitted).

39. Mezei, F., in *Neutron Scattering—1981*, p. 379, Ed. Faber, Jr., J., Am. Inst. Phys., New York (1982).

40. Gerlach, P., Schärpf, O., Prandl, W. and Dorner, B., *J. de Physique 43*, C7-151 (1982).

41. Oberthür, R.C., Rawiso, M., and Schärpf, O.; Higgins, J., Gabrys, B. and Schärpf, O. (work in progress).

42. Fermi, E., Amaldi, E., Pontecorvo, B., Rasetti, F. and Segre, E., *Ric. Scientifica 5(2)*, 282 (1934).

43. Fender, B.E.F., Hobbis, L.C.W. and Manning, G., Phil. *Trans. R. Soc. Lond. B290*, 657 (1980).

44. Windsor, C.G., *Pulsed Neutron Scattering*, Taylor & Francis, London (1981).

Structure & Motion: Membranes, Nucleic Acids & Proteins,
Eds., E. Clementi, G. Corongiu, M. H. Sarma & R. H. Sarma,
ISBN 0-940030-12-8, Adenine Press,   Copyright Adenine Press, 1985.

Investigation of Protein Dynamics by Mössbauer Spectroscopy: The Time Dependence of Protein-specific Motions

Fritz Parak

Institut für Physikalische Chemie der Westfälischen Wilhelms-
Universität Münster, Schloßplatz 4-7, 4400 Münster, Fed. Rep. Germany

Abstract

The iron atom of iron containing proteins can be used as a label for the dynamics of the molecule. Mössbauer spectroscopy allows the determination of the mean square displacement of the iron as well as the time scale at which the motion occurs.

Investigation of deoxygenated myoglobin crystals shows that well below 200 K the molecules are frozen into conformational substates. At higher temperatures a gain in entropy drives the molecules into a "transition state" where slow protein specific motion occur with a characteristic time longer than 10^{-9}s. These motions are strongly overdamped and can be understood as restricted Brownian diffusion controlled by the internal viscosity of the molecule and the viscosity of the environment.

A comparison with theoretical investigation shows that computer simulations give correct amplitudes but fail in the description of the time dependence of the motions.

Introduction

A complete description of the dynamics of a protein molecule requires the knowledge of the coordinates of all atoms and the velocities and accelerations of the atoms at any time. Such a description, if possible, would however, not focus on the peculiarities of protein specific motions correlated to functional properties. For a more practicable approach one may look for types of motions present also in small molecules and for additional modes which are protein-specific. The different modes of motions can be described by the amplitudes $<x^2>^{1/2}$ of the moving atoms and by characteristic times, τ, necessary to move over the full amplitude.

The upper limit for stretching vibrations of small molecules lies at frequencies up to $10^{14}s^{-1}$ yielding characteristic times of 10^{-14}s. Vibrations of H-atoms against the whole molecule represent the upper limit. The lower limit of vibrations are usually modes determined by the dimensions of the molecule and the dispersion relation correlating wavelengths with frequencies via the velocity of sound. The diameter of the molecule (about 30 A for myoglobin) yields $\lambda/2$ of the longest mode of motion. With a velocity of sound between $6.4 \cdot 10^5$ and $6.9 \cdot 10^4$ cm/s(1) one obtains frequencies between 10^{12} and 10^{11} s^{-1}. Assuming that the longest quasi acoustical mode follows the 8 α-helices in myoglobin the value of $\lambda/2$ increases to about 190 A yielding as the lowest frequency $1.8 \cdot 10^{10}$s^{-1}.

From this estimation it is not surprising that a normal mode analysis of pancreatic typsin inhibitor gives as the lower limit frequencies of about $2 \cdot 10^{10}$s^{-1}(2). The molecular dynamics simulations using potentials which are much more realistic give practically the same lower limit (3). Does this mean that the essential motions in a protein cannot occur with a characteristic time longer than 10^{-10}s?

Numerous experimental investigations show the importance of protein specific motions on a much longer time scale. NMR investigation yield jump rates for the flip of tyrosin side chains between two possible positions between 10 and 10^6s^{-1} (4). In the case of myoglobin, CO photoflash experiments prove the functional significance of slow structural relaxations (5). At temperatures below 200 K the myoglobin molecules are frozen in conformational substates representing slightly different configurations of the protein. Above 200 K structural relaxations of the molecule occur. Relaxation rates of 10 to 10^4s^{-1} at 230 K determine the rebinding of CO after the photoflash. Extrapolation of the relaxation rates to room temperature are consistent with 10^8s^{-1} as the order of magnitude.

Theoretical and experimental results give an inconsistent picture of the timescale of protein specific motions. In this contribution Mössbauer measurements on myoglobin crystals are used to explain the discrepancy and to give a physical picture of protein specific motions with characteristic times longer than 10^{-10}s.

Materials and Methods

The discussion is based on Mössbauer experiments and X-ray structure investigation of myoglobin crystals.

For the Mössbauer absorption spectroscopy the heme iron of myoglobin had to be replaced by the Mössbauer isotope ^{57}Fe (6). Absorption experiments were performed on deoxygenated myoglobin between 4.2 K and 293 K (7). The data analysis used two features of the Mössbauer spectra: i) In the whole temperature range the mean square displacement, $<x^2>$, of the iron can be obtained from the logarithm of the "absorption area" of the Lorentzian energy distribution of the spectrum. Note that only displacements are measured which have a characteristic time of motion faster than 10^{-7}s. ii) Above 200 K the energy distribution of the Mössbauer spectra changes. The line width increases and indicates slow diffusive modes with a characteristic time of about 10^{-7}s. Moreover, additional broad absorption lines appear, labeling pronounced motions with a characteristic time varying from $2 \cdot 10^{-8}$s to $3 \cdot 10^{-9}$s between 200 K and 283 K.

Mean square displacements of the atoms of met myoglobin obtained from X-ray structure analysis are available in the literature. Values at 300 K, 275 K and 250 K are given in (8) and values at 80 K in (9). Additionally, X-ray structure analysis of myoglobin was performed at 300 K, 190 K, 165 K and 120 K. The analysis of the data is still not finished but preliminary $<x^2>$-values of these investigations were partly used.

Results and Discussions

Labeling protein dynamics at one atom in the molecule yields a displacement s(t) which is a superposition of contributions from various modes of motions.

$$\vec{s}(t) = \sum_i a_i \vec{x}_i(t) \tag{1}$$

$\vec{x}_i(t)$ is the amplitude of the mode i coupled with the constant a_i to the atom under consideration. A quite general equation of motion for isotropic amplitudes using a one dimensional description for $\vec{x}_i$ is given by:

$$\ddot{x}(t) + 2\beta\dot{x}(t) + c\ fkt(x) = F(t) \tag{2}$$

Under the influence of driving force F(t) the atom performs a damped motion characterized by the friction β and a restoring force depending on the displacement x from an average position. The type of motion depends on the nature of β, c fkt(x) and F(t).

In the system under consideration the driving force F(t) comes from the temperature. Considering a system containing protein molecules and water at room temperature, F(t) is a random fluctuating force or torque due to the Brownian motion of the water molecules (compare for instance (1)). In order to get an analytic solution of equ. (2) one starts with a harmonic backdriving force:

$$c\ fkt(x) = \omega^2 x \tag{3}$$

where ω gives the frequency of the oscillations (10).

A solution of equ. (2) together with equ. (3) allows the calculation of the mean square displacement, $<x^2>$, of the atom under consideration. One can now apply this scheme to the analysis of the motion of iron in myoglobin characterized by the mean square displacement as obtained from Mössbauer experiments, $<x^2>^\gamma$, or X-ray structure analysis, $<x^2>^x$.

In the simplest case the damping is rather small ($\beta \ll \omega$) (11). The Mössbauer spectrum is determined by a well-defined Lorentzian corresponding to elastic inter-action of the ^{57}Fe nucleus with the γ-radiation. Inelastic processes reducing the area of Mössbauer spectrum give rise to energy shifted lines not visible in the spectrum (phonon interaction).

The $<x^2>^\gamma$-values determined by Mössbauer spectroscopy between 4.2 K and 200 K can be understood from this case. In a picture of normal mode analysis or protein dynamics simulations this corresponds to relative high frequency modes. The corresponding amplitudes are abbreviated $<x^2_v>^{1/2}$.

Above 200 K the $<x^2>$-values increase drastically with temperature yielding 0.064 A^2 at 298 K (7). The additional modes of motions contribute the value $<x^2_t>$

$$<x^2>^\gamma = <x^2_v> + <x^2_t> \tag{4}$$

This large value at physiological conditions is only measured if the myoglobin molecule is surrounded by water. In dry protein the amplitudes are much smaller (6,12) and follow the temperature dependence obtained between 4.2 K and 200 K for $<x^2_v>$. The large $<x^2_t>$ at physiological conditions comes, therefore, from protein-specific modes only present in the active molecule.

An analysis of the spectral distribution of the Mössbauer experiments shows that $<x^2_t>$ comes from strongly overdamped motions ($\omega \ll \beta$) (11). For this case the coefficients of equ. (2) and (3) are correlated by:

$$\frac{\omega^2}{2\beta_t} = \alpha_t \tag{5}$$

Frequency and friction of the motion can no longer be separated. The motion is described by the relaxation rate α_t. The amplitude $<x^2_t>^{1/2}$ is no longer reached by simple oscillations, but by a diffusive type of motion. The atom under consideration gets from its neighbours an accelleration due to a stochastic fluctuation. The corresponding motion is, however, damped out rather quickly and yields only a small displacement. The next fluctuation may move the atom in the same or in the opposite direction. In contrast to a true diffusion the mean square displacement $<x^2_t>$ is, however, limited due to the restoring force $\omega^2 x_t$. The atom needs about α^{-1} seconds to move over the distance $<x^2_t>^{1/2}$. Protein specific dynamics can be understood by overdamped Brownian diffusion in restricted space. Obviously it can be viewed that the Brownian motion of the water determines the internal motions of the protein molecule.

One has now to compare the characteristic time of protein specific motions (around $\alpha^{-1} = 3\cdot10^{-9}$s at room temperature) with theoretical calculations (2,3,13). Assuming that the calculations yield the essential frequencies in the molecule the pronounced mode of motions labeled by Mössbauer spectroscopy is certainly correlated with maxima in the low frequency regime. Taking from (2,3) a value of about 16 cm^{-1} as the order of magnitude gives $\omega \approx 3\cdot10^{12}$s^{-1}. Together with $\alpha = 3\cdot10^8$s^{-1} one obtains $\beta \approx 10^{16}$s^{-1} as the order of magnitude for the friction. This large value cannot be understood from the usual model of hard collisions. An appropriate picture for large damping comes from the laminar flow in a liquid of high viscosity. In the theory a correlation length L is introduced measuring the distance from the moving particle where the viscous liquid is still not at rest (14). For myoglobin one obtains a length of about 300 A (15). This is about 10 times the diameter of the molecule. Protein dynamics are essentially influenced by the cellular medium surrounding the molecule. The important influence of the viscosity of the environment on the kinetics of CO-rebinding in myoglobin was described in (16).

In order to visualize the time dependence of protein specific dynamics the adiabatic contribution to the friction can be put into a potential as shown in Fig. 1. The one dimensional drawn potential has a parabolic envelope accounting for harmonic restoring forces. Deep and shallow potential wells represent the damping. Falling into a trap stops the fluctuation in the molecule for some time and allows a change of the direction of motion if the stochastic force F(t) of eqn. (3) pushes the molecule again out of the trap. Two types of traps are drawn in Fig. 1. The deep potential wells c represent conformational substates where the molecule can be

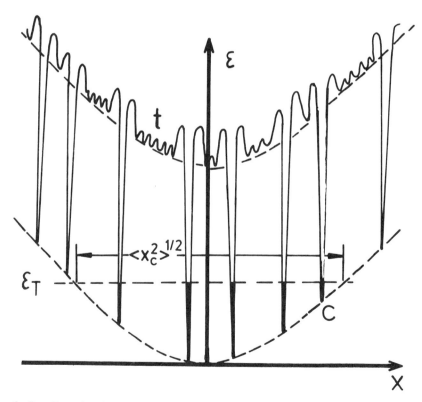

Figure 1. One dimensional representation of the potential describing the adiabatic part of the friction slowing down protein specific dynamics. The conformational substates C are represented by deep traps in which the molecule can be frozen in at low temperature. At the temperature T all conformational substates below $\mathcal{E}_T$ are populated. The distribution of conformational substates is measured by $<x_c^2>$.

frozen in at low temperatures. The existence of these states follows from CO flash experiments (5) and from X-ray structure investigation on myoglobin (8,9,17). Obviously a fluctuation of a molecule from one conformational substate to another goes via the top of the barrier between these states. As usual, this top of the barrier is called transition state.

Mössbauer experiments in comparison with X-ray structure analysis show that at room temperature a molecule is most of the time in the transition state (18). Therefore, the transition state contains a much larger phase space volume as the conformational substates. This means that the transition state represents a large gain of entropy with respect to a conformational substate. Protein dynamics can be understood as an entropy driven process (19).

A problem arises if one compares the mean square displacements $<x^2>$, measured by X-ray structure analysis, Mössbauer spectroscopy and calculations from a normal mode analysis (2) and from protein dynamics simulations (3). As for instance shown in (20) calculations agree reasonably well with X-ray structure results which suggest that the displacements come from motions with a frequency around $3 \cdot 10^{12} s^{-1}$ and larger. However, Mössbauer experiments show that at least at the position of iron only about 16% of the mean square displacement determined by X-ray structure analysis comes from fast vibrations. About 43% is due to strongly overdamped motions as described before and the remaining 41% stem from motions with characteristic times longer than 10^{-7}, including also static disorder. How can this discrepancy between theoretical values and Mössbauer results be removed?

One may assume that the measurements at the iron does not reflect protein dynamics but local motions of the iron or motions of the molecule as moiety. An exotic motion of the iron only with respect to its nearest neighbours above 200 K would reflect itself in the hyperfine interactions measured by the quadrupole splitting of the Mössbauer spectrum. As shown in (21) no anomalies occur at 200 K. A motion of the entire molecule would influence all atoms of the molecule. Comparison of X-ray and Mössbauer data (7,8) rules out this possibility. A comparison of Mössbauer measurement on deoxy hemoglobin and hemochrome where the iron is linked additionally to the nitrogen of the distal histidine shows that the iron behaves similar as the helix to which it is bound (22). Recently the second order Doppler shift of iron in myoglobin crystals was determined by Mössbauer spectroscopy which is a measure of the mean square velocity $<v^2>$, of the iron. No change in the temperature dependence was found above 200 K (23). The determined velocities have to be attributed to the fast vibrations which yield $<x^2_v>$. This shows in turn that $<x^2_t>$ comes from displacements with a significantly longer characteristic time of motion. Summarizing, there is at present no experimental hint that the interpretation of the Mössbauer results as given before is wrong.

One has now to look on the limitations of the theoretical calculations. Mean square displacements are obtained from a time average performed over sampling times long enough to allow the atom under considerations to travel over the distance

$<x^2>^{\frac{1}{2}}$. Only if the sampling time fulfills this condition can displacements be compared with the values obtained by an ensemble average determined from X-ray structure analysis. The agreement between X-ray results and the calculations shows that the sampling times of $5 \cdot 10^{-11}$s as used in protein dynamics calculations are indeed long enough. A potential according Fig. 1 can, however, not be explored in this time. Protein dynamics simulations obviously use potentials where the damping which slows down the characteristic time of motion is drastically underestimated. This is also shown by the fact that protein dynamics simulations get a protein interior viscosity of about $0.3 \cdot 10^{-2}$P (13) while Mössbauer data yield values of 40 P and more. The potentials used in the calculations are closer to the envelope of the potential obtained from Mössbauer experiments. This envelope is of course not necessarily always harmonic as shown in Fig. 1 at the position of the iron. In this context one has to remember that the $<x^2>$-values obtained from X-ray measurements depend on ω but not on β_t. For a harmonic envelope one obtains:

$$<x^2_c> = \frac{k_B T}{m\omega^2} \qquad (6)$$

k_B is the Boltzman constant, T the temperature and m the mass of the moving particle. X-rays make a flash photograph of the mean square displacements in the different molecules yielding results insensitive to the characteristic times of motions.

The underestimation of the damping in the theoretical calculations comes mainly from the fact that the experimentally determined high friction β_t is not only due to structural relaxations occurring within the molecule but also to the damping in the surroundings at distances up to 10 molecule diameters away. At present such a large environment is not included in the calculations.

Conclusions

Motion in protein includes characteristic times from 10^{-14}s to several seconds and even unmeasurably long times. Rebinding experiments of CO in myoglobin show that the kinetics in this molecule are governed by rather slow structural relaxations with characteristic times between 10^{-9} and 10^{-2}s. The structural relaxation depends on the viscosity of the surrounding medium. Mössbauer experiments prove protein specific motions to be strongly overdamped. They can be described by restricted Brownian diffusion with a characteristic time longer than 10^{-9}s. The large damping is due to molecular internal relaxations and to relaxations with a correlation length up to 300 A. At present protein dynamics simulations do not describe the time dependence of protein specific motions. The calculations are, however, capable to obtain the amplitudes of the displacements, analyzing the envelope of the potential which governs the motion.

This work was supported by the Deutsche Forschungsgemeinschaft and the Bundesministerium für Forschung und Technolgie. I gratefully acknowledge valuable discussions with H. Frauenfelder, E. W. Knapp and L. Reinisch.

References and Footnotes

1. See for instance, Peticolas, W. L., in *Methods in Enzymology, Vol. 61,* p. 425; ed.: C. H. W. Hirs and S. N. Timasheff, Academic Press (1979).
2. Go, N., Noguti, T. and Nishikawa, T., *Proc. Natl. Acad. Sci. USA 80,* 3696 (1983).
3. Levy, R. M., Perahia, D. and Karplus, M., *Proc. Natl. Acad. Sci. USA 79,* 1346 (1982).
4. Wagner, G., DeMarco, A. and Wüthrich, K., *Biophys. Struct. and Mechanisme 2,* 139 (1976).
5. Austin, R. H., Beeson, K. W., Eisenstein, L., Frauenfelder, H. and Gunsalus, I. C., *Biochem. 14, 5355* (1975).
6. Parak, F. and Formanek, H, *Acta Cryst. A 27,* 573 (1971).
7. Parak, F., Knapp, E. W. and Kucheida, D., *J. Mol. Biol. 161,* 177 (1982).
8. Frauenfelder, H., Petsko, G. A. and Tsernoglou, D., *Nature (London) 280,* 588 (1979).
9. Hartmann, H., Parak, F., Steigemann, W., Petsko, G. A., Ringe Ponzi, D. and Frauenfelder, H., *Proc. Natl. Acad. Sci. USA 79,* 4967 (1982).
10. A mathematical treatment which allows also to handle unharmonic backdriving forces was recently given by Nadler, W. and Schulten, K., *Phys. Rev. Lett. 57,* 1712 (1983).
11. The detailed analysis is given in: Knapp, E. W., Fischer, S. F. and Parak, F., *J. Chem. Phys. 78,* 4701 (1982).
12. Krupyjanskii, Yu, Parak, F., Goldanskii, V. I., Mössbauer, R. L., Gaubmann, E. E., Engelmann, H. and Suzdalev, I. P., *Z. Nat. Forsch. 37c,* 57 (1982).
13. Swaminathan, S., Ichiye, T., van Gunsteren, W. and Karplus, M., *Biochemistry 21,* 5230, 1982.
14. Landau, L. D. and Lifeschitz, E. M. *Lehrbuch d. theoretischen Physik, Vol. 6,* Hydrodynamik, Academie Verlag Berlin (1966).
15. Knapp, E. W. and Parak, F. (1984), personal communication.
16. Beece, D., Eisenstein, L., Frauenfelder, H., Good, D., Marden, M. C., Reinisch, L., Reynolds, A. H., Sorensen, L. B. and Yue, K. T., *Biochem, 19,* 5147 (1980).
17. Parak, F., et al (1984), *X-ray structure investigation of myoglobin between 40 K and room temperature,* in progress.
18. Knapp, E. W., Fischer, S. F. and Parak, F., *J. Physic, Chem. 86, 5042* (1982).
19. Parak, F. and Knapp, E. W. (1984) *Proc. Natl. Acad. Sci. USA 81,* in press.
20. Northrup, S. H., Pear, M. R., McCommon, J. A., Karplus, J. A. and Takano, T., *Nature (London) 287,* 659 (1980).
21. Parak, F., Finck, P., Kucheida, D. and Mössbauer, R. L., *Hyperfine Interactions 10,* 1075 (1981).
22. Mayo, K. H., Kucheida, D., Parak, F. and Chien, J. C. W., *Proc. Natl. Acad. Sci. USA 80, 5294* (1983).
23. Reinisch, L., Heidemeier, J. and Parak, F., in progress.

Structure & Motion: Membranes, Nucleic Acids & Proteins,
Eds., E. Clementi, G. Corongiu, M. H. Sarma & R. H. Sarma,
ISBN 0-940030-12-8, Adenine Press,   Copyright Adenine Press, 1985.

Topological Aspects of Conformational Transformations in Proteins—Preferential Pathways of Protein Folding

P. De-Santis, A. Palleschi, S. Chiavarini
Dipartimento di Chimica, Universita di Roma
La Sapienza—00100 Rome, Italy

Abstract

The paper addresses the problems of general trends in protein folding and try to establish such trends with respect to a peptide-ribbon topological properties. The topological parameters evaluated on a set of 49 proteins show striking regularities that extend beyond the secondary structures actually present and are interpreted as a manifestation of the topological invariance of conformational transformations in globular proteins.

Introduction

It is generally accepted that the primary structure of a protein is designed not only to ensure the appropriate stability to the tertiary structure, but also to direct the folding processes along preferential pathways short cutting the majority of metastable conformations before settling in the final native structure.

In fact, protein folding involves, apart the overcoming of energy barriers, entangled movements of the polypeptide chain which become progressively hampered by the growing in importance of long range interactions and then only concerted transformations can take place. In addition, cooperativity in conformational transformations could be forced by the interactions between protein regions and ribosomes, as well as simply by excluded volume effects.

We have shown that such a class of transformations could be selected on the basis of their topological characters and recognized "a posteriori" using a suitable representation of the tertiary structure (1,2).

The present paper investigates the topological aspects of the conformations in polypeptides to get an insight into the protein folding process and put forward some hypotheses about possible conformational pathways the polypeptide chain undertakes to reach the tertiary structure.

Energy constraints in polypeptide chain transconformations.

The representation of the tertiary structure in terms of conformational parameters, namely the rotation angles ϕ and ψ around the N-C$_a$ and C$_a$-C bounds respectively, allowed us to recognize van der Waals interactions as the main factor responsible of the thermodynamic stability of the protein structures. In fact, the representative points of the local conformations of the amino acids residues lie into regions of the energy conformational map nearby minima of the van der Waals interactions evaluated using semi-empirical energy functions.

A feature of the energy maps relevant to the considerations will be made in the present paper, is that complete rotations around N-C$_a$ and C$_a$-C bonds, which periodically reproduce the same conformations, appear to meet necessarily with high energy barriers. As an example Fig. 1 shows the conformational energy surface in terms of the rotation angles ϕ and ψ in the case of the l-alanine dipeptide as evaluated using our best set of van der Waals potential functions (3). The gride corresponds to increments of 10° on the ϕ and ψ axes; the maxima are cut off at the value of 10 Kcal/mole over the deepest minimum of the right-handed α-helix (α_R).

The saddles between α_R, β and α_L conformations are easily recognized.

Whilst higher energy values would be considered with less confidence than the lower values in the conformational energy calculations, nevertheless, it appears highly probable that the local conformational transitions between the sterically allowed conformations follow just one of the two geometrically equivalent pathways: e.g. the $\alpha_R \Rightarrow \beta$ transition should occur by increasing ψ in counterclockwise direction of about 180°, because the geometrically equivalent rotation in the opposite

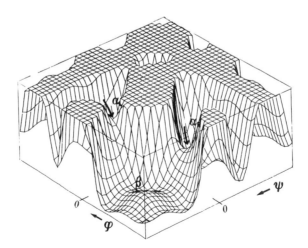

Figure 1. Van der Walls conformational energy surface of l-alanine dipeptide. The grid is 10° for both the rotation angles ϕ and ψ; the maxima are cut off at the value of 10 Kcal/mole over the deepest minimum corresponding to the right-handed α-helix.

direction is obstructed by an energy barrier at least 10 Kcal/mole high. Therefore the majority of conformational transformations in a polypeptide chain would occur within the conformational space represented in Fig. 1; in this case in fact the boundary energy barriers are expected to be higher than in isolated dipeptide units.

It is easy to realize that hindering cyclic changes of both ϕ and ψ imposes stringent correlations in the conformational motions and transformations of the polypeptide chain, when restrained by long range intramolecular forces as well as constrained by interactions with the environments.

Topology of the conformational transformations in a polypeptide chain.

Restrictions on the single rotations around the skeletal bonds as well as the long range interactions within the polypeptide chain and the "viscosity" of the site where conformational transformations occur, brake the chain motions, causing a strong reduction of the conformational freedom. As a result, the number of accessible metastable conformations decreases, and the transformation process follows optimized conformational pathways where internal rotation correlations satisfy these constraints.

In order to select such constrained conformational pathways in a stretch of polypeptide chain when conformational transformations occur, we have adopted an useful representation of the polypeptide chain as a continuous ribbon enveloping the peptide groups, so that one edge fits the oxygen atoms and the other one the hydrogen atoms, as shown in Fig. 2; thus if one starts from a polypeptide standard

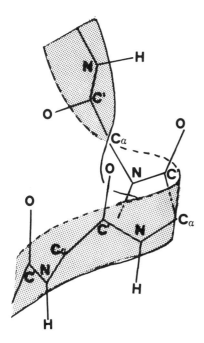

Figure 2. Representation of the polypeptide chain as a continuous ribbon enveloping the peptide groups.

conformation (we have chosen α_R), its transformation into whichever chain conformation is associated to a definite change of the twisting and writhing of the peptide-ribbon (if the minimum conformational pathway is followed) namely, to the ribbon topology.

In fact, the ribbon topology (4,5) is characterized by three parameters: the twisting number T, the writhing number W and the linking number L. T is the twist of the ribbon around its axis: when the axis is a straight one, T is given by the integral of the azimuthal angle in number of cycles; when the axis writhes in the space, it can be obtained by the related curvilinear integral along the axis.

W represents the writhe of the ribbon axis: when it follows a cylindrical helix,

$$W = \nu - T = \nu(1 - \sin \epsilon)$$

where ν is the number of helical turns and ϵ is the pitch angle.

The linking number L is the number of times one edge of the ribbon is linked to the other edge: for a closed ribbon L is an integer number and a "topological invariant", namely, it does not change for any deformation of the ribbon; it is related to the twisting and writhing numbers by the relation:

$$L = T + W$$

In the case of a polypeptide chain we assumed the ribbon as a linear mosaic of peptide groups and adopted the above equation, strictly valid for continuous ribbons, in order to evaluate the "topological parameters of a polypeptide conformation". Thus, we assumed as a measure of T the sum of the angles of rotation T_i around the i-th virtual bond after this bond was made collinear to the next one by rotating through the virtual bond angle θ_i.

In order to evaluate W, we considered the ribbon as folded around an ideal line given by the sequence of local axes (h_i) allowing the coincidence of the i-th peptide group on the nearest neighbour one by a simple rotation of the angle ϕ around h_i(6). In the case of helical structures this line coincides with the helical axis. With this approximation the linking number L of a polypeptide chain is given by

$$L = \sum_1^u L_i = \sum_1^u T_i + \sum_1^u \phi_i(1 - d_i/l \sin(\theta_i/2))$$

where d_i is the projection of the virtual bond on the axis h_i, l and θ_i the virtual bond length and angle respectively, and n the number of amino acid residues; this is, in fact, the writhing number per peptide group of a helical structure. The value of L_i can be defined in terms of the conformational parameters ϕ_i and ψ_i. If the α_R conformation is assumed as the standard structure of a polypeptide chain, L-L$_a$ represents the relative rotation (in number of turns) of the terminals when the chain

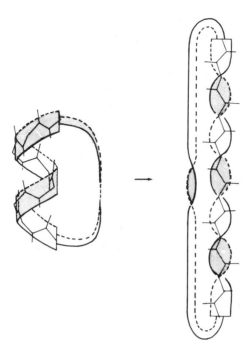

Figure 3. Pictorial drawing of the transformation of 6 amino acid residues from α_R to β(pleated) structure.

assumes a final structure (characterized by the linking number L), starting from the α-helical structure.

In fact, let's consider the polypeptide chain in α_R conformation formally closed using, for example, a relaxed elastic ribbon connecting the terminal peptide groups; meanwhile conformational transformations occur, the mutual rotations of terminals accumulate into the ribbon and are easily evaluated since the operation is eventually made of eliminating the interfolding between the ribbon and the polypeptide chain (always possible). As an example, the transformation of 6 residues from α_R to β (pleated) structures is pictorially illustrated in Fig. (3).

The transformation in a β (twisted) structure, normally found in proteins, requires a further twisting of the peptide ribbon and, obviously, an equivalent twisting increment (in the opposite way) of the elastic ribbon of about 180°.

Since the result is independent of the order of the rotations, L-L$_a$ is a "property" of the polypeptide conformation because independent of the particular pathway followed in the conformational transformation, providing that it lies within the energy borders of Fig. (1).

As a matter of fact, we have tested the value of the linking number given analitically in many cases, using mechanical models and an elastic ribbon, and always found a good agreement. Of course, our formula gives correct values in the case of transformations between helical conformations; it works however satisfactorily also in the cases of non regular conformations, at least for those sterically allowed.

Therefore, polypeptide conformations having equal linking numbers can be defined as "topologically equivalent or homeomorphic" in that they could be interconverted without rotational movements of the terminals when conformational energy minimum pathways are followed. As a consequence, if the terminals of the polypeptide chain are linked, as in the cyclopeptides, the linking number difference of two structures represents the minimum number of boundary energy barriers which must be overcome in order to have their conformational interconversion.

As examples of homeomorphic structures, Fig. (4) illustrates three skeleton conformations proposed by different authors (7,8,9) for the cyclopeptide Gramicidin S. The conformations in Fig. 4a and 4b are characterized by very close values of L-L$_a$ (4.2 and 4.1, $\Delta L \cong O$) in spite of their different structures in Fig. (4c), whereas the last

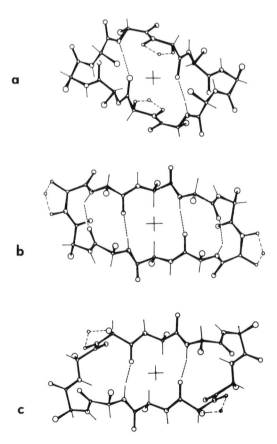

Figure 4. Skeleton conformations of the cyclopeptide Gramidicin S proposed by different authors; a (ref. 7); b (ref. 8); c (ref. 9) as examples of homeomorphic structures.

conformation, whilst apparently similar to the structure in Fig. (4b), has a different L-L$_a$ value (6.0, $\Delta L \cong 2$).

Topological constraints in protein folding

Although the polypeptide chain of the proteins is not closed, nevertheless, the linking number could turn out an useful parameter to investigate the mechanisms of folding. In fact, the intramolecular interactions which increase during the pathway towards the native tertiary structure, brake the chain motions, generating a network of topological domains where homeomorphic conformational transformations are favoured. In fact changes of linking numbers in protein transformations require an energy cost either for crossing the boundary barriers or for breaking the network of long-range interactions. Furthermore, it is plausible that exist, during the protein synthesis in the ribosome, interactions with the proteic and nucleic moieties which further limit the movements of the polypeptide chain.

It is thus evident that the concept of homeomorphic conformations introduces a topological correlation in the conformational transformations which could become an important factor aside the neighbourhood correlation (as in α-helix formation) capable to direct the folding process along optimized pathways towards the final tertiary structure. An important aspect of the topological correlation is its high cooperativity. In fact, in elastic ribbons, homeomorphic transformations correspond to transitions between twist and writhe which result in catastrofic processes (10) because slight deviations from the uniformity of the mechanical properties along the ribbon become amplified, generally resulting in a definite "tertiary structure" corresponding to the best coupling between twist and writhe. With the aim of providing informations about the protein folding process; owing to the postulated invariance of the linking number along the kinetic pathways towards the native structure, we have analyzed the topological parameters of the tertiary structures of 49 proteins obtained by X-ray studies (11) and evaluated the trends of the linking number at each position along the sequence. Some examples of such "topological diagrams" are shown in Fig. (5). They illustrate the topological features of Myoglobin, Adenylate Kinase, Rubredoxin, Ferredoxin, Cytochromes C551 and B5. As can be easily observed the experimental points are generally fitted by straight line segments having different lengths but only two alternative pitch angles, with high accuracy (the correlation coefficients were always higher than .998).

In order to quantify the relative dispersion of the pitch angles, the histogram was evaluated of the least square linear fitting of all the segments containing more than 3 amino acid residues; in the topological diagrams the units were chosen in order to have proportionality between the segment lengths and the extension of the related secondary structures in the physical space. The histogram is shown in Fig. (6) and represents 684 segments corresponding to 5633 amino acid residues.

The topological character of the segments with pitch angle $\sim 0°$ coincides with that of the right-handed α-helix as well as the segments with $\sim 60°$ have topological

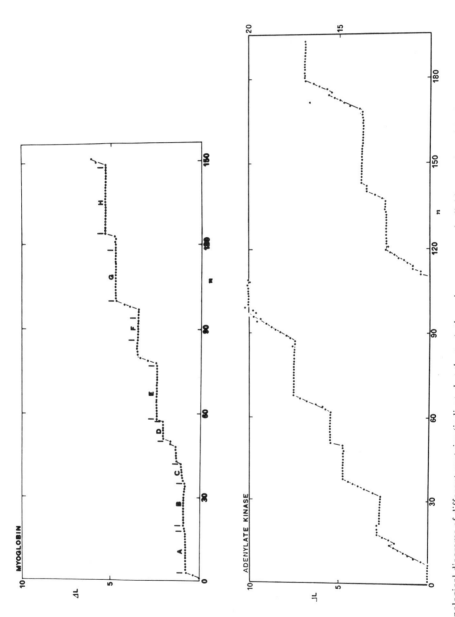

Figure 5. Topological diagrams of different proteins (indicated at the top): the points represent the linking number of the first n amino acid residues along the sequence; the interpolating segments represent the topology of the peptide-ribbon.

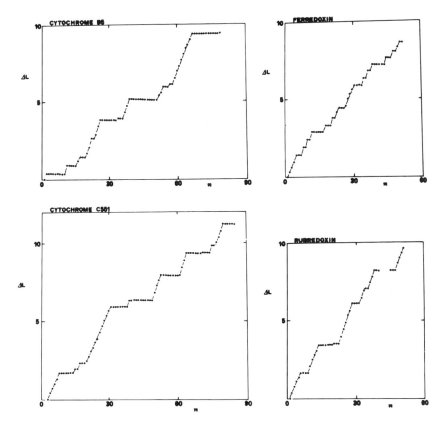

Figure 5 continued from previous page.

features similar to β (twisted) structures. For this reason, we called α and β the two topologies existing in proteins.

Actually, the topological diagrams localize the regions of the amino acid sequence with α or β topological character, namely, with a "virtual" α or β secondary structure;

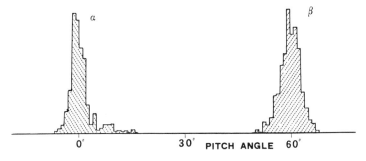

Figure 6. Histogram of the least square pitch angle distribution of the segments containing more than 3 residues in the topological diagrams of 49 proteins.

they, however, generally possess a so high degree of agreement with the trend expected for the α and β structures to feel confident that these regions actually have or had before the latter steps of the folding process, those secondary structures.

This consideration appears rather convincing looking, for example, at the topological diagram of Myoglobin in Fig. (5). As it can be seen all the α-helix regions are easily recognized but the persistence of the α topological character over the A . . . AB . . . B . . . C− and partly CD regions, suggests that the whole sequence from the amino end to the 48th amino acid residue should have been a long α-helix in early steps of folding. The same is worth for the G . . . GH . . . H regions, where the GH corner, which actually extends over 6 amino acids residues in the tertiary structure, was plausibly before in the α-helix conformation; thus the F helix should have been 8 amino acid residues (belonging to EF and FG corners) longer.

Fig. (7) pictorially illustrates the topological equivalence of the GH corner with an α-helix. The topological diagram of Rubredoxin in Fig. 5 which is considered an all-β protein, shows few regions of α topological character not corresponding to actual α-helices.

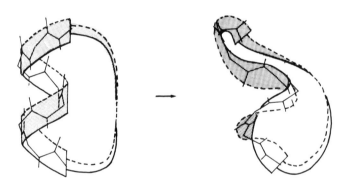

Figure 7. Pictorial drawing illustrating the topological equivalence of the GH corner of Myoglobin with an α-helix.

The case of BPTI, which is often used as a model protein in literature, is illustrated in greater details in Fig. (8); thus, the scale of ΔL is amplified in order to evaluate the degree of accuracy of the topological diagrams.

The existence of topological correlations along the polypeptide chain convincingly springs out from the comparison of the topological diagram with the upper diagram which illustrates (in the same scale) the sequence of the local linking number contributions per amino acid residue. It is evident that the large deviations from the standard Lα and Lβ contributions are mutually compensated along the chain.

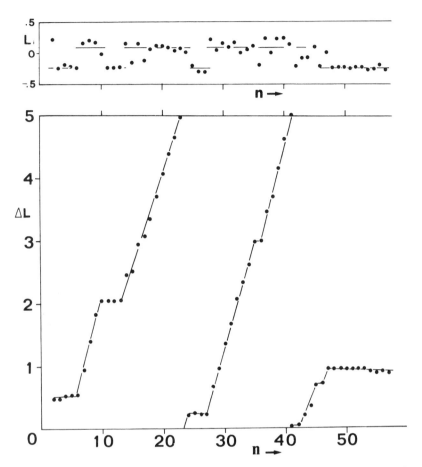

Figure 8. Amplified topological diagram of BPTI. The fitting lines represent the topological trend of the continuous peptide-ribbon. The upper diagram shows the contributions to the linking number by the single amino acid residues along the sequence.

The presence of a such topological compensation is a general feature of the topological diagrams and characterizes the secondary as well as the non regular structures; whilst this is obvious in the former case where it has the effect of conserving the structural integrity and the network of hydrogen bonds, in the other case it is plausibly the consequence of a previous conformational regularity. Thus, such concerted changes of conformational parameters which are rather hard to rationalize in the conformational diagrams, are easily realized in the topological diagrams; particularly where the conformational distortions are large enough to fully destroy the regularity of the original structure.

Regular conformational distortions which are related to the presence of super secondary structures, as twisted β sheets and coiled coil α-helices, result in pitch changes in the topological diagrams as quantified by the dispersion of the two maxima of the histogram in Fig. (6): e.g., in the case of twisted β strands, the coil angle, τ, is linearly related to the pitch angle, p. as:

$$\tau = 5.0p - 280°$$

In few cases, the topological regularity shows certain significant deviations from the linear trends; often, these singularities occur at glycine residues on account of their larger conformational versatility; when, however, different amino acid residues are involved, their conformations are often not sterically allowed rising the suspect of their misdetermination by X-ray analysis. Examples are Val 99, Lys 100, Ala 175 of Adenylate Kinase in Fig. (5) which conformational parameters assigned by crystal structure analysis are $(27°, 87°)$, $(161°, -32°)$, $(113°, -120°)$ for ϕ and ψ, respectively, out of the sterically allowed regions of the energy conformational maps.

Finally, topological diagrams allow an easy detection of topological homologies in a protein and between different proteins, deriving from a virtual conformational equivalence which could reflect distant philogenetic relations (2): e.g. the gene duplication of ferredoxin appears evident in the pertinent topological diagram in Fig. (5).

Concluding Remarks

The results presented in this paper allow us to put forward an attractive hypothesis about the folding processes occurring in the proteins.

It is known that proteins undergo a wide variety of conformational transformations ranging from small changes occurring also in the crystals and having either dynamical or statistical characters (12,13), to large changes of the entire globular structure in denaturation processes.

Whilst the first kind of conformational transformations belongs to the class of the homeomorphic transconformations, the larger conformational changes which break the compactness of protein structure should be in general not topologically invariant. Nevertheless, since the rate limiting energy barrier between the folded and unfolded states has been shown to lie very close to the fully folded state for the protein studied so far (12), the kinetics of the protein folding should be determined by conformational transformations topologically invariant.

The lack of dependence on the bulk viscosity of the solvent observed in the final folding during some denaturation processes, (15,16), argues infact against large chain motions which would be rate limited by external frictions; thus, the occurring conformational transformations are, plausibly, topologically constrained as a consequence, of limited chain diffusional motions; (translational as well rotational) into the solvent.

In the case of proteins growing in the ribosome it is likely that also the intermediate steps of the protein folding could be mainly represented by conformational transformations topologically invariant because of the additional interactions with the ribosome, which further brake the chain movements, and excluded volume effects.

If this hypothesis is assumed, the folding process would follow those conformational pathways which are topologically invariant starting from polypeptide conformations topologically "prepared" in the early steps of the synthesis.

It is noteworthy that the success in predicting the structure of PTI by a computer simulation of protein folding (7,8) when starting from a β extended conformation except for the terminal amino acid residues in α-helical conformation as in the tertiary structure, could be in relation with the fact that such a starting structure is "topologically prepared" (except for few amino acid residues) as the pertinent topological diagram in Fig. (8) clearly shows.

Finally, starting from the topological representations of proteins it is easy to evaluate the "a priori" propensity of each amino acid residue to occur with α or β character as well as the correlation matrices, in other words, the basic statistical parameters for trying predictions on the topology of a protein based on the knowledge of the primary structure. We are attempting such statistical approach, with the aim to derive the propensities of the amino acid residues to occur as α or β conformer in the early unfolded protein, where the conformational correlation were plausibly limited to the nearest neighbour amino acid residues.

Further researches need to gain supports on the general conclusions of this paper.

References and Footnotes

1. De Santis, P., Morosetti, G., Palleschi, A., *Biopolymers, 22,* 37-42 (1983).
2. De Santis, P., Chiavarini, S., Morosetti, S., Palleschi, A. *Biopolymers 23,* 1547-1562 (1984).
3. Liquori, A. M. in Symposium on Principles of Biomolecular Organization, Ciba Foundation, J. A. Churchill Ltd, London (1966), pp. 40.
4. Fuller, F. B., *Proc. Nat. Acad. Sci., USA, 68,* 815-819 (1971).
5. Crick, F. H. C., *Proc. Nat. Acad. Sci., USA, 73,* 2639-2643 (1976).
6. Damiani, A., De Santis, P., *J. Chem. Phys., 48,* 4071-4075 (1968).
7. Liquori, A. M., De Santis, P., Kovacs, A. L., Mazzarella, M., *Nature* (London), *211,* 1039 (1966).
8. De Santis, P., Liquori, A. M., *Biopolymers, 10,* 699 (1971).
9. Leach, S. J., Nemethy, G., Scheraga, H. A., *Biopolymers, 4,* 369 (1966).
10. Le Bret, M., *Biopolymers, 18,* 1709-1725 (1979).
11. Protein Data Bank, Department of Chemistry, Brookhaven National Laboratory, Associated Universities, Inc., Upton, L. I. New York 1973.
12. Karplus, M., in *Dynamics of Proteins, Proceedings of the Second SUNYA Conversation in the Discipline Biomolecular Stereodynamics,* Volume II, Ed., Sarma, R. H., Adenine Press, New York (1981).
13. Delepierre, M., Dobson, C. M., Hoch, J. C., Olejniczak, E. T., Poulsen, F. M., Ratcliffe, R. G., Redfield, G., in *Structure and Dynamics of Proteins by Proton NMR Applications of Nuclear Overhauser Effects to Lysozyme, Proceedings of the Second SUNYA Conversation in the Discipline Biomolecular Stereodynamics.,* Volume II., Ed., Sarma, R. H., Adenine Press, New York (1981).

14. Creighton, T. E., Conformational Flexibility in Proteins in *Structural Aspects of Recognition and Assembly in Macromolecules,* Balaban, M., Sussman, J. L., Trand, W., and Yonath, A. (1981), Balaban ISS, Rehovot and Philadelphia.
15. Epstein, H. F., Schechter, A. N., Chen, R. F., Anfinsen C. B., *J. Mol. Biol. 60,* 499-508 (1971).
16. Tsong, T. Y., Baldwin, R. L., *Biopolymers, 17,* 1669-1679 (1978).
17. Levitt, M., Warshel, A., *Nature, 253,* 694-698 (1975).
18. Robson, B., Osguthorpe, D. J., *J. Mol. Biol., 132,* 19-51 (1979).

Structure & Motion: Membranes, Nucleic Acids & Proteins,
Eds., E. Clementi, G. Corongiu, M. H. Sarma & R. H. Sarma,
ISBN 0-940030-12-8, Adenine Press, Copyright Adenine Press, 1985.

Proton NMR Studies of Protein Folding and Unfolding

Christopher M. Dobson, Philip A. Evans and Robert O. Fox

Inorganic Chemistry Laboratory, University of Oxford
South Parks Road, Oxford OX1 3QR

Abstract

Methods by which nuclear magnetic resonance spectroscopy may be used to investigate the folding and unfolding of proteins are discussed. The potential of one approach, that involving magnetization transfer techniques, is illustrated with results from studies of hen lysozyme. The significance of the results and the wider applicability of the methods are considered.

Introduction

Nuclear magnetic resonance spectroscopy (NMR) has been applied extensively to the study of proteins in solution (1,2). The most detailed studies have been carried out with small proteins where the resonances of many nuclei (mainly protons but also carbons) can now be resolved and assigned with confidence. Once resonances have been assigned, the structure and dynamics of individual groups in the protein can be investigated (3). Most of the studies have been concerned with the nature of globular proteins in their native states and such studies have been described elsewhere in detail (4-6).

It is well established that the compact globular structure of native proteins is lost upon denaturation (7). This is seen clearly in NMR spectra by the absence of resonances with chemical shifts grossly different from those observed in small peptides. In the spectra of native proteins, chemical shifts are strongly perturbed by local interactions between residues in close proximity (9). It is not clear, however, to what extent unfolded proteins conform to the simple model of a random coil. Indeed, it has been pointed out that since no single solvent condition will be the ideal solvent for the entire polypeptide chain, some non-random behaviour is to be expected (7). NMR spectra have shown some evidence for non-random behaviour in denatured proteins. If all residues experienced an equivalent solvent-like environment it would be expected that all chemically equivalent nuclei in residues of the same type would have very similar NMR parameters such as chemical shift or relaxation rates, with only small perturbations arising from interactions between neighbouring residues. ^{13}C NMR spectra of denatured lysozyme and ribonuclease show heterogeneity in chemical shifts principally associated with resonances of hydrophobic residues (10,11). There is less heterogeneity in less polar solvents and it is suggested that the denatured proteins have a loosely bound structure involving hydrophobic residues (10). The difference in the ¹H NMR spectra of lysozyme denatured in dimethyl sulphoxide and aqueous solutions has also been suggested as evidence for significant non-random behaviour of the protein in aqueous solution (12). Other NMR evidence for non-random behaviour of denatured proteins includes the observation of unequal T_1 relaxation times and pK_a values for the different histidine residue of ribonuclease (13).

A major problem with the study of denatured proteins by NMR is that their spectra tend to be so crowded that it is difficult to observe, assign and study individual resonances. The parameters that provide conformational information in native proteins, such as nuclear Overhauser effects and coupling constants, cannot easily be extracted. An approach to this problem is to examine simpler model systems which can be studied using conventional methods. One example involves the peptide hormone glucagon which has been shown to have structured elements in its monomeric state (14), and is of additional interest because of conformational changes resulting from specific association reactions (15). Recently, the C- and S- peptide fragments of ribonuclease have been studied (16). These small peptides can form stable α-helices at low temperatures and these have been characterized by NMR methods. It has also been shown that certain features characteristic of this helix formation can be discerned in the spectrum of denatured ribonuclease itself under similar conditions (17). Another approach to the problem of observing and identifying resonances in the spectrum of a denatured protein is to relate them to the resonances of the corresponding nuclei in the spectrum of the native protein (12). Techniques by which this can be achieved, based on magnetization transfer effects, are discussed later in this article.

A detailed description of residual structure or of localised conformational preferences in denatured proteins would be of profound significance for studies of protein folding. The probabilty of the formation of elements of structure which constitute folding nuclei may be a crucial determinant of the folding rate. Before returning to this point, we review briefly approaches to studying folding and unfolding of proteins by ¹H NMR; the insensitivity of ¹³C NMR has resulted in few studies using this nucleus.

NMR Studies of Protein Folding and Unfolding

The most straightforward approach, in principle, is to follow in real time the course of folding or unfolding as manifest in changes in the NMR spectrum of the protein. This method has, for example, been used to study the refolding of ribonuclease at low temperatures (18,19). In these studies the transient appearance of resonances attributed to kinetic folding intermediates was observed. The limitation of this type of approach is that it is not possible to accumulate an NMR spectrum with an acceptable signal-to-noise ratio in a very short time. A reaction can thus only be followed if conditions can be found in which it proceeds sufficiently slowly. One way of avoiding this difficulty is to trap transiently formed species so that they may be observed at leisure. An example of this approach is that free thiol groups can be alkylated during the refolding of reduced disulphide-containing proteins, thereby trapping species with an incomplete complement of disulphide bonds. The folding pathway of BPTI has been mapped out in detail in this way (20), and several of the intermediates have been studied by NMR (21).

A variation on the trapping theme makes use of the protection from solvent exchange afforded to labile protons by folded structure in a protein (22). A protein, unfolded in H_2O, is allowed to refold in a substantial excess of D_2O. In the case of those labile protons that are slow to exchange from the native state, the extent to which they are exchanged for solvent deuterons will depend on the competition between the folding and exchange reactions. If the folding pathway involves partially folded intermediates it is apparent that some protons will become protected from solvent at an earlier stage than others. The proportion of protons in these sites will therefore be correspondingly higher. NMR can be used to determine relative amide proton concentrations, and hence identify regions of structure formed early in folding. Studies of this type have been reported for ribonuclease (22).

These experiments are all carried out under conditions where the native state is thermodynamically strongly favoured over the denatured state. An alternative approach is to choose conditions in which folded and unfolded states are in equilibrium. This may be achieved by variation of pH, increase of temperature or addition of denaturants. In some circumstances, for example the 50 residue lac-repressor headpiece fragment, the interconversion between the folded and unfolded states is fast enough under the conditions studied that chemical shift values vary continuously through the denaturation transition (23). These values represent averages of the chemical shifts in the two states weighted according to their relative populations. In cases where the rates of interconversion are slow on the NMR timescale, discrete resonances are observed for protons in the different states. The relative areas of these resonances depend on the relative populations of the different states, which can then be followed as a function of the denaturation conditions. An intermediate with a significant equilibrium population could therefore be observed and characterized by NMR under these conditions. One case where distinct resonance of different species are observed is staphylococcal nuclease; in this case a minor folded component persists over a wide range of conditions (24,25).

Intermediates can also be searched for by comparing the intensities of the resonances of different residues in the spectrum of a native protein as the degree of denaturation is changed. If the mid-points of denaturation detected for the different residues are different, this provides evidence that an intermediate exists. In studies of some proteins, such as lysozyme, no such differences have been detected under the conditions examined (26,27). This suggests that a two-state model adequately describes their folding equilibria. For the thermal unfolding of ribonuclease at low pH, however, a difference of approximately 2°C in the denaturation midpoints of different residues has been reported (28). Similarly for ribonuclease in the presence of urea or guanidinium chloride the mid-points of denaturation detected for different residues were found to be significantly different (29).

A particularly important aspect of these equilibrium studies concerns the rates at which the interconversions of different species occur. If accurate rate data were available it would, for example, enable the kinetic behaviour of different regions of a protein to be compared. The kinetic significance of any minor species observed at equilibrium could also be investigated. In the slow exchange region, where distinct resonances are observed from the different states of the protein, the technique of magnetization transfer has the potential to provide this information (12). In the next section of this paper, studies of lysozyme are described which investigate the feasibility of this approach.

Magnetization Transfer Studies of Lysozyme

Lysozyme from hen egg-white has 129 amino acid residues. The ^{1}H NMR spectrum of the protein in its native state has been studied in detail (30,31). At present at least one resonance of nearly 50 amino acid residues has been identified and assigned, and much information concerning the structure and dynamics of the protein in solution has been obtained (3). Figure 1 shows part of the ^{1}H NMR spectrum of lysozyme at temperatures above and below that at which denaturation occurs. The resonances between 6.0 and 8.8 ppm are from aromatic protons, the others from H$^\alpha$ protons. These spectra illustrate two points made earlier. There are large differ-

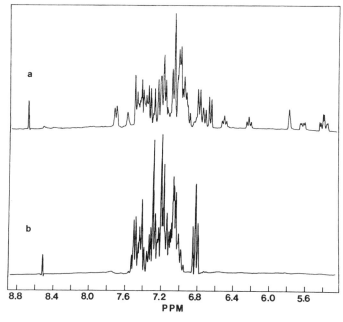

Figure 1. Low-field region of 300 MHz ^{1}H NMR spectra of lysozyme. Resolution has been enhanced by Gaussian multiplication, and labile protons exchanged for deuterons. The top spectrum, at 65°C, is of native lysozyme and the bottom spectrum, at 82°C, is of denatured lysozyme. All the data shown in this article are for 5mM lysozyme in D$_2$O at pH 3.8.

ences in the spectra of the native and denatured states and the spread of chemical shifts is smaller in the spectrum of the denatured protein (12).

Thermal denaturation has been followed in detail by recording spectra at different temperatures (27,32). The exchange rate between the native and denatured species is sufficiently slow that when both forms exist in significant concentrations the NMR spectrum is a summation of the resonances from the two states. Figure 2 shows this for the fully resolved H$^\epsilon$ resonance of His 15. As the temperature is increased, the intensity of the resonance from the native protein can be seen to decrease whilst that of the denatured protein increases. The ratio of the areas of the two peaks provides a value for the equilibrium constant as a function of temperature. Data of this type have been obtained for a number of residues of lysozyme (27,32); some examples are shown in Figure 3. The residues involved are found in quite different regions of the lysozyme molecule. The crystal structure shows that the molecule in its native state is divided into two parts by a deep cleft (33). One part contains a hydrophobic region in which Leu 17, Trp 28 and Trp 108

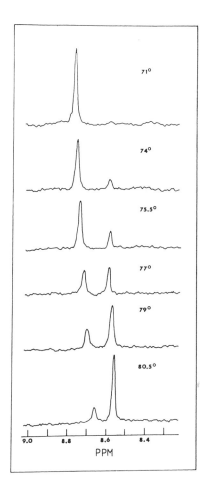

Figure 2. 300 MHz ^{1}H NMR spectra of lysozyme at various temperatures showing the resonance of His 15 H$^\epsilon$ in the native and denatured forms of the protein.

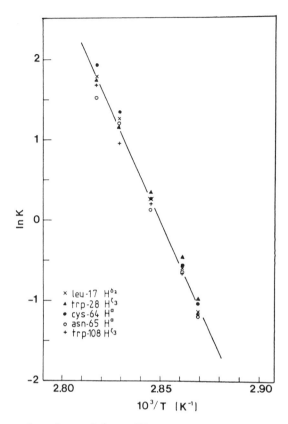

Figure 3. Temperature dependence of the equilibrium constant K determined from resonances of several residues of lysozyme. In each case K is the ratio of the intensity in the denatured state to that in the native state.

are found. Cys 64 and Asn 65 are situated in the other part, which contains a region of β-sheet structure. As Figure 3 shows, within experimental error the denaturation temperature (77.5°C) and the enthalpy of denaturation (110 kcal mol^{-1}) are the same for each of these residues.

In its simplest form, a magnetization transfer experiment involves a situation where interconversion takes place between two or more chemical forms of a molecule in which a given nucleus has a different resonant frequency in each form (34,35). Irradiating selectively at the frequency of one resonance prior to accumulating the NMR spectrum causes saturation of that resonance, and this can be transfered by the exchange process to the resonance of the same nucleus in the other form. The results of experiments of this type for lysozyme are given in Figure 4. This shows the upfield region, that containing resonances of methyl groups, of a spectrum of lysozyme at the denaturation temperature. The effects of irradiating resonances in this spectrum are revealed by difference spectroscopy (12). When the resonance at -0.5 ppm, that of one of the methyl groups of Leu 17 in the native protein, is

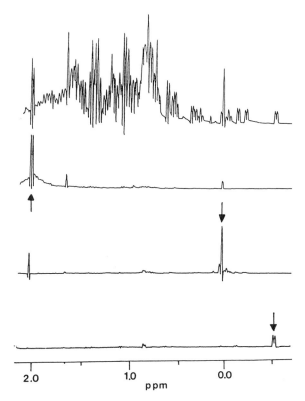

Figure 4. The top spectrum shows the high-field region of the 300 MHz ¹H NMR spectrum of lysozyme recorded at 77°C, close to the mid-point of denaturation. Below this are difference spectra showing the effects of irradiating at the positions indicated (12).

irradiated the difference spectrum shows not only this resonance but also one close to 0.8 ppm. The latter corresponds to the resonance of the Leu 17 methyl group in the denatured state. Similarly irradiation of the resonance at 0.0 ppm, that of the methyl group of Met 105 in the native protein, results in clear transfer of magnetization to a resonance close to 2.0 ppm. This is then identified as that of the Met 105 methyl group in denatured lysozyme. A complementary experiment is also shown in Figure 4. Irradiation at about 2.0 ppm reveals resonances at 0.0 ppm and at 1.7 ppm. This occurs because the irradiation has saturated the nearly overlapping peaks of the methyl group of both Met 12 and Met 105 in the denatured state. The resonance at 1.7 ppm is that of Met 12 in the native protein.

The correlation of peaks in the native and denatured spectra provides a method by which assignments may be made in the spectrum of the denatured protein for those residues with resonances assigned in the native protein. The selective irradiation of resonances throughout the spectrum would permit this to be achieved. An alternative approach is to make use of two-dimensional NMR techniques (12). In these

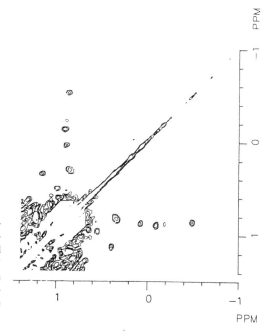

Figure 5. Contour map of the high-field region of a 470 MHz ^{1}H two-dimensional chemical exchange spectrum of lysozyme recorded at 78°C. This was obtained using the NOESY pulse sequence (37,38); further details are given elsewhere (12). The off-diagonal features represent cross-peaks at the chemical shifts of the native and denatured resonances.

experiments each pair of resonances, corresponding to a nucleus in the native and denatured states, gives rise to a cross-peak in the spectrum. Figure 5 shows part of a contour plot from such an experiment. The correlation of the upfield methyl group resonances of the native protein (see Figure 4) with the corresponding resonances of the denatured state can be observed clearly in this diagram.

The experiments described above have not required a quantitative analysis of the magnitude of magnetization transfer effects. It is, however, possible to carry out this analysis and to use it to obtain kinetic information (34,35). In the case of the irradiation experiments, increasing the length of time of irradiation increases the extent of magnetization transfer. This is shown in Figure 6 for the H$^\epsilon$ resonance of His 15 of the native form following irradiation of the resonance of the denatured form. The intensity, I, of the native resonance decreases towards a limiting value, I$^\infty$, which is determined, according to the conventional two-spin treatment, by

$$\frac{I^o}{I^\infty} = 1 + k_1 T_{1N}$$

where I^o is the unperturbed intensity, k_1 is the unfolding rate and T_{1N} the apparent spin-lattice relaxation time of the resonance of the native form. The time-development of the effect also depends on k_1 and T_{1N}, according to

$$\frac{d \ln |I - I^\infty|}{dt} = -(k_1 + 1/T_{1N})$$

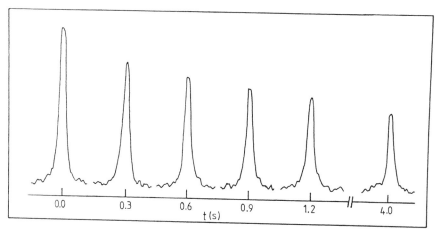

Figure 6. Time-dependence of the transfer of magnetization from denatured to native lysozyme at 77°C observed from the His 15 H$^\epsilon$ resonance. The intensity of the resonance in the native form (see Figure 2) is shown as a function of the length of time irradiation was applied to the resonance of the denatured form.

Hence, determination of the limiting magnetisation transfer effect and its rate of development leads to values for k_1 and T_{1N} by solution of these equations. That the behaviour in the real protein system corresponds well to this description is shown in figure 7. No evidence for kinetic heterogeneity under these conditions has been obtained (32). The folding rate, k_2, can be determined independently by an analogous experiment in which a resonance of the native form is saturated.

These and related experiments have permitted rates of interconversion of native and denatured lysozyme to be determined for residues in different regions of the lysozyme molecule (32). The rates measured for the different residues are the same

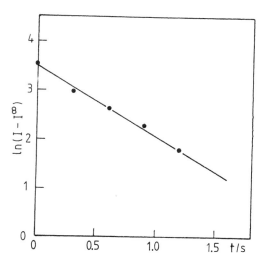

Figure 7. Semi-logarithmic plot of the data from Figure 6. The ratio I^o/I$^\infty$ = 2.0, and the slope d ln(I−I$^\infty$)/dt = 1.4s^{-1}. Hence, using the analysis described in the text, T_{1N} = 1.4s and the unfolding rate, k_1 = 0.7s^{-1}.

within experimental error, and taken with the thermodynamic data (Figure 3) demonstrate the highly cooperative nature of the folding and unfolding processes under the conditions used here. It has also been possible to determine apparent activation energies from temperature dependence studies (32). That for the unfolding process is large and positive, but that for the folding process is negative. The observation is in accord with a number of other studies of protein folding kinetics using different techniques, including a study of lysozyme in LiBr solution (39). The combined activation energies are close to the thermodynamic enthalpy of unfolding, and the temperature at which the folding and unfolding rates were equal was found to be identical to the mid-point of denaturation obtained from the data of Figure 3. These results give considerable confidence in the application of the magnetization transfer method to protein folding and unfolding.

Further Applications

The results on lysozyme demonstrate that the magnetization transfer technique can provide a method of assigning the NMR spectrum of an unfolded protein, and that it can provide consistent values for the rates at which the folded and unfolded states interconvert.

Assignment of the spectrum of an unfolded protein is an essential step towards a detailed characterisation of its structure and dynamic properties. The unfolded state undoubtedly comprises a large number of significantly different conformations in rapid equilibrium (12). NMR parameters such as chemical shifts and coupling constants reflect this average and could in principle permit a definition of the various structures involved. Another approach is to combine the kinetic measurements with results from NMR studies of hydrogen exchange. At temperatures close to the unfolding transition, labile hydrogens buried in the native structures may exchange with solvent via the unfolded state (27,40). In conjunction with the rates of folding and unfolding determined from the magnetization transfer experiments, the exchange kinetics of individual hydrogens can provide information about the lability of hydrogens in the unfolded state (32). This information should enable regions of persistent structure in the unfolded state to be identified. These may represent localised elements of secondary structure which subsequently combine leading to more extensive folding (41). Greater populations of localised structures at lower temperatures could be related to the increase in the folding rate as the temperature is reduced.

The approach demonstrated here for lysozyme should be applicable to other proteins provided that conditions can be chosen such that the interconversion between folded and unfolded states is fast enough to compete with spin-lattice (T_1) relaxation. Rates in the range of $0.1s^{-1}$ to $10s^{-1}$ are likely to be within these conditions. Magnetization transfer has been observed in studies of the bovine pancreatic trypsin inhibitor (42,43), and of staphylococcal nuclease (25). No effects could, however, be observed for ribonuclease folding intermediates stabilised in low temperature experiments (44).

Magnetization transfer studies of staphylococcal nuclease are of particular interest because equilibrium concentrations of more than one folded state exist. The interconversion of the different folded states and the unfolded state is fast enough, at least close to the denaturation temperature, to be measured by magnetization transfer techniques (25). Furthermore the unfolding rates determined by magnetization transfer differ substantially (25). Detailed kinetic studies of the interconversion of these species are in progress.

Acknowledgements

This is a contribution from the Oxford Enzyme Group, which is supported by the Science and Engineering Research Council. R. O. F. acknowledges the receipt of a Fellowship from the Jane Coffin Childs Memorial Fund. We thank M. Delepierre and F. M. Poulsen for valuable discussions.

References and Footnotes

1. Campbell, I. D. and Dobson, C. M., *Methods Biochem. Anal., 25,* 1 (1979).
2. Jardetzky, O. and Roberts, G. C. K., *NMR in Molecular Biology,* Academic Press, New York (1981).
3. Dobson, C. M., in *Structure and Dynamics: Nucleic Acids and Proteins* (Clementi, E. and Sarma, R. H., Eds.) Adenine Press, New York, p. 451 (1983).
4. Boswell, A. P., Eley, C. G. S., Moore, G. R., Robinson, M. N., Williams, G. and Williams, R. J. P., *Eur. J. Biochem., 124,* 289 (1982).
5. Wagner, G. and Wüthrich, K., *J. Mol. Biol., 155,* 347 (1982).
6. Delepierre, M., Dobson, C. M. and Poulsen, F. M., *Biochemistry, 21,* 4756 (1982).
7. Tanford, C., *Adv. Protein Chem., 23,* 121 (1968).
8. McDonald, C. C. and Phillips, W. D., *J. Am. Chem. Soc., 91,* 1513 (1969).
9. Perkins, S. J., in *Biological Magnetic Resonance* (Berliner, L. J. and Reuben, J., Eds.) Plenum, New York, vol. 4, p. 193 (1982).
10. Howarth, O. W., and Lian, L. Y., *J. Chem. Soc. Chem. Commun.,* 258 (1981).
11. Howarth, O. W. and Lian, L. Y., *Biochemistry 23,* 3522 (1984).
12. Dobson, C. M., Evans, P. A. and Williamson, K. L., *FEBS Lett., 168,* 331 (1984).
13. Matthews, C. R. and Westmoreland, D. G., *Biochemistry, 14,* 4532 (1975).
14. Braun, W., Wider, G., Lee, K. H. and Wüthrich, K., *J. Mol. Biol., 169,* 921 (1983).
15. Wagman, M. E., Dobson, C. M. and Karplus, M., *FEBS Lett., 119,* 265 (1980).
16. Kim, P. S. and Baldwin, R. L., *Nature, 307* 329 (1984).
17. Bierzynski, A. and Baldwin, R. L., *J. Mol. Biol., 162,* 173 (1982).
18. Blum, A. D., Smallcombe, S. H. and Baldwin, R. L., *J. Mol. Biol., 118,* 305 (1978).
19. Biringer, R. G. and Fink, A. L., *Biochemistry, 21,* 4748 (1982).
20. Creighton, T. E., *Prog. Biophys. Mol. Biol., 33,* 231 (1978).
21. States, D. J., Dobson, C. M., Karplus, M. and Creighton, T. E., *J. Mol. Biol. 174,* 411 (1984).
22. Kuwajima, K., Kim, P. S. and Baldwin, R. L., *Biopolymers, 22,* 59 (1983).
23. Wemmer, D., Ribeiro, A. A., Bray, R. P., Wade-Jardetzky, N. G. and Jardetzky, O., *Biochemistry, 20,* 829 (1981).
24. Markley, J. L., Williams, M. N. and Jardetzky, O., *Proc. Natl. Acad. Sci. USA, 65,* 645 (1970).
25. Fox, R. O., Evans, P. A. and Dobson, C. M., to be published.
26. McDonald, C. C., Phillips, W. D., and Glickson, J. D., *J. Am. Chem. Soc., 93,* 235 (1971).
27. Wedin, R. E., Delepierre, M., Dobson, C. M. and Poulsen, F. M., *Biochemistry, 21,* 1098 (1982).
28. Westmoreland, D. G. and Matthews, C. R., *Proc. Natl. Acad. Sci. USA, 70,* 914 (1973).
29. Benz, F. W. and Roberts, G. C. K., *J. Mol. Biol., 91,* 345 (1975).
30. Poulsen, F. M., Hoch, J. C. and Dobson, C. M., *Biochemistry, 19,* 2597 (1980).

31. Redfield, C., Poulsen, F. M., and Dobson, C. M., *Eur. J. Biochem., 128.* 527 (1982).

32. Dobson, C. M. and Evans, P. A., *Biochemistry.* in press.

33. Blake, C. C. F., Johnson, L. N., Mair, G. A., North, A. C. T., Phillips, D. C. and Sarma, V. R., *Proc. Roy. Soc. London, Ser. B, 167,* 378 (1967).

34. Forsén, S. and Hoffman, R. A., *J. Chem. Phys., 39,* 2892 (1963).

35. Campbell, I. D., Dobson, C. M., Ratcliffe, R. G. and Williams, R. J. P., *J. Magn. Reson., 29,* 397 (1978).

36. Wider, G., Macura, S., Kumar, A., Ernst, R. R., and Wüthrich, K., *J. Magn. Reson., 56,* 207 (1984).

37. Boyd, J., Moore, G. R. and Williams, G., *J. Magn. Reson. 58,* 511 (1984).

38. Macura, S., Huang, Y., Suter, D. and Ernst, R. R., *J. Magn. Reson., 43,* 259 (1981).

39. Segawa, S., Husimi, Y. and Wada, A., *Biopolymers, 12,* 2521 (1973).

40. Woodward, C. K. and Hilton, B. D., *Biophys. J., 32,* 561 (1980).

41. Go, N., *Ann. Rev. Biophys. Bioeng., 36,* 183 (1983).

42. Wüthrich, K., Wagner, G., Richarz, R. and Perkins, S. J., *Biochemistry, 17,* 2253 (1978).

43. Wüthrich, K., Roder, H. and Wagner, G., in *Protein Folding. Proc. 28th Conf. German Biochem. Soc.* (Jaenicke, R. Ed.) Elsevier/North Holland, Amsterdam, p. 549 (1980).

44. Biringer, R. G. and Fink, A. L., *J. Mol. Biol., 160,* 87 (1982).

Structure & Motion: Membranes, Nucleic Acids & Proteins,
Eds., E. Clementi, G. Corongiu, M. H. Sarma & R. H. Sarma,
ISBN 0-940030-12-8, Adenine Press, Copyright Adenine Press, 1985.

Functional Aspects of the Neutral Patterns in Protein Evolution

B. Borstnik* and G. L. Hofacker

Theoretische Chemie
Technische Universitat Munchen
8046 Garching, W-Germany

Abstract

In view of growing evidence in favour of a "neutral" evolution strategy we analyzed the observed frequencies of point mutations under functional aspects. Our objective was to find out to which extent the neutral patterns in the evolution of proteins were dominated by the changes in physico-chemical properties with the replacement of amino acid residues. To describe overall properties of proteins we introduced a 20-dimensional characteristics space spanned by eigenvectors of a property preservation matrix closely related to the Dayhoff matrix of mutation frequencies. A subspace of the characteristics space related to the physico-chemical residue properties polarity, hydrophobicity and residual volume can be shown to be most strongly preserved in mutations. We then demonstrate that the genetic code may already be adapted to the preservation of amino acid properties, thus contributing to the observed neutralistic patterns. The degree of neutralistic adaption of the genetic code is assessed by computing the mutation probability matrix of nucleotides optimally fitted to the empirical amino acid exchange rates. We also showed by a maximum entropy analysis that modelling protein evolution as a random process, while preserving just the sum of the three most relevant amino acid properties, yields some of the most salient features of the observed amino acid replacement frequencies.

1. Introduction

Evidence for the existence of neutral mutations found in the amino acid exchange patterns of protein evolution (1) was hard to discard but the question remained what was their meaning, what their functional definition. In particular, it had to be explained how neutral mutations could get accepted and in which sense this concept suited the recognized principles of evolution.

There is probably no other basic principle underlying evolution phenomena than Darwin's, through there may be something to be understood regarding its exact interpretation. Under these auspices one should look for the deeper rational of

---

*Permanent address: Boris Kidric Institute of Chemistry, 61000 Ljubljana, Yugoslavia

neutral mutations by analyzing mutation and selection processes for proteins from a functional point of view. In this paper we examine the hypothesis that the primary purpose of neutral mutations is the preservation of protein functions, some of which are easily characterized by crude overall physico-chemical properties like polarity, residual volume, hydrophobicity etc. Despite of the simplicity by which we formalize these quantities and the small number of overall properties which turn out to be relevant, a concept of neutral mutations emerges which appears to be a direct consequence of Darwinistic selection of mutations submitted by a genetic code adapted on its part to the preservation of function. In a forthcoming paper (2) we will further elaborate on the rational of neutral mutations within a Darwinistic paradigm by showing optimal properties of this very process.

2. A vector space of amino acid properties

In an earlier paper (3) it was argued that Dayhoff's matrix of amino acid exchange probabilities provided in fact a stochastic description of neutral mutations. Dayhoff et al. (4) noted already that their replacement probabilities strongly reflected the likely exchange of amino acids with similar physico-chemical properties which makes their data an appropriate basis for a study of neutral strategies.

The question of how amino acid properties correlate with their mutation frequencies and their codon multiplicities has been studied very well (5,6,7). A large number (up to 20) of amino acid properties were quantified and attributed to the individual molecules. Suppose a list of 20 amino acid properties were given for the 20 amino acids. Then, naively, one could think of characterizing each amino acid by its 20 component property vector or, inversely, each property by its components in the amino acids. If the property vectors span the 20 dimensional composition space, there should be a way to derive at least some of the most characteristic ones from the Dayhoff matrix. To find the property vectors predominantly preserved in neutral mutations, we follow the subsequent line of arguments.

The property vectors ($\vec{G}^{(m)}$) be represented by average values $\overline{G}^{(m)}$ and the root mean squares (rms) of their variations, $\vec{G}_{rms}^{(m)}$

$$\overline{G}^{(m)} = \sum_{i=1}^{20} G_i^{(m)}/20 \quad (1a) \qquad G_{rms}^{(m)} = \left\{ \sum_{i=1}^{20} (G_i^{(m)} - \overline{G}^{(m)})^2 \right\}^{1/2} \quad (1b)$$

The vector of variations is then normalized and given by

$$g_i^{(m)} = (G_i^{(m)} - \overline{G}^{(m)})/G_{rms} \quad (2)$$

In the first three columns of Table I a list of three vectors is given. The degree of independence of these property variation vectors is then measured by their scalar products ($\vec{g}^{(m)}, \vec{g}^{(n)}$), which are the correlation coefficients. A glance at Table II, listing inner products of the empirical property vectors volume (8), hydrophobicity

Table I

Normalized variations of three empirical amino acid property vectors and five significant eigenvectors of the operator M^{Day}. In the last row the corresponding F values, resp. also eigenvalues, are presented.

Amino Acid	Properties Volume	Hydro-phobicity	Polarity	$\vec{h}^{(4)}$	$\vec{h}^{(5)}$	$\vec{h}^{(6)}$	$\vec{h}^{(7)}$	$\vec{h}^{(8)}$
Gly	−0.43	−0.23	0.06	0.10	0.09	−0.09	−0.18	−0.02
Ala	−0.28	−0.08	−0.02	0.08	0.05	−0.09	−0.14	−0.01
Pro	−0.28	0.28	−0.03	0.11	0.10	−0.09	−0.16	0.10
Ser	−0.28	−0.24	0.07	0.09	0.07	−0.05	−0.11	0.05
Thr	−0.12	−0.24	0.02	0.08	0.04	−0.09	−0.14	0.03
Gln	0.01	−0.25	0.19	0.12	0.13	0.13	0.13	−0.12
Asn	−0.15	−0.23	0.28	0.10	0.12	0.03	−0.04	−0.06
Glu	−0.01	−0.12	0.34	0.12	0.14	0.02	−0.09	−0.13
Asp	−0.16	−0.25	0.40	0.13	0.15	0.01	−0.11	−0.13
Lys	0.19	−0.12	0.25	0.11	0.06	0.10	0.05	0.31
Arg	0.21	0.06	0.19	0.13	0.10	0.21	0.26	0.73
His	0.06	−0.09	0.18	0.08	0.20	0.34	0.56	−0.51
Val	0.00	−0.08	−0.21	0.03	−0.12	−0.26	−0.13	−0.10
Ile	0.14	0.11	−0.27	0.02	−0.17	−0.35	−0.11	−0.11
Met	0.11	0.35	−0.22	0.01	−0.84	0.40	0.01	−0.01
Leu	0.14	0.07	−0.29	−0.05	−0.23	−0.22	0.03	−0.10
Cys	−0.16	0.04	−0.24	0.07	−0.06	0.02	0.05	−0.00
Phe	0.26	0.30	−0.26	−0.55	−0.03	−0.45	0.52	0.12
Tyr	0.28	0.26	−0.18	−0.74	0.20	0.42	−0.39	−0.02
Trp	0.46	0.47	−0.25	0.02	−0.01	0.00	−0.01	−0.02
F/λ_0	0.44	0.40	0.36	0.21	0.32	0.48	0.54	0.60

(9) and polarity (10, 11) shows that empirical property vectors cannot be used to span a high dimensional property space, as even these 3 fundamental residuum properties turn out to be strongly correlated, witnessed by their correlation coefficients. We note that one of these vectors deviates only by 35° from the plain

Table II

Correlation coefficients for the three amino acid property vectors (first three rows) and their decomposition in terms of five significant eigenvectors of the M^{Day}-matrix.

Scalar product	Volume	Hydrophobicity	Polarity
Volume	1	0.638	−0.345
Hydrophobicity	0.638	1	−0.709
Polarity	−0.345	−0.709	1
$\vec{h}^{(4)}$	−0.473	−0.501	0.489
$\vec{h}^{(5)}$	−0.20	−0.44	−0.56
$\vec{h}^{(6)}$	0.159	0.050	0.323
$\vec{h}^{(7)}$	0.338	0.102	0.051
$\vec{h}^{(8)}$	0.162	0.163	0.053

of the two others, what is close to the average angle of 30° for a random vector. The only sensible way thus seems to be the deduction of preserved properties from the observed mutation frequencies.

Intuitively one would expect that the replacement of amino acid i by j imposes a risk of rejection on the mutant which is somehow proportional to the change in the m-th amino acid property $|g_i^{(m)} - g_j^{(m)}|$ or some power of it, $|g_i^{(m)} - g_j^{(m)}|^r$. If multiplied by the number of accepted j $\Rightarrow$ i point mutations, A_{ij}, and summed over all amino acid pairs, this functional F should be a measure for the rejection risk induced by point mutations changing average amino acid properties:

$$F(A, g^{(m)}) = \sum_{i>j} A_{ij} |g_i^{(m)} - g_j^{(m)}|^r / \sum_{i>j} A_{ij} \tag{3}$$

We adopt $r = 2$ as a convention. By standard variational procedure we introduce a Lagrange multiplier to take care of the normalization condition $\sum_i (g_i^{(m)})^2 = 1$ arriving at the eigenvalue problem

$$M\vec{g} = \lambda\vec{g} \tag{4a}$$

with

$$M_{ij} = -A_{ij} / \sum_{k>l} A_{kl} \qquad i \neq j$$

$$M_{jj} = \sum_{i=1}^{20}{}' M_{ij} \tag{4b}$$

This yields 20 property vectors $\vec{h}^{(m)}$ with nonnegative eigenvalues λ_m and the peculiarity that

$$F(A, \vec{h}^{(m)}) = \lambda_m \tag{5}$$

One of the λ's must be zero due to (4b). For a uniform A matrix, making all amino acids equivalent ($A_{ij}^{uni} = 1$) it can easily be seen that there is a 19-fold degeneracy with $\lambda_i = \lambda_o = 2/19$ for $i = 2, .., 20$ and $\lambda_1 = 0$. This means that each vector g satisfying (1), (2) and (3) leads to $F(A^{uni}, \vec{g}) = \lambda_o$. For the Dayhoff (12) matrix of transition frequencies A^{Day}, on the other hand, we obtain eigenvalues which are quite evenly distributed over the interval $[0, 2.66\lambda_o]$. As noted above, the first eigenvalue vanishes and the corresponding eigenvector must be $h_i^{(1)} = 1/20$. The next two eigenvectors are also very special, having only one coefficient different from zero, namely at the place of tryptophan and cysteine. This expresses the fact that their mutation frequencies are very low such that $F(A^{Day}, \vec{h}^{(2)})$ and $F(A^{Day}, \vec{h}^{(3)})$ turns out very low as well ($\lambda_2 = 0,057 \lambda_o$; $\lambda_3 = 0,18 \lambda_o$). The following eigenvectors now are what we are looking for—they span the vector space of the most significant property vectors. This is illustrated in Tables I and II. Table I shows three vectors representing the empirical amino acid characteristics residual volume, hydrophobicity and polarity, as well as five significant eigenvectors of the M matrix. Comparing the F measures of the 3 empirical property vectors with the eigenvalues λ_4 to λ_8 suggests that volume, hydrophobicity and polarity belong to the subspace spanned by the M-eigenvectors with the lower eigenvalues. An even clearer evidence of this fact can be found in Table II where the scalar products of the three property vectors with these M-eigenvectors are presented. The descreasing tendency of the projections with rising eigenvector indices is evident. Specifically we can conclude that polarity is preserved the most in neutral mutations and that residual volume and hydrophobicity also belong to the subspace of amino acid properties which are rather well preserved.

We analysed several other amino acid property vectors besides the ones presented in Table I, but found none with an F measure nearly as low as polarity. The property subspace spanned by the vectors $\vec{h}^{(4)}$ to $\vec{h}^{(8)}$ comprises (besides residual volume and hydrophobicity) also notable components of the vectors: average non-bonded energy per residuum (13), bulkiness (6), propensity to form a β structure (14), pk value of carboxyl group (15) and pk value of α-amino group (15). Among the property vectors for which one can detect no tendency of being preserved in mutations are composition (7) and propensity to form an α-helix (14).

3. Characteristics of the existing genetic code within the distribution of all conceivable codes

The genetic code has the definite property to favour amino acid exchanges which preserve physico-chemical properties. This was known for some time (1), but Dayhoff et al. (4) could show moreover that the most frequent interchanges occur between codons differing by only one nucleotide. With the results obtained in the preceding section, one can analyse these observations in a quantitative way.

We now turn to a detailed analysis of the Dayhoff mutation probabilites. They must be determined by three significant factors which are in hierarchical order:

1. The coding of amino acids.
2. The mutation probability matrix of nucleotides.
3. The acceptance of mutations on the protein level.

Each of these three stages are characterized by an effective amino acid replacement matrix A and a corresponding F-measure. Their adaption to a neutral strategy can then be judged by comparing F values for the three empirical property vectors volume, hydrophobicity and polarity as well as the "inherent" properties $\vec{h}^{(4)}.....,\vec{h}^{(8)}$. Given the effective A matrix, F is computed via equ. (3); the results are collected in Table III. We can thus focus on the determination of the effective A matrices.

A matrix A^{nat}, soleley determined by the coding of amino acids, can easily be constructed with sufficient accuracy. For simplicity we adopt the hypothesis of a uniform mutation rate on the DNA level and neglect two and three nucleotide

Table III

Values of the $F(A, \vec{g})$ measure relative to λ_o, the measure due to homogeneous A matrix. The measures are evaluated for three property vectors and five eigenvectors $\vec{h}^{(i)}$ and with four different A matrices: The observed interchange frequencies (A^{Day}); amino acid submission frequencies due to one nucleotide interchanges with homogeneous rate using natural code (A^{nat}); the same as A^{nat} except that the elements of nucleotide mutation probability matrix were optimized (A^{opt}); amino acid submission frequencies due to one nucleotide interchanges with most probable random code (A^{rand}).

Vector A	Volume	Hydrophobicity	Polarity	$\vec{h}^{(4)}$	$\vec{h}^{(5)}$	$\vec{h}^{(6)}$	$\vec{h}^{(7)}$	$\vec{h}^{(8)}$
A^{Day}	0.44	0.40	0.36	0.21	0.32	0.48	0.54	0.60
A^{opt}	0.79	0.64	0.60	0.50	0.45	0.70	0.80	1.20
A^{nat}	0.82	0.71	0.57	0.54	0.44	0.68	0.88	1.25
A^{rand}	1.01	0.91	0.78	0.82	0.58	0.87	0.89	1.37

interchanges. Within one codon the A_{ij}^{nat} elements can then be determined as the number of ways in which two residua can be connected by single nucleotide interchanges. According to the structure of the existing genetic code this can be achieved for 75 out of 190 possible distinct codon pairs; for the remaining 115 pairs one gets $A_{ij}^{nat} = 0$. The nonzero elements can have the values 1, 2, 3, 4 and 6. An example for $A_{ij}^{nat} = 1$ is the pair ile, arg; for $A_{ij}^{nat} = 2$ the pair gln, pro, for $A_{ij}^{nat} = 3$ there are 3 pairs only: ile, thr; ile, val and ile, met. $A_{ij}^{nat} = 4$ has, among others, the pair ala, gly and the pairs which belong to $A_{ij}^{nat} = 6$ are ser, thr; gly, arg; arg, ser and phe, leu. The results for the $F(A^{nat}, \vec{g})$ measure are given in the third row of Table III. As can be expected, the F measures evaluated with A^{nat} for our three empirical property vectors are well above the ones generated by the Dayhoff matrix A^{Day}. The same is true for the inherent properties $\vec{h}^{(m)}$. On the other hand, it is worth noting that the measures $F(A^{nat}, \vec{g})$ for $\vec{g}$ being residual volume, hydrophobicity, polarity or $\vec{h}^{(4)}...\vec{h}^{(8)}$ are still below λ_o, the eigenvalue in the case of uniform mutation rates on the amino acid level. Yet, this finding is still not sufficient to judge the conjecture that the attribution of codons to amino acids is itself the result of adaption to a neutral strategy. A crucial comparison can only be made to the F-numbers of a randomly generated code. The idea of comparing the natural code with a random code in terms of changes of amino acid properties due to single nucleotide substitutions was first brought up by Alf-Steinberger (16).

When generating realizations of the genetic code at random, we must be aware of the established fact that only two nucleotides were coding the amino acids in some early stage of evolution while the third functioned as a spacing (17,18). If we conceive the genetic code as representing permutations of 21 elements on a 3-dimensional grid of 4 x 4 x 4 = 64 sites it is plausible that the evolution of the code proceeded predominantly in the two dimensions corresponding to the first two places of the codon. We thus used an algorithm which generated random configurations of the code without destroying doublets and quadruplets of codons coding one amino acid due to degeneracy in the third place.

Starting out from the natural code the following operations were performed:

a) Permutations of all sixteen quadruplets in two dimensions due to the first and the second position of the codon; b) Interchanges of the doublets in the boxes where the degeneracy in third place is twofold (words ending with U and C coding one amino acid and the words ending with A or G coding another one). The pair of "stop" codons and the pair ("stop", tyr) were also subjected to this kind of interchange; c) met and tryptophan were also interchanged.

In a typical calculation up to 10 interchanges were generated. This yielded the corresponding A^{rand} matrices which in turn determined the F measures. The relative number of realizations of the code as a function of the values is given in fig. 1 in the case of polarity of amino acid residua. The distribution has the shape of slightly distorted Gaussian with the natural code belonging to the low F value wing of the distribution. We note that the maximum of the distribution is close to λ_o, the F value for uniform amino acid replacement frequencies. Other g vectors, such as residual volume, hydrophobicity and some eigenvectors h give pictures similar to the one shown for polarity, except that the positions of $F(A^{Day}, \vec{g})$ and $F(A^{nat}, \vec{g})$ vary according to Table III.

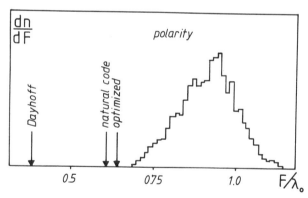

Figure 1. The histogram shows the distribution of $F(A^{rand}, \vec{g})$ measure with $\vec{g}$ being the variation of polarity of amino acid residua and A^{rand} the amino acid submission frequency matrix due to randomly generated genetic codes. The arrows show the values of the F measure with A matrix corresponding to the observed amino acid interchange frequencies (Dayhoff); submission frequencies due to natural code and due to optimized nucleotide mutation probability matrix.

Qualitatively we can distinguish 3 different cases. In the first one, referring to the property vectors within the space spanned by $\vec{h}^{(4)},...,\vec{h}^{(7)}$, we can conclude that these were recognized in the adaption of the genetic code to a neutral strategy. This is also borne out by Fig. 2 which gives direct evidence that one nucleotide exchanges tend to connect similar residua. The second case is the inherent property represented by $\vec{h}^{(8)}$. Properties of this kind seem to be recognized at a higher stage

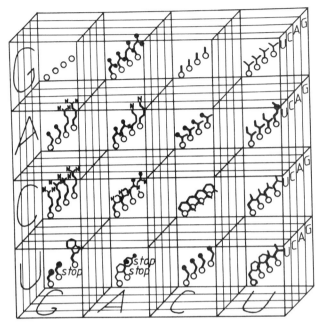

Figure 2. The natural genetic code (represented in a scheme due to Eigen and Schuster (22)). The vertical and horizontal directions correspond to the first, resp. second place in the codeword; nucleotides in reset perspective are in third place. Only the backbone of amino acid residua is symbolized, hydrogen atoms being omitted. The empty circles represent C atoms, small black circles oxygen atoms, big black circles are the sulphurs and the N is nitrogen.

of neutralistic adaption since their $F(A^{Day}, \vec{g})$ value is below the main peak of the F-measure distribution of random codes, whereas $F(A^{nat}, \vec{g})$ is located on top of it. As a third case we refer to the subspace of property vectors with $\lambda_i > \lambda_o$, which play no role whatsoever in a neutral strategy.

4. Optimization of the mutation probability matrix

In the preceding section we evaluated, in terms of the F measure, to which degree the coding of amino acids contributed to the neutralistic mutational pattern. Our conclusions were based on the assumption of equally probable nucleotide exchanges. If this were not true in general, it could be that the existing code was even further adapted in the sense that the submissions of mutations to the protein level fitted even closer the observed probabilities. Since it is hard to tackle this question inductively, we felt it was appropriate to start from the empirical Dayhoff exchange probabilities and determine from it the 4 x 4 nucleotide exchange matrix which predicts a mutational submission pattern as close as possible to the Dayhoff matrix.

To that end we must write down explicitly the relationship between the nucleotide exchange matrix $w^{(n)}$ of rank 4 and the matrix of submissions on the protein level

$w^{(a)}$ of rank 20. Consider a pair of amino acids a_i and a_j which are unambiguously coded by the nucleotide triplets

$$a_i = \{\alpha, \gamma, \pi\} \tag{6a}$$

and $\quad a_j = \{\beta, \delta, \varphi\} \tag{6b}$

The elements of the transition probability matrix $\overline{w}_{ij}^{(a)}$ can then be expressed as a product of elements from the $w^{(n)}$ matrix:

$$\overline{w}_{ij}^{(a)} = w_{\alpha\beta}^{(n)} \, w_{\gamma\delta}^{(n)} \, w_{\pi\varphi}^{(n)} \tag{7}$$

To take into account the fact that some amino acids are coded by more than one triplet, we have to sum the expression (7) over the pairs of triplets which code amino acids a_i and a_j. For simplicity of description we interpret indices i and j to identify partial components of amino acids u and v. The elements of the complete 20 x 20 mutation probability matrix then read

$$\overline{w}_{uv}^{(a)} = \sum_{i,j} \overline{f}_j^{(a)} \, \overline{w}_{ij}^{(a)} \, / \, \overline{f}_v^{(a)} \tag{8}$$

where

$$\overline{f}_v^{(a)} = \sum_j \overline{f}_j^{(a)} \tag{9a}$$

$$\overline{f}^{(a)} = f_\beta^{(n)} \, f_\delta^{(n)} \, f_\varphi^{(n)} \tag{9b}$$

The matrix of submissions to the protein level is completely determined by equs. (7)-(9), i.e. by the nucleotide replacement probabilities.

If there would be no selection on the protein level, the submission matrix $\overline{w}^{(a)}$ and the Dayhoff matrix would be identical. We can now solve the problem to determine the nucleotide replacement matrix $w^{(n)}$ which produces a submission pattern as close as possible to Dayhoff's. From here on, this is only a numerical problem. The results of our calculation may, yet, warrant some extra interest when more information on nucleotide exchange probabilities becomes available.

The solution of the question stated above can be achieved by the minimization of some non-negative measure in terms of the difference between submission frequencies and observed interchange frequencies. In constructing such a functional, we have to decide which quantity to compare. We chose to minimize the sums of the squares of differences between the elements of the relatedness odds matrix as defined by Dayhoff (12). This neutralizes the excessive influence of frequent residua which can occur in other minimization schemes. Our functional thus had the form:

$$H(w^{(n)}) = \sum_{u \neq v} (\overline{w}_{uv}^{(a)} \, / \, \overline{f}_u^{(a)} - w_{uv}^{(a)} \, / \, f_u^{(a)})^2 \tag{10}$$

The equations resulting from the minimization of (10) are nonlinear in $w^{(n)}$ but could be solved by a brute force Monte Carlo procedure.

The mutation probability matrix for the nucleotides and the corresponding composition vector constitute a system of 8 degrees of freedom. The 4 x 4 matrix and its composition vector require 20 numbers, but are constrained by 12 relationships. These are four equs. (A5), six equs. (A7), the normalization condition and the setting of the PAM value for the mutation matrix. To minimize the functional (10) we started from the uniform matrix $w^{(n)}$ as a zeroth approximation and varied all six independent off diagonal elements by random steps within a prescribed range while the 3 independent components of the composition vector were derived from the overdetermined equs. (9). After some tenthousand steps no further decrease in the functional could be achieved and the possibility of a local minimum was ruled out by standardized state of the art procedures. This was done for $PAM^{(n)} = 2$. $w^{(n)}$ for higher $PAM^{(n)}$ values was constructed according to (A4) and used further to define $w^{(a)}$ via equ. (8) which then entered the functional (10). The results are collected in Table V.

The results of the optimization procedure now allow a more subtle judgment on the adaption of the genetic code to the neutral strategy. Fig. 1 shows that after optimization of the nucleotide matrix the F measure of polarity does not improve. However, as can be seen from Table III, for the subspace of the significant property vectors the F measure is lower on average after optimization of the nucleotide mutation matrix.

Furthermore, the correlation coefficient between the observed amino acid interchange frequencies by A^{Day} and the corresponding calculated submission frequencies increases from 0.403 for A^{Day} / A^{nat}, to 0.623 for the pair A^{Day} / A^{opt}. Table IV demonstrates how the various amino acid composition vectors relate to Dayhoff's.

Table IV
The correlation coefficients of the Dayhoff composition vector with respect
to various calculated composition vectors

composition vector based on codon multipicity (examples: f(gly)=4/61; f(ser)=6/61 ...)	0.632
composition vector based on optimized nucleotide replacement probabilities	0.657
Maximum entropy formalism with one constraint (residual volume)	0.738
Maximum entropy formalism with three constraints (residual volume, polarity and hydrophobicity)	0.776

Table V

The results of the optimization procedure

Mutation probability matrix of nucleotides and the corresponding composition vector

	U	C	A	G		
U	.9855	.0049	.0078	.0007	.242	U
C	.0041	.9779	.0064	.0058	.199	C
A	.0096	.0096	.9753	.0122	.300	A
G	.0008	.0075	.0105	.9813	.259	G

The rise of the correlation coefficient when going from the simple model, with the amino acid compositions being entirely determined by the degeneracy of the genetic code, to the scheme with optimized nucleotide replacement probabilities is small with respect to before mentioned comparison.

The conclusion is that some further adaption of the coding of amino acids to a neutral strategy beyond the A^{nat} stage is possible. There is hope that the true nucleotide replacement probability matrix can soon be derived from mutational studies on nucleotides where the silent mutations (nucleotide exchanges within degenerate codons) allow an estimate of mutational submission frequencies to the protein level. We will then be able to say precisely to which degree the coding of amino acids is adapted to the neutral strategy expressed by the Dayhoff matrix. There seem, yet, to be many factors involved in the determination of the nucleotide replacement matrix (see e.g. (19)) which are not completely understood at this stage.

5. A neutralistic interpretation of the control of mutations on the protein level

Our investigation of neutral strategies cannot be concluded without some hypothesis to which extent the neutralistic pattern of mutations might be due to nucleotide replacement probabilities or control of acception on the protein level. We have given convincing evidence that the genetic code favours neutralistic submissions but we also concluded that the submissions might support the actual amino acid replacement probabilities to the extent, witnessed by the correlation coefficients between A^{Day}, A^{nat} and A^{opt}. Is there a way to tell which functional requirements determine acception or rejection of a submitted mutation on the protein level?

In sec. 2 we found that the preservation of the amino acid properties polarity, residual volume and hydrophobicity were of primary importance according to Dayhoff's data. If we imagine that the proper retainment of these quantities cannot be accomplished completely on the nucleotide level with only six replacement probabilities to adapt, it will be interesting to see whether imposition of such functional restrictions combined with random selection on the protein level will lead to a mutation probability matrix closer to the empirical one.

This problem can be solved within the maximum entropy formalism (see, e.g. (20)). It was noted by R. Coutelle and G. L. Hofacker (21) that the gain in compositional information of an amino acid distribution $\vec{f}$ over a prior one, $\vec{f}^{\,\circ}$ was given by

$$\Delta S = \sum_{i=1}^{20} f_i \ln(f_i/f_i^\circ) \tag{11}$$

S is a non-negative quantity which vanishes only for $\vec{f} = \vec{f}^{\,\circ}$. If we identify f_i°, the "prior expectation", with the optimized composition vector (9a) of sec. 3, the composition resulting from a random process subjected to certain restrictions is, in the spirit of information theory, obtained by maximizing ΔS under given constraints. For computational simplicity we demand that the physico-chemical properties of amino acids are preserved in mutations in the average, i.e. we have constraints of the general form

$$\sum_i f_i G_i^{(r)} = \overline{G}^{(r)} \tag{12}$$

where $G_i^{(r)}$ are the components of the r-th property vector. The solution of the extremal procedure for f has the form

$$f_i = f_i^\circ \exp\left(-\sum_r G_i^{(r)}\right) / \left\{ \sum_j f_j^\circ \exp\left(-\sum_r \lambda_r G_j^{(r)}\right)\right\} \tag{13}$$

with Lagrange parameters λ_r which we determined by a Monte Carlo procedure.

The extremal procedure was carried through in 2 cases: a) one constraint only which was residual volume; b) three constraints being residual volume, hydrophobicity and polarity.

The conserved averages $\overline{G}^{(r)}$ were the empirical ones. Fig. 3 shows a marked average improvement of the composition vectors calculated in this way with respect to Dayhoff's. The correlation coefficients of the resulting composition vectors with respect to the Dayhoff ones are given in Table IV. We consider these results quite remarkable in view of the simplistic assumptions made about the constraints. It lends further support to the idea that the adaption of the genetic code to a neutral strategy stems from an earlier stage of evolution but has been superceded by a rather strict functional control on the protein level.

Discussion

In the preceding sections we tried to analyze the apparent neutral strategy of molecular evolution. We first gave quantitative support to the conjecture that neutral mutations served the retainment of amino acid properties in mutations. It

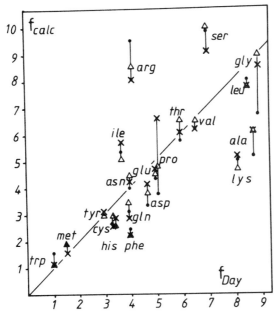

Figure 3. Plot of the calculated amino acid composition vectors (in %) obtained in three different ways, versus the experimentally determined ones. The calculated composition vectors were obtained by 1) Optimizing the nucleotide mutation probability matrix and on its basis the amino acid mutation probability matrix and the corresponding composition vector were calculated (circles). 2) Applying in addition the maximum entropy formalism with the constraints being the residual volume conservation (triangles) resp. residual volume, hydrophobicity and polarity conservation (crosses).

was the whole space of properties with low F values which was preserved, not the individual physico-chemical properties polarity residual volume and hydrophobicity which altogether spanned it. This may be an indication that the control of the mutational pattern "sees" amino acid properties somewhat different from our own physico-chemical frame of reference.

On the basis of these more precisely defined amino acid properties we could take up the question to which extent the neutral mutational pattern was due to the distribution of mutations submitted to the protein level. The results depicted in fig. 1 and Table III demonstrated a remarkable adaption of the genetic code to the conservation of the 3 most essential amino acid properties polarity, residual volume and hydrophobicity. These findings alone would, after random selection on the protein level, support the conjecture of a neutral strategy. We then showed that the adaption of the genetic code to produce neutral submissions could go much further. Though we cannot compare these nucleotide replacement probabilities with reliable empirical data, our "educated guess" regarding the kind of mutational control executed on the protein level turned our surprising well. As borne out by fig. 3, the control on the protein level again seems to favour retainment of the 3 most important amino acid properties.

It is the simple property-preserving control of submissions on the protein level which needs further explanation. If one does not want to give up the Darwinistic principle one must find a process in this scheme involving only mutations and selection which leads to a fitter mutant. The solution may lie in the fact that the rate by which a species produces acceptable mutants is essential for its survival. In other words, if essential amino acid properties are changed in a mutation, its chance to get accepted becomes very small and the possibility to produce a superior mutant in a later generation is wasted. This is why we see a prevalent pattern of neutral mutations emerge by any statistics of past accepted mutational events. We will show in a forthcoming paper that neutral mutations actually emerge from a principle of optimal search (for a "survivor") by a mutational random walk in the phase space of protein sequences.

Acknowledgement

This work was supported by the Deutsche Forschungsgemeinschaft and the Fonds der Chemischen Industrie. We would like to thank Dr. J. Manz for discussions and the suggestion to optimize the nucleotide probability matrix.

Appendix 1

The stochastic description of mutational nucleic acid and protein replacements.

The stochastic formalism as it can be applied to neutralistic evolutionary changes can be found in the protein atlas by M. O. Dayhoff (12), a more comprehensive version is given by Coutelle and Hofacker (3). We briefly review some of the relevant formulas to show that they apply to the evolution of nucleotide as well as protein compositions.

The master equation

$$\frac{d}{dt} f_i = \sum_{j \neq i} W_{ij} f_j - f_i \sum_{j \neq i} W_{ji} \tag{A1}$$

is defined in terms of probabilities (resp. relative compositions) f_i and mutation rate probabilities W. In the protein case, the indices run from 1 to 20 in the case of RNA nucleotides from 1 to 4.

The time evolution operator of (A1) called mutation probability matrix describes the change in the nucleotide or protein composition in time dt by

$$w(dt) \vec{f}(t) = f(t+dt) = f(t) + \frac{f}{t} dt \tag{A2}$$

Comparison with (A1) yields

$$w_{ij} = \begin{cases} W_{ij}\, dt & i \neq j \\ 1 - dt \sum_{k \neq j} W_{kj} & i = j \end{cases} \tag{A3a}$$

The quantities

$$dt \sum W_{ki} \tag{A3b}$$

give the probability that the i-th element will mutate (the suitably normalized sum is called mutability). The expected number of elements dN hit by mutations in dt is then proportional to the average value of the sums in (A3b)

$$\frac{dN}{N} = \frac{dt}{R} \sum_{i \neq j} W_{ij} \tag{A3c}$$

with R = 20 or 4.

The quantity (A3c) multiplied by 100 is introduced as the percentage of accepted point mutations (PAM's), (4),

$$d(PAM) = (100/R)\, dt \sum_{i \neq j} W_{ij} \tag{A3d}$$

and PAM's can be looked at as proporational to the time variable on the mutational clock,

$$PAM = (100/R)t \sum_{i \neq j} W_{ij} \tag{A3e}$$

The mutation probability matrix for a finite PAM value can be obtained by repeated application of the operator (dt)

$$w(\tau) = \lim_{\substack{N \Rightarrow \infty \\ Ndt \Rightarrow \tau}} \{w(dt)\}^N \tag{A4}$$

There is a special state $\vec{f}^o$ with a stationary property, i.e. $d\vec{f}^o/dt = 0$, which implies that the mutations leading from an element of a certian type to any other one are cancelled by the reverse mutations. Obviously, this f must belong to the eigenvalue $\lambda = 1$ and

$$w(t)\vec{f}^o = \vec{f}^o \tag{A5}$$

Furthermore, as the r.h.s. of (A1) vanishes, we have

$$\sum_{j \neq i} (W_{ij}\, f_j^o - W_{ji}\, f_i^o) = 0 \tag{A6}$$

The determinant of the linear equations (A6) must be zero to assure conservation of probability and a solution f therefore exists. The question whether detailed balance, namely

$$W_{ij}/W_{ji} = f_i^\circ / f_j^\circ \qquad (A7)$$

is a reasonable assumption is still open; it would assure a simple relaxation type development of the composition vector $\vec{f}$ (t) towards $\vec{f}^\circ$. In either case, the neutral mutations tend to establish a stationary composition.

Note that $w_{ij}f_j$ is proporational to the number of accepted j $\Rightarrow$ i point mutations (see sec. 2),

$$A_{ij} \propto w_{ij} f_j.$$

References and Footnotes

1. Woese, C. R., *The Genetic Code*, New York, Harper and Row (1967).
2. Borstnik, B. and Hofacker, G. L., to be published.
3. Coutelle, R. and Hofacker, G. L., *J. Theor. Biol. 15*, 615 (1982).
4. Dayhoff, M. O., ed., *Atlas of protein sequence and structure, Vol. 5*, Natl. Biomed. Res. Found., Washington, D. C., (1972).
5. Kubota, Y., Takanashi, S., Nishikawa, K., and Ooi, T., *J. Theor. Biol. 91*, 347 (1981).
6. Zimmerman, J. M., Eliezer, N., and Simha, R., *J. Theor. Biol. 21*, 170 (1968).
7. Grantham, R., *Science 185*, 862 (1974).
8. Goldsack, D. E., and Chalifoux, R. C., *J. Theor. Biol. 39*, 645 (1973).
9. Jones, D. D., *J. Theor. Biol. 50*, 167 (1975).
10. Woese, C. R., *Naturwissenschaften 60*, 447 (1973).
11. Aboderin, A. A., *Int. J. Biochem. 2*, 537 (1971).
12. Dayhoff, M. O., ed., *Atlas of protein sequence and structure Vol. 5, Suppl. 3*, Natl. Biomed. Res. Found., Washington, D. C., (1978).
13. Oobatake M. and Ooi, T., *J. Theor. Biol. 67*, 567 (1977).
14. Chou, P. Y. and Fasman, G. D., *Biochem. 13*, 222 (1974).
15. Sober, H. A., ed., *Handbook of Biochemistry, Selected Data for Molecular Biology, 2nd ed.*, The Chemical Rubber Co. (1970).
16. Alf-Steinberger, C., *Proc. Nat. Acad. Sci. 64*, 584 (1969).
17. Jukes, T. H., *Molecules and Evolution*, Columbia Univ. Press, New York (1966).
18. Yockey, H. P., *J. Thero. Biol. 67*, 345 (1977).
19. Kimura, M. *Evolution of Genes and Proteins*, Eds., Nei, M. and Koehn, R. K., Sinauer Associates Inc. (1983).
20. Levine, R. D. and Tribus, M., ed., *The maximum entropy formalism*, MIT Press (1979).
21. Coutelle, R., Hofacker, G. L., and Levine, R. D., *J. Mol. Evol. 13*, 57 (1979).
22. Eigen, M. and Schuster, P., *The Hypercycle*, Springer Verlag (1979).

Structure & Motion: Membranes, Nucleic Acids & Proteins,
Eds., E. Clementi, G. Corongiu, M. H. Sarma & R. H. Sarma,
ISBN 0-940030-12-8, Adenine Press, Copyright Adenine Press, 1985.

Metal Ions in Biochemistry

Ivano Bertini and Claudio Luchinat
Department of Chemistry
University of Florence, via G. Capponi 7
50121 Florence, Italy

Abstract

A broad survey is given of the role of metal ions in living organisms, and the interaction of metal ions with biomolecules is discussed. Among redox reactions the studies on Ru-cytochrome c and on azurin reacted with chromium are surveyed in order to appreciate some problems related to electron transfer processes. The state of knowledge on cytochrome c oxidase and on its function is summarized; then erythrocyte superoxide dismutase and liver alcohol dehydrogenase are discussed in some detail. Among Lewis acids carbonic anhydrase and carboxypeptidase A are presented and their function and reactivity compared. Metal transport is also discussed with particular attention devoted to the structure of siderophyllins.

1. Introduction

In the first half of 19th century the importance of iron for human life was already established when hemoglobin was isolated and analyzed. Its central place in cell respiration became clear when cytochromes were discovered by MacMunn and by Keilin in 1925(1). Since 1930 more and more metal ions have been recognized as essential;(2) however, there is an increasing consensus on dating(3) around 1950 the start of a specific interest in the relation between chemical structure and biological function of metal protein systems,(4,5) within the frame of exploding advances in both biochemistry and inorganic chemistry. This new field, still lacking of sharp boundaries, is known today either as bioinorganic chemistry or inorganic biochemistry.

Metallic elements, or elements commonly considered by inorganic chemists, found in a healthy human adult of 70 Kg of weight are listed in Table I(6). Asterisks indicate that the element is essential to either animals, plants, or both. Sodium, potassium and magnesium are major components, together with chlorine, of body fluids and cytoplasm. They are present as free ions. Calcium is a component of the skeletal material and is also found, together with sodium, as ion outside cellular membranes. The less abundant transition metals are generally associated with proteins, or with smaller ligands. Manganese (II) is somewhat similar also to magnesium (II). Other elements like aluminum, strontium, barium, lead, cadmium, and tin are present ubiquitously but they are not known to be essential to any organism.

Table I

Inorganic Elements in a healthy human adult of 70 Kg weight (ppm on a wet basis)
Asterisks indicate established essentiality for animals, plants, or both(6)

Element	Quantity (ppm)	Element	Quantity (ppm)
*Ca	1.4×10^4	*Se	0.4
*S	6.4×10^3	Cd	0.3-0.4
*P	6.3×10^3	Ba	0.3
*Na	2.6×10^3	*I	0.2-0.3
*K	2.2×10^3	*Mn	0.17-0.3
*Cl	1.8×10^3	*Mo	0.2
*Mg	4.0×10^2	*B	0.2
*Fe	60-70	*As	0.05
*Si	40	*Cr	0.04
*Zn	20-33	*Ni	0.04
Rb	5.1	Li	0.03
Sr	4.6	*V	0.03
Br	1.7-7.5	*Co	0.016
Pb	1.7		
*Cu	1.1		
Sn	0.07-1.5		
Al	0.2-0.6		

In this contribution we will discuss the role of the various metal ions in biological functions and then will focus the interest on the role of metals in catalysis: the structural features of the metal environment will be analyzed together with the investigation methodologies, and some examples will be presented in detail. Finally, the metal transport processes will be briefly reviewed.

2. Role of Metal Ions

A living system continuously performs a huge amount of chemical transformations which occur in different compartments of the body, and yet are complicatedly related in chains of chemical events spanning different tissues and organs. The body as a whole is not isolated from the environment and continuously exchanges chemicals with it through respiration, nutrition and excretion. It is probably better not to visualize the living system as a reaction vessel where reagents evolve towards products until chemical equilibrium is reached but rather as an industrial plant where reagents enter and products are removed in a non-equilibrium steady state, the goal being not to obtain the products but to keep the plant working.

Given this "Philosophy" of living systems, it is no surprise that even the simplest chemical reactions are often carried out *in vivo* by chemical routes different from those a chemist would follow in the laboratory; nor that metallic elements are utilized for so many purposes, and to so high a degree of sophistication, in tissues that are ninety-nine per cent (99%) made of "organic" molecules (excluding water). Indeed, the very same chemical complexity needed by a living system to be what it is has surely required the exploitation of the great variety of chemical properties and reactivities displayed by metal ions since the beginning of the evolution of life. As an example, duplication of polynucleotide chains, one of the key steps in the origin of life, has been shown to occur *in vitro* with a reasonable fidelity in the presence of divalent metal ions(7).

The utilization of the various metal ions is of course related to their different chemical properties. Alkaline ions, like Na^+ and K^+, usualy exist as free aquaions in aqueous media and show little tendency to firmly interact with ligands, with a few exceptions. Therefore, they are mainly used in animal tissues as electric carriers to regulate the electrochemical potential across the cell membrane; often the potential difference between the inside and the outside of the cell is as high as 0.1 V, the inside potential being negative with respect to the outside. The build-up and discharge of this potential allows for instance the nerve fibers to conduct impulses, and therefore to transfer information and muscle commands across the body. Under the conditions for such a potential difference to exist, the interior of the cell is found to contain large amounts of K^+ but relatively little Na^+, while the reverse holds for the outside. Therefore the cell has a means of discriminating between K^+ and Na^+, which are so similar from a chemical point of view, and uses this capacity to create a potential difference. This is done at the expenses of the energy liberated by hydrolysis of ATP, in turn catalyzed by a metal-activated enzyme (ATPase) embedded in the cell membrane(8,9).

The discharge of the electric potential is probably accomplished by ionophores, generally multidentate oxygen- containing ligands especially suited to coordinate alkaline ions and to carry them through the hydrophobic inner section of the membrane(10,11). In this case, the driving force is simply due to the concentration gradient of the ion.

Mg^{2+} and Ca^{2+} have a weak tendency to firmly interact with donor atoms of proteins or of other biological molecules; according to this property, which is also partially common to Mn^{2+}, these ions are used as enzyme activators: the transient interaction of the ion with a protein may facilitate binding and proper orientation and activation of a substrate through formation of a ternary complex, or trigger a conformational change which induces an active conformation at the catalytic site. Typical in this respect are the functions of Mg^{2+} and Mn^{2+} ions as activators of kinases, or of Mg^{2+} as activators of ATPase. Owing to its role of ion pump, the latter enzyme also requires K^+ and Na^+ as cofactors; the detail of its mechanism are still controversial(9), but it is instructive to note that the ATPase activity, *i.e.* the efficiency of pumping is regulated both by the concentration of the "wrong" monovalent cation on each side of the membrane and by the concentration of Mg^{2+} in the *interior* of the cell. By regulating the amount of Mg^{2+} in the cellular fluid the cell has therefore a means of controlling the build-up of the potential difference.

Mg^{2+} and Ca^{2+} transport across the membrane is also controlled by a specific ATPase(12). Calcium regulation in particular is important for muscle contraction. Ca^{2+} is pumped by the appropriate ATPase in the sarcoplasmic reticulum surrounding the muscle fiber, where it is temporarily stored as a complex with calcium-binding proteins. Upon nervous impulse, the isolated ATPase molecules on the sarcoplasmic reticulum membrane aggregate(13,14), opening up hydrophobic channels for the rapid efflux of Ca^{2+} driven by its concentration gradient. Ca^{2+} then interacts with the protein troponin-C, which in turn triggers another ATPase present in the muscle fiber; this results in locking together the actin and myosin parts of the fibers causing macroscopic muscle contraction(15).

Calcium also plays a structural role in proteins, where it is prevalently bound to oxygen atoms of carboxylate groups. Finally the low solubility of calcium salts is exploited for the formation of bones, teeth, and shells. The controlled precipitation of calcium phosphate crystals in the formation of bones(16,17) is another example of the fine tuning of processes, such as nucleation, that are of difficult control in chemical laboratory.

The other transition metal ions are firmly bound to specific proteins and explicate some functions compatible with the properties of the metal ions. For example, zinc (II) acts as a Lewis acid in a series of enzymatic functions(18,19). It is well known that transition metal ions are good Lewis acids, their Lewis acidity following the Irwing-Williams series ($Mn^{2+} < Fe^{2+} < Co^{2+} < Ni^{2+} < Cu^{2+} > Zn^{2+}$). In biochemical processes only zinc(II) and nickel(II) are used, and for different substrates. Cobalt(II) is as good as zinc(II); however, nature prefers the latter, presumably owing to the higher stability constant of its complexes(20). The best Lewis acid in inorganic chemistry, *i.e.* copper(II) is never used as such, probably because its redox properties may damage biopolymers. As outlined by Williams(20), the Lewis acid properties of zinc(II) are enhanced by the binding mode to the protein. Zinc(II) favors lower coordination numbers with respect to nickel(II) and copper(II); indeed, when a metal is tetracoordinated, it is more acidic than when the coordination number is larger. This property is accompanied by a large coordination flexibility. Furthermore, OH^- and H_2O ligands which are often involved in the catalytic process exchange better than in nickel(II) complexes.

Dioxygen is carried by metalloproteins and in particular by metal ions with π donor capacities. Iron(II) heme proteins are the natural carriers in mammalians. The heme moiety, besides favoring the low spin configuration upon oxygen binding, decreases the acidity of the complex, so that hemoglobin has low affinity for water at the axial position. Pairs of adjacent copper(I) ions are also effective as dioxygen carriers. Iron(II) and copper(I) are also oxygen activators. The transport of oxygen is obviously a primary requirement for living systems; again, it is evident that only metal ions whose chemical properties are appropriately tuned by particular ligands are suited for the purpose.

Perhaps the most striking example of the essentiality of metal ions is the requirement to perform redox reactions common to all living systems. The metallic elements mainly used for redox reactions are iron, copper, cobalt, molibdenum, and nickel. Redox reactions can be classified as electron transfer, oxygen and dioxygen insertion, nitrogen fixation, and hydrogen transfer. Iron is generally associated with reactions involving hydrogen and oxygen, molibdenum with nitrogen (N_2, NO_3^-), cobalt with carbon (CH_3^-). Electron transfer processes usually occur between couples with close reduction potentials and with centers within 1.5 nm. Again, the particular heme group and axial ligands are capable of modulating the potential of the Fe^{3+}/Fe^{2+} couple and, likewise, the coordination geometry and nature of the donor atoms are capable of determining the potential of the Cu^{2+}/Cu^+ couple. The change in oxidation number following the redox reaction may cause some change in the coordination sphere with some consequent protein rearrangement, possibly up to the external part of the protein in order to signalling the oxidation state to reactants(21). Oxygen insertion reactions are performed by the heme-FeO group when iron is in the oxidation state +4 (and even formally +5), and by MoO groups. Molibdenum is a 4d element which can afford an electron transfer capability of three electrons with relatively small potentials. Regarding the Fe-heme group, systems with positive reduction potential are good oxygen carriers, whereas systems with negative potentials are good oxygen activators. Again, this depends on the relative stability of the $M^{n+}O_2$ moiety with respect to $M^{(n+1)+}O_2^-$, $M^{(n+2)+}O_2^{2-}$, and $M^{(n+2)+}O^{2-}$.

Hydrogen transfer reactions are generally catalyzed by organic molecules with reduction potentials close to zero. Hydride transfer reactions often occur through NAD^+ and are catalyzed by several enzymes among which a zinc(II) enzyme (liver alcohol dehydrogenase). Some hydrogen transfer occurs through radicalic reactions. For example, coenzyme B_{12} undergoes the following reaction:

$$Co^{II}-CH_2-R \rightleftarrows Co^{II} + \ ^{\cdot}CH_2-R$$

and then is the radical the active species at the enzyme site(22). The cobalt ligand of coenzyme B_{12} is corrin, which produces such an in-plane strong field to make the Co^{II}/Co^I and Co^{III}/Co^{II} potentials close to zero.

Almost fifty per cent (50%) of the known enzymes firmly contain metal ions or require metal ions in order to perform their function. The different uses of different metals which apparently could carry out the same reactions is a fascinating feature of biological systems. There is some overlap of function for metallo and non-metalloenzymes, but rare. The use of a particular element with respect to a similar one for a given function, its selective binding through suitable ligands, its transportation in fluids and membranes, and finally the explication of the biological function in which the element is involved, require highly sophisticated chemical processes which we must try to understand.

3. Coordination Chemistry of Metal Ions

Alkaline metal ions do not give rise to stable metal complexes with ligands having donor atoms other than oxygen. In aqueous solutions even oxygen-containing mono- or bi-dentate ligands are not able to displace water molecules from the aquaions.

However, these ions are efficiently complexed by certain classes of natural or artificial multidentate ligands. The long-standing interest(23) in their coordination chemistry is due to the problem of rationalizing the apparent high discriminating power of biological systems towards K^+ or Na^+(24). Extensive studies on synthetic cyclic(25,26) or bicyclic(27-30) ethers (crown ethers or cryptands, respectively) show that a major way of distinguishing between Na^+ and K^+ is simply the size of the cavity, and reflects the difference in ionic radii of the two ions (95 and 133 pm, respectively); on the other hand, ions of similar size but different charge like, for instance, Na^+ and Ca^{2+} (ionic radius 99 pm) can be discriminated by changing the side substituents on the rings. These properties can be usefully compared with those observed on natural macrocyclic ionophores(31,32) which are able to selectively carry monovalent cations across the cell membranes. Related to the above issue is of course the fascinating problem of the active transport efficiency of membrane proteins. In general, these proteins (probably including some ATPase's) are dimeric or oligomeric aggregates disposed across the entire membrane section in such a way to allow, when properly triggered, the passage of cations by opening hydrophylic channels amidst the subunits(33). The detailed mechanisms by which channels open up and force cations against concentration gradients are still largely unknown, although they have to be ultimately based on the transient formation of coordination bonds between the cation and oxygen donors on the surface of the subunits facing the channel. A beautiful crystal structure of a macrocyclic polyether has been recently reported(34) where the cycles are piled one on top of the other, in such a way to form a channel displaying mixed-site K^+ binding. The structure (Figure 1) can be looked at as a multiple exposure photograph of a single K^+ ion *migrating* through the channel.

Alkaline earth metal ions are relatively more suited to give rise to coordination compounds also with mono- and bi-dentate ligands. While Ca^{2+} and the heavier members of the group still prefer oxygen ligands, Mg^{2+} may also be coordinated by nitrogen donors, like in porphyrines, chlorophylls, and related compounds. In addition, Ca^{2+} has a greater tendency to be accommodated in distorted coordination polyhedra, sometimes reaching coordination numbers higher than six. These two properties provide the chemical ground for the differentiation of their biological role. As an example of the structural function of Ca^{2+} the X-ray structure of thermolysin(35) is shown in Figure 2. The presence of the four calcium ions prevents self-digestion and confers a high thermal stability to the protein.

A large share of the biochemistry of magnesium is connected with its interactions in the cell cytoplasm and nucleus with nucleotides, nucleosides, and their derivatives. With a bit of oversimplification, the general picture can be that Mg^{2+} acts as a bridge between specific enzymes and the above molecules, with formation of ternary complexes. This is the case of ATPase, where Mg^{2+} activates the hydrolysis of ATP(36,37), or RNA polymerase, where it has been proposed(38) that Mg^{2+} bridges the enzyme to the appropriate nucleoside triphosphate, the elongation substrate, catalyzing dephosphorylation of the latter (Figure 3).

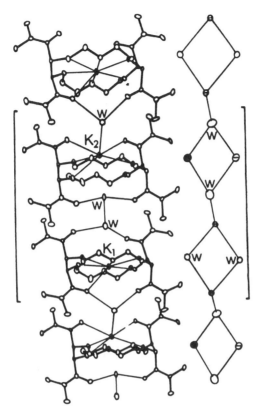

Figure 1. Crystal structure of the polyfunctional macrocyclic K^+ complex $([(M,K)_2,3H_2O]^{2+} \cdot [3KBr,4H_2O]^{2-})$, where $M = 2,3,11,12$ tetracarboxamide-18-crown 6. W stands for water and K_1 and K_2 indicate two non-equivalent K^+ sites in the channel(34).

Furthermore, Mg^{2+} ions are known to confer additional stability to DNA, RNA and polyribonucleotides. In every case the interaction is with the oxygens of the phosphoryl groups, as show in Figure 4(39).

The coordination chemistry of manganese (II) is somewhat in between that of alkaline earth ions and of the other transition ions; therefore manganese (II) can either act as an activator or form stable complexes with proteins. The other transition metal ions most often are found as stable complexes.

It is customary to use the formation constant

$$K = \frac{[PM]}{[P][M]}$$

where P is the protein and M is the metal, to define the borderline between the two

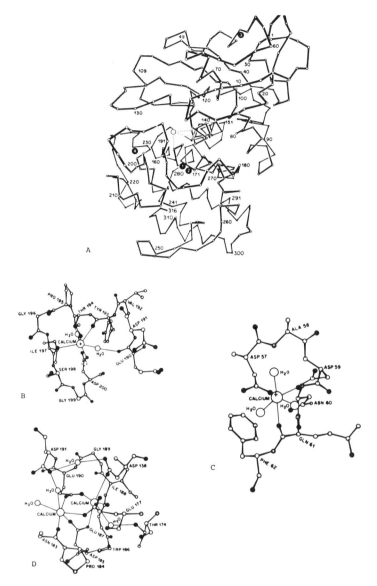

Figure 2. Backbone conformation of thermolysin showing the four calcium ions (●) and the zinc (○) ion (A); close-up views of the calcium (O) binding sites 1-2 (D), 3 (C) and 4 (B).

types of the interactions. When K is smaller than, say, 10^8 M^{-1} the PM adduct is defined as metal-protein complex and the protein a metal-activated enzyme. When K is larger than 10^8 M^{-1} the adduct is defined as metalloprotein. The affinity of proteins for the various metal ions follows the series Mg^{2+}, Ca^{2+} < Mn^{2+} < Fe^{2+} < Co^{2+} < Ni^{2+}, Cu^{2+}, Zn^{2+}. By following the previous statement, Ca^{2+}, Mg^{2+} and sometimes Mn^{2+} give rise to metal-protein complexes and act as protein activators or triggers.

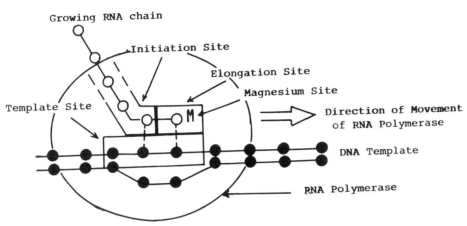

Figure 3. Proposed(38) activating role of Mg²⁺ in the function of *E. coli* RNA polymerase: Mg²⁺ binds the enzyme and coordinates the phosphate oxygen of the appropriate nucleoside triphosphate in the elongation site. The latter then reacts with the *α*-phosphate group of the nucleotide (initiation site) in the growing RNA chain.

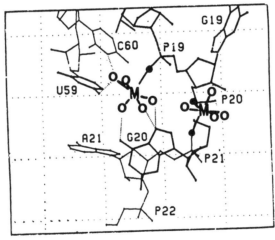

Figure 4. Schematic drawing of two of the coordination sites of Mg²⁺ (M) in the D loop of tRNA. One Mg²⁺ ion is coordinated to a phosphate (P19) oxygen (●) and to five water molecules (○), the second to the phosphate P20 and P21 oxygens and to four water molecules(39).

The distinction between the two types of interactions is also of geometric type. Activation generally occurs at the surface of the protein, involves few donor atoms from the protein (even only one), and allows the metal to complete the coordination sphere with water molecules most often to reach pseudooctahedral coordination. Stable binding often occurs inside the protein, more donor atoms from the protein are involved, and the coordination number and geometry are variable.

If the metal although bound inside, is exposed to solvent, *i.e.* the metal is in a cleft or cavity, it may perform direct catalytic activity. Therefore metalloenzymes are a subset of metalloproteins. Non exposed metal ions in metalloproteins may be used for electron transfer reactions, may have a role in stabilizing the ternary structure, or may be transported in metal-carrier proteins.

The metal ligands suitable for coordination bonds available in proteins are reported in Table II together with their pKa values. Upon coordination the apparent pK_a can be lowered by several units; for example, histidines with pK_a of 6-7, when coordinated to a metal ion can be titrated with pK_a of ~4. The pK_a values of Table II, with the latter consideration in mind, give the pH range for a residue to act as a ligand. Furthermore, there are groups with lone pairs which are essentially pH independent and which may be relevant as donor atoms (particularly methionine). Finally, among exogenous ligands we can list water and its anions OH^- and O^{2-}, anions in general, buffers, substrates and inhibitors.

Table II
Potential metal-binding groups available in proteins, and range of pK_a values
for those having acid-base properties

Group	pK_a
α-COOH	3.1-3.5
β,γ-COOH (Asp, Glu)	4.4-4.7
Imidazole (His)	6.4-7.2
α-NH$_2$	7.8-8
ϵ-NH$_2$ (Lys)	9.6-10.5
Phenolic-OH (Tyr)	9.6-9.8
Sulphydryl (Cys)	9.0-9.4
NH$_2^+$ (Peptide bond)	10.2-10.8
Guanidinium (Arg)	11.6-12.6
Aliphatic-OH (Thr, Ser)	>14
Thioether S (Met)	

The coordination chemistry of the resulting metalloproteins may be very unusual in the sense that there is a large variety of metal-donor atoms sets which, when discovered, had nothing to share with the known inorganic compounds. We can say that inorganic chemists have been pushed towards the synthesis of complexes similar in the chromophore to the naturally occurring chromophores in order to understand their peculiar structural and electronic properties. Typical examples are the blue proteins(40) (Figure 5) and the iron sulfur proteins(41) (Figure 6).

As a typical example of blue protein we can examine plastocyanin(40) (Figure 5). The chromophore is constituted by copper(II) (when the protein is in the oxidized state), two sulfur atoms (one from a cysteine and the other from a methionine) and two histidine nitrogens. The resulting geometry is pseudotetrahedral. An enormous number of small models have been prepared with the aim of reproducing the electronic and redox properties of the natural proteins. Of course, the strain in the

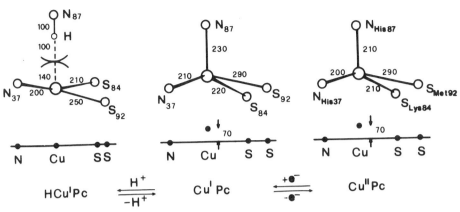

Figure 5. Copper site in plastocyanin (Pc) with copper-donors bond lengths (pm): HCu^{I}Pc is the reduced, redox inactive form; Cu^{I}Pc is the reduced, active form; Cu^{II}Pc is the oxidized, active form. The lower portion of each diagram shows the deviation of the copper atom from the N^{37}-S^{84}-S^{92} plane(40).

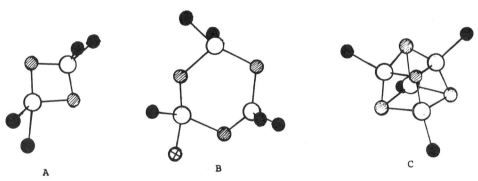

A   B   C

Figure 6. Selected examples of iron-sulfur clusters in proteins. A:2Fe-2S(Cys)^4 cluster in S. *platensis* ferredoxin. B: 3Fe-3S (Cys)^5(Glu)cluster in *A. Vinelandii* Ferredoxin. C: 4 Fe-4S(Cys)^{4}cluster in *C. Vinosum* HIPIP(41). (○=Fe; ○=S; ●=S(Cys); ⊗=O(Glu)).

bonds and angles(40) due to the protein structure cannot be successfully mimicked. The particular strain in the coordination polyhedron around the metal ion is referred to as "entatic state"(42,43) at the metal site, and is thought to be the way through which nature modulates the chemical properties of metal ions in order to accomplish a certain function. The iron sulfur proteins are typical electron carriers; formally one iron of the cluster changes its oxidation state from +2 to +3 or vice versa. The clusters may contain 2, 3, or 4 iron atoms. Each cluster provides a delocalized, highly polarizable redox center. Again the work in the field of inorganic chemistry has been quite useful in the understanding of the chemical bonds and of the chemical properties of the ligands(44).

In order to appreciate the unicity of the chromophores in metalloproteins we would like to recall the reader's attention to the fact that even a chromophore of the type

$$\left[\begin{array}{c} N \\ N \longrightarrow Zn \longrightarrow OH_2 \\ N \end{array} \right]^{2+}$$

as found in carbonic anhydrase (see section 6) has not been synthetized by inorganic chemists.

Finally, in a survey of the chromophores encountered in metalloproteins a special consideration should be devoted to porphyrinato, dihydroporphyrinato and corrinato anions (Figure 7).

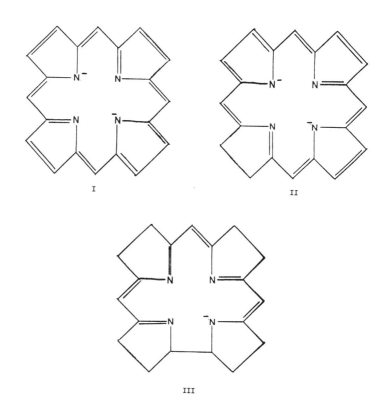

Figure 7. Schematic structure of porphyrinato(I), dihydroporphyrinato(II) and corrinato(III) anions.

It should be recalled that metal ions bound in the interior of the protein may be exposed to the solvent, in such a way that they can accomplish a catalytic function. The idea of entatic state of the metal ion and its environment has been stressed even for these systems(45). It is the authors' opinion that when the metal ion is exposed to solvent and the protein residues continue to be coordinated even in the presence of exogenous ligands, then the entatic effect has to be both less apparent and less important and the coordination compound, even if difficult to be reproduced by models, better fits with the ideas coordination chemists have about coordination compounds. The surface of the cavity, its geometrical contours, its hydrophobicity or hydrophylicity, and its charge, are powerful filters in order to select the proper substrate. With this respect it is instructive to analyze the affinity of anions as it depends on the charge in the cavity. Carboxypeptidase A contains in the cavity a Glu(46) residue which presumably has a pK_a around 5-6 (Table II). Indeed, k_{cat} displays a pK_a of ~6(47) indicating that a glutamate ion is necessary for the catalytic process. K_M of the substrate depends on pH with a pK_a of about 9(47). A further proton is thus released by the system constituted by the coordinated water and a Tyr residue. Carboxypeptidase A is soluble in high salt concentration, *e.g.* 1M NaCl. At pH > 6 chloride apparently does not bind the metal ion, while it does so at lower pH values(48). The presence of the negative glutamate ion is thus capable to disfavor anion binding and to definitely prevent chloride binding. NCO^- and N_3^- bind also in the pH range 6-9, and their affinity drops to zero with a pK_a of about 9(49). A further negative charge also therefore prevents strong ligands from binding. Similar to carboxypeptidase A is carbonic anhydrase: the metal ion of the latter enzyme binds anions with an affinity which decreases with increasing pH, and the pH dependence can be ascribed to the deprotonation of two groups(50). There is agreement now that one of them is the coordinated water molecule. Direct competition exists between anions and OH^-. Apparently OH^- is capable to prevent binding of all anions.

4. Metal Ion Substitution as Investigation Tool

Investigation of metalloprotein systems and of metals interacting with moieties of biological relevance in general is performed with all kinds of spectroscopies. However, a given spectroscopy or any investigation tool is more suitable for certain metal ions than for others. On this consideration, researchers have been trying to substitute the native metal ion with a different one and then ascertain the biological function of the new derivative. A typical example is provided by the zinc enzymes. Zinc(II) is not particularly suitable for electronic spectroscopy, is diamagnetic, and has a low-abundance isotope with low NMR sensitivity. However, zinc(II) can be substituted by many other divalent metal ions, often with maintenance of catalytic activity (Table III). Cobalt(II) is the best ion both as spectroscopic probe and biological substituent. Cobalt(II) can be investigated by electronic spectroscopy, MCD, EPR, NMR of protein ligands. The enzyme interaction with inhibitors and substrates can easily be monitored. In Table IV the metal ions that have been used to substitute the native ion are shown. The ionic radii are also reported. Potassium can be studied directly through NMR although it is easier on sensitivity grounds to

Table III

Substitution of zinc (II) in metalloenzymes with other divalent metal ions
and relative catalytic activity of the resulting derivatives(135)

Enzyme	Donor Set	Substituted Metals
Alcohol dehydrogenase	S_4^b; S_2NO^b	$Co^{II}(70)$, $Cu^{II}(1)$, $Cu^I(8)$, $Cd^{II}(30)$, $Ni^{II}(12)$
Superoxide dismutase	N_3O^b	$Co^{II}(100)$, $Cu^{II}(50)$, $Cd^{II}(100)$
Aspartate Transcarbamylase	S_4	$Mn^{II}(100)$, $Ni^{II}(100)$, $Cd^{II}(130)$
Transcarboxylase	?	$Co^{II}(100)$, $Cu^{II}(0)$
RNA polymerase	?	$Co^{II}(100)$
Carboxypeptidase	$N_2O_3^b$	$Mn^{II}(28)$, $Fe^{II}(29)$, $Co^{II}(215)$, $Ni^{II}(50)$, $Cu^{II}(0)$, $Cd^{II}(0)$, $Hg^{II}(0)$, $Co^{III}(0)$
Thermolysin	$N_2O_2^b$	$Mn^{II}(60)$, $Co^{II}(100)$, $Cu^{II}(105)$, $Cd^{II}(185)$, $Hg^{II}(100)$
Alkaline Phosphatase	N_4O; NO_5; O_6	$Mn^{II}(0)$, $Co^{II}(20)$, $Ni^{II}(0)$, $Cu^{II}(0)$, $Cd^{II}(0)$, $Hg^{II}(0)$
β-Lactamase II	N_3S	$Mn^{II}(3)$, $Co^{II}(11)$, $Ni^{II}(0)$, $Cu^{II}(0)$, $Cd^{II}(11)$, $Hg^{II}(4)$
Carbonic Anhydrase	N_3O^b	$Mn^{II}(4)$, $Co^{II}(56)$, $Ni^{II}(5)$, $Cu^{II}(1)$, $Cd^{II}(4)$, $Hg^{II}(0)$, $Co^{III}(0)$
Aldolase	$N_3O + ?$	$Mn^{II}(15)$, $Fe^{II}(67)$, $Co^{II}(85)$, $Ni^{II}(11)$, $Cu^{II}(0)$, $Cd^{II}(0)$, $Hg^{II}(0)$
Phosphomannose Isomerase	?	$Mn^{II}(50)$, $Co^{II}(90)$, $Ni^{II}(0)$, $Cu^{II}(130)$
Pyruvate Carboxylase	?	$Co^{II}(100)$

[a] Percent activities relative to the native enzyme in parentheses.
[b] From x-ray data.

study Tl^+ and even Cs^+. Unfortunately, Tl^+ is a good NMR probe but not as good as far as biochemistry is concerned. Tl^{3+} has been used as substituent for Fe^{3+}.

Table IV

Metal Substitution

Native Cation	Ionic Radius (pm)	Probe with Ionic Radius (pm)	Biochemistry
K^+	133	$Tl^+(140)$	moderate
		$Cs^+(169)$	good
Mg^{2+}	65	$Mn^{2+}(80)$	good
		$Ni^{2+}(69)$	poor
Ca^{2+}	99	$Mn^{2+}(80)$	poor
		$Eu^{2+}(112)$	redox reactions
		$Ln^{3+}(115-93)$	moderate
Zn^{2+}	69	$Co^{2+}(72)$	excellent
		$Cu^{2+}(72)$	poor
		$Mn^{2+}(80)$	moderate
		$Cd^{2+}(97)$	moderate
Cu^{2+}	72	$Co^{2+}(72)$	poor
Cu^+	96	$Ag^+(126)$	moderate
Fe^{3+}	64	$Gd^{3+}(102)$	moderate
		$Tb^{3+}(100)$	moderate
		$Eu^{3+}(102)$	moderate
		$Ho^{3+}(97)$	moderate
		$Ga^{3+}(62)$	moderate
		$Tl^{3+}(95)$	redox reactions

Mg^{2+} is successfully substituted by Mn^{2+}, which is commonly studied by EPR and NMR. Ca^{2+} is often substituted by lanthanides, which are investigated with typical techniques (NMR, fluorescence). Cu^{2+} can be directly investigated through EPR spectroscopy, although it has also been substituted with Co^{2+}.

5. Redox Reactions

We will pick up in this section some examples to illustrate the accomplishments in the area, and the problems which still deserve further investigation efforts. First of all we will discuss the pathways of electron transfer in metalloproteins and then the mechanism of cytochrome oxydase which reduces the dioxygen molecule to water. Then attention will be devoted to better known enzymes like copper-zinc superoxide dismutase and liver alcohol dehydrogenase.

5.1 Electron Transfer Pathways

One problem in electron transfer reactions is the understanding of the relationship between the rate of electron transfer within any pair of protein redox centers and their separation. Gray et al.(51) have synthesized a pentaamineruthenium(III) derivative of horse heart cytochrome c. The sixth ligand of ruthenium is the protein residue histidine 33, which is located 1.5 nm away from iron. The reduction potentials of the two redox sites are: $Fe^{3+}/Fe^{2+} = .26$ V; $Ru^{3+}/Ru^{2+} = .15$V . $Ru(bipy)_3^{2+}$ in the luminescent excited state is a strong reductant ($-.8$ V), and reduces ruthenium(III) in Ru-Cytochrome c faster than iron(III). Then the following intramolecular electron transfer takes place

$$Ru^{2+}\text{-cytochrome c}(Fe^{3+}) \Rightarrow Ru^{3+}\text{-cytochrome c}(Fe)$$

The intramolecular rate constant is about 20 s^{-1}, and is in the same general range as the turnover rates of several multisite redox enzymes. Such rate constant is surprisingly temperature independent implying that solvation and inner sphere reorganizational barriers are negligibly small, or somehow cancel each other. In natural systems intramolecular electron transfer rate constants may be temperature dependent.

The pathway for intermolecular electron transfer is another point to be understood. Pecht(52) has allowed hexaaquachromium(II) to interact with oxidized azurin. $Cr(OH_2)_6^{2+}$ reduces azurin and the resulting chromium(III) forms a kinetically inert complex with the protein. By recognizing the residues interacting with chromium(III) the reductant binding site can be identified, and the chromium(II)$\Rightarrow$copper(II) electron transfer pathway can be figured out. It is suggested that chromium(III) is coordinated to Lys-85 and Glu-91. A site which allows chromium to simultaneously bind Lys-85 and Glu-91 can be located $\sim$1nm away from the copper site. At this point it is assumed that the metal ion remains coordinated to the very same site where it donates the electron, and then the pathway of the electron from chromium(II) to copper(II) can be considered. Lys-85 and Glu-91 define an opening into the protein's interior exposing the N_δ of the imidazole ring of His-35, which is coplanar with that of His-46, the latter being coordinated to copper(II) (Figure 8). The N_δ-C_ϵ edge of His-35 is overlapping the C_δ-N_ϵ edge of His-46, with a separation of ~380 pm between the planes. Chromium(II) could interact with Lys-85 and Glu-91, and have a coordinated water interacting via hydrogen bond with the N_δ of His-35. The latter atom belongs to a π system which includes His-35, His-46 and d_π orbitals of copper(II). This could be the electron transfer pathway.

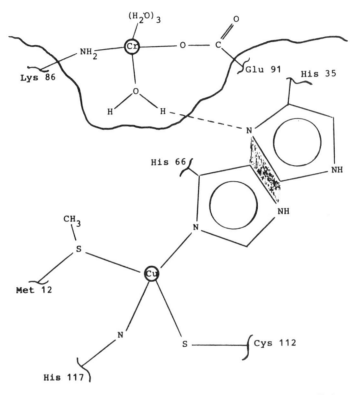

Figure 8. Proposed electron transfer pathway from chromium(II) to copper(II) in azurin(52).

The effect of chromium(III) binding on the reactivity of azurin with two of its presumed partners in the redox system of the bacterium has also been investigated(53). The electron exchange reaction between azurin and cytochrome c_{551} occurs with the same mechanism, but is significantly smaller in the presence of chromium(III). This suggests the involvement of the labeled region in the electron transfer reaction. On the other hand, the cytochrome oxidase-catalyzed oxidation of azurin by O_2 is unaffected by the presence of chromium(III), indicating the existence of a second active site involved in the reduction of cytochrome oxidase. It has been proposed that in cytochrome c there are different binding sites for oxidants and reductants, even for very similar reagents like bis(dipicolinato) cobaltate(III) and bis(dipicolinato) ferrate(II)(54). The protein can discriminate among different oxidants or different reductants as well(55). Therefore, several possible electron pathways exist inside the protein.

5.2 Cytochrome c Oxidase

The enzyme is located in the mitochondrial membrane, where it takes care that the reaction

$$O_2 + 4 \text{ Ferrocytochrome c} + 4H^+ \Rightarrow 2H_2O + 4 \text{ Ferricytochrome c}$$

occurs rapidly and without any toxic intermediates of oxygen reduction. Cytochrome c oxidase contains four inequivalent metal centers: two are heme A, *i.e.* iron porphyrins with a formal group and a hydrophobic isoprenoid side chain, and the other two are unusual copper(II) ions (Cu_A and Cu_B) which are different from one another. One iron and one copper ions (Fe_{a3} and Cu_B) are in the site capable of binding O_2 and exogenous ligands, whereas the other two metal ions (Fe_a and Cu_A) are buried into the protein and serve to mediate the transfer of electrons between cytochrome c and the oxygen reduction site. Fe_a accepts one electron from ferrocytochrome c and transfers it to the Cu_A center, *i.e,* the other buried copper site. The reduction potential of Fe_a has to be similar to that of ferrocytochrome c. Fe_a is six coordinated and low spin, with two imidazoles as axial ligands(56). Cu_A transfers the electron to the O_2 reduction site; it is bound to one histidine nitrogen and a cysteine sulfur(57). Two other sulfur atoms have been suggested as donors(58). The electronic relaxation rates as estimated by ENDOR spectroscopy indicate that there is some weak interaction between Fe_a and Cu_B, thus suggesting an upper limit for their distance of about 1.5 nm(59). Fe_{a3} and Cu_B are the oxygen binding site. When they are in the reduced form iron(II) is high spin with $S=2$(60) and copper(I) is diamagnetic. When they are fully oxidized, iron(III) is high spin with $S=5/2$ and copper(II) is $S=1/2$; since the two ions are strongly antiferromagnetically coupled the resulting system is again $S=2$(61).

Both fully oxidized and fully reduced Fe_{a3}-Cu_B systems are EPR silent. Upon partial reduction of the oxidized enzyme at neutral pH, an EPR spectrum is observed typical of high spin heme with a g value of 6, whereas at high pH the EPR spectrum indicates a low spin heme (a g value at 2.6)(62). Therefore Fe_{a3} at low pH is assigned as Fe-heme-OH_2, whereas at high pH is assigned as Fe-heme-OH. The magnetic coupling has disappeared upon Cu_B reduction. The magnetic coupling can be disrupted also by exogenous ligands. An EPR investigation on the system reduced Fe_{a3}-NO with isotopically labelled nitrogens both on NO and protein histidines has lead to the conclusion that a histidine ligand completes the coordination in Fe_{a3}(63). The ligands to Cu_B remain uncertain. Possibly copper is bound to three histidine ligands(59); a bridging cysteine ligand has also been suggested(64).

The distance between Fe_{a3} and Cu_B is about 0.5 nm and probably varies with the state of the enzyme. This proximity is critical for efficient electron transfer and for

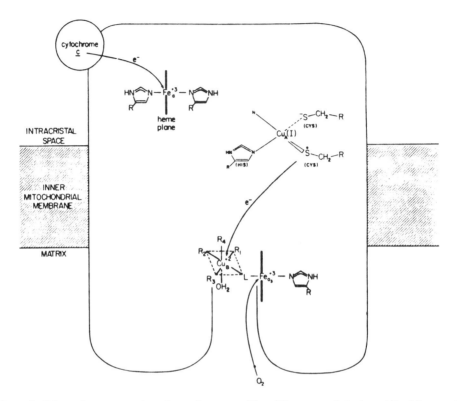

Figure 9. Schematic representation of cytochrome c oxidase. The enzyme is in the oxidized form, and the proposed pathway of electrons from reduced cytochrome c to Fe_a, and from Cu_A to the oxygen-reduction site, is shown(59).

stabilizing O_2 reduction intermediates. Cu_A and Fe_a are relatively far from the other two metal sites (Figure 9).

While it seems that the structural coupling between the Cu_B-Fe_{a3} site and Cu_a site is little if any, there is evidence that the reduction potential of Fe_a may be strongly influenced by the redox state of Fe_{a3}(65).

Fe_a and Cu_A serve to accept electrons from cytochrome c which binds at the cytosol side of the mitochondrial membrane; probably Fe_a and Cu_A are located near that side of the membrane (Figure 9). The electrons are then transferred to the O_2 reduction site. The binding of O_2 raises the reduction potential of Fe_{a3} above that of Cu_B in order to keep O_2 anchored (Figure 10). Cu_B would then serve to transfer one electron into the site to yield a peroxide bound to the oxidized site (Figure 10, step I). Then by further reduction the O-O bond would break to give rise to a ferryl species and a Cu_B^{2+} (step II). Another electron completes the reduction (step III). Since in the cycling enzyme Fe_a and/or Cu_A are reduced, a further electron transfer would reduce Cu_B to give the "half reduced" species with g=6.

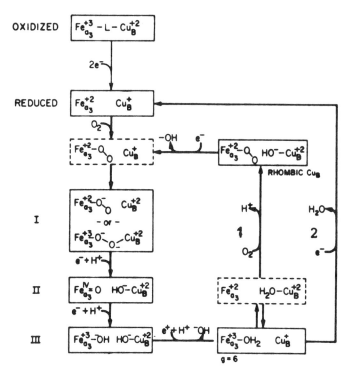

Figure 10. Proposed mechanism for the reduction of dioxygen by cytochrome c oxidase (see text). Electrons are supplied at the various stages by the Fe_a and Cu_A centers (not shown). O_2, H_2O, OH^- and H^+ participating in the reaction all come from the matrix side of the membrane(59).

Then, depending on the availability of electrons and dioxygen, the cycle may be completed by either pathway 1 or 2 (Figure 10)(59).

There is increasing evidence that the enzyme can undergo a conformational change between two states, named "resting" and "pulsed"(66-68). The enzyme evolves towards the latter state, which is more active, during turnover. The existence of two interconvertible conformations may be a means of regulating the overall process in vivo(68).

5.3 Copper-zinc Superoxide Dismutase

Copper-zinc superoxide dismutase is a dimeric molecule, each subunit containing both copper(II) and zinc(II). It shows a large catalytic efficiency towards the dismutation of the superoxide ion although its role "in vivo" is not completely settled(69-71). The catalyzed reaction can be written as

$$O_2^- + Cu^{2+} \rightleftharpoons Cu^+ + O_2$$

$$O_2^- + Cu^+ + 2H^+ \rightleftharpoons Cu^{2+} + H_2O_2$$

where the copper ion is alternately reduced and oxidized during each catalytic cycle. The redox potential of a copper ion depends on the oxidation state of copper in the other subunit(72), but this behavior is not clearly related to the occurrence of cooperative or anticooperative effects. The X-ray structure at 200 pm resolution has definitely shown(73,74) that a copper atom is bound to four histidine ligands, one of which is bridging copper(II) with the zinc atom. The four nitrogen donors of copper(II) are approximately arranged at the vertices of a flattened tetrahedron, with a water molecule semicoordinated along an axial position, as shown in Figure 11. On the other side of the water molecule copper(II) faces hydrophobic residues of the protein. The zinc(II) ion, which is coordinated to two additional histidines and to an aspartate residue, has probably a structural role(75).

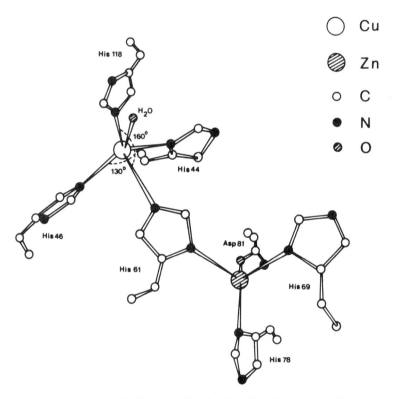

Figure 11. Active site of superoxide dismutase. The four histidine nitrogens coordinated to copper(II) are arranged at the vertices of a flattened tetrahedron; the N-Cu-N angles are also shown(73).

As shown by the reaction scheme, copper is essential for the catalytic activity. The superoxide ion can reduce copper(II) through outer sphere or inner sphere redox reactions. There is a tendency to believe that the reaction is inner sphere(69) on the ground of the existence of superoxide adducts of model copper(II) complexes(75,76) and on the competitive inhibition pattern of anions(77), which are known to bind the copper(II) ion in SOD, and of consideration of the charge distribution in the active site cavity(78).

In this case some steps in the catalytic reaction would be

$$(1)$$

In particular, it is proposed that in species (4) the Cu-Zn bridge is broken, since reduction of Cu^{2+} to Cu^+ requires a proton(72,79), and that the bridging histidine remains coordinated to Zn^{2+} because of the similarity of the X-ray absorption edges of Zn^{2+} in both oxidized and reduced species(80). It may be noted that all the proton transfers involved have to be very fast, since the overall reaction rate is close to the diffusion limit. This suggests the possibility of a buffer- mediated proton transfer at least around neutral pH. Zn- deprived SOD is substantially as active as native SOD(74). If, however, SOD is treated with phenylglyoxal (PHG) a large decrease in activity is observed(81). From a spectroscopic point of view we notice that the native enzyme shows an unusually rhombic EPR spectrum with three resolved g values(82) (Table V) whereas the other two derivatives show essentially tetragonal EPR patterns(83).

Table V
EPR parameters of native,(81) zinc-deprived,(82) and PHG-modified(83) superoxide dismutase

	g_x	g_y	g_z	$A_{||}$ (cm^{-1} x 10^4)
Native[a]	2.02	2.11	2.26	143
Zinc-deprived[a]	2.07[b]		2.26	158
PHG-modified[c]	2.07[b]		2.27	155

[a] bovine erithrocyte enzyme
[b] the spectrum is assumed to be axial
[c] yeast enzyme

In our opinion the rhombicity of the EPR spectrum of the native enzyme shows deviation from a tetragonal geometry; for instance, strong rhombic components are observed when the chromophores are intermediate between trigonal bipyramidal and square pyramidal geometry(84). Removal of zinc, or addition of PHG, releases some strain in the protein chromophore, thus providing a pseudoaxial EPR spectrum. However, the EPR pattern is not related to the activity of the enzyme, as demonstrated by the zinc- deprived SOD(75,83). Water proton ^{1}H NMR measurements have shown that water is present in all three cases at substantially equal distances (85-87). ^{17}O measurements on the native and PHG derivatives are consistent with a water still coordinated in apical position(87,88), the latter chromophore being closer to tetragonal symmetry. Despite the electronic relaxation time, as estimated from variable field ^{1}H NMR data, increases on passing from the native to the PHG modified SOD, ^{17}O NMR relaxation slightly decreases(87), consistent with the decrease in A/ℏ expected for a more regular axial coordination. Anionic inhibitors bind essentially with the same affinity native and zinc-deprived SOD(86), whereas the affinity of anions for the PHG-modified protein is at least one order of magnitude smaller(83-87). This confirms a parallel behavior of anions and O_2^-; however, the mechanism which would account for this parallel behavior is still controversial. It would be tempting to believe that the charge inside the cavity regulates the affinity of anions and O_2^-. In the active cavity an arginine is present with a pK_a presumably around 12, which maintains a positive charge within the cavity. PHG would react with it giving a covalent adduct(81). The modified arginine side chain could then swing out of the cavity, removing the positive charge and decreasing the affinity of anions and O_2^-(78,83). The competition between anions and SOD would thus be due to the different charge of the cavity when anions are bound to the metal. Indeed, other metallo-proteins often bind one anion but rarely two. Whereas there is agreement that anion inhibitors cause a detachment of a histidine from binding to copper(II)(78,83,89-91), it is not yet demonstrated which histidine is removed, nor the mechanism causing the detachment.

5.4 Liver Alcohol Dehydrogenase

Mammalian alcohol dehydrogenases are present in large quantity in the liver. Liver alcohol dehydrogenases (LADH) have a broad substrate specificity, and in man are mainly involved in the metabolism of ingested ethanol(92,93). They are dimeric enzymes, the subunits being not active in the monomeric state. Each subunit is divided in two domains joined by a narrow neck region and separated by the deep active site cleft(94,95). One domain binds the coenzyme and the other two zinc atoms (catalytic domain). Two coenzyme molecules can bind per enzyme molecule independently of each other (96). One zinc atom is bound to two cysteines and one histidine, the fourth position being open to the solvent in a cavity 2.5 nm deep. The other zinc is bound to four sulfur atoms from cysteine residues. The two catalytic zinc atoms of the molecule are separated by 4.7 nm. Thus, there is no direct interaction between the two catalytic centers. Indirect interaction mediated by the conformational change induced by the coenzyme(96) is, however, quite possible since the subunits are bound together through the domains at which coenzymes bind. The substrate binding occurs through aminoacids of both subunits. The zinc(II) ion exposed to the solvent is necessary for catalysis(92) although it is not necessary for coenzyme binding(97,98). The other zinc ion is referred to as non-catalytic metal ion, and may have a structural role(99). Inhibitors can bind either in presence and/or in absence of coenzyme(92,95,100). Anions of the type of $Pt(CN)_4^{2-}$, $Au(CN)_2^-$, halides, etc. bind at the general anion binding site where the phosphate groups of the coenzyme are bound(101). Probably this site involves Arg-47 and Arg-369. Therefore they are coenzyme-competitive. This behavior is different from that with carbonic anhydrase. Also 2,2'-bipyridine and 1,10-phenanthroline are coenzyme-competitive, although they bind the zinc atom and for this reason are partially competitive also with the substrate(102,103). The aromatic moiety of the ligand partially overlaps with the nicotinamide moiety of the coenzyme. Imidazole type ligands(103), as well as pyrazole(104), form both binary complexes with LADH and ternary complexes in the presence of NAD^+. The ligands bind to the catalytic zinc atom displacing the zinc bound water(95). Pyrazole has the suitable geometry to allow interaction between its NH nitrogen and the C_4 carbon of NAD^+(104). The enzyme has a pK_a of about 9.2 which regulates the association rate of NAD^+ and NADH(105). The dissociation rate constant of NAD^+ depends on a pK_a of 7.6, whereas the dissociation rate constant of NADH depends on a pK_a of 11.2(105).

According to Pettersson(105-107), the coordinated water would have a pK_a of 9.2, which would shift to 7.6 upon NAD^+ binding and to 11.2 upon NADH binding. The same pK_a values are found for the enzyme substituted with cobalt at the catalytic site(108). However, ^1H NMR studies on the latter derivative have shown that these acid-base processes affect the coordination geometry around the metal ion, and that the coordinated histidine proton is also somehow involved(109).

The optimum pH for the alcohol oxidation is around 8 and the pH region of activity is 6-10(92). The principal steps involved in the catalytic pathways, and their probable sequence, are:

1) binding of NAD^+ to the enzyme, accompanied by release of a proton (water deprotonation according to several authors (92,105-107,110)).

2) formation of a ternary complex with the alcohol substrate; it is proposed that the alcoholic proton is transferred to the coordinated hydroxide with release of water and coordination of the alcoholate oxygen(92,100,111).

3) Direct transfer of a hydride ion from the alcoholic αCH_2 group to the C_4 carbon of NAD^+, resulting in a weakly coordinated aldehyde(112).

4) Dissociation of the aldehyde from the coordination sphere by addition of a water molecule.

5) Dissociation of NADH. This is the rate limiting step of the overall reaction(92). X-ray data(95,96) shows clear indication of a major conformational change on passing from the binary complex with NADH (closed form) to the unligated enzyme (open form). The closed form is required to properly orient the nicotinamide ring towards the substrate(112).

Steps 2) and 3) might occur simultaneously; the concomitant hydride transfer from the alcohol to NAD^+ would surely lower the energy barrier, which is very high for the formation of a true alcoholate intermediate. However, this is in contrast with isotope effect data which seem to indicate that proton and hydride transfer do not occur simultaneously(113,114).

A conformational change similar to that found for the dissociation of NADH could also take place upon binding of NAD^+. Unfortunately, no crystalline binary complex with NAD^+ could be isolated due to the tendency of NAD^+ to reduce to NADH in the active site.

6. The Metal as Lewis Acid

Already in LADH the role of zinc has been shown to be that of a Lewis acid for binding of alcohol and then for activation of proton and hydride transfer. Zinc probably plays a similar role in the many zinc enzymes. They may provide a more active nucleophile like ZnOH with respect to H_2O, or may modify the electronic structure of the substrate upon binding with the metal ion. We will discuss here in some detail two typical enzymes, *i.e.* carbonic anhydrase which is a lyase and carboxypeptidase which is a hydrolase. Zinc(II) in carbonic anhydrase has been shown by X-ray analysis to be bound to three histidine nitrogens and to be exposed to solvent(115). At least another histidine is always present in the cavity. The enzyme contains therefore two acidic groups, the histidine and the coordinated water(116). The acid base equilibria can be summarized as follows with four microconstants.

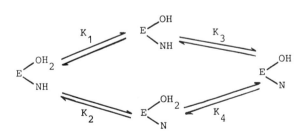

The values for the various pK_a's for the high activity bovine carbonic anhydrase II are:(117) $pK_1=6.1$ $pK_2=6.3$ $pK_3=7.9$ $pK_4=7.7$. The effect of the presence of the histidine in the cavity is to lower the pK_a of the coordinated water to pK_1: otherwise it would be equal to pK_4. The reaction catalyzed is

$$CO_2 + H_2O \rightleftharpoons HCO_3^- + H^+$$

and CO_2 is attacked by the coordinated OH^- (50,118-120).

At this point a five coordinated intermediate is formed with a further water molecule. Evidence for an equilibrium between four and five coordinated species

has been provided by the cobalt(II) derivative(121). The association of a water molecule provides a pathway for bicarbonate detachment.

The high turnover of the process required H^+ to exchange at a rate which is larger than the proton diffusion limit. Indeed, it is found that proton release is buffer assisted(122). This mechanism requires that the active species of the enzyme are those containing the hydroxide ion. One of these, *i.e.* the monodiprotonated species, is in equilibrium with the species indicated as

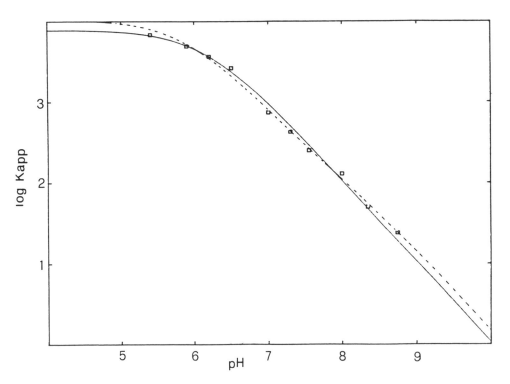

i.e. with the proton transferred from histidine to OH^-. It is this proton transfer which is the rate limiting step of the mechanism (123,124). Anionic inhibitors bind the diprotonated species, and to a lesser extent the monoprotonated species. The pH dependence of affinity of NO_3^- for cobalt(II) substituted bovine carbonic anhydrase II is shown in Figure 12(117).

Figure 12. pH dependence of the apparent affinity constant of NO_3^- for cobalt(II) bovine carbonic anhydrase II as measured on unbuffered solutions through spectrophotometric titrations. The dashed line is a fit with two pK_a's regulating anion affinity(117).

Carbon dioxide possibly has a binding site in the cavity well distinct from the metal site. This is shown by ^{13}C NMR on the CO_2-HCO_3^- system in the presence of the copper(II) substituted derivative. ^{13}C NMR spectra of water solutions containing CO_2 and HCO_3^- show two signals which coalesce in the presence of the active cobalt(II) enzyme(125). However in the presence of copper (II) carbonic anhydrase the two signals are well separated and contain information on their independent interaction with the protein (126). It results that ^{13}C of CO_2 is about 600-800 pm away from the paramagnetic center whereas HCO_3^- is directly bound to it. Analogous results are obtained on the manganese(II) derivative(127).

Such data may not be directly transferable to the native enzyme; nevertheless they may be helpful in understanding the overall properties of the enzyme. For example, a water molecule could coordinate to the native enzyme-substrate system in an intermediate state. Analogously, anions bind the metal in CoCA giving rise to either tetra- or pentacoordinated adducts(120,121).

With the native enzyme the derivatives would be only tetracoordinated(128), but the step represented by the five coordinate adduct with the substrate HCO_3^- is now well accepted in the catalytic process. It is worthy to notice here that Monte Carlo calculations have shown the existence of two potential wells for water molecules in the first coordination sphere of the metal in CA(129).

As pointed out earlier, carbonic anhydrase and carboxypeptidase are similar in that both have two acidic groups. The latter enzyme displays activity only in the pH range between the first ($pK_a \simeq 5.5$) and the second ($pK_a \simeq 9$)deprotonation(47). From X-ray it results that zinc(II) in CPA is bound to two histidine nitrogens, a Glu residue, and a water molecule(130). The first comment regards the binding mode of Glu; it may behave as bidentate: the reported O-Zn distances are 220 and 230 ± 20 pm(130). These distances are longer than usual; the possibility that these figures are the result of long and short distances statistically disordered may not be ruled out.

Again by X-ray analysis the substrate GlyTyr has been found to interact with several residues(130). The peptidic NH interacts via hydrogen bond with Tyr 248, and the terminal COO^- with Arg 145. This possibly is the kind of interaction with the protein of every terminal carboxyl aminoacid to be cleaved. Whereas the carbonyl group interacts with zinc, the amino group is 50% bound to zinc and 50% bound to Glu 270. The catalytic efficiency of the enzyme depends on two pK_a's: one at about 6 and the other at about 9. The acidic groups in the cavity are: Glu 270, H_2O, and Tyr 248 which, however, in the free enzyme is possibly outside the cavity. Also, the cobalt(II) substituted enzyme has pK_a's around 6 and 9. The affinity of N_3^- and NCO^- for the metal decreases with increasing pH with the same pK_a values(49,131); at low pH there is competition between the above anions and Cl^-(48), which is necessary to keep the enzyme in solution. We feel that the low pK_a is due to Glu 270 (or to the Glu-H_2O system). Surely Glu is expected to have lower pK_a and therefore the equilibrium will be

$$Glu \underset{}{\overset{pK_{a1}}{\rightleftharpoons}} Glu^- + H^+$$

The high pK_a could be consistent with the coordinated water; it is higher than in CA but here the metal is bound to a negative anion (Glu 72); also Tyr 248 may, however, have a similar pK_a but it should be close enough to the metal to affect its electronic spectrum. Indeed, the electronic spectra of CoCPA are pH-dependent with a pK_a of 8.8(132).

There is agreement that peptide substrates bind the metal, apparently substituting the water molecule of the native enzyme(133). This very same step is believed to occur in CoCPA. In our opinion, however, the active cobalt(II) and nickel(II) derivatives have room for another ligand in the coordination sphere(134,137), and substrate could be bound to the metal simultaneously with a residual coordinated water. We would interpret the electronic spectra of CoCPA and its derivatives with β phenylpropionate and GlyTyr as due to five coordinated chromophores(136,137) whereas the electronic spectra of the N_3^- and NCO^- derivatives as due to tetracoordination (Figure 13). The interpretation is mainly based on intensity considerations since five coordinated chromophores have been found to provide less intense spectra than tetracoordinated chromophores(134). The nuclear relaxing properties of cobalt(II) ions have also been related to the coordination number, and proton T_1^{-1} of coordinated histidines indicate a change in coordination number on passing from CoCPA to the N_3^- and NCO^- derivatives(49). These data do not appear consistent with a recent X-ray report on Ni and CoCPA(138). The cobalt(II) derivative appears essentially identical to the native enzyme and the nickel(II) derivative shows evidence for an empty coordination site. Of course there may be structural differences between the enzyme solid and the enzyme solution. Otherwise, we may suggest that mono- and bi-dentate behavior of Glu-72 changes in presence of inhibitors.

A coordinated water in presence of substrate can be important for the catalytic process. In an intermediate state it may pass to OH^- and attack the carbonyl carbon. Lipscomb(130) and, independently, Nagakawa(139) have predicted that a coordinated hydroxide would sterically be suited for such an attack. This is also the mechanism of model compounds in hydrolytic functions(140,141).

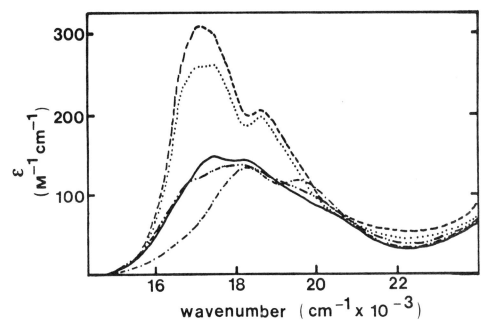

Figure 13. Electronic absorption spectra of cobalt(II) carboxypeptidase A(——) and of its adducts with GlyTyr(- - - - -), β-phenylpropionate (- · · · · · -), azide (- - - -), and cyanate(· · · · · · ·).(132,137).

7. Metal Transport

Metal ions have a proper flow in biological space and time in living organisms(142). The space in a living organism can be divided into compartments like inner side of cells, the walls of membranes, or the places where metal salts as microcrystalline solids are present. The distribution of metal ions among the various compartments is time dependent; furthermore, the compartments themselves may not be fixed in space.

Thus, both time and space require mechanisms for the movement of the ion. In the case of ions which at pH ~ 7 are stable as aquaions a mechanism for movement can be a concentration gradient and, as previously shown, particular ionophores allow the passage through membranes. For other ions, in particular those of transition metals, transportation has to be ensured by chelation and then a given protein or the ligand in general becomes the carrier. In animals trace metals are initially complexed in the intestinal lumen by small ligands which allow their controlled absorption through the intestinal wall and their entrance in the blood stream. Stronger ligands present in the plasma sequester the ions and deliver them to the cells of the appropriate tissues. These ligands are summarized in Table VI, together with some known storage systems. With the possible exception of histidine for copper(143), sulphur-containing aminoacids for cadmium(144), and citrate for nickel(145), the metal transport systems are in general macromolecular proteins; these can be recognized by specific membrane proteins causing the incorporation

Table VI

Biological Systems for Metal Ion Transport and Storage in Mammalians and in Other Living Species[a]

Metal Iron	Transport	Storage
Iron	Transferrin	Ferritin Hemosiderin
	Hydroxamate siderophores[b] (ferrichromes, ferrioxamines) phenolate siderophores[b] (Enterobactin, Itoic acid . . .)	
Copper	Serum albumin, histidine, Caeruloplasmin(?)	Caeruloplasmin(?)
Cadmium	S-Aminoacids and Peptides Albumin, Macroglobulin	Metallothionein
Nickel	Macroglobulin, Transferrin(?) Cytrate(?) Citrate[c]	
Vanadium	Transferrin, $H_2VO_4^{-}$[d]	Ferritin
Zinc	Transferrin, Macroglobulin Serum Albumin	Metallothionein
Cobalt	Transferrin (?)	Metallothionein(?)
Chromium	Transferrin (?)	
Manganese	Transmanganin (?)	

[a]Where not otherwise indicated, the systems are referred to mammalians
[b]In Microorganisms
[c]In Plants
[d]In Tunicate Ascidia Nigra

of the metal in the appropriate cells. In the case of copper, it is proposed that the metal is initially complexed by serum albumin or histidine, and brought through the portal vein to the liver, where it is incorporated into caeruloplasmin(146). The latter protein, which accounts for about 95% of the total copper in plasma, may contribute to carry the metal to the places of synthesis of the functional copper proteins. Similarly, it has been proposed that zinc is carried by transferrin or serum albumin to the liver, and then released to the carrier protein α_2 macroglobulin(147-150).

From Table VI, the general role of transferrin in metal transport is apparent although, as suggested by the name, its main role has been usually considered to be the transport of iron. Serum transferrins belong to a class of proteins, siderophyllins, generally associated with iron-complexation and bacteriostatic requirements; other similar siderophyllins are lactoferrins from mammalian milk and conalbumins from avian egg white. They are cigar-shaped monomeric proteins of molecular weight 80,000 containing two similar metal binding sites at the two ends of the molecule (N-terminal and C-terminal site) and two identical carbohydrate chains linked to the protein in the C-terminal domain(151). The proteins have high affinity for trivalent metal ions, especially for iron(III), but also bind bipositive cations. Metal binding requires the presence of one bicarbonate (or carbonate) ion per site as synergistic anion(152). The X-ray structure has not yet been refined enough to allow the identification of the metal ligand residues(153,154). The presence of tyrosyl residues has been proposed from UV difference spectroscopy on the iron(III) and other metal substituted derivatives;(155) analogous studies on model complexes suggest that two is the number of such residues actually coordinated in each protein site(156). Histidine residues have been proposed on the ground of ESR data on the copper(II) derivatives(157); however, superhyperfine splitting only reveals one nitrogen for sure(158). The synergistic anion and solvent ligand (H_2O or OH^-) are believed to complete the metal coordination sphere.

The two binding sites are not equivalent, as indicated by the different kinetics of metal binding(159) and the difference in stability at low pH(160-161), but often indistinguishable from a spectroscopic point of view. A notable exception is constituted by the dithallium (III) transferrin, that is relatively stable despite the high reduction potential of tallium(III): the ^{205}Tl NMR spectra show two separate signals which could be individually assigned to the two sites on the ground of the different pH dependence of the affinity and with the aid of mixed-metal derivatives(162).

In a recent NMR study on the cobalt(II) conalbumin the isotropically shifted signals (Figure 14) could be successfully assigned as arising from *two* histidines and *two* tyrosines in each metal site(163). A qualitative analysis of the dipolar contributions to the shifts suggests that the two tyrosines are *trans* to each other on the z axis of a distorted octahedral chromophore. Reexamination of the data on the

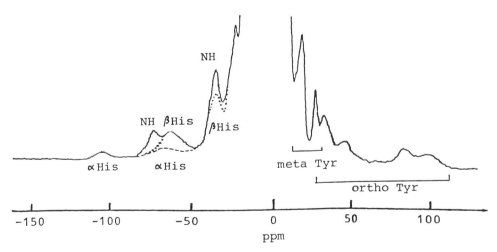

Figure 14. 60 MHz ^1H NMR spectrum of cobalt(II) substituted conalbumin in H_2O. The dotted lines refer to the spectrum obtained on a deuterium-exchanged sample. The dashed line indicates a resonance observed upon saturation of the upper signals. The assignment of the signals to histidine and tyrosine protons is also shown(163).

copper(II)(158) and oxovanadium(IV) derivatives(64,165) allows us to draw the tentative picture of the active site shown in Figure 15. The *cis* position of the two histidines, one of which is along the z axis (coincident with the V-O axis and Cu-OH$_2$ axis respectively) may account for the observation of the coupling with only one histidine nitrogen in the copper derivative, since the unpaired electron responsible for the coupling mainly resides in the xy plane, and with the prevalence of oxygens in the xy plane indicated by the g and A values of the VO^{2+} derivative.

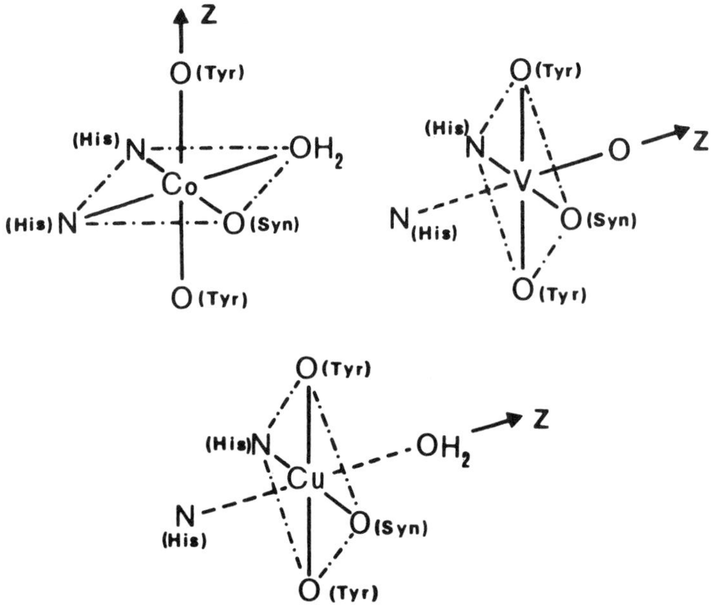

Figure 15. Proposed coordination geometry and ligand arrangement in cobalt(II)-, oxovanadium(IV)-, and copper(II)-substituted siderophyllins.

A breakthrough in the understanding of siderophillins would be represented by the characterization of the specific role of the synergistic ion.

8. Conclusion

A quite large share of research efforts is today conveyed towards the understanding of the properties of metal ions modulated by the great artist which is Nature. These efforts answer the question of why a given metal and its donors are chosen for a certain function. Metalloproteins provide a wealth of examples of peculiar properties of metal ions hardly encountered in inorganic chemistry; they furthermore can perform many functions and each of them with a unique fascinating mechanism: from electron transfer, to metal transport and to catalysis of very many reactions. At the moment each metal-containing biochemical system represents a world by itself to be discovered and understood; the comparison among the working mechanisms of the different worlds will surely be quite instructive and will provide a great progress in chemical and biochemical sciences.

References and Footnotes

1. Keilin, D., 1966, *The History of Cell Respiration and Cytochrome*, University Press, Cambridge.
2. Schwarz, K., 1977, *Clinical Chemistry and Chemical Toxicology of Metals*, Elsevier, Amsterdam.
3. Malmström B. G., 1983, *Chemica Scripta, 21*, 7.
4. Lehninger, A. L., 1950, *Physiol. Rev. 30*, 393.
5. Williams, R. J. P., 1953, *Biol. Rev., 28*, 381.
6. Underwood, E. J., 1977, *Trace Elements in Human and Animal Nutrition*, 4th. Ed., Acad. Press, N.Y.
7. Inoue, T. and Orgel, L.E., 1983, *Science, 219*, 859.
8. Kagawa Y., 1978, *Biochim. Biophys. Acta, 505*, 45.
9. *Membrane Transport*, Martonosi, A. N., ed., 1982, Plenum, N.Y.
10. Lehn, J. M., 1983, *Stud. Phys. Theor. Chem., 24*, 181.
11. Lehn, J. M., 1979, *Pure Applied Chem., 1*, 979.
12. Berman, M. C., 1982, *Biochim. Biophys. Acta, 694*, 95.
13. Le Maire, M., Jørgensen, K. E., Roigaard-Petersen, H., and Møller, J. V., 1976, *Biochemistry, 15*, 5805.
14. Chyn, T. and Martonosi, A.N., 1977, *Biochim Biophys. Acta, 468*, 114.
15. Birnbaum, E. R. and Sykes, B. D., 1978, *Biochemistry, 17*, 4965.
16. Williams, R. J. P., 1970, *Quart, Rev. Chem. Soc., 24*, 331.
17. *Calcium in Biological Systems*, Duncan, C. J., ed., 1976, Cambridge Univ. Press.
18. *Metal Ions in Biological Systems*, Sigel, H., ed., vol. 15,1983, Marcel Dekker, N. Y.
19. *Metal Ions in Biology*, Spiro, T. G., ed., vol. 5, 1983, John Wiley and Sons, N.Y.
20. Williams, R. J. P., 1982, *Pure Appl. Chem., 54*, 1889.
21. Moore, G. R. and Williams, R. J. P., 1980, *Eur. J. Biochem., 103*, 503.
22. Johnson, A. W., 1980, *Chem. Soc. Rev., 9*, 125.
23. Williams, R. J. P., 1953, *Analyst, 78*, 586.
24. Lehn, J. M., 1979, *Pure Appl. Chem., 51*, 979.
25. Izatt, R. M., Eatough, D. J. and Christensen, J. J., 1973, *Struct. Bonding, 16*, 161.
26. Coxon, A. C., Laidler, D. A., Pettman, R. B. and Stoddart, J. F., 1978, *J. Am. Chem. Soc., 100*, 8260.
27. Lehn, J. M., 1973, *Struct. Bonding, 16*, 1.
28. Lehn, J. M., 1978, *Acc. Chem. Res., 11*, 49.
29. Lehn, J. M., 1978, *Pure Appl. Chem., 50*, 861.
30. Lehn, J. M., 1983, *Inorg. Chim. Acta, 79*, 67.
31. Simon, W., Morf, W. E. and Meier, P. C., 1973, *Struct. Bonding, 16*, 113.
32. Fenton, D. E., 1977, *Chem. Soc. Rev., 6*, 325.
33. Klingenberg, M., 1981, *Nature, 290*, 449.
34. Behr, J. P., Lehn, J. M., Dock, A.-C. and Moras, D., 1982, *Nature, 295*, 526.
35. Colman, P. M., Jansonius, J. N., Matthews, B. W., 1972 *J. Mol. Biol., 70*, 701; Matthews, B. W. and Weaver, L. H., 1974, *Biochemistry 13*, 1719.
36. Albers, R. W., 1976, in *The Enzymes of Biological Membranes*, Martonosi, A.N., ed., vol. 3, Wiley, N. Y., pp. 283.
37. *Na⁺, K⁺ ATPase Structure and Kinetics*, Skoy, J. C. and Nørby, J. G., eds., 1979, Academic Press, N. Y.
38. Bean, B. L., Koren, R., and Mildvan, A. S., 1977 *Biochemistry, 16*, 3322.
39. Teeter, M. T., Quigley, G. J., and Rich, A., 1980, in *Metal Ions in Biology* Spiro, T. G., ed., vol. 1, John Wiley and Sons, N. Y., pp. 145.
40. Freeman, H. C., 1980, in *Coordination Chemistry-21*, Laurent, J. P., ed., Pergamon Press, Oxford, pp. *29*
41. Stout, C. D., 1982, in *Metal Ions in Biology*, Spiro, T. G., ed., vol. 4, John Wiley and Sons, N. Y., pp. 97.

42. Vallee, B. L. and Williams, R. J. P., 1968, *Proc. Natl. Acad. Sci., U. S. A., 59,* 498.

43. Vallee , B. L. and Williams, R. J. P., 1968, *Chem. in Britain, 4,* 397.

44. Berg, J. M. and Holm, R. H., 1982, in *Metal Ions in Biology,* Spiro, T. G., ed. Vol. 4, John Wiley and Sons, N. Y. pp. 1

45. Vallee, B. L, 1983, *Metal Ions In Biology,* Spiro, T. G., ed., vol 5., John Wiley and Sons, N. Y., pp. 1

46. Lipscomb, W. N., Hartsuch, J. A., Quiocho, F. A., and Reeke, G. N., 1969, *Proc. Natl. Acad. Sci., U. S. A. 64,* 28.

47. Auld, D. S, and Vallee, B. L., 1970, *Biochemistry, 9,* 4352.

48. Geoghegan, K. F., Holmquist, B., Spilburg, C. A., and Vallee, B. L., 1983, *Biochemistry, 22,* 1847.

49. Bertini I., Lanini, G., Luchinat, C. and Monnanni, R., submitted for publication.

50. Lindskog, S., 1983, in *Metal Ions in Biology,* Spiro, T. G., ed., vol. 5, John Wiley and Sons, N. Y. pp. 79.

51. Yocom, K. M., Winkler, J. R., Nocera, D. G., Bordignon, E., and Gray, H. B., 1983, *Chemica Scripta, 21,* 29.

52. Farver, O. and Pecht, I., 1981, *Israel J. Chem. 21,* 13.

53. Farver, O., Blatt, Y. and Pecht, I., 1982, *Biochemistry, 21,* 3556.

54. Grant Mauk A., Coyle, C. L., Bordignon, E., and Gray, H. B., 1979, *J. Am. Chem. Soc., 101,* 5054.

55. Butler, J., Davies, D. M., Sykes, A. G., Koppenol, W. H., Osheroff, N., and Margoliash, E., 1981, *J. Am. Chem. Soc., 103,* 469.

56. Babcock, G. T., Vickery, L. E., and Palmer, G., 1976, *J. Biol. Chem, 251,* 7907; Babcock, G. T., Callahan, P. M., Ondries, M. R. and Salmeen, I., 1981, *Biochemistry, 20,* 959.

57. Stevens, T. H., Martin, C. T., Wang, H., Brudwig, G. W., Scholes, C. P., and Chan, S. I., 1982, *J. Biol. Chem, 257,* 12106.

58. Chan, S. I., Martin, C. T., Wang, H., Brudwig, G. W., and Stevens, T. H., 1983, in *The Coordination Chemistry of Metalloenzymes,* Bertini, I., Drago, R. S. and Luchinat, C., eds., D. Reidel, Dordrecht, pp. 313.

59. Blair, D. F. Martin, C. T., Gelles, J., Wang, H., Brudwig, G. W., Stevens, T. H., and Chan, S. I., 1983, *Chem. Scripta, 21,* 43.

60. Kent, T. A., Young, L. J., Palmer, G., Fee, J. A., and Münck, E., 1983, *J. Biol. Chem., 258,* 8543.

61. Tweedle, M. F., Wilson, L. J., Garcia Iñiguez, L., Babcock, G. I., and Palmer, G., 1978, *J.Biol. Chem., 253,* 8065.

62. Hartzell, C. R. and Beinert, H., 1974, *Biochim. Biophys. Acta, 368,* 318.

63. Stevens, T. H. and Chan, S. I., 1981, *J. Biol. Chem., 256,* 1069.

64. Powers, L., Chance, B., Ching, Y., and Angiolillo, P., 1981, *Biophys. J., 34,* 465.

65. Wikstrom, M. K. F., Harmon, H. J., Ingledew, W. J., and Chance B., 1976, *Febs Lett, 65,* 259.

66. Brunori, M., Antonini, E., and Wilson, M. T., 1981, in *Metal Ions in Biological Systems,* Sigel, H., ed, vol. 13, Marcel Dekker, N. Y., pp. 187.

67. Brunori, M., Colosimo, A., Wilson, M. T., Sarti, P., and Antonini, E., 1983, *Febs Lett, 152,* 75.

68. Brunori, M., Colosimo, A., and Silvestrini, M. C., 1983, *Pure Appl. Chem., 55,* 1049.

69. Fee, J. A., 1981, in *Metal Ions in Biological Systems,* Sigel, H., ed., vol. 13, Marcel Dekker, N. Y., pp. 259.

70. Valentine, J. S and Pantoliano, M. W., 1981, in *Metal Ions in Biology,* Spiro, T. G., ed., vol 3, John Wiley and Sons, N.Y., pp. 291

71. Fee, J. A., 1980, in *Metal Ions in Biology,* Spiro, T. G., ed., vol. 2, John Wiley and Sons, N.Y., pp. 209

72. Lawrence, G. D. and Sawyer, D. T., 1979, *Biochemistry, 18,* 3045.

73. Tainer, J. A., Getzoff, E. D., Beem, K. M., Richardson, J.S. and Richardson, D. C., 1982, *J.Mol. Biol., 160,* 181.

74. Tainer, J. A., Getzoff, E. D., Richardson, J. S., and Richardson, D. C., 1983, *Nature, 306,* 284.

75. Pantoliano, M. W., Valentine, J. S., Burger, A. R., and Lippard, S. J., 1982, *J. Inorg. Biochem., 17,* 325.

76. Nappa, M., Valentine, J. S., Miksztal, A. R., Schugar, H. J., Isled, S. S., 1979, *J. Am. Chem. Soc., 101,* 7744.

77. Rigo, A., Viglino, P., and Rotilio, G., 1975, *Biochem. Biophys. Res. Commun, 63,* 1013.

78. Getzoff, E. D., Tainer, J. A., Weiner, P. K., Kollman, P. A., Richardson, J. S., and Richardson, D. C., 1983, *Nature, 306,* 287.

79. Fee, J. A. and Di Corleto, P. E.,1973, *Biochemistry, 12,* 4893.

80. Blumberg, W. E., Peisach, J., Eisenberger, P., and Fee, J. A., 1978, *Biochemistry, 17,* 1842; Blumberg, W. E., Peisach, J., and Fee, J. A., 1980, *Fed, Proc., 39,* 1976.

81. Malinowski, D. P. and Fridovich, I., 1979, *Biochemistry, 18,* 5909; Borders, C. L., Jr., and Johansen, J. T., 1980, *Carlsberg Res. Commun., 45,* 185; 1980, *Biochem. Biophys. Res. Commun., 96,* 1071.

82. Rotilio, G., Morpurgo, L., Giovagnoli, C., Calabrese, L., and Mondovi, B., 1972, *Biochemistry, 11,* 2187.

83. Pantoliano, M. W., Valentine, J. S., Mammone, R. J., and Scholler, D. M., 1982, *J. Am Chem. Soc., 104,* 1717; Bermingham-McDonough, O, Valentine, J.S., et al, 1982, *Biochem. Biophys. Res. Commun., 108,* 1376.

84. Bencini, A., Bertini, I., Gatteschi, D, and Scozzafava, A., 1978, *Inorg. Chem., 17,* 3194.

85. Gaber, B. P., Brown, R. D., Koenig, S. H., and Fee, J. A., 1972, *Biochim. Biophys. Acta, 271* 1.

86. Bertini, I., Luchinat, C., Monnanni, R., Scozzafava, A., and Borghi, E., 1984, *Inorg. Chim. Acta, 91* 109.

87. Bertini, I., Lanini, G., and Luchinat, C., 1984, *Inorg. Chim. Acta, 93,* 000.

88. Bertini, I., Luchinat, C., and Messori, L., 1981, *Biochem. Biophys. Res. Commun., 101,* 577.

89. Bertini, I., Luchinat, C., and Scozzafava, A., 1980, *J. Am. Chem. Soc., 102,* 7349.

90. Bertini, I., Borghi, E., Luchinat, C., and Scozzafava, A., 1981, *J.Am. Chem. Soc., 103,* 7779.

91. Strothkamp, K. G. and Lippard, S. J., 1981, *Biochemistry, 20,* 7488

92. Bränden, C.-I., Jörnvall, H., Eklund, H. and Furugren, B., 1975, in *The Enzymes,* vol. 11, 3rd. ed., Boyer, P. D., ed., Academic Press, N. Y. pp. 103.

93. Rossmann, M. G., Liljas, A., Bränden, C.-I., and Banaszak, L. J., 1975, in *The Enzymes,* vol. 11, 3rd ed., Boyer, P. D., ed., Academic Press, N. Y., pp. 61.

94. Eklund, H., Nördstrom, B., Zeppezauer, E., Söderlund, G., Ohlsson, I., Boiwe, T., Söderberg, B.-O., Tapia, O., Bränden, C.-I., and Åkeson, Å., 1976. *J. Mol. Biol., 102,* 27.

95. Eklund, H. and Bränden, C.-I., 1983, in *Metal Ions in Biology,* Spiro, T. G., ed. vol 5, John Wiley and Sons, N. Y. pp. 123.

96. Eklund, H., Samama, J.-P., Wallen, L., Bränden, C.-I., Åkeson, Å., and Jones, T. A., 1981, *J. Mol. Biol., 146,* 561.

97. Schneider, G., Eklund, H., Cedergren-Zeppezauer, E., and Zeppezauer, M., 1983, *EMBO J., 2,* 685.

98. Dietrich, H., MacGibbon, A. K. H., Dunn, M. F., and Zeppezauer, M., 1983, *Biochemistry, 22,.* 3432.

99. Drum, D. E., Li, T.-K. and Vallee, B., L., 1969, *Biochemistry, 8* 3792.

100. Cedergren-Zeppezauer, E., 1983, *Biochemistry, 22,* 5761.

101. Norne, J.-E., Lilja, H., Lindman, B., Einarsson, R., and Zeppezauer, M., 1975, *Eur, J. Biochem, 59,* 463.

102. Eklund, H., Nordström, B., Zeppezauer, E., Söderlund, G., Ohlson, I,. Boiwe, T., and Bränden, C.-I., 1974, *Febs Lett., 44,* 200.

103. Boiwe, T., and Bränden, C.-I., 1977, *Eur. J. Biochem, 77,* 173.

104. Eklund, H., Samama, J.-P., and Wallen, L., 1982, *Biochemistry, 21,* 4858.

105. Kvassman, J. and Pettersson, G., 1979, *Eur. J. Biochem., 100,* 115.

106. Kvassman, J. and Pettersson, G., 1978, *Eur. J. Biochem, 87,* 417.

107. Anderson, P., Kvassman, J., Lindström, A., Olden, B., and Pettersson, G., 1981, *Eur. J. Biochem, 113,* 425.

108. Dietrich, H., and Zeppezauer, M., 1982, *J. Inorg. Biochem., 17,* 227.

109. Bertini, I., Gerber, M., Lanini, G., Luchinat, C., Maret, W., Rawer, S., and Zeppezauer, M., 1984, *J. Am Chem. Soc. 106.* 1826.

110. Dworschack, R. T., and Plapp, B. V., 1977, *Biochemistry, 16,* 2716.

111. Kvassman, J. and Pettersson, G., 1980, *Eur. J. Biochem, 103,* 557.

112. Eklund, H., Samama, J.-P., Plapp, B. V., and Bränden, C.-I., 1982, *J. Biol. Chem., 257,* 14349.

113. Schmidt, J., Chen, J., De Traglia, M., Minkeo, D., and McFarland, J. T., 1979, *J. Am Chem. Soc., 101,* 3634.

114. Klinman, J. P., 1981, *CRC Crit. Rev. Biochem., 10,* 39.

115. Kannan, K. K., Notstrand, B., Fridborg, K., Lövgren, S., Ohlsson, A., Petef, M., 1975, *Proc. Natl. Acad. Sci., U.S.A. 72,* 5l; Liljas, A., Kannan, K. K., Bergsten, P. L., Waara, I., Fridborg, K., Strandberg, B., Carlbom, V., Jarup, L., Lövgren, S., Petef, M., 1972, *Nature New Biol., 235,* 131.

116. Simonsson, I., Lindskog, S., 1982, *Eur. J. Biochem, 123,* 29.

117. Unpublished results from our laboratory.
118. Lindskog, S., Coleman, J. E., 1973, *Proc. Natl. Acad. Sci., U.S.A. 70,* 2505.
119. Pocker, Y. and Sarkanen, S., 1978, *Adv. Enzymol, 47,* 149.
120. Bertini, I., Luchinat, C., 1983, *Acc. Chem. Res., 16,* 212.
121. Bertini, I., Canti, G., Luchinat, C., Scozzafava, A., 1978, *J. Am. Chem. Soc., 100,* 4873.
122. Jonsson, B. H., Steiner, H., Lindskog, S., 1976, *Febs Lett., 64,* 310.
123. Silverman, D. N., Vincent, S. H., 1982, *CRC Crit. Rev. Biochem. 14,* 207; Venkatasubban, K. S., Silverman, D. N., 1980, *Biochemistry, 19,* 4984.
124. Steiner, H., Jonsson, B. H., Lindskog, S., 1975, *Eur. J. Biochem., 59,* 253.
125. Yeagle, P. L. Lochmüller, C. H., and Henkens, R. W., 1977, *J. Am. Chem. Soc., 99,* 3194.
126. Bertini, I. Borghi, E., and Luchinat, C., 1979, *J. Am. Chem. Soc, 101,* 7069; Bertini, I., Borghi, E., Canti, G., and Luchinat, C., 1983, *J. Inorg. Biochem, 18,* 221.
127. Led, J. J., Neesgaard, E., Johansen, J. T., 1982, *Febs Lett., 147,* 74.
128. Yachandra, V., Powers, L., and Spiro, T. G., 1983, *J. Am. Chem. Soc., 105,* 6596.
129. Clementi, E., Corongiu, G., Jönsson, B., and Romano, S., 1979, *Febs Lett, 100,* 313; E. Clementi, Corongiu, G., Jönsson, B., and Romano, S., 1979, *Gazz, Chim. Ital., 109,* 669.
130. Lipscomb, W. N., 1974, *Tetrahedron, 30,* 1725; Rees, D. C., Lewis, M., Honzatko, R. B., Lipscomb, W. N. and Hardman, K. D., 1981, *Proc. Natl. Acad. Sci, U.S.A., 78,* 3408.
131. Bertini, I., Lanini, G., Luchinat, C. and Monnanni, R., 1983, in *The Coordination Chemistry of Metalloenzymes,* Bertini, I., Drago, R. S. and Luchinat, C., eds., Reidel, D., Dordrecht, pp. 93.
132. Latt, S. A. and Vallee, B. L., 1971, *Biochemistry, 10,* 4263.
133. Rees, D. C. and Lipscomb, W. N., 1981, *Proc. Natl. Acad. Sci., U.S.A., 78,* 5455.
134. Rosenberg, R. C., Root, C. A., Wang, R.-H., Cerdonio, M. and Gray, H. B., 1973, *Proc. Natl. Acad. Sci., U.S.A., 70,* 161.
135. Bertini, I. and Luchinat, C., 1983, in *Metal Ions in Biological Systems* Sigel, H., ed., vol. 15, Marcel Dekker, N. Y., pp. 101.
136. Bertini, I., Canti, G., and Luchinat, C., 1982, *J. Am Chem. Soc., 104,* 4943.
137. Bertini, I., 1983, in *The Coordination Chemistry of Metalloenzymes,* Bertini, I., Drago, R. S. and Luchinat, C., eds., Reidel, D., Dordrecht, pp. 1.
138. Hardman, K. D. and Lipscomb, W. N., 1984, *J. Am. Chem. Soc., 106,* 463.
139. Nakagawa, S., Umeyama, H., Kitaura, K. and Morokuma, K., 1981, *Chem. Pharm. Bull., 29,* 1.
140. Buckingham, D. A., Keene, F. R., and Sargeson, A. M., 1974, *J. Am. Chem. Soc., 96,* 4981.
141. Groves, J. T. and Dias, R. M., 1983,in *The Coordination Chemistry of Metalloenzymes,* Bertini, I., Drago, R. S., and Luchinat, C., eds. D. Reidel, Dordrecht, pp. 79.
142. Williams, R. J. P., 1983, *Pure Appl. Chem., 55,* 1089.
143. Sarkar, B., and Kruck, T.P.A., 1966, in *Biochemistry of Copper,* Peisach, J., Aisen, P., and Blumberg, W. E., eds. Academic Press, N. Y. pp.183.
144. Cherrian, M. G., 1979, *Experientia, Suppl., 34,* 339.
145. Nielsen, F. H., 1980, *J. Nutr., 110,* 965.
146. Frieden, E., 1978, in *Trace Element Metabolism in Man and Animals,* Kirchgessner, M., ed., Tech. Univ. Munich Freising, Weihenstephen, pp. 8.
147. Henkin, R. I., 1974, in *Protein-Metal Interactions,* Friedman, M., ed., Plenum, N. Y., pp. 299.
148. Evans, G. W., 1976, *Proc. Exp. Biol. Med., 51,* 775.
149. Charlwood, P. A., 1979, *Biochim. Biophys. Acta, 581,* 260.
150. Smith, K. T., Failla, M. L., and Cousins, R. J., 1979, *Biochem. J., 184,* 627.
151. Metz-Boutigue, M.-H., Jolles, J., Jolles, P., Mazurier, J., Spik, G. and Montreuil, J., 1980, *Biochim, Biophys. Acta, 622,* 308.
152. Schade, A. L. and Reinhart, R. W., 1966, *Protides Biol. Fluids Proc. Colloq. Bruges, 14,* 75.
153. Gorinsky, B., Horsburgh, C., Lindley, P. F., Moss, D.S., Parkar, M. and Watson, J. L., 1979, *Nature, 281,* 157.
154. DeLucas, L. J., Suddath, F. L., Gams, R. A., and Bugg, C. E., 1978, *J. Mol. Biol., 123,* 285.
155. Gelb, M. H. and Harris, D. C., 1980, *Arch. Biochem. Biophys. 200,* 93.
156. Pecoraro, V. L., Harris, W. R., Carrano, C. J., and Raymond, K. N., 1981, *Biochemistry, 20,* 7033.
157. Aasa, R. and Aisen, P., 1968, *J. Biol. Chem., 243,* 2399.
158. Froncisz, W. and Aisen, P., 1982, *Biochim. Biophys. Acta., 700,* 55.

159. Huebers, H., Josephson, B., Huebers, E., Caiba, E., and Finch, C., 1981, *Proc. Natl. Acad. Sci. U.S.A., 78,* 2572.
160. Harris, D. C., and Aisen, P., 1976, *Nature(London), 257,* 821.
161. Zapolski, E. J., and Princiotto, J. V., 1977, *Biochem, J., 166,* 175.
162. Bertini, I., Luchinat, C., and Messori, L., 1983, *J. Am. Chem. Soc., 105,* 1347.
163. Bertini, I., Luchinat, C., Messori, L. and Scozzafava, A., 1984, *Eur. J. Biochem., 141,* 375.
164. White, L. K. and Chasteen, N. D., 1979, *J. Phys. Chem., 83,* 279.
165. Casey, J. D. and Chasteen, N. D., 1980, *J. Inorg. Biochem, 13,* 111.

Structure & Motion: Membranes, Nucleic Acids & Proteins,
Eds., E. Clementi, G. Corongiu, M. H. Sarma & R. H. Sarma,
ISBN 0-940030-12-8, Adenine Press,   Copyright Adenine Press, 1985.

Iron Atom Displacement and Ligand Reactivity in Ferrous Heme Proteins: Effects of Proximal Bond

Paolo Ascenzi[a], Maurizio Brunori[a†],
Massimiliano Coletta[a] and Teddy G. Traylor[b]

[a]C. N. R., Center for Molecular Biology,
Institutes of Chemistry and Biochemistry, Faculty of Medicine,
University of Rome "La Sapienza", Piazzale Aldo Moro 2,
00185 Rome, Italy

[b]Department of Chemistry, University of California, San Diego,
La Jolla, California 92093, U. S. A.

Abstract

The effect of pH on the spectral properties and second-order rate constants for CO binding to several monomeric hemoproteins is reported. As the pH is lowered, the second-order rate constant of sperm whale, horse, *Dermochelys coriacea, Coryphaena hippurus* and *Aplysia limacina* myoglobins (Mb) (the latter only in the presence of 0.1 M acetate) increases, attaining a value at least one order of magnitude higher than that at pH 7.0. This behaviour, which resembles that reported for chelated heme model compounds (J. Geibel, C. K. Chang and T. G. Traylor, *J. Am. Chem. Soc. 97.* 5924 (1975)), is not however observed for CO binding to *Aplysia limacina* Mb in the absence of acetate and *Chironomus thummi thummi* erythrocruorin (Ery), which are pH-independent.

The lowering of pH brings about in all these monomeric hemoproteins, with exception of *Aplysia limacina* Mb in the absence of acetate, a change of the visible absorption spectrum of the deoxy form, similar to that observed in chelated heme model compounds (T. G. Traylor, C. K. Chang, J. Geibel, A. Berzinis, T. Mincey and J. Cannon, *J. Am. Chem. Soc. 101,* 6716 (1979)). By analogy with these model compounds, such spectroscopic and kinetic effects have been attributed to the protonation of the proximal HisF8, which leads to the progressive predominance, as the pH is lowered, of a ferrous deoxy tetracoordinated form of the heme's iron, characterized by a faster second-order rate constant for CO binding.

On the basis of this interpretation, the energetics involved in the movement of the deoxy Fe atom toward the heme plane can be considered a main determinant of the activation free energy barrier for the binding of carbon monoxide.

[†]To whom correspondence should be addressed (phone 06/4950291).

Introduction

The binding of gaseous ligands, such as O_2 and CO to hemoproteins occurs through the formation of a bond between the ligand and the Fe atom located at the center of the active site, *i.e.* the heme.

The reactivity for heme's ligands has been shown in several hemoproteins to be modulated by: (i) the conformation of the globin in the vicinity of the metal atom, especially at the level of the distal histidine (1); (ii) the geometry of the heme (2,3); and (iii) the heme-globin interactions, occurring either through the covalent $HisF8N_r$-Fe bond and van der Waals contacts between globin and the side groups of the heme's pyrroles (4).

In the case of CO binding, lowering of pH down to $\simeq 3$ brings about in sperm whale Mb (5) a 40-fold increase of the association second-order kinetic constant, *e.g.* from pH 7.0 ($k = 5x10^5$ $M^{-1}s^{-1}$) to pH 2.7 ($k = 2.0x10^7$ $M^{-1}s^{-1}$), with a $pK_a = 3.45$. This value is similar to that ($pK_a = 3.60$) reported spectroscopically for "chelated" mesoheme and attributed to the progressive predominance, as the pH is lowered, of a tetracoordinated form which also displays a higher second-order rate constant for the binding of CO (6). By analogy, the same event was thought to take place in sperm whale Mb to explain the observed pH- rate profile (7).

However, subsequent extension to other monomeric hemoproteins, namely *Aplysia limacina* Mb and *Chironomus thummi thummi* monomeric Ery (Ery III) gave different results: (i) *Aplysia limacina* Mb, although characterized at pH 7.0 by a CO binding rate similar to sperm whale Mb ($k = 5x10^5$ $M^{-1}s^{-1}$), maintains a slow rate even going to a pH as low as 3.5; and (ii) *Chironomus thummi thummi* Ery displays a pH-independent second-order rate constant ($k \simeq 3x10^7 M^{-1}s^{-1}$) between pH 4 and 9. The different behaviour between these three hemoproteins arose the question of whether a general kinetic mechanism accounting for this variable pH-dependence may be proposed (8).

Further extension of the kinetic investigation to these and to other monomeric hemoproteins, namely horse, *Coryphaena hippurus* and *Dermochelys coriacea* Mbs, shows that: (i) as the pH is lowered, every monomeric hemoprotein undergoes the protonation of the $HisF8N_r$, and the consequent transition from penta- to tetracoordinated heme takes place; (ii) effects of residues on the distal side of the heme pocket play a minor role in the pH-dependence of CO binding kinetics; (iii) the limiting step in the binding pathway of CO seems to be the energy involved in the movement of the Fe atom toward the heme plane.

Materials and Methods

Sperm whale Mb (type II) and horse Mb (type I) were purchased from Sigma Chemical Co. (St. Louis, USA). Both hemoproteins were purified as reported elsewhere (2). *Aplysia limacina* Mb was prepared according to Rossi Fanelli and Antonini

(9). *Dermochelys coriacea* Mb was purified according to Ascenzi et al. (10). *Coryphaena hippurus* Mb and *Chironomus thummi thummi* Ery (Ery III) were generous gifts of Dr. J. V. Bannister and Prof. R. Huber, respectively.

CO was obtained from Caracciolo (Rome, Italy).

All the other reagents (from Merck AG, Wuppertal, FRG) were of analytical grade and used without further purification.

The following buffers were used for stopped-flow experiments (all concentrations referring to before mixing): 0.3 M phosphate buffer (in the presence or absence of a 0.2 M acetic acid/acetate mixture), pH 2-5.5; 0.3 M acetate and 0.2 M citrate buffers, pH 3.5-5.5; 1 mM and 0.3 M phosphate buffer, pH 5.5-9.5 (all sodium salts). In flash photolysis experiments, for pH above 3.5, 0.1 M acetate and phosphate, in the same range as in stopped-flow experiments, and 2% borate buffer, pH 8-9 were used (all sodium salts).

Rate constants for CO binding were obtained by flash photolysis (pH 3.5-9.5) and stopped-flow (pH 2-8).

The spectrum of the transient deoxy form at low pH, before the protein denaturation has time to occur, was determined by pH-jump experiments as reported by Giacometti et al. (7).

All experiments were performed between 21 and 23°C.

Results and Discussion

The unliganded form of the pentacoordinated "chelated" protoheme (pH 7.0) is characterized by a visible absorption spectrum very similar to that of different deoxy Mbs, but completely different from that of protoheme in the tetracoordinated form (see Fig. 1A) (11).

Figure 1B shows the spectra of the deoxy *Coryphaena hippurus* Mb at pH 2.4 and 7.0 (12). The similarity between the spectra of *Coryphaena hippurus* Mb at low pH and of the deoxy tetracoordinated protoheme (see Figure 1) is immediately apparent. Every hemoprotein investigated displays a similar transient spectrum at low pH. The characteristic two-banded visible spectrum represents an unequivocal evidence that, under these conditions, the proximal histidine becomes protonated in the deoxygenated form at low pH, leading to the breakage of the proximal fifth axial bond of the Fe atom (13).

Furthermore, the pH-profile of CO binding rates to sperm whale, *Coryphaena hippurus*, *Dermochelys coriacea* and horse Mbs are similar (see Figure 2 and Table I) in terms of absolute values of the rates, whereas pK_a of the transition, although essentially similar to that of "chelated" mesoheme ($= 4.0$), are slightly different

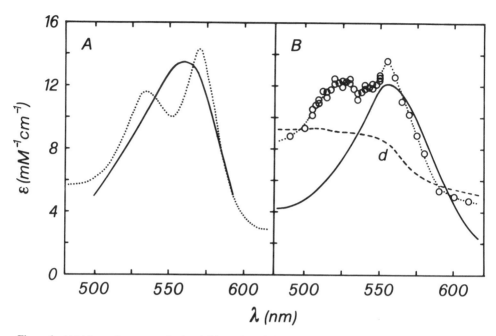

Figure 1. *(A)* Absorption spectra in the visible region of the ferrous deoxy pentacoordinated "chelated" protoheme (continous line; from Ref. 11) and of the tetracoordinated protoheme (dotted line) in 2% cetyltrimethylammonium bromide at pH 7.0, 0.1 M phosphate buffer. *(B)* Absorption spectra in the visible region of the ferrous deoxy *Coryphaena hippurus* Mb at pH 7.0, 0.5 mM phosphate buffer (continous line) and of the denatured protein at acid pH (= 2.40) in 0.15 M phosphate buffer (dashed line, marked by *d*; no effects of ionic strength and buffer composition on the absorption spectra have been detected). The experimental points (○) represent the values of optical density derived from pH-jump of *Coryphaena hippurus* Mb (see Ref. 12).

among the various proteins investigated (see Table I), probably reflecting variations in the heme-protein interactions. However, *Chironomus thummi thummi* Ery does not display any rate variation between pH 9.0 and 2.7 (see Figure 2), even though over this pH range the protein undergoes a spectral transition from the penta-coordinated to the tetracoordinated form, similarly to the other monomeric hemoproteins investigated. It may suggest that either: (i) the observed rate of CO binding in *Chironomus thummi thummi* Ery is not controlled by the Fe-CO bond formation, contrary to the other hemoproteins, but it is limited by ligand diffusion to the active site; or (ii) the breaking of the fifth axial bond of the Fe atom does not affect the CO binding rate to this protein. Case (i) appears very unlikely, since laser photolysis of CO-*Chironomus thummi thummi* Ery indicates that the geminate rebinding of carbon monoxide is slower than ligand diffusion outward in the bulk solution (J. Hofrichter et al., unpublished results). However, since in deoxy *Chironomus thummi thummi* Ery the distance of the Fe atom from the heme plane is smaller (0.17 A) (14) than in sperm whale Mb (0.55 A) (15) and human HbA (0.63 A) (16), the faster CO binding rate displayed by *Chironomus thummi thummi* Ery could be related to a smaller free energy barrier. It leads to propose a model in

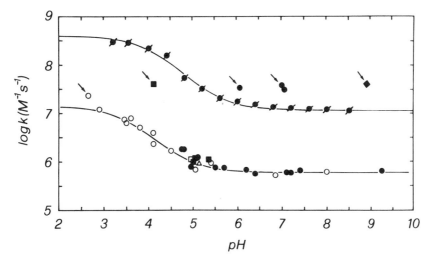

Figure 2. pH-dependence of the second-order rate constant for the binding of CO to *Chironomus thummi thummi* Ery (symbols indicated by arrows), sperm whale Mb and 3-[1-imidazolyl]-propylamide monomethylester mesoheme (symbols indicated by slashes; from Ref. 18), the latter in 2% cetyltrimethyl-ammonium bromide. Open and closed symbols refer to data obtained respectively with stopped-flow and flash photolysis. Different symbols correspond to the buffers used, i.e. phosphate (○,●), acetate (□,■), citrate (△) and borate (◆). The continous lines are the best fits to theoretical curves assuming a single ionizing group with a $pK_a = 3.45$ and 4.00 for sperm whale Mb and 3-[1-imidazolyl]-propilamide monomethylester mesoheme, respectively.

Table I

Values of the rate constants for CO binding at neutral and acid pH along with pK_a values for the penta- to tetracoordinate transition in monomeric hemoproteins

Proteins	pH	k $(M^{-1}s^{-1})$	pK_a
Sperm Whale Mb	7.10	6.0×10^5	3.45
	2.90	1.4×10^7	
Horse Mb	7.55	5.9×10^5	2.68
	2.55	2.7×10^7	
Coryphaena Hippurus Mb	6.95	1.2×10^6	3.55
	2.72	1.6×10^7	
Dermochelys coriacea Mb	7.00	1.3×10^5	2.59
	2.75	1.3×10^7	
Aplysia limacina Mb	7.10	6.0×10^5	2.98^a
	2.40^a	7.5×10^6	
	2.42^b	1.0×10^6	
Chironomus thummi thummi Ery	7.05	3.1×10^7	
	2.65	2.3×10^7	

[a]0.15 M phosphate buffer and 0.1 M acetate/acetic acid mixture.
[b]0.15 M phosphate buffer.

which the movement of the iron toward the heme is the crucial step in controlling the height of the activation barrier for CO binding to hemoproteins. In this perspective, the pH-independence of CO binding to *Chironomus thummi thummi* Ery would simply indicate that the position of the iron atom with respect to the heme plane is not affected in this protein by the protonation of the proximal histidine and the consequent cleavage of the proximal bond. Therefore, in the other hemoproteins tetracoordination enhances the rate of binding to the deoxy form mainly making energetically less unfavorable a conformation characterized by the iron closer to the heme plane, leading to a decrease of the activation barrier for CO binding by 2 Kcal/M.

In the case of *Aplysia limacina* Mb, the penta- to tetracoordinate transition occurs only if acetic acid/acetate is added to the phosphate buffer, displaying a pK_a quite similar to the other monomeric hemoproteins (see Table I). It suggests that in this protein weakening of the HisF8N$_{\epsilon}$-Fe bond is dependent on the binding of the carboxylate, which raises the pK_a of the penta- to tetracoordinate transition, thus acting as a rate-enhancing factor. As a matter of fact, this phenomenon has been shown also for the hexa to pentacoordinate transition of the nitric oxide derivative of *Aplysia limacina* Mb (17).

In spite of the similar pH-rate profile, the CO binding rate of the "chelated" mesoheme is ten to twenty times faster than in these monomeric hemoproteins at all pH values (see Figure 2) (18) which is likely to be attributed to steric effects related to the aminoacid residues present on the distal side of the heme pocket. In this regard, histidine deserves a special mention, since its substitution by smaller residues has been shown to bring about a decrease in the free energy kinetic barrier and an enhancement of the binding rate constant. However, such an expected rate increase is not observed in *Aplysia limacina* Mb, which lacks the distal histidine replaced by a solvent-exposed lysine (19). This is possibly due to a compensation by a larger value of the activation free energy for the binding of CO, resulting in a cancellation of the kinetic enhancing factor represented by the aminoacid substitution.

In conclusion, from the data reported here it is evident that distal steric effects, represented by the aminoacid residues of the heme pocket, and heme-globin interactions (proximal effects) both contribute to determine the free energy activation barrier for CO binding. However, distal determinants seem responsible mainly for reactivity differences among various hemoproteins while proximal determinants are mostly involved in the modulation of reactivity within the same protein and for the same ligand, as linked to external conditions, related to binding of protons or other small molecules like acetate.

This statement, even though proposed only for the ligand CO, seems of general application to heme proteins, since also in human HbA the difference in the kinetic activation barrier for CO binding to the two quaternary conformations ranges around 2 Kcal/M (20). Moreover, steric constraints from the distal side do not look dramatically different between R and T structural conformations (21), indicating that such a decrease in the free energy barrier is mostly attributable to the different heme geometry in the two conformations.

References and Footnotes

1. K. Moffat, J. F. Deatherage and D. W. Seybert, *Science 206,* 1035 (1979).
2. Antonini, E. and Brunori, M. in *Hemoglobin and Myoglobin in Their Reaction with Ligands,* North-Holland Publishing Co., Amsterdam (1971).
3. Perutz, M. F., *Nature (London) 228,* 726 (1970).
4. Seybert, D. W., Moffat, K., Gibson Q. H. and Chang, C. K., *J. Biol. Chem. 252,* 4225 (1977).
5. Abbreviations are: Mb, myoglobin; Ery, erythrocruorin; "chelated" proto or mesoheme, 3-[1-imidazolyl]-propylamide monomethyl ester proto or mesoheme.
6. Cannon, J., Geibel, J., Whipple, M. and Traylor, T. G., *J. Am. Chem. Soc. 98,* 3395 (1976).
7. Giacometti, G. M., Traylor, T. G., Ascenzi, P., Brunori, M. and Antonini, E., *J. Biol. Chem. 252,* 7447 (1977).
8. Traylor, T. G., Deardurff, L A., Coletta, M., Ascenzi, P., Antonini, E. and Brunori, M., *J. Biol. Chem. 258,* 12147 (1983).
9. Rossi Fanelli, A. and Antonini, E., *Biokhimija 22,* 336 (1957).
10. Ascenzi, P., Condò, S. G., Bellelli, A.,Barra, D., Bannister, W. H., Giardina, B. and Brunori, M., *Biochim. Biophys. Acta 788,* 281 (1984).
11. Traylor, T. G., Chang, C. K., Geibel, J., Berzinis, A., Mincey, T. and J. Cannon, *J. Am. Chem. Soc. 101,* 6716 (1979).
12. Because of the protein denaturation which occurs at low pH, the deoxy spectra were obtained by a pH-jump in the stopped-flow, keeping the protein unliganded. From the final spectrum of the denaturated protein and the amplitude of the absorption change, which follows the mixing, it is possible to construct the transient spectrum of deoxygenated hemoprotein immediately after the pH-jump (see Ref. 7).
13. The appearance of tetracoordinate spectrum immediately after mixing the deoxy form of the monomeric hemoprotein, kept at neutral pH, with the acid buffer (say pH 2.5) indicates that the protonation of the proximal histidyl residue, and the consequent tetracoordination of the heme, occurs at a rate much faster than the dead time of the rapid mixing apparatus (3 ms.). Moreover, the reversibility of the pH effects on the denaturation has been controlled spectrophotometrically for sperm whale Mb and *Aplysia limacina* Mb.
14. Steigemann, W. and Weber, E. *J. Mol. Biol. 127,* 309 (1979).
15. Takano, T. *J. Mol. Biol. 110,* 569 (1977).
16. Fermi, G. *J. Mol. Biol. 97,* 237 (1975).
17. Ascenzi, P., Giacometti, G. M., Antonini, E., Rotilio, G. and Brunori, M., *J. Biol. Chem. 256,* 5383 (1981).
18. Geibel, J., Chang, C. K. and Traylor,T. G. *J. Am. Chem. Soc. 97,* 5924 (1975).
19. Bolognesi, M., Coda, A., Gatti, G., Ascenzi, P. and Brunori, M.,*J. Mol. Biol.* to be submitted (1984).
20. Sawicki, C. A. and Gibson,Q. H. *J. Biol. Chem. 251,* 1533 (1976).
21. Baldwin, J. and Chothia, C. *J. Mol. Biol. 129,* 175 (1979).

Structure & Motion: Membranes, Nucleic Acids & Proteins,
Eds., E. Clementi, G. Corongiu, M. H. Sarma & R. H. Sarma,
ISBN 0-940030-12-8, Adenine Press,   Copyright Adenine Press, 1985.

Heuristic Studies of Structure-Function Relationships in Enzymes: Carboxypeptidase and Thermolysin

Michael N. Liebman and Harel Weinstein
Department of Pharmacology
Mount Sinai School of Medicine
of the City University of New York
New York, New York 10029

Abstract

We present an approach to the study of structure-function relationships in proteins that are fundamental to the design of proteins, peptides and ligands with predetermined properties. The approach is based on macromolecular structural analysis and theoretical chemistry, and is illustrated by a comparative study of the crystal structures of the metalloenzymes carboxypeptidase A and thermolysin, and of their complexes with inhibitors of different sizes and molecular structures. The analysis follows a general protocol proposed for heuristic studies that is outlined in detail. The techniques of structural superposition, linear distance plot analysis, distance matrix and partitioned distance matrix analysis, are described and used to analyze the secondary and tertiary structures of the native enzymes, to identify the sites of ligand-protein interaction in these enzymes, the structural consequences of the complexation, and the structural basis for recognition, specificity and reactivity. The results reveal similarities in the patterns of ligand-contacts, and the resulting structural perturbations that are independent of similarities in structural details in these two proteins and may therefore constitute general mechanisms of ligand-protein interaction in this class of metalloenzymes.

Introduction

The accurate and quantitative description of molecular structure and its relation to biological activity is a fundamental requirement for the understanding of molecular mechanisms of action in biological systems. Yet, only a few biological systems have been isolated and purified enough to be suitable for structural analyses able to yield the detail necessary for such quantification. This shortcoming is particularly evident in the lack of structural information on the proteins that constitute the biological receptors for hormones and neurotransmitters, and in the relatively small number of enzymes that have been studied in detail. Consequently, generalizations (e.g., see (1)) and heuristic modeling (e.g., see (2)) based on functionally or structurally related proteins constitute a major source of information. *In heuristic modeling, one system is studied in detail in order to infer the properties of another, functionally and/or structurally homologous system.* This approach should be especially

339

well suited for the study of enzymes belonging to classes in which functional mechanisms are well characterized, and structures are known at atomic resolution for several members. Analysis of the known structural details in these enzymes leads to an elucidation of the principles of macrostructural organization in these classes of proteins; the analysis of the structures of complexes between the enzymes and their substrates, inhibitors, or transition-state analogs, affords insight into structure-function relationship in that particular class of enzymes. The principles resulting from such studies should be applicable to the structurally unknown members of the class: hypotheses regarding the structural origin of differences between the individual members of the class, as well as insight concerning the functional consequences of such differences should emerge from these studies. For example, results from the preliminary comparison of carboxypeptidase A (CPA) and thermolysin (TLN) presented here, indicate how structural and functional similarity in the active site region of this class of zinc enzymes can be achieved in the absence of full structural homology, and point to the probable structural source of some of the observed differences between the specificities of the two enzymes. The hypotheses resulting from such studies can be probed experimentally by measuring and comparing the specificities and kinetic characteristics of the enzymes acting on a variety of substrates and inhibitors, and new substrates and inhibitors can be designed to test specific inferences.

In a documented example (2) of a successful use of the heuristic approach in experimental studies, even in the absence of a detailed theoretical analysis of the structure, the hypotheses for recognition and complementarity obtained from biochemical studies of CPA have led to the design of specific inhibitors for another metalloendopeptidase, angiotensin converting enzyme, ACE, (3). The inhibitor was designed to interact with the major determinants for specificity in the active site of CPA: a carboxyl-binding group, a group with affinity for the carboxy terminal peptide bond, and a tightly bound zinc ion that coordinates with the carboxyl of the scissile bond. A validation of the working hypothesis underlying this design is found in the recent X-ray crystallographic analysis (4) of the structure of the complex formed when the inhibitor, 3-mercaptopropanoyl-L-proline, modified to fit the specificity requirements, was diffused into the crystal of another metallopeptidase, thermolysin. This study provided the first determination of the mode of binding of a mercaptan inhibitor to a zinc peptidase, and confirmed the geometry that had been assumed for the active site complex in ACE (4) on the basis of its relation to CPA.

The advanced theoretical tools used in heuristic studies must combine the approaches and techniques used to analyze the three-dimensional structure resulting from X-ray crystallography, with the computational methods of quantum chemistry and molecular mechanics used to evaluate the physico-chemical properties of the systems and to study the energies and reactivities of the proteins interacting with the ligands (i.e., substrates, inhibitors, transition-state analogs). A variety of experimental approaches, including X-ray crystallography, can be used to probe the conclusions from the theoretical studies in which the various methods were used (5-8). To extend such studies to systems in which only limited structural data is available, the modeling approaches take advantage of the details of structural or functional homology. Specific examples of systems that can be studied on the basis of structural homology (i.e., homologous amino acid sequences or three-dimensional structures) include the modeling of sickle-cell hemoglobin dysfunction based on the properties of normal oxy-deoxy-carbomonoxy hemoglobin (9,10), the modeling of thrombin activity based on specificity data from other serine proteases such as trypsin and chymotrypsin (11), and the modeling of immunoglobin specificity based on that of other antibodies (12). In itself, functional homology presents a less defined guideline than structural homology for heuristic studies, because it could include modeling of functionally related proteins that bear no homology in their amino acid sequences, or their three-dimensional structures, (e.g., carboxypeptidase and thermolysin, as described below). However, inclusion of *functional homology considerations* in the heuristic modeling expands the scope of the analysis from the boundaries of the ligand recognition site (active site, or binding pocket) to regions external to this site. The importance of this extension is underlined by the observation that active sites and binding regions typically comprise less than 15% of the total volume of a protein, and yet the remainder of the structure is conformationally conserved through evolution, apparently for its functional role (13).

As illustrated below, we utilize the methods of macromolecular structural analysis and theoretical chemistry to study the known interaction of substrates, inhibitors, and transition-state analogs with CPA and TLN in order to examine the relationship between structure and function in these metalloenzymes, and to develop a heuristic model for the exploration of functional and stereomolecular specificity in related macromolecules. In addition to the approaches presented here, the molecular basis for the differences between the specificities of this pair of related enzymes is also sought from theoretical calculations of structure, and from simulations of interactions with ligands in order to infer on structural modifications of substrates (and inhibitors) that would improve their performance. Such guided synthesis of compounds designed to meet new specificity requirements, and to be tested experimentally, yields the necessary probes to test the hypotheses emerging from the theoretical analysis.

General Approach to Heuristic Studies

To define the source of reactivity and specificity determinants in the enzymes we study their macrostructural organization and conformational perturbation using our system of computer programs for macrostructural analysis, STRAP (STructural Representation and Analysis Package). The major steps in the protocols we follow in our heuristic studies are illustrated below for the case of CPA and TLN, and are followed by a description of the methods used in the analysis.

Step 1. Characterization of the Binding of Active Site Inhibitors:

The atomic coordinates from the X-ray diffraction are studied to characterize the binding of active site inhibitors. Coordinates of the native structure and of complexes with a variety of inhibitors, substrates or transition-state analogs are analyzed with the methods included in the STRAP package, as described below.

The structural perturbations induced by complexation are analyzed by comparing structures of the native and complexed enzymes. The methods of analysis (described below) include Distance Matrix analysis and Partitioned Distance Matrix analysis to reveal perturbations within the active site region and beyond it, while Structural Superposition and Linear Distance Plot methods can be used to evaluate quantitatively those segments of the molecule that remain unperturbed upon ligand binding. As illustrated below for the analysis of CPA and TLN, these methods enable the analysis of structural sub-domains and the *separation* of those conformational perturbations which occur within the secondary structure of the protein from those in the tertiary structure, or between folding domains. It is important to examine the influence of structural domains on macromolecular flexibility that could affect both the recognition and binding process, especially because the active sites of those multidomain enzymes that are known structurally have been shown to occur at inter-domain interfaces.

Step 2. Analysis of van der Waals Surfaces:

a. Separate molecules:

The full van der Waals surfaces of the enzyme and the complexed inhibitors are generated and examined for each enzyme-inhibitor pair to assess the tightness of the fit between the interacting molecules. This leads to inferences on the probablity that solvent remains trapped in the complexes, and identifies likely sites for solvent accumulation; this analysis also reveals possible access channels through which solvents could reach the active site region in the complex.

b. Complexes:

The van der Waals surfaces of the crystallographically defined enzyme complexes are analyzed to learn about possible sites of steric hindrance that could contribute to differences in the specificities of the two enzymes, and the groups within the active sites of CPA and TLN that could be responsible for these differences.

Step 3. Examination of Electrostatic Complementarity:

The electrostatic complementarity between each enzyme and its inhibitors is examined by comparing the electrostatic potential inside the active site with the electrostatic potential generated by the separate inhibitors in the conformations found in the complex, as well as in energetically optimized conformations (see Steps 4 and 5, below). Regions of optimal matching between the enzyme and the ligand in its bound conformation are identified, and the role of electrostatic complementarity in the stabilization of the complex is inferred. Comparisons with results obtained from energy optimized structures lead to inferences regarding the importance of electrostatic forces in the recognition, orientation, and anchoring of the ligands in the active site. We also identify separately the contribution of the active site and the entire protein to the electrostatic forces in the active site in order to infer on the role of the entire enzyme structure in the electrostatic stabilization of the complex.

Step 4. Definition of Molecular Reactivity Characteristics of the Enzyme:

The recognition of a substrate or inhibitor by an enzyme is not a single-event process but a series of interactions, so that examination of the final binding product evidenced in the X-ray crystallographic structure of the complexed enzyme might not fully define all the molecular reactivity characteristics exhibited by the enzyme. It is therefore important to explore the molecular arrangements that are known to occur, as well as those that are energetically likely to occur in the enzyme and the ligands.

The conformations and energetics of the arrangements of the ligands (substrates and inhibitors) in the active site as well as the energies involved in transitions from optimal conformations of the free ligand to the bound conformations of the complexes can be evaluated with a variety of molecular mechanics methods.

Step 5. Construction of Improved Substrates and Inhibitors:

Results from the theoretical analysis of recognition and specificity described above, in combination with results from biochemical studies, are used to predict improved substrate or inhibitor properties. Structures that incorporate these improvements can be constructed computationally to simulate their interactions with the active site. A variety of theoretical methods, including molecular mechanics and quantum chemical calculations are used for these simulations. These calculations could suggest alternative binding modes, i.e., different binding orientation in the two compared enzymes, which can be analyzed by superposition of the appropriate conformers to explore the possibility that they occur due to small structural differences in the two enzymes. Such structural differences could recruit secondary interaction sites in the active site and thereby induce changes in specificity, thus indicating a structure-function relationship that may also differentiate other members of this class of enzymes—an important clue for the heuristic modeling.

The Methods of Macrostructural Analysis

Many of the methods we use for structural analysis, for molecular modeling, and for the exploration of the specificity and reactivity determinants of enzymes, were developed only recently (14-17) but were already used in our own laboratory to study several protein structures and their interactions with ligands (18-21). These methods include the Linear Distance Plot analysis, Partitioned Distance Matrix analysis and Topological Structure searching. Other methods that we utilize include modifications to existing techniques (e.g., Distance Matrix analysis and Structural Superposition methods) which permit them to operate at a higher degree of resolution in the analysis of structure-function relationships in macromolecular systems. *The emphasis in the molecular modeling is not simply on the determination of similarities between two structures,* say that of an enzyme in its native form and in its complexed form with an inhibitor in the active site. *Rather, we analyze in detail the similarities between discrete parts of the secondary and tertiary structure, the extent of the differences, their structural nature, and their potential relation to the function of the macromolecule.*

a. Structural Superposition

We use an extensively modified version of the classical approach to structural superposition, to define a "best" transformation of one molecule to another by minimizing the root-mean-square difference between two sets of equivalent atomic positions: for the superposition of the atomic coordinates of one molecule, A, onto a second molecule, B, we use an algorithm based on that of Rao and Rossmann (14). The 'best' transformation is defined as that which minimizes the root mean square difference between two sets of equivalent atomic positions, $[X_i]'_A = [RT][X_i]_A$. The merit function, F, that is minimized is the root mean square difference $F = [\Sigma([X_i]_B - [X_i]'_A)^2/N]^{1/2}$ between the positions of the equivalenced atoms of molecules A and B.

The degree of significance of this metric has been investigated (21) and its strength was shown to be in establishing similarity of two homologous structures known at low resolution. This becomes a drawback in the higher-resolution analysis of the detailed differences between two structures because this metric is a single statistic which is averaged over the entire set of equivalenced atoms. We have modified the superposition algorithm, to incorporate the distribution of Δr, (where $\Delta r = (\Delta x^2 + \Delta y^2 + \Delta z^2)^{1/2}$) a one-tailed, skewed distribution. This distribution is normal in three-dimensions as a "chi-squared" (X^2) distribution with three degrees of freedom and we analyze this distribution of the fit of the equivalenced atom pairs according to a chi-squared statistic. The selection of outlyers from this statistic is preset at a conservative confidence level of 9.999%. The outlyers are removed from the equivalenced atom pairs and the translation and rotation matrices are redetermined, statistically refined as described above, and the entire procedure is iteratively applied until convergence of the equivalenced atom pairs is achieved upon application of the rotation and translation matrix. Convergence is defined as the the inability to distinguish outlyers from the theoretical distribution at the prescribed confidence level using a maximum likelihood statistic.

b. Linear Distance plot

The linear distance plot is a computer-assisted technique for rapid analysis of the secondary structure of a protein from its X-ray crystallographic coordinates. As illustrated below in the analysis of CPA and TLN, its *pattern recognition capabilities* can be used to: 1) identify components of secondary structure and define their boundaries; 2) locate repetitive motifs of secondary structure; 3) compare protein structures by identifying regions of structural insertions and deletions, as well as conformational changes, without considering amino acid sequences. To perform this analysis, the structure is partitioned into segments containing N amino acids following each α-carbon, from residue 1 in the sequence to (Ntot$-$N), where Ntot is the total number of amino acids in the polypeptide. The sum(s) of the distances between the α-carbon origin of this neighborhood and each of its (N$-$1) subsequent neighbors yields a characteristic value that reflects the conformation of the polypeptide chain within that segment. For example, for an ideal α-helix containing 10 amino acids (Ntot$=$10) and a selected segment size of 4(N$=$4), the sum of distances (S) will be identical for residues 1 through 6 because the components i.e., the distances between α-carbons in the α-helix, are constant. If the helix ends were extended such that the end-to-end distance increased significantly without a change in the number of residues, the component distances would also increase yielding a larger S. The parameter S thus reflects the overall curvature of a polypeptide chain because the conformation of a segment of the chain may be considered to have a set of helical parameters which are unique to a certain folding pattern constructed from 'virtually bonded' α-carbon atoms.

A spectrum-like decomposition of the overall conformation that reflects the curvature of the polypeptide chain is obtained by plotting S calculated for each neighborhood versus the residue number of the atom of origin of the segment (see Fig. 1). This spectrum-like representation can be interpreted by comparison to S values obtained from structures of ideal homopolymers. In Figure 1, the horizontal lines indicate the S values obtained for twenty ideal homopolymers superimposed on the linear distance plot of CPA. The neighborhood size in the figure was N$=$4, chosen to optimize the discrimination power of this technique; the probability of occurrence of a given conformation diminishes with increasing N. One limitation of this technique is its inability to distinguish mirror images: in the linear distance plot, a left-handed helical conformation will appear identical to a right-handed helix possessing similar helical parameters. Its strengths lie, however, in *comparisons of topological folding in two related proteins where it removes the need for either structural superposition or alignment of the amino acid sequence.* Where homologous proteins are involved, or when comparing two different conformations of a protein, *difference linear distance plots* can be generated by pointwise subtraction of one plot from the other to reveal regions of secondary structural change, and patterns indicative of the form of these conformational perturbations, e.g., helix bending, etc.

Pattern recognition analysis of sequences of structures identified by the "frequencies" in the plot is used to reveal ordered groupings of elements of nonclassical secondary structure that might be recognizable in other, non-homologous protein structures. These patterns provide the basis for comparison of structural organization among functionally related proteins, based on criteria transcending the classical identification of helices, sheets and turns.

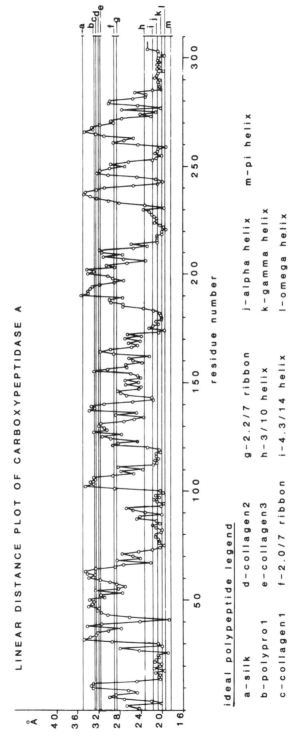

Figure 1. Linear distance plot of native carboxypeptidase A from its crystal structure (3CPA). Horizontal lines indicate the values for the ideal polypeptides.

c. Distance Matrix analysis

The distance matrix is a convenient form of representation of the conformation of polypeptides that has proven useful in describing the organization of secondary and tertiary structure into structural domains. It was originally conceived by Phillips (15), and Ooi and Nishikawa (16) and is most commonly used to represent the conformation of a protein structure determined crystallographically in comparison to conformations generated by protein folding techniques. With the use of this technique, elements of secondary and supersecondary structure have been identified by Kuntz (17), and Liebman described its use in the identification of tertiary structure, supersecondary structure, and folding domains (18). Recent extensions of this technique have rendered it useful for comparisons of homologous or even non-homologous protein structures, providing structural evidence of gene duplication and also structural palindromes, revealing patterns of macromolecular recognition by extensions to quaternary structural representation (19).

The algorithm for the construction of a distance matrix (20) is based on the list of all α-carbons numbered from 1 to N (where N is the total number of amino acids in the protein) in order of the amino acid sequence, from amino to carboxy terminus. The matrix is square, symmetrical, and of order N, with each ij-th element being the distance between the α-carbon (i) and the α-carbon (j), i.e. $r_{ij}=r_{ji}$; all the diagonal elements are equal to zero. A unique representation of all the distance pairs within the N amino acids in the protein is contained in either the upper or the lower half of the matrix.

A graphic version of the matrix (see Fig. 2) is achieved by generating graphic elements (symbols) representing equi-distance contours within certain preset distance ranges, i.e., a different symbol or a different shading intensity for each range of r_{ij} values. Typical boundaries for the ranges of these values are 5.0A, 10.0A and 15.0A. This results in a contouring that highlights regions of contact within the structure that are close together in three-dimensional space although they may be distant in amino acid sequence and thus in placement along the polypeptide chain. In general, contours for the distance ranges given above will concentrate near the diagonal, with i and j of approximately equal magnitude; the patterns formed by the shading of these areas are therefore indicative of secondary structural elements. Such short distances between distant residues result from the tertiary folding of the protein. The contours that appear more distant from the diagonal identify sequence-distant i and j residues that are geometrically close, and are therefore indicative of the tertiary structure and the folding of the protein. The elements of secondary structure in CPA, as well as their interactions (i.e., tertiary structural features), are identified in Fig. 2.

Certain classical shading patterns have been identified as corresponding to well defined structures such as α-helices, parallel and anti-parallel β-sheet structures, contiguous β-sheet folding domains and β-barrels, as well as tertiary structural patterns resulting from the interactions of sheets, turns and helices. Examples of

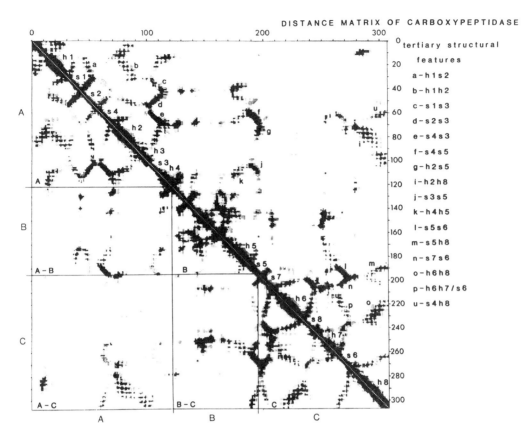

Figure 2. Distance matrix of native carboxypeptidase A from its crystal structure (3CPA), contoured at 5.0, 10.0, and 15.0 A (decreased shading), showing definition of structural domains A, B and C. The interactions between the elements of secondary structure identified in the distance matrix are listed as tertiary structural features.

these patterns are identifiable in Fig. 2. Some general patterns can also be recognized as indicative of folding domains. These regions, contoured with intermediate density and located in contiguous positions along the diagonal, represent interactions between folding domains. Such features reflect the organization of structural elements that can be parallel, i.e., structures which repeat within the protein and are oriented with the same directionality of amino-to-carboxy terminus (e.g., those originating from gene duplication) or anti-parallel, i.e., structures which repeat within the proteins and are oriented with opposing directionality with respect to the amino-to-carboxy polarity of the protein chain (e.g., those occurring in the structural palindromes present in the serine proteases) (17,18).

An important advantage of the distance matrix technique for structural representation is that it belongs to a group of *methods that make possible structural comparisons without the need for direct superposition* of partial or complete structures. The distance matrix representation, which is based on internal distances and is therefore not related to any specific origin, will not be altered by rotation and/or translation of the coordinates. Two macromolecules may therefore be compared directly by simple examination of their respective distance matrices, without need for superposition. The algorithm for generating the distance matrix renders it invariant to a mirror inversion of the structure, i.e., a totally inverted protein and its mirror image would present the same distance matrix.

d. Partitioned distance matrix analysis

This is a rapid survey technique for the comparison of a series of structurally related molecules; it utilizes the pattern recognition power of the distance matrix described above to compare elements of secondary and tertiary structure, and folding domains that are common even in proteins with different three-dimensional conformations. The method, described in detail and illustrated both in its simple and extended applications (19,20,21), is presented here only in the simpler form that was applied to the direct comparison of the complexes of CPA and TLN, as illustrated below.

The distance matrix is partitioned on the basis of results from secondary structure analysis obtained .from the linear distance plot by marking regions of secondary structural elements, e.g., α-helices, β-sheets, β-turns, etc., or some of the more complex patterns of protein folding that are identifiable in the linear distance plot. The number of partitions is of the order M, where M<N(N is the total number of amino acids in the protein). To survey a series of structures, the distance between the i-th and j-th elements of a given partition in one molecule is subtracted from that between corresponding pairs in the other molecules; both the absolute sum of the differences (APDM) and the signed (algebraic) sum of the differences (SPDM) are tabulated within each partition of the two compared molecules. Both sums are normalized to the number of component distance pairs to make possible a comparison of average distances among partitions containing different numbers of contributors to the sums; the averages are also weighted relative to the experimental errors within the partitions. (The errors are computer generated from the positional errors assigned to the individual atomic coordinates used in the computation, based on the original crystallographic report, or as determined by a statistical analysis of the protein structure). In this manner, it becomes possible to determine the relative size of the differences in conformation, i.e., relative to secondary structural partitions or the tertiary structural partitions, and relative to the potential error in the measurements of these distances. The contrast between the SPDM analysis and the APDM analysis is significant because the algebraic sum may vanish if many large terms are present but are equal in magnitude and opposite in sign. Such compensation could be due to a specific form of conformational change, e.g., from symmetrical conformational changes. The APDM analysis would reveal the presence of such compensated differences by exhibiting large terms in elements that have relatively small values in SPDM. Concerted conformational changes, i.e., expansions of structure (or contractions), are expressed by large differences that are approximately equal in absolute value in both APDM and SPDM.

To probe for the site and form of the conformational perturbation, this analysis can also be applied using partitionings consisting of all atoms of the individual sections rather than the α-carbon virtual bonds. The strength of partitioned distance matrix analysis in macrostructural comparison is *the ability to simultaneously survey secondary, tertiary and quaternary structural features and also to qualitatively examine the source of conformational perturbations.*

e. Energy calculations for macromolecular structures

The quantum mechanical methods for the calculation of molecular structure and properties that are used in the study of enzyme structure and activity have been described in detail and reviewed extensively (22). Of special interest are the methods for the calculation of electrostatic properties of macromolecules, e.g., the molecular electrostatic potential (MEP). The use of MEP to study chemical properties, molecular interactions, and biological activities of molecules has been documented and reviewed critically in a recent symposium (23), specific applications to the study of the properties of CPA have been illustrated (24,25) and also recently expanded to include MEP studies of interactions of CPA with ligands (26,27). Studies of MEP maps for a variety of other macromolecular systems, including enzymes (28,32), have demonstrated the power of this analysis and the importance of electrostatic considerations in the study of interaction of a variety of ligands with biological macromolecules (33,34).

Molecular Modeling of the Structure and Specificity Determinants of CPA and TLN.

Both enzymes are zinc-containing peptidases which have been thought to have undergone possible convergent-evolution to provide the necessary zinc center for catalytic activity (35). A high degree of structural homology could be anticipated for the two enzymes, since both hydrolyze peptide bonds on the amino side of large hydrophobic residues, and in view of their similar molecular weight (i.e. about 34000). No significant structural homology appears to exist. However, examination of the active sites of the two molecules, particularly in reference to the zinc ligands and the binding site for peptide inhibitors, revealed a surprising degree of similarity (35). These observations indicate that the results of the present preliminary report that aims to exemplify the power of the techniques used in this type of analysis, should also reveal some of the aspects of structural organization of zinc-containing peptidases that will be useful in the identification of general properties that may be shared by other enzymes in this class.

In a detailed report (36), we present the results of a comparative macrostructural analysis of the structures of native carboxypeptidase A (CPA), carboxypeptidase A complexed with GLY-L-TYR (CPA-GY), with an inhibitor protein isolated from potato (CPA-PI), and carboxypeptidase B (CPB). Three-dimensional structures of these complexes were available from X-ray crystallographic studies in Lipscomb's group and deposited in the Protein Data Bank (37); we performed a structural analysis of the protein organization of the enzyme, and of the interaction of its secondary, tertiary and domain structure (e.g., see Figs. 1 and 2). Only a limited analysis of the direct specificity requirements of the active site was possible as a result of the low occupancy (i.e. only 40%) of the GLY-L-TYR in the CPA-GY crystal structure. We examined the independent behavior (response to inhibitor binding) of the three folding domains of this enzyme, determined regions of secondary and tertiary structural perturbation which accompany direct ligand contact within the active site region, and described the potential *propagation* of these conformational perturbations due to contacts between the active site region of the enzyme and the other elements of the secondary and tertiary structure that are external to the active site (27,36). In addition, we studied the electrostatic complementarity of the active site (27), and performed quantum chemical calculations on models representing molecular interactions in the active site (26,28). Below we present a brief summary of results.

Some of the observations were summarized recently and led to the following conclusions (27):

1) The enzyme consists of three independent folding domains which exhibit conformational flexibility upon inhibitor binding;

2) A possible role of the overall protein organization in the recognition process is revealed by concerted changes induced in the enzyme structure outside the contact region upon binding of inhibitors;

3) Although the two inhibitors compared in their complexes with CPA (CPA−GY and CPA−PI) differ greatly in size and complexity, they are very similar in the elements of recognition and complementarity to the enzyme as well as in the qualitative nature (but not locus) of the major conformational perturbations induced upon formation of enzyme inhibitor complexes. This suggests that the conclusions reached from the present analysis should be generalizable to other substrates and inhibitors of carboxypeptidase A;

4) The electrostatic potential surfaces reveal the complementarity between enzyme and inhibitor within the active site region, although the molecular structural components responsible for this complementarity differ in the two inhibitors;

5) The MEP map calculated in the active site region from the contribution of the active site model is very similar to that generated by the entire protein; it differs only in its lower polarity, i.e., the smaller absolute values of the positive and negative regions;

6) Cavities observed within the active site in both enzyme-inhibitor complexes could contain solvent that may be involved in binding or catalytic activity.

For the analogous study of TLN, the data available from the Protein Data Bank (37) consisted of native thermolysin (3TLN), thermolysin complexed with L-LEU-NHOH (4TLN), thermolysin complexed with HONH-Benzylmalonyl-ALA-GLY-pNitroanilide (5TLN) and thermolysin complexed with $ClCH_2CO-DL-(NOH)LEU-OCH_3$ (7TLN). The distance matrix of thermolysin shown in Fig. 3, readily reveals the presence of two folding domains, containing residues 1-159, and 160-314. The high content of helix within the secondary structure, and the strong interactions between helices, both within each of the individual folding domains and in the interaction between the folding domains, is also evident in the contoured distance matrix (Fig 3).

THERMOLYSIN

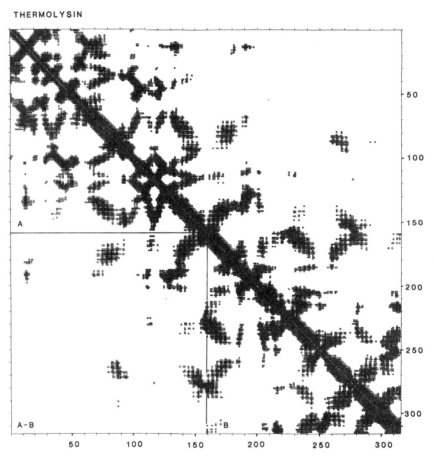

Figure 3. Distance matrix of native thermolysin from its crystal structure (3TLN), contoured at 5.0, 10.0 and 15.0 A (decreased shading). Structural domains A and B are identified.

Analysis of the distances between atoms in the inhibitors and in the enzyme was used to define the contacts and delineate the binding pocket. The total number of contacts was found to be 49 in 4TLN, 89 in 5TLN, and 63 in 7TLN, reflecting the

Table I

Enzyme-Inhibitor and Enzyme-Ca^{2+} Contacts

	4TLN	5TLN	7TLN
	112 ASN	112	112
	113 ALA	113	113
	114 PHE	114	114
	—	130 PHE	—
	—	133 LEU	—
	139 VAL	139	139
	142 HIS	142	142
	143 GLU	143	143
	146 HIS	146	146
	166 GLU	166	166
	169 SER	—	—
	188 ILE	188	188
	189 GLY	189	189
	202 LEU	202	202
	203 ARG	203	203
	—	226 ASP	—
	231 HIS	231	231
CA1	ASP138, THR174, GLU177, ASP105, ILE188, GLY189, GLU190		
CA2	GLU177, ASN183, PRO184, ASP185, GLU190, ASP191		
CA3	ASP57, ALA58, ASP59, PHE62, ASP67		
CA4	TYR193, THR194, ILE197, GLY199, ASP200		

comparative size and mode of binding of these inhibitors. The list of contacts is given in Table 1.

In an analogous computation we defined the list of contacts for the four Ca^{2+} ions in the TLN structures; the list of these contacts is also given in Table 1.

Preliminary applications of the Structural Superposition procedure to TLN and its complexes with inhibitors were also carried out, without reference to the domain structure, and including the four calcium ions and the zinc in the comparisons. The results of these comparisons are given in Table 2. In these comparisons to 3TLN, several residues were declared as outliers by the superposition technique in all the compared pairs: 1I,14L,51P,79V,85K,89N,96N,101R,119E,128Q,156I,168I,197I,198S, 289S,316K. In addition, in the comparison of 4TLN with 3TLN, the following residues were also declared as outliers: 91L,108Q,118S,158Q,175L,187E,210K,225Q. In the comparison of 5TLN with 3TLN the additional outliers included: 2T,9V,25S, 33N,120M,129T,201S,202L,211Y,227N. These outliers indicate the amino acids that were most perturbed by the interaction with the individual inhibitors, either through direct contact or by conformational changes induced in more distant parts of the structure. As we observed in the analysis of carboxypeptidase (see above, and (27)), there are many regions of conformational perturbation that are common to all inhibitors that bind to the active site, but *not all conformational perturbations are*

Table II
Results of Superposition of Thermolysin Complexes with Native Thermolysin (deviation in A)

	Initial		Final		Calcium Sites				Zinc Site
	# atoms	rms	# atoms	rms	CA1	CA2	CA3	CA4	Zn
4TLN-3TLN	2437	.22	2379	.19	.09	.10	.18	.11	.07
5TLN-3TLN	2437	.24	2378	.21	.11	.08	.22	.15	.21
7TLN-3TLN	2437	.14			.07	.13	.05	.09	.08

restricted to this common region, indicating that conformational perturbation outside the active site is linked to the specificity determinants of the enzyme.

These results are preliminary and further analysis of the nature and consequences of the perturbation is required, but analysis of the superposition already indicates a high degree of similarity in the regions of residues 10-20, 125-135, 145-160, 195-205, 265-275, and 305-315. This similarity is reflected either in the lack of change induced by inhibitor binding or in very similar patterns of change. A more refined, and hence more reliable, representation of these similarities is provided by the analysis of the linear distance plots.

Linear distance plot analysis was performed on thermolysin and its complexes, comparing the 3TLN, 4TLN, 5TLN, and 7TLN structures (only the linear distance plot of 3TLN is shown in Fig. 4). The results suggest differences in the definitions of structural boundaries and structural types as compared with features published in the Protein Data Bank, as was also the case with carboxypeptidase (27,36). Different linear distance plots were also computed for the 4TLN-3TLN, 5TLN-3TLN and 7TLN-3TLN comparisons, and are also shown in Fig. 4. As noted in the carboxypeptidase analysis (27,36), structural differences were observed with this technique that were not readily apparent from the superposition analysis. As with carboxypeptidase, all regions of the final superposition profile in which conformational perturbation within thermolysin is apparent (see above), are also identified by the difference linear distance plot analysis (Fig. 4). However, the changes revealed by this analysis only reflect perturbations of contiguous regions of the protein, e.g. secondary structure. *Combining the information from all the methods of analysis it also becomes possible to infer on changes in the tertiary structure.* For example, the contact analysis reveals that one of the Ca^{2+} ions (CA3) is in close contact with the region of the protein in which conformational changes are induced in 4TLN and 5TLN but not in 7TLN. For 4TLN and 5TLN this Ca^{2+} site is shown in Table 2 to exhibit the largest deviation in the superposition analysis which is based on the protein, not on the Ca^{2+} coordinates. These are also inhibitor contact regions which distinguish 4TLN and 5TLN from 7TLN (including residues 130 and 169). In the partitioned distance matrix analysis shown below, in which seven major partitions of the structure defined by the linear distance plot are compared to examine

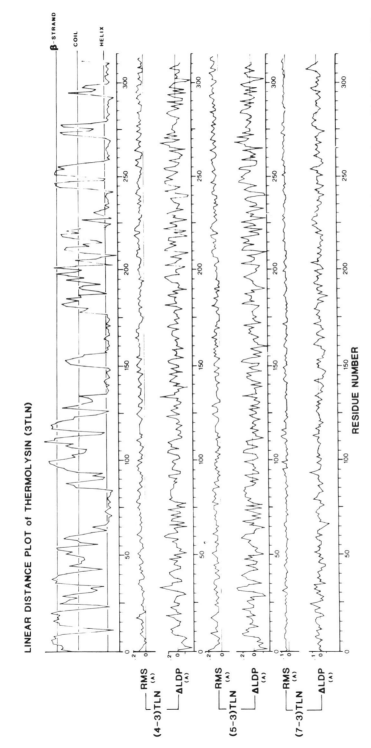

Figure 4. Linear distance plot of thermolysin (3TLN—top strip) and structural comparisons to the structures of three complexes with inhibitors: 4TLN, 5TLN and 7TLN. Each pair of strips in the comparison presents the root mean square (RMS) deviation (in Å) calculated from the superposition analysis, and the deviation (in Å) calculated from the difference linear distance plots (delta LDP) of each pair of structures (i.e., the enzyme-inhibitor complexes—4TLN, 5TLN, or 7TLN—minus the native enzyme structure, 3TLN).

the tertiary structure changes in the complexes, the main changes observed (expressed in A) involve partitions D and E which include residues 136-159 and 160-178, respectively.

4TLN

	A	B	C	D	E	F	G
A	.08						
B	.08	.09					
C	.09	.10	.08				
D	.12	.12	.11	.08			
E	.13	.13	.11	.09	.07		
F	.10	.10	.10	.09	.10	.10	
G	.12	.11	.11	.09	.10	.10	.10

Partition	Residues
A	1- 64
B	65- 85
C	86-135
D	136-159
E	160-178
F	179-259
G	260-316

5TLN

	A	B	C	D	E	F	G
A	.10						
B	.09	.09					
C	.10	.11	.09				
D	.14	.14	.13	.09			
E	.15	.15	.13	.09	.09		
F	.12	.12	.11	.10	.10	.11	
G	.14	.13	.13	.10	.10	.11	.10

Partitions in Domain I (domain A in Fig 3): A,B,C,D

Partitions in Domain II (domain B in Fig. 3) E,F,G

7TLN

	A	B	C	D	E	F	G
A	.04						
B	.04	.05					
C	.05	.04	.05				
D	.05	.04	.05	.04			
E	.05	.04	.06	.04	.04		
F	.04	.04	.06	.04	.04	.04	
G	.05	.04	.06	.05	.04	.04	.04

These two portions, D and E, are at the interface of the two domains in the protein structure of TLN, revealed by the distance matrix (Fig. 3). Large numerical values of the A and B vs D and E elements in the matrices appear only in 4TLN and 5TLN, but not in 7TLN, emphasizing the larger changes in tertiary structure that occur between those regions even though the secondary structural changes were not large. The only other major change revealed by the partitioned distance matrix is between the amino and carboxy terminals (i.e., partitions A and G).

Preliminary Conclusions

The analysis of the structure of thermolysin and complexes with inhibitors is not yet complete. The conclusions reached from the comparison with the carboxy-peptidase are therefore preliminary. Several salient points emerge, however, from the results discussed above.

1) In contrast to CPA where we observed frequent transitions between elements of secondary structure in two of the three domains (i.e., A and C), TLN is characterized by an alternation between extended regions containing helices and regions of β-sheet conformation. Also in contrast to CPA, which contains a domain (B) that exhibits little structure categorizable as helix or sheet, both domains of TLN exhibit predominantly periodic structures;

2) The domain interface in TLN consists of helix-helix interactions whereas in CPA the three domain interfaces, A to B, B to C, and A to C, consist of interactions between β-strands. Because the active sites of these enzymes, as of other multidomain proteins, reside at domain interfaces, this constitutes an organizational difference in the environment of the active sites of TLN and CPA which may entail functional consequences in interactions with substrates and inhibitors;

3) As in the complexes of CPA, most of the ligand-protein contacts in the complexes of TLN are similar for all the ligands although the ligands differ in size;

4) As observed for CPA, the regions of conformational change induced by complexation in TLN are not limited to the regions of contact, and they vary with the individual inhibitors;

5) Within the complexes of TLN, an orientational difference of the positioning of inhibitors within the active site is immediately revealed by the large structural changes induced in 4TLN and 5TLN, but not 7TLN. Although 4TLN and 5TLN differ in their orientation, both affect sites in contact with CA3, a calcium binding site that is not in contact with the inhibitor in 7TLN, pointing to a possible causal relationship between major structural changes in the Ca^{2+} position, and perturbation of the protein structure.

6) Comparison of the groups of structures of CPA and TLN complexes leads to the general conclusion that structural perturbation occurs in regions of the enzyme outside the active site (binding pocket), and not all regions of contact with ligand exhibit conformational perturbation.

Comparison with the conclusions obtained for CPA reveals that *these inferences appear to be independent of structural organization of the protein, both inside and outside the binding pocket, and are fundamental to the definition of recognition, specificity, and reactivity in ligand-protein interaction.* Specific molecular determinants for recognition, specificity, and reactivity should be revealed in a subsequent modeling stage in which we will compare the functional properties of the proteins and ligands (e.g., molecular electrostatic potentials, electric fields, interaction energies) to identify the basis for their similarity, and the reasons for the differences in their characteristics.

Acknowledgements

M. L. Liebman is a Revson Fellow supported by an award from ᴄ.ᴊe Revson Foundation. H. Weinstein is recipient of a Research Scientist Development Award (K02 DA-00060) from the National Institute on Drug Abuse. Generous grants of computer funds from the University Computer Center of the City University of New York and the Computer Center of the Graduate School of CUNY are gratefully acknowledged.

References and Footnotes

1. Bunning, P., Holmquist, B. and Riordan, J. F., *Biochem. Biophys. Res. Commun. 83,* 1442 (1978).
2. Ondetti, M. A. and Cushman, D. W., *Ann. Rev. Biochem. 51,* 283 (1982).
3. Cushman, D. W., Cheung, H. S., Sabo, E. F. and Ondetti, M. A. *Biochemistry 16,* 5485 (1977).
4. Monzingo, A. F. and Matthews, B. W., *Biochemistry 21,* 3390 (1982).
5. Wippf, G., Dearing, A., Weiner, P. K., Blaney, J. M. and Kollman, P. A., *J. Amer. Chem. Soc. 105,* 997 (1983).
6. Blaney, J. M. and Langridge, R., *J. Med. Chem. 25,* 777 (1982).
7. Veber, D. F., Freidinger, R. W., Perlow, D. S., Palevda, W. M. Holly, F. W., Strachan, R. G., Nutt, R. F., Arison, B. H., Hommick, C., Randall, W. C., Glitzer, M., Saperstein, R. and Hirschman, R., *Nature 292,* 55 (1981).
8. Kuo, L. C., Fukuyama, J. M. and Makinen, N. W., *J. Mol. Biol. 163,* 63 (1983).
9. Baldwin, J. and Chothia, C., *J. Mol. Biol 129,* 175 (1979).
10. Amato, S. V., Liebman, M. N. and Brinigar, W. S., *J. Protein Chem.,* submitted (1984).
11. Mann, K. G., Nesheim, M. E., Hibbard, L. S. and Tracey, P. B., *Ann. N. Y. Acad. Sci. 370,* 378 (1981).
12. Novotny, J., Bruccoleri, R., Newel, J., Murphy, D., Haber, E. and Karplus, M., *J. Biol. Chem. 258,* 14433 (1983).
13. Rossmann, M. G., Liljas, A., Branden, C. I. and Banaszak, L. J., in *The Enzymes,* Ed., P. D. Boyer, Academic Press, New York Vol. II, p. 62, (1975).
14. Rao, S. T. and Rossmann, M. G., *J. Molec. Biol. 76,* 241 (1973).
15. Phillips, D. C. in *British Biochemistry Past and Present,* Ed., T. W. Goodwin, Academic Press, London, p. 11 (1970).
16. Nishikawa, K., Ooi, R. T., Isogai, Y. and Saito, N., *J. Phys. Soc. (Japan) 32,* 1331 (1972).
17. Kuntz, I. D., *J. Amer. Chem. Soc. 97,* 4362 (1975).
18. Liebman, M. N., Ph.D. Thesis, Dept. of Chemistry, Michigan State University (1977).

19. Liebman, M. N., *Biophys. J. 32*, 213, (1980).
20. Liebman, M. N. in *Molecular Structure and Biological Activity,* Eds. J. F. Griffin and W. L. Duax, Elsevier Science Pub. Co. New York, p. 193 (1982).
21. Liebman, M. N. in *Structural Aspects of Recognition and Assembly in Biological Macromolecules,* Eds. M. Balaban, J. Sussman and A. Yonath, Balaban, ISS, Philadelphia, p. 147, (1981).
22. Weinstein, H. and Green, J. P. (Eds.), *Quantum Chemistry in Biomedical Sciences,* Ann. N. Y. Acad. Sci. Vol. 367 (1981).
23. Politzer, P. and Truhlar, D. G.(Eds.), *Chemical Applications of Atomic and Molecular Electrostatic Potentials,* Plenum Press, New York (1981).
24. Hayes, D. M. and Kollman, P. A., *J. Amer. Chem. Soc. 98*, 3335 (1975); *ibid* 98, 7811 (1976).
25. Allen, L. C., *Ann. N. Y. Acad. Sci. 367*, 383 (1981).
26. Weinstein, H., Osman, R., Topiol, S. and Venanzi, C. A. in *Quantitative Approaches to Drug Design,* Ed., J. C. Dearden, Elsevier Science Publishers B.V., Amersterdam, p. 81 (1983).
27. Weinstein, H., Liebman, M. N. and Venanzi, C. A. in *New Methods in Drug Research-I,* Ed., A. Makriyannis, J. R. Prous Publ. Co., Barcelona, 1984, in press.
28. Osman, R., Weinstein, H. and Topiol, S., *Ann. N. Y. Acad. Sci. 367*, 356 (1981).
29. Sheridan, R. P. and Allen, L. C., *J. Amer. Chem. Soc. 103*, 1544 (1981).
30. Etchebest, C., Lavery, R. and Pullman, A., *Stud. Biophys. 90*, 7 (1982).
31. Lavery, R., Pullman, A. and Pullman, B., *Theoret. Chim. Acta. 62*, 93 (1982).
32. Weiner, P. K., Langridge, R., Blaney, J. M., Schaefer, R. and Kollman, P. A., *Proc. Natl. Acad. Sci. USA 79*, 3754 (1982).
33. Koppenol, W. H. and Margoliash, E., *J. Biol. Chem. 257*, 4426 (1982).
34. Warshel, A., *Proc. Natl. Acad. Sci. USA 81*, 444 (1984).
35. Kester, W. R. and Matthews, B. W., *J. Biol. Chem. 252*, 7704 (1977).
36. Liebman, M. N., Venanzi, C. A. and Weinstein, H., 1984, *Biopolymers,* submitted.
37. Bernstein, F. C., Koetzle, T. F., Williams, G. J. B., Meyer Jr., E. F. and Tasumi, M., *J. Mol. Biol. 112*, 535 (1977).

Structure & Motion: Membranes, Nucleic Acids & Proteins,
Eds., E. Clementi, G. Corongiu, M. H. Sarma & R. H. Sarma,
ISBN 0-940030-12-8, Adenine Press, Copyright Adenine Press, 1985.

Modeling The Activation Energy and Dynamics
of Electron Transfer Reactions in Proteins

Antonie K. Churg and Arieh Warshel
Department of Chemistry
University of Southern California
Los Angeles, California 90089

Abstract

A microscopic treatment of electron transfer (ET) between proteins is described. This treatment involves static and dynamic models. The static model evaluates the contribution of the reorganization energy, α, to the rate of ET. This model is applied to the case of electron exchange between the heme groups of two cytochrome c molecules. It is shown that the observed X-ray structural differences between reduced and oxidized cytochrome c can be used to obtain an estimate of α. This estimate depends on the electrostatic interactions of the heme and the protein dipoles and is rigorously independent of the folding energy of the protein. The dynamic model is a semiclassical trajectory approach that simulates the molecular dynamics of electron exchange in a system of two cytochrome c's. This semiclassical approach, which is designed to treat realistic anharmonic systems, gives a rate constant quite similar to that obtained using the static approach. It is found that the time dependent energy gap obtained in the semiclassical approach can provide a basis for a microscopic harmonic approximation of ET in proteins. That is, the Fourier transform of the energy gap can be used in a "dispersed polaron" model. This model allows one to explore the temperature dependence of the rate constant within the harmonic approximation. A preliminary study using this model gives results which may be relevant to experimentally observed ET rates in photosynthetic bacteria.

I. Introduction

Understanding the relation between the three-dimensional structure of electron transfer (ET) proteins and the energetics and rates of their reactions has been a central problem in biochemistry. The various factors in the energetics are difficult to isolate experimentally. Moreover, a crucial factor determining the rate of ET, the so-called reorganization energy (1,2), cannot be measured even though it is directly related to the structure. Thus, a theoretical treatment of ET in proteins is potentially insightful.

While the spatial coordinates of the "reacting groups" usually describe the progress of a reaction, in ET reactions, the reaction coordinate is not the electron's position in space. According to the formal theory of ET, the proper reaction variable is the dielectric polarization of the medium around the electron donor and acceptor charge distributions (1-3). At the outset, it is not obvious how the polarization fluctuations of proteins differ in magnitude or frequency from those of isotropic solutions. The frequency spectrum of the dielectric polarization is of significant interest in the cases of low temperature ET, where tunneling is the only accessible channel. In particular there is a great interest in the temperature dependence of ET reactions of photosynthesis, whose rates are constant over a wide temperature range (4). In principle, the nature of these fluctuations can be obtained from trajectory calculations of the classical motion. The trajectory simulation approach also offers an intriguing possibility of explicitly calculating the ET rate by semiclassical quantum mechanics.

This paper describes preliminary results of calculations simulating the dynamics of electron exchange between the reduced and oxidized forms of cytochrome c. In Section II the Marcus-Levich theory of ET (derived for a dielectric continuum solvent) is applied to construct a static model for the relation between protein X-ray structure and the intrinsic ET rate. Here, the important quantity is the reorganization energy, α, which is the electrostatic energy required to switch the oxidation states of the heme groups. The X-ray structures of the oxidized and reduced forms of the protein give an estimate of α.

Section III evaluates the rate constant for ET reactions using a semiclassical trajectory approach. This simulation uses the calculated motion of the protein atoms to evaluate the time dependent energy gap which determines the rate of electron exchange. The relation between the semiclassical results and the static results is examined. This section also presents a "dispersed polaron" model which lets one explore the temperature dependence of the rate constant within the harmonic approximation.

II. Static Model for ET Activation Energy in Proteins

Figure 1 introduces several important features of ET reactions in protein environments. In these systems, the electron donor (D_1) and acceptor (A_2) are molecules such as heme or flavin in their reduced and oxidized forms, respectively. These molecules are inside a protein, which, in turn, is surrounded by water. Thus, protein 1 and the surrounding water constitute the solvent for D_1 and protein 2 and surrounding water constitute the solvent for A_2. As far as the ET reaction energetics are concerned, the interaction of the protein dipoles and charges and the surrounding water dipoles with the charge distribution associated with D_1 and A_2 is crucial. The ET process is a redistribution of charge, and the distance between electron donor and acceptor is typically quite large ($>5\text{Å}$) in biological ET. Therefore the electron can hop from D_1 to A_2 only if the system reaches the dipole configurations for which the total energy of the system is unchanged when the electron jumps. Calculating the probability of reaching these special configurations is necessary in order to calculate the overall rate.

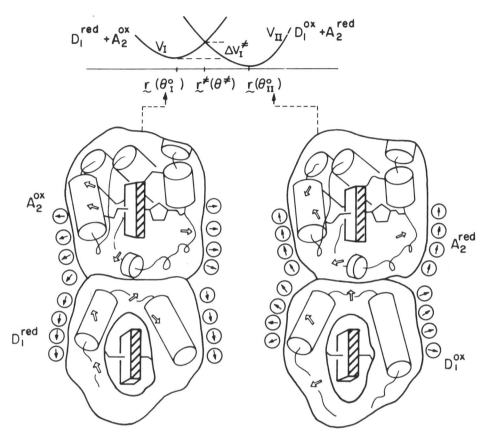

Figure 1. A sketch relating the structure to the energetics of electron transfer from a donor D_1^{red} in protein 1 to an acceptor A_2^{ox} in protein 2. The protein dipoles and water dipoles (encircled arrows) reorient from their equilibrium configuration, $\underline{r}(\theta_1^o)$ around reactants to that around products, $\underline{r}(\theta_{II}^o)$. The schematic of protein 2 is as diagrammed by Adman (5). ET occurs at the special configurations $\underline{r}^*(\theta^*)$ for which the potential surfaces V_I and V_{II} intersect.

As indicated at the top of Figure 1, the special dipole configurations correspond to the intersection of two potential surfaces. The reactants, D_1^{red} and A_2^{ox}, are associated with charge distribution I and the products, D_1^{ox} and A_2^{red}, are associated with charge distribution II. The potential surfaces V_I and V_{II} are traced out by varying the orientation of the dipoles in the system, and the special configurations that allow electron transfer to occur (if at all) correspond to intersection of the potential surfaces. The height of the intersection above the minimum of V_I corresponds to the activation (potential) energy for reaching the special configurations. The figure also indicates that the difference in equilibrium structures, $\underline{r}(\theta_I^o)$ and $\underline{r}(\theta_{II}^o)$, is directly related to the height of the intersection. That is, the further "apart" $\underline{r}(\theta_I^o)$ and $\underline{r}(\theta_{II}^o)$ are, or the steeper the surfaces V_I and V_{II} are, the larger is $\Delta V^\#$.

With these considerations, one can construct potential surfaces for the contribution of protein reorganization to the barrier for electron exchange between two cytochrome c molecules. The X-ray structures (6) of the reduced and oxidized forms represent the most probable configurations of the protein dipoles around reduced and oxidized heme respectively. Since the only difference between the two forms of the protein is the net charge on the heme (0 in the reduced, +1 in the oxidized form), and the observed structural differences are small, we assume that the configurations relevant to the electron exchange reaction are intermediate between the observed structures of the two forms. Furthermore, since the electrostatic interaction of the protein charges and dipoles with the partial atomic charges of the heme atoms varies slowly with atomic coordinates compared to the other interactions, the electrostatic potential of the intermediate configurations can be approximated by linearly interpolating the coordinates of the protein atoms. Essentially, the proteins are assumed to have so many degrees of freedom that the same set of electrostatic potentials would be obtained if one correctly constrained bond lengths and angles as done in Section II. Note that as far as the requirement for intersection of the potential surfaces is concerned, only the electrostatic interaction of the heme charges, Q, and the solvent (protein) dipoles and charges, q, is relevant. The potential surfaces so constructed (7) are shown in Figure 2. The height of the intersection above the minima of V_I and V_{II} is small: 0.6 kcal/mol, indicating that the protein tertiary structure favors configurations that allow electron transfer to occur.

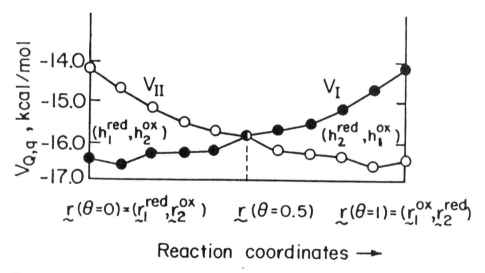

Figure 2. A direct interpolation of the protein potential surface for electron exchange using the observed X-ray structures of cytochrome c. The reaction coordinates are given by: $\underline{r}_1(\theta) = \underline{r}_1^{red} + \theta(\underline{r}_1^{ox} - \underline{r}_1^{red})$, $\underline{r}_2(\theta) = \underline{r}_2^{ox} + \theta(\underline{r}_2^{red} - \underline{r}_2^{ox})$ where subscripts 1 and 2 refer to molecules "1" and "2", θ is a number between 0 and 1, and $\underline{r}^{ox}$ and $\underline{r}^{red}$ are the corresponding observed X-ray coordinates. The potentials V_I and V_{II} are the electrostatic interaction, $V_{Q,q}$, between the heme partial atomic charges, Q and the protein dipoles, q: "I" labels the reactants (heme$_1^{red}$, heme$_2^{ox}$) and "II" labels the products (heme$_1^{ox}$, heme$_2^{red}$).

Figure 2 shows an interesting correlation with the formal theory of ET derived for dielectric continuum solvents (where free and potential energy are equivalent, and the free energy surfaces are modelled as parabolic functions of a formal coordinate). In the Marcus-Levich theory the height of the intersection above the minimum V_I is given by

$$\Delta g^{+} = (\alpha + \Delta G^{\circ})^2 / 4\alpha \tag{1}$$

where α is the solvent reorganization energy. In the case of electron exchange reactions ΔG° is 0 and the reorganization energy is the energy to switch the charges when the reactants are in the equilibrium cofigurations. That is, for electron exchange reactions in a dielectric continuum solvent $\Delta g^{+} = \alpha/4$. In the cytochrome case α is seen to be 2.4 kcal/mol, and $\Delta V^{+} = \alpha/4$. Such parabolic potential surfaces would not have been expected for a discrete structure.

III. Dynamic Model

Although the static approach considered above is a reasonable approximation, it is insightful to examine electron transfer reactions by simulating the dynamics of the reacting system. Such a study can test the assumptions of the static model, provide a molecular realization of the parameters of formal electron transfer model theories and help in analyzing the temperature dependence of the rate constant.

Figure 3 introduces concepts useful in treating ET dynamics. The reactant and product charge distributions, I and II, correspond to two quantum mechanical electronic states, whose energies, V_I and V_{II}, fluctuate along the classical trajectories of the nuclei in the system. ET from one cytochrome molecule to another corresponds to a transition from state I to II when the system passes through an intersection of V_I and V_{II}. The vibrations of the (protein) solvent, the "reaction coordinate" in Figure 3 produce fluctuations in the quantity $\Delta V(\underline{r}) = V_{II}(\underline{r}) - V_I(\underline{r})$, and ET can occur at the special configurations $\underline{r}^{+}$ where $\Delta V(\underline{r}^{+}) = 0$. Note that $\Delta V(\underline{r})$ is the energy change on switching the charges when the system is in configuration $\underline{r}$, and this is the quantity ΔV_{Qq} in Figure 2. Note also that because the separation between electron donor and acceptor is large, quantum mechanical coupling between states I and II is weak, and time-dependent perturbation theory of non-adiabatic transitions is applicable.

This work analyzes the fluctuations in $\Delta V(\underline{r})$ using a semiclassical trajectory approach and compares the calculated rate constant to that obtained by the formal theory of ET. Section A outlines our semiclassical trajectory approach, which provides the rate constant by numerically integrating the transition probability along the classical trajectory of the nuclei. Section B decomposes the fluctuations of ΔV by Fourier analysis. The characteristic frequencies and their amplitudes are interpreted in terms of "origin shifts" of harmonic vibrations. These effective shifts can be used in a harmonic model which provide the approximate temperature dependence of the rate constant.

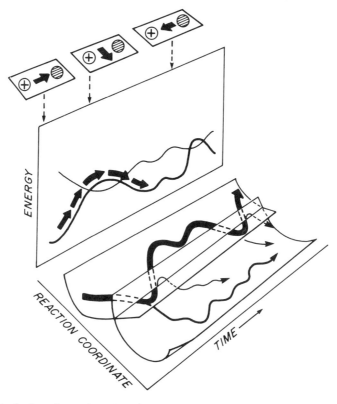

Figure 3. Sketch of solvent fluctuations that permit ET to occur. The solvent "moves" in the field of the reactants (hypothetically a neutral donor and a positively charged acceptor). The solvent trajectory is drawn as a thick black curve in the reaction coordinate vs. time plane, and as thick arrows in the energy vs. time plane. Three orientations of a hypothetical dipole are shown above the energy vs. time sketch: When the dipole points away from the + charge the potential energy is low; with the dipole perpendicular to the line joining the donor and acceptor the potential surfaces for the reactants and products intersect, and if the dipole points toward the charge, the reactant potential surface (the darker of the two curves in the energy vs time plane) is at a maximum. At the intersections of the two potential surfaces there is a probability for ET—sketched as thin, wavy "trajectories" breaking away from the main trajectory.

A. Semiclassical Trajectory Approach

The Marcus-Levich theory gives the rate constant for ET reactions by the well-known equation

$$k_{I,II} = (\pi / \hbar^2 k_b T \alpha)^{1/2} \sigma_{I,II}^2 \exp\{-\Delta g^+ / k_b T\} \tag{2}$$

Here α is the solvent reorganization energy, and $\sigma_{I,II}$ is the coupling between states I and II. Δg^+ in this equation is evaluated by eq. (1) without a rigorous derivation for realistic anharmonic systems. Marcus-Levich theory is valid within the harmonic approximation. However, refinements are needed to relate eq. (1) to the microscopic structure of real systems. Levich's approach uses essentially ΔV° rather than ΔG° in eq. (1); Marcus' treatment is purely classical, and therefore not applicable to ET involving tunneling. Nevertheless, the one-dimensional treatment gives a starting point. The rate constant of eq.(2) can be derived using the probability distribution for a (one-dimensional) reaction coordinate x and its conjugate velocity, $\dot{x}$:

$$k_{I,II} = \int P(x^\neq)\, P(\dot{x})\dot{x}\, \Gamma(\dot{x})\, d\dot{x} \tag{3a}$$

The quantity $\Gamma(\dot{x})$ is the Landau-Zener probability for hopping from I to II on passage through the intersection region (8):

$$\Gamma_{LZ}(\dot{x}) = 2\pi(\sigma_{I,II})^2/|\{\hbar\dot{x}\partial\Delta V/\partial x)\}|_{x^\neq} \tag{3b}$$

In the harmonic approximation one obtains eq.(2) from eqs.(3a) and (3b). Substituting a typical value $\sigma^\circ_{I,II}=.01$ kcal/mol (9), which should be scaled by the ratio $[\sigma(R)/\sigma^\circ]^2$, for a given value of the distance R between the donor and acceptor and $a=2.4$ kcal/mol (from Section I) in eq. (2), one obtains $k_{I,II}$ (harmonic)$=(\sigma/\sigma^\circ)^2$ $\times 4.22\times 10^9$ sec^{-1}.

We proceed to calculate $k_{I,II}$ by a semiclassical trajectory approach which has been developed to simulate photoisomerization reactions (10) and to explore the validity of eq. (2) for anharmonic polar solvents (11). Before describing our results we outline the main points of this semiclassical approach. The time dependent wavefunction of the system is written as:

$$\psi(t) = \sum_k b_k(t)\phi_k(\underline{r},\underline{R},) \exp\{-(i/\hbar)\int^t V_k[\underline{r}(t'),\underline{R}(t')]dt'\} \tag{4}$$

where $\underline{r}$ are the coordinates of the solvent (protein) and $\underline{R}$ are the coordinates of the donor and acceptor. The ϕ_k are the diabatic electonic wavefunctions, satisfying $H(r,R)\phi_I=V_I(r,R)\phi_I+\sigma(R)\phi_{II}$ where $\sigma(R)$ is the coupling constant, $(\sigma_{I,II}$ in eq. (2) and (3)). Substituting $\psi(t)$ in the time dependent Schrodinger equation gives:

$$\dot{b}_{II}(t) = -(i/\hbar)\sigma(\underline{R}(t))b_I(t) \exp\{-(i/\hbar)\int^t \Delta V[\underline{r}(t'),\underline{R}(t')]dt'\} \tag{5}$$

where b denotes db/dt and $\Delta V=V_{II}-V_I$. Given the initial conditions $|b_{II}(0)|^2=0$, $|b_I(0)|^2=1$, it is possible to determine the probability $|b_{II}|^2$ of being in state II for any path $(\underline{r}(t),\underline{R}(t))$ of the solute solvent system by integrating eq. (5). This is done by starting a trajectory near the minimum of V_I, adjusting the velocities until the system equilibrates with an average kinetic energy that corresponds to a given temperature, and then integrating eq. (5) along the classical trajectory $(\underline{r}(t),\underline{R}(t))$. Our semiclassical approach implies that the trajectories are split at points t_i where $|b_{II}|^2$ increases by a significant amount (these points occur when $\Delta V(t)=0$). A fraction $(|b_I|^2-|\delta b_{II}|^2$ stays on V_I while a fraction $|\delta b_{II}|^2$ is trapped in state II (See Fig. 3). Dealing with a multidimensional system we neglect interference between trajectories that cross to state II at different times, because such trajectories are likely to be trapped in different vibrational states. Thus the overall probability that the system is in state II is calculated by propagating a long-time trajectory on state I and summing the crossing possibilities $|b_{II}|^2$ (evaluated by integrating eq. (4) around the individual crossing points, t_i, where $\Delta V = 0$.

The overall rate is given by:

$$k_{I,II} = \lim_{\Delta t \Rightarrow \tau} \{ \frac{\Delta}{\Delta t} |b_{II}(\Delta t)|^2 \} = \lim_{\Delta t \Rightarrow \tau} \{ \frac{\Delta}{\Delta t} \sum_i |\delta b_{II}|^2 \} \tag{6}$$

$$= \lim_{\Delta t \Rightarrow \tau} \frac{\Delta}{\Delta t} \sum_i |\int_{t_i^-}^{t_i^+} \{ (i\sigma/\hbar) \exp[-i/\hbar) \int_0^t \Delta V(t')dt'] \} dt|^2$$

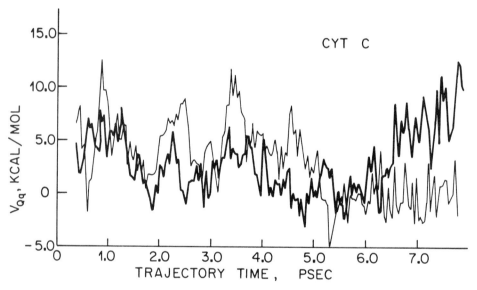

Figure 4a. The fluctuations of the potentials V_I (dark tracing) and V_{II} (lighter tracing) as the protein atoms move in the field of charges $Q_I = (Q_1^{red}, Q_2^{ox})$. $V_I(t) = V_{I1}(t) + V_{I2}(t) + C$; $V_{II}(t) = V_{II1}(t) + V_{II2}(t)$, where $V_{I1}(t) = V(Q_1^{red}, \underline{r}_1^{red}(t))$, $V_{I2}(t) = V(Q_2^{ox}, \underline{r}_2^{ox}(t))$, $V_{II1}(t) = V(Q_1^{ox}, \underline{r}_1^{red}(t))$, $V_{II2}(t) = (Q_2^{red}, \underline{r}_2^{ox}(t))$; $\underline{r}_i^{ox}(t)$ denotes "atoms of protein i moving in field of Q^{ox}". The energy required to switch the charges at time t is the energy gap, $\Delta V(t) = V_{II}(t) - V_I(t) - C = \Delta V_1(t) + \Delta V_2(t) - C$, where $\Delta V_1(t) = V_{II1}(t) - V_{I1}(t)$, $\Delta V_2(t) = V_{II2}(t) - V_{I2}(t)$. The constant C was defined by the following procedure: For the separate trajectories, distribution functions $P_1(\Delta V_1)$ and $P_2(\Delta V_2)$ were calculated (12). $P_i(\Delta V_i) \cdot \delta(\Delta V_i)$ gives the probability that ΔV_i is the range $(\Delta V_i, \Delta V_i + \delta(\Delta V_i))$. The maxima of P_1 and P_2 were at $\Delta V_1^* = 4$ kcal/mol, $\Delta V_2^* = 11$ Kcal/mol. For these simulations the most probable charge switching energy is $\Delta V_1^* + \Delta V_2^* = 15$ kcal/mol, in disagreement with the static model which gives 2.5. Assuming that the fluctuations of the system around the incorrect minimum are similar to those around the correct one, we can correct the energy gap by using C=12.5 kcal/mol. The discrepancy that requires the introduction of the correction C is probably due to the fact that the water molecules surrounding the protein were not included in the present simulation. Proper simulation of the water molecules or introduction of an effective constraint is needed to prevent protein 2 from contracting around the partial charges of the oxidized heme. Note also that the water contribution should be included in the reorganization energy (7).

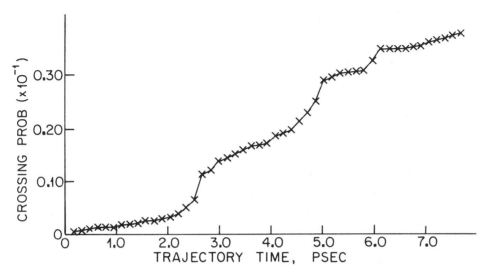

Figure 4b. The simulated ET probability accumulated over the course of the cytochrome trajectory. ET probability accumulates at the crossings of V_I and V_{II} in Figure 4a.

where $t_i^{\pm} = (t_{i\pm 1} - t_i)/2$ and τ is the trajectory time after which the computed k reaches a constant value. As shown in ref. (11), the analytical continuation of $|\delta b_{II}|^2$ can be approximated by the Γ_{LZ} of eq. (3b) and eq. (6) can be approximated by eq. (2).

The results of modeling the rate of electron exchange between reduced and oxidized cytochrome c at 300°K are presented in Figure 4. The quantities labelled V_{Qq} in the upper panel are the electrostatic potentials V_I (dark) and V_{II} (light). The two protein conformations fluctuate (independently) under the influence of bond stretching, angle bending, Van der Waals, and electrostatic forces. The trajectories propagate under the influence of charges of state I only. That is, V_{II} is the computed value of V_{Qq} if the heme charges were to be switched, but the trajectory is propagated with the reactant charges. ET probability accumulates near the crossings of V_I and V_{II}, which correspond to $\Delta V(\underline{r}(t),\underline{R}(t))=0$. The number of crossings is related to the reorganization energy which is given by the most probable value of ΔV. Using this principle, the curve $V_I(t)$ was shifted relative to $V_{II}(t)$ using a calibration described in the caption of Figure 4. The numerical values of V_I and V_{II} shown in the figure were then used to compute the ET rate according to eq. (5), setting $\sigma_{I,II}^0 = .01$ kcal/mol. The resulting transition probability, $|b_{II}|^2$ as a function of time is shown in the lower panel of Figure 4. Accordingly, the ET rate is $(\sigma/\sigma^0)^2$ $\times(3.5\times 10^{-2}/8\times 10^{-12}$ sec$) = (\sigma/\sigma^0)^2 \times 4.4\times 10^9$ sec^{-1}, in excellent agreement with $(\sigma/\sigma^0)^2 \times 4.2\times 10^9$ sec^{-1} from eq. (2).

B. The Dispersed Polaron Model

The ET rate law of eq. (2) is based on the heuristic assumption that the many degrees of freedom in the discrete protein structure can be described by a single effective harmonic oscillator. The close agreement between the rate calculated by eq. (2) and by eq. (5) suggests that a harmonic model of ET might be applicable to proteins. Therefore, we try to extract from $\Delta V(t)$ the microscopic parameters of the harmonic (polaron) theory of ET.

The time dependent energy gap $\Delta V(t)$ can be decomposed by evaluating its Fourier transform

$$\Delta V(t) = \sum_s A_s \cos(\omega_s t + \phi_s) + \overline{\Delta V(t)} \qquad (7)$$

where the A_s are the Fourier amplitudes of frequency ω_s, ϕ_s are the phases, and the last term in eq. (7) is the time average of $\Delta V(t)$. The moduli, $|A_s|$, of the Fourier coefficients calculated for the trajectory data are shown in Figure 5. Apparently the characteristic frequencies of the fluctuations in $\Delta V(t)$ are in the 60-80 cm^{-1} range.

The relation between the Fourier coefficients A_s and the reorganization energy

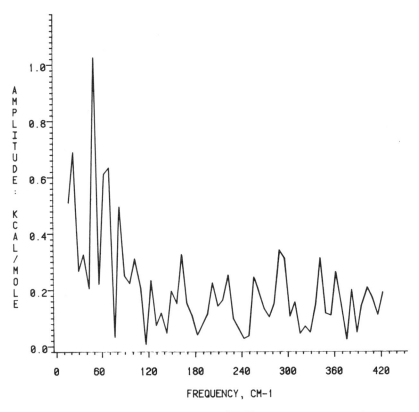

Figure 5. Moduli if the Fourier components of $\Delta V(t) - \overline{\Delta V(t)}$; $\Delta V(t)$ is obtained by subtracting the dark from the light tracing in Figure 4a.

can be described as follows. The potential surfaces V_I and V_{II} are approximated by harmonic potentials:

$$V_I = \sum_s \hbar\omega_s q_s^2$$

$$V_{II} = \sum_s \hbar\omega_s (q_s - \Delta_s)^2 \tag{8}$$

$$\Delta V(t) = \sum_s \hbar\omega_s q_s(t)\Delta_s + \overline{\Delta V(t)}$$

where q_s is the dimensionless normal coordinate associated with ω_s, and Δ_s is the "origin shift" for ω_s; the set of Δ's describe the lateral displacement of the equilibrium coordinates (at the minima of the potential surfaces), and the total reorganization energy is $\sum_s \hbar\omega_s \Delta_s^2/2$. The average of the q_s obtained by the classical trajectories of thermally equilibrated harmonic systems is given (for $\hbar\omega_s \ll k_b T$) by:

$$q_s(t) = \cos(\omega_s t + \phi_s)[\overline{n}_s(T)]^{1/2} \tag{9a}$$

$$\overline{n}_s(T) = [\exp(h\omega_s/k_b T) - 1]^{-1} \tag{9b}$$

Substituting eq. (9) in (8) and equating with (7) one obtains

$$|\Delta_s| = |A_s|/[(\hbar\omega_s)(\overline{n}_s(T))^{1/2}] \tag{10}$$

The origin shifts calculated from eq. (10) are shown in Figure (6).

The origin shifts obtained from the Fourier transform of $\Delta V(r)$ can be used to estimate the rate constant and its temperature dependence, that is, the rate constant of an electron transfer reaction, or any other radiationless process, is given (within the harmonic approximation) by the well known expression (13-16)

$$k = (\sigma^2/\hbar^2) \int_{-\infty}^{\infty} \exp\{-G\} \exp\{G_+(t) + G_-(t)\}dt \tag{11}$$

$$G_+ = (1/2)\sum_s \Delta_s^2(\overline{n}_s + 1)\exp\{i\omega_s t\}$$

$$G_- = (1/2)\sum_s \Delta_s^2 \overline{n}_s \exp\{-i\omega_s t\}$$

$$G = (1/2)\sum_s \Delta_s^2 (2\overline{n}_s + 1)$$

This harmonic expression can be evaluated directly (see related treatment in ref. (17)), although its stationary phase solution is well known (16). Here we only point out the behavior at the low and high temperatures. For $T \Rightarrow 0$ the rate constant is proportional to the square of the Franck-Condon factor, $C(0, n^\circ)$ of the multi-

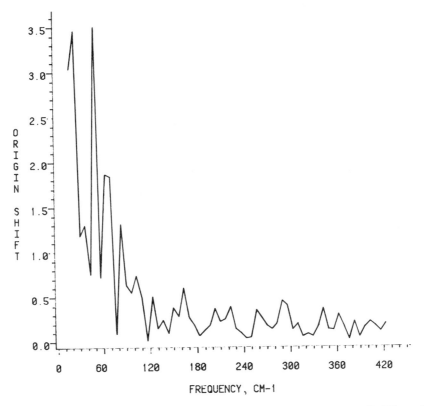

Figure 6. The origin shifts calculated by eqs. (9) and (10) from the Fourier moduli of Figure 5.

dimensional system, where $\underline{n}^\circ$ is the quantum number that satisfies the relation $(\Sigma \hbar\omega_s n_s^\circ - \Delta E_o \Rightarrow \min)$. For the experimentally relevant cases (e.g. photosynthetic systems) where $\Delta E \neq 0$

$$k = (\sigma^2/\hbar^2\overline{\omega}) \, \Pi_s \, [C_s(0,n_s^\circ)]^2 = (\sigma^2/\hbar^2\overline{\omega})\exp\{-\Sigma_s \Delta_s^2\} \, \Pi_s \, \Delta_s^{2n}/n_s^\circ! \qquad (12)$$

where $1/\overline{\omega}$ is the effective density of states for the system. In the high temperature limit where $k_b T \gg \hbar\omega_s$ (for the relevant modes) the "hot progressions" start to contribute, and the system behaves according to eq. (2) where the activation barrier is given by

$$\Delta g^+ = \Sigma \, \alpha_s/4$$

$$\alpha_s = (1/2)\hbar\omega_s \Delta_s^2 \qquad\qquad (13)$$

In the case of cytochrome c simulated here we find from Fig. 5 that the calculated $|\Delta(\omega)|$ have a peak near 60-80 cm^{-1}. If the system is described by an effective mode with this ω_s (in equilibrium with the low frequency modes of the system), then the rate constant will change from that of activated process eq. (2) to tunneling (eq.(12)) when $\bar{n}\Rightarrow0$, for the 80 cm^{-1} mode. The correspondence between this finding and the experimentally observed temperature dependence of the oxidation of cytochrome c in the reaction center of photosynthetic bacteria (for review see ref. (4)) is clearly interesting. Note, however, that the calibration approach presented in the caption of Figure 4 did not correct for the overestimate of the origin shifts associated with the relaxation of the oxidized cytochrome from its X-ray position. Thus our preliminary finding is meaningful only if the overestimated origin shifts are associated with low frequencies with $\omega < 40$ cm^{-1} and not with the peak at 60-80 cm^{-1}.

IV. Concluding Remarks

This work explored the miscroscopic nature of biological ET reactions using static and dynamic models. These preliminary microscopic studies indicated the following points: (i) The key quantity in ET reactions, the "charge switching potentials", (ΔV), defines the reaction coordinates; (ii) The energetics and rates of electron exchange, calculated for fixed (rather large) donor-acceptor separation, are consistent with the harmonic approximation; (iii) The characteristic frequencies relevant to ET rates are low ($\hbar\omega \lesssim 0.2$ kcal/mol), and at room temperature vibrational tunneling involving these modes is not an important mechanism. The fluctuations of $\Delta V(t)$ give the characteristic frequencies and origin shifts which can be used to explore the reaction rate within the harmonic approximation.

This work did not treat the energetics of forming biological electron donor-acceptor complexes (which would require more structural information than is currently available). Availability of the relevant structural information may let one estimate the actual value of $\sigma(R)$ and obtain the rate constant by scaling the results of the present calculations by $(\sigma(R)/\sigma^\circ)^2$. However, a more complete calculation should also consider the effect of thermal fluctuation in $\sigma(R)$. These fluctuations reflect the changes in the distance between the donor and acceptor, which seems to play an important role in nonbiological ET processes. In biologically optimized systems one expects the donor and acceptor to be assembled in an optimal binding. In this case the effect of fluctuations of $\sigma(R)$ is likely to be small but studies of its actual magnitude are clearly needed.

The temperature dependence of the rate of electron transfer reactions in biological molecules has been the subject of intensive studies in recent years (4). Microscopic simulations of the observed temperature dependence provides a key check on any molecular model of ET reactions. The temperature dependence can be obtained by using our semiclassical trajectory approach for several temperatures. However, a much simpler approximate result can be obtained by the "dispersed polaron model" presented in this work. It is encouraging to note that this model predicts a transition from an activated reaction to a tunneling process in the temperature range found experimentally in photosynthetic bacteria.

Acknowledgement

This work was supported by Grant GM-24492 from the National Institutes of Health and Grant PCM-8303385 from the National Science Foundation.

References and Footnotes

1. (a) Marcus, R. A., *J. Chem. Phys., 24,* 966-978 (1956). (b) *Ibid, 24,* 979-980 (1956). (c) Marcus, R. A., *Annu. Rev. Phys. Chem. 15,* 155-96 (1964).
2. Levich, V. G., *Adv. Electrochem. Electrochem. Eng. 4,* 249-371 (1966).
3. Kestner, N. R., Logan, J. and Jortner, J., *J. Phys. Chem. 78,* 2148-66 (1974).
4. Blakenship, R. E. and Parson, W. W., in: *Photosynthesis in Relation to Model Systems,* J. Barber, ed., Elsevier-North-Holland (1979) pp. 71-114.
5. Adman, E., *Biochem. Biophys. Acta 549,* 107-44 (1979).
6. (a) Takano, T. and Dickerson, R. E., *J. Mol. Biol. 153,* 79-94 (1981). (b) *Ibid. 153,* 95-115 (1981).
7. Churg, A. K., Weiss, R. M. , Warshel, A. and Takano, T., *J. Phys. Chem. 87,* 1683-94 (1983).
8. (a) Zener, C., *Proc. R. Soc. London, Ser. A137,* 363-40 (1932). (b) Landau, L. D., *Phys. Z. Sowjetunion 2,* 46-48 (1932).
9. (a) Marcus, R. A. and Siders, P., *J. Phys. Chem. 86,* 622-30 (1982) and references therein. (b) Hopfield, J. J., *Proc. Natl. Acad. Sci. USA 71,* 3640-4 (1974). (c) Brunschweig, B. S., Logan, J., Newton, M. D. and Sutin, N., *J. Amer. Chem. Soc. 102,* 5798-5809 (1980). (d) Beitz, J. V. and Miller, J. R., *J. Chem. Phys. 71,* 4579-95 (1979).
10. Warshel, A. and Karplus, M., *Chem. Phys. Lett. 32,* 11-17 (1975).
11. Warshel, A., *J. Phys. Chem. 86,* 2218-24 (1982).
12. Margenau, H. and Murphy, G. M., *The Mathematics of Physics and Chemistry, Vol 2.,* Van Nostrand-Reinhold, N. Y. (1964), Chapter 3.
13. Kubo, R. and Toyozawa, Y., *Progr. Theor. Phys. 13,* 160-182 (1955).
14. Lax, M., *J. Chem. Phys. 20,* 1752-60 (1952).
15. Englman, R. and Jortner, J., *Mol. Phys. 18,* 145-64 (1970).
16. Brailsford, A. D. and Chang, T. Y., *J. Chem. Phys. 53,* 3108-13 (1970).
17. a. Mukamel, S., *J. Chem. Phys. 77,* 173-81 (1982). b. Warshel, A., Stern, P. and Mukamel, S., in: *Time Resolved Vibrational Spectroscopy,* G. H. Atkinson, ed., Academic Press, Inc., N. Y. (1983) pp. 41-51.

Structure & Motion: Membranes, Nucleic Acids & Proteins,
Eds., E. Clementi, G. Corongiu, M. H. Sarma & R. H. Sarma,
ISBN 0-940030-12-8, Adenine Press, Copyright Adenine Press, 1985.

Ab-Initio Calculations in the Study of Enzyme Inhibition

Colin Thomson and Colin Edge
National Foundation for Cancer Research Project,
Department of Chemistry, University of St. Andrews,
St. Andrews, KY16 9ST, Scotland

Richard Brandt
National Foundation for Cancer Research Project,
Department of Biochemistry, Medical College of Virginia,
Virginia Commonwealth University, Richmond, Virginia
U.S.A.

Abstract

Theoretical calculations on inhibitors or Glyoxalase I and on model reactive intermediates are reported. The role played by molecular electrostatic potential calculations in such a study is discussed and used to suggest possible inhibitors.

Introduction

Enzyme inhibition is of undoubted importance as a basis for drug action (1), and there are currently many clinically useful drugs such as, for example, penicillins, digitalis, methotrexate and aspirin which are enzyme inhibitors.

The design of potent and selective enzyme inhibitors has advanced enormously in recent years, as more detailed information has become available about the physiological and pathological role of the enzymes themselves, and especially as more detailed structural data continues to be obtained from X-ray diffraction experiments (2) and other physical methods (3).

Furthermore, advances in the understanding of enzyme mechanisms and their chemical models (4) have changed the emphasis in the design studies of enzyme inhibitors, leading to the development of mechanism based inhibitors.

One particularly attractive approach is the design of inhibitors which are transition state analogues.

Transition state inhibitors

The transition state of an enzymatically catalyzed reaction is bound to the enzyme much more tightly than its substrate(s), a fact first pointed out by Pauling (5), and therefore might be used as a template for the design of more potent and specific enzyme inhibitors (6-9). These inhibitors are known as transition state analogues, and may have considerable pharmacological potential. For instance, they might be designed specifically to inhibit enzymes which are vital to tumour cells, but not to normal cells, to take an extreme example, or they may inhibit enzymes for which there are quantitative differences between normal and abnormal cells. In the case of inhibitors which act by mimicking the substrate of an enzyme reaction, the structure of the substrate, and indeed of the enzyme-substrate complex is frequently known, and inhibitors can be designed accordingly.

Unlike the substrate(s) and product(s) of a reaction, however, transition state structures cannot be studied directly and structures involved in analogue studies were commonly derived from reasonable interpolations between the structures of reactant(s) and product(s), based on some assumed mechanism for the reaction.

Recent advances in theoretical chemistry have, however, made it possible to calculate theoretically the structure and energy of the transition states (and other intermediates) in chemical reactions, and therefore we now have an additional tool for use in the design of analogues of transition state structures.

Theoretical studies of transition states and intermediates

Reactant(s) and product(s) in any reaction correspond to minima on the multidimensional potential energy surface of the reaction. Location of these minima is relatively straightforward, but there may during the course of the reaction be other relative minima on the surface, corresponding to intermediates in the reaction. These are frequently short lived and not isolable. Separating the minima there are transition states. These stationary points are characterized by the following (10).

The potential energy for a system of $f = 3N-6$ degrees of freedom, where N is the number of atoms, can be expressed as a function of the 3N-6 generalized coordinates

$$V = V(q_1, q_2, ... q_f) \tag{1}$$

Furthermore, the stationary points q° satisfy the conditions

$$\frac{\partial V}{\partial q_i} = 0 \ i = 1, 2, ... f \tag{2}$$

However, V(q) can also be expressed as a power series about an extremum point q°

$$V(q) = V(o) + \tfrac{1}{2} \sum_i^f \sum_j^f F_{ij}q_iq_j + \text{higher terms} \tag{3}$$

where

$$F_{ij} = \left(\frac{\partial^2 V}{\partial q_i \partial q_j}\right) q^\circ = F_{ji} \qquad (4)$$

are the usual force constants, which form a symmetric matrix **F**.

The second order term is usually the only one considered and hence

$$V(q) = \tfrac{1}{2}\mathbf{q}^T\mathbf{F}\mathbf{q}$$

where **q** is the column vector of the q_i.

The shape of the surface about the q° is determined by the eigenvalues and eigenvectors of **F**, i.e.

$$\mathbf{F} U_i = x_i U_i$$

Minima correspond to positive values of x_i, and transition states have all values of $x_i > 0$ except one which is negative.

Until quite recently the calculation of the energy derivatives was difficult and time consuming, but since Pulay (11,12) showed how gradients could be calculated for SCF wave functions, there has been much effort devoted to the development of fast and efficent methods for the evaluation of these energy derivatives, and it is now possible to systematically locate the minima (and hence calculate theoretically geometries of reactants, products and intermediates on the P.E. surface) and the transition states, and for the latter, of course, the theoretical structural data is information not readily obtained experimentally (13).

Thus these sophisticated quantum chemical techniques can be of great use in studying reaction pathways in biochemistry, and in particular can be applied to the design of transition state analogues. Andrews (14,15) has used these methods to calculate semi-empirical (MINDO/3) geometries for the transition state structure for the reaction catalysed by the mitochondrial GABA-transaminase, and then to suggest, via calculations, transition state analogues which might inhibit this enzyme, with some success.

It is clear that this method has great potential, and we have investigated the use of ab-initio SCF calculations in the design of transition state inhibitors of an enzyme system glyoxalase-I which may be important in controlling cell division. Our long term aim is the design, via ab-initio calculations, of transition state analogues which have anti-tumour activity and we report the results of our work to date in the present paper.

The glyoxalase enzyme system and cell proliferation

There is currently considerable interest in the cellular metabolism of methylglyoxal (MG), stimulated by the investigations of Szent-Györgyi and co-workers (16-19), who showed that α-β dicarbonyls were growth inhibitory, and that methylglyoxal in particular is a normal constituent of tissue (20).

Early in this century, the glyoxalase enzyme system responsible for the catabolism of methylglyoxal was discovered (21,22). This system comprises two separate enzymes: glyoxalase-I (S-lactoyl-glutathione methylglyoxal lyase, isomerizing; EC4.4.1.5) and glyoxalase-II (S-2 hydroxyacylglutathione hydrolase; EC3.1.2.6) and glutathione (GSH) is required as a co-factor (23). The final product of the reaction is D-lactate (not the L-lactate of glycolysis) (24).

The currently accepted reaction scheme is depicted in Figure 1. The initial reaction between MG and GSH results in the formation of the α-ketohemithiol acetal which in the presence of glyoxalase-I rearranges to an α-hydroxythiol ester. The mechanism of this rearrangement is of considerable interest, since it has been shown that MG inhibits tumour growth (25), and if it is involved in the control of cell division, as suggested by Szent-Györgyi (19), the inhibition of the glyoxalase enzyme may be a potential means of regulating growth. There is much experimental support for the inhibition of cellular growth by a variety of inhibitors of

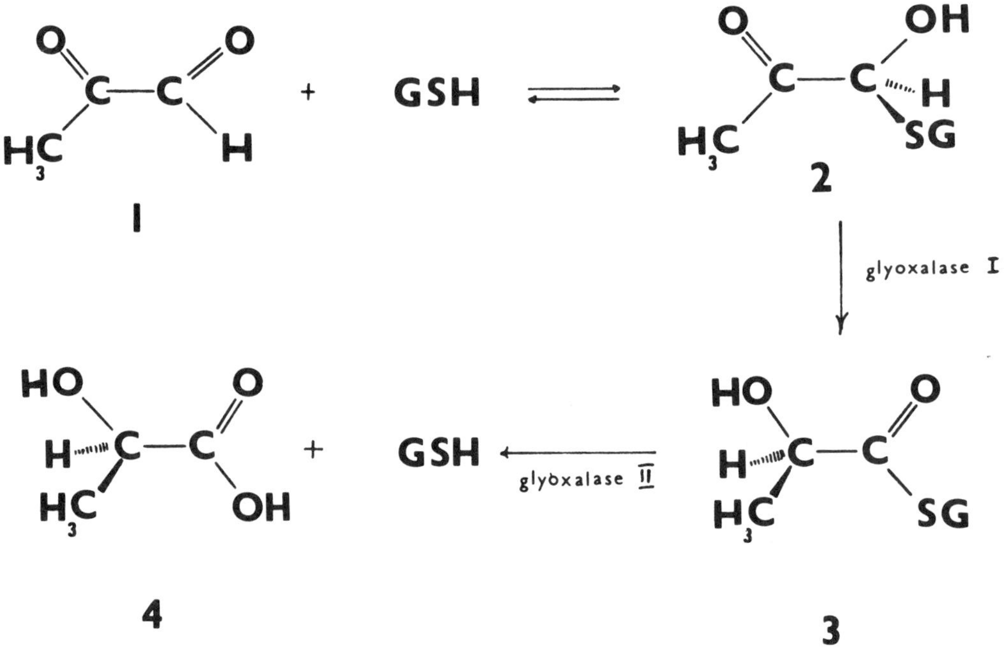

Figure 1. Steps in the metabolism of methyl glyoxal (1) to lactic and (4) by the glyoxalase enzyme system.

glyoxalase-I. For instance, Vince et al showed that a variety of S-alkylglutathiones act as competitive inhibitors (26) and several other competitive inhibitors have also been studied (27-30). There have recently been reports of other inhibitors which are not substrate analogues (31-33).

In order to understand the mode of action of the various inhibitors, and hopefully to design new and more effective inhibitors for possible use as chemotherapeutic agents, it is necessary to investigate in more detail the mechanism of the reaction involving glyoxalase-I.

There have been many experimental studies on this system, and the nature of the enzyme itself has been reviewed by Mannervik (34). It has been detected in various organisms, prokaryotes as well as eukaryotes, and seems to be as ubiquitous as glutathione itself. The purified enzyme has been obtained from a variety of different sources, but in terms of the inhibitor studies described in our work, the human red blood cell is the source of this enzyme. The general molecular properties are known, but not the detailed structure of the active site. However, it has been demonstrated that Zn^{2+} is the metal co-factor of the native enzyme (35), and also that a basic amino acid residue is present, possibly histidine (36). Other divalent metal ions, such as Mg^{2+} (37), and Mn^{2+} (38), can replace Zn^{2+}, without substantial loss of activity, and in recent work using flourescence and NMR relaxation studies, it has been shown that two water molecules are involved in the metal binding (39).

Whatever the precise structure of the active site, from a mechanistic viewpoint the pathway from this adduct of MG (2 in Figure 1) to the thioester of lactic acid could go through a variety of possible intermediates, and it is clearly of considerable importance to investigate these, since it should shed more light on the mechanism of inhibition by known inhibitors and suggest other possible inhibitors.

Of particular promise is inhibition by transition state analogues, and investigations of such inhibitors was first reported by Douglas and Nadvi (32) who showed that several compounds containing the enediol structure (Figure 2) act as inhibitors, since one of the possible intermediates or transition states has this structure.

However, there are other possible intermediates in this reaction scheme and one obvious way to investigate these molecules is via theoretical calculations.

Lavery and Pullman (40) in 1979 carried out the first ab-initio quantum mechanical calculations on this system and studied the energetics of the reactions involving a variety of postulated intermediates, including the enediol, with the scheme depicted in Figure 3. These authors used CH_3SH as a model system for GSH, and also investigated the effect of Mg^{2+} on the energetics, since Mg^{2+} or other divalent ions are essential for enzymatic activity (37,38).

The present paper deals with a more extensive study of this reaction scheme, with a broader objective, namely to suggest additional criteria for evaluation of the inhibi-

MOLECULE	FORMULA

R = H   Esculetin
    CH$_3$ 4-methyl esculetin

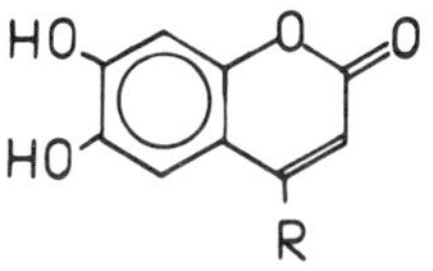

Iso esculetin

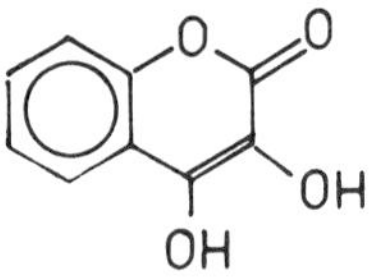

Squaric Acid

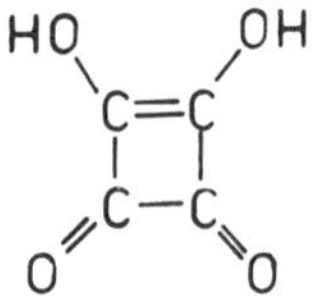

Maltol

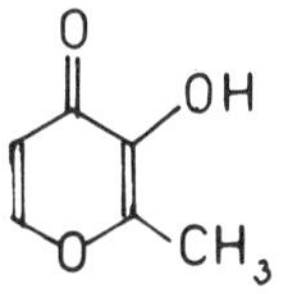

2,3 Dihydroxypyridine

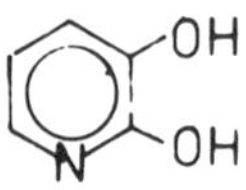

2,3 Dihydroxybenzoic Acid

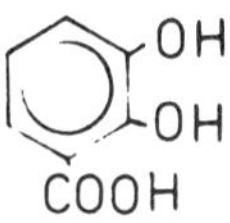

Figure 2. Inhibitors of glyoxalase-I.

tors of glyoxalase-I, studies of enzyme inhibition, and quantum chemical studies of the various inhibitors and intermediates.

We have previously described the experimental study of several potential transition state inhibitors of human red blood cell glyoxalase-I, using an in-vitro spectro-photometric analysis to determine their 50% inhibition concentration (I_{50}), (41,42) and also our initial results of an attempt to correlate the inhibitory potential of the compounds investigated with "reactivity indices" obtained from ab-initio quantum

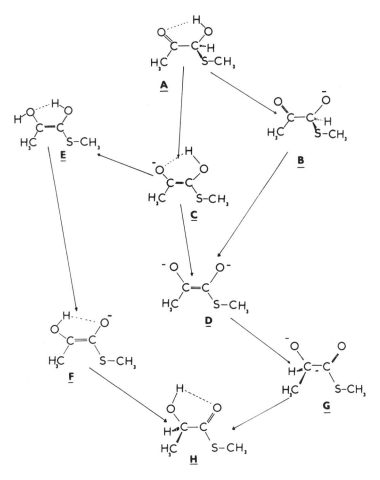

Figure 3. Possible reaction pathways between reactant A and product H (after Lavery and Pullman (40)).

chemical calculations on the inhibitors and various models for the transition state and intermediates in the reaction scheme of Figure 3 (43). The present paper describes our more recent experimental and theoretical work on inhibitors of this system.

The molecular electrostatic potential in inhibition studies

Quantum chemical calculations are being increasingly applied to calculate the geometry and conformation of molecules of biological interest, especially now that energy minimisation using the force or gradient method (11-13) is available in many standard SCF programme packages. Of course, the precise details of such geometries are dependent on the quality of the basis set, but in the investigations described in this work, we believe that the results of minimal basis set calculations will yield essentially the same information as more extended basis calculations for the following reasons.

In the usual model for enzyme-substrate reactions, during the binding of substrate (structure known or calculable) to the active site (structure unknown or only partly known), the initial recognition will involve relatively long range non-covalent interactions in which the conformation of the substrate will match the active site conformation (lock and key model).

However, since this interaction takes place over distances greater than normal bonding distances, it is likely that a more realistic criteria for mutual affinity between the enzymes active site and the substrate is not just a conformational match but a complementary match in the total field generated by the substrate. This field is directly calculable from the total charge density $\rho(r)$ of the molecule and the electrostatic potential maps $V(r)$ have proved of enormous use in the investigation of the reactivity criteria for drug receptor interactions (44,45).

One major advantage of such V maps is that they are much less sensitive to the *precise* geometry of the molecule, and to the basis set, than are other criteria such as atomic populations, binding energies etc. We have found that if the geometries are only partially optimised so that the forces on any internal coordinates are $< \sim 0.07$ hartree, the V maps are qualitatively very similar to those from more accurate wave functions, particularly with regard to the positions of minima on the V maps. Similar conclusions concerning the utility of the STO-3G basis set in such studies have been reported by Osman et al. (45) Clearly any attempt to investigate theoretically the enzyme reaction must use a model system, and we have followed the work of Lavery and Pullman who modelled glutathione by methanethiol CH_3SH. The model depicted in Figure 3 which represents a suitable model for the substrate.

There are a variety of possible intermediates in the transformation of the α-ketohemithiol acetal to the thioester of lactic acid, and several of these have previously been studied by Lavery and Pullman and the energetics of the various processes computed. These authors also studied the energetics of the various processes in the presence of Mg^{2+}, since this is believed to be involved in the active site, and indeed is essential to stabilise some of the anionic intermediates. These authors conclude that the most likely reaction path is via the enediol with a less likely alternative involving the dianion (40).

The possible involvement of the enediol has already been suggested from the other experimental studies of the inhibition of glyoxalase-I, and it was to this intermediate that we first turned our attention in the present study.

Strategy in the theoretical calculations

Our work uses the structural data of Lavery and Pullman as a starting point for calculations on the possible intermediates. We have calculated the wave functions and electrostatic potential maps for the model enediol intermediate using the standard STO-3G basis set, both with and without the Mg^{2+} ion, whose position was optimized, starting from the geometry used in Ref. 40. No attempt was made to exhaustively re-optimise the geometry using the force method, although computation of the forces showed these to be low for the wave functions used. As pointed out above, the qualitative features of the maps are essentially independent of the *precise* geometry. We carried out similar computations for the other intermediates.

We then compared the V maps with those calculated for a wide variety of known inhibitors of glyoxalase-I which differ significantly in their inhibition (41,42). These compounds differ quite substantially in their chemical structure.

Calculational details

The calculation of the electronic wave functions for all the molecules studied were carried out with the GAUSSIAN 80 program system (46). The majority of the calculations were carried out using the STO-3G basis set (47) and the Restricted Hartree Fock (RHF) procedure for the SCF calculations. Limited geometry optimizations were carried out (mainly for the OH groups) using the force method, although for some systems a more extensive optimization study was attempted (see below). However, providing the forces on the atoms were <0.07 hartree, there was very little difference between the results.

The calculations of the electrostatic potential maps were carried out using the same basis set and our version of GAUSSIAN 80, extensively modified to improve the convergence, to include Morukuma's energy decomposition analysis (48) and the electrostatic potential programme developed by Tomasi et al. (49) We have also added the electrostatic potential calculation as an option to the population analysis links of GAUSSIAN 80.

All calculations were carried out on the VAX 11/780 at the University of St. Andrews.

The STO-3G basis set was used throughout all geometry optimization calculations, including the same quality basis set for Mg and S. These are internal to the Gaussian 80 program, and the S and Mg basis sets in particular differ from those used by Lavery and Pullman. The problem of the description of ions with minimal basis sets is well known (50), and we have for these species, investigated the use of additional basis sets, including diffuse functions, although this work is incomplete. However, it is clear that the qualitative results with the STC 3G basis set agree with those obtained with more extensive basis sets. In particular, although the STO-3G calculations on the anions usually do not give the HOMO as bound, we have found that the electrostatic molecular potential maps are not changed significantly in appearance from more accurate calculations. Put another way, such maps reflect the main features of the potential near the molecule even with rather crude basis sets.

In the particular case of the dianion derived from squaric acid $C_4O_4^{2-}$, we have optimized the geometry using the recently developed 4-31G+ basis set (52), which has an additional added diffuse function. The results are not substantially different, despite the large number of unbound orbitals in the STO-3G calculation, and the qualitative features of the map remains the same, although there are of course large variations in the depths of the minima.

With regard to the molecular geometry of the inhibitors, we have started from a conformation with standard bond lengths and evaluated the forces on the nucleii. In many cases, these were substantial, but a limited optimization, particularly of the OH group conformations, usually reduced the forces to <0.07 hartree. A detailed comparison of the maps, however, showed that once the forces were <0.07 hartree the qualitative appearance of the maps were virtually identical, and so for the larger systems in particular, we have not carried out optimizations if the forces satisfied the above criteria.

It has been pointed out several times that for complexes, the basis set extension error (BSEE) may be important for an STO-3G basis set (53). However, its main effect will be on energies and not on the qualitative features. Work is in progress, however, to check this supposition.

The electrostatic potential maps were generated in the molecular plane, and also in plane 1.6A above and below the molecular plane, and in a variety of planes perpendicular to the molecular plane. Apart from the CH_3-groups, all these molecules are planar, and non planar analogues are not inhibitory (41). In attempting to correlate the molecular electrostatic potential maps with the inhibitory effects of the various molecules, we have used a variety of criteria to which we now turn.

Results and Discussion

a) Structure of the intermediates

There is good evidence that the reaction catalysed by glyoxalase-I is primarily a one substrate reaction involving the hemimercaptal adduct (A) as substrate (34). Model system studies suggest that the enediol intermediate is involved in the rearrangement, with a fast proton transfer occurring in a protected active site on the

Table I

Total and Relative Energies of the intermediates A—F of Figure 3 calculated using the STO-3G basis set

Intermediate	Energy*	Relative Energies[†] This Work	Ref. 40
A	−695.11630	0	0
B	−694.29781	513	394
C	−694.32025	499	354
D	−693.30043	1139	877
E	−695.101073	+9.5	−11
F	−694.34722	483	345

*Values in hartree
[†]Values in kcal mol^{-1}

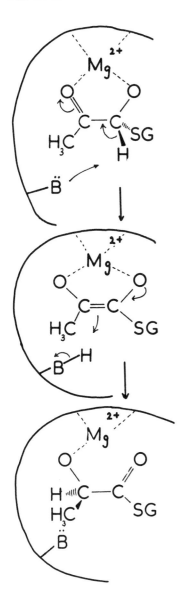

Figure 4. One possible mechanism for the catalytic mechanism at the active site of glyoxalase-I.

enzyme (36,53). It was also concluded that the acidity of the hemimercaptal is due to the electron withdrawing effect of the acyl group, and that the carbanion is stabilised by the thiol group, which is why glutathione is a co-factor (54). Figure 4 depicts the possible sequence of events.

However, despite the attraction of this hypothesis, there are other possible intermediates such as those depicted in Figure 3 and the careful study by Lavery and Pullman examined some of these (40). Our calculations are in broad agreement with their results as is shown in Table I which compares our total and relative energies of the intermediates with the values obtained by Lavery and Pullman. Our geometrical parameters were, however, partly optimized and differ appreciably from those of Ref. 40. We also carried out parallel calculations using a chelated Mg^{2+} ion in the same configurations described by Lavery and Pullman. Table II gives the relevant energies and binding energies compared with Ref. 40. Agreement is quite satisfactory. Our conclusions therefore are the same regarding the likely intermediacy of the enediol intermediate, probably via the pathway $A \Rightarrow C \Rightarrow E \Rightarrow F \Rightarrow H$, involving two successive deprotonations and the anionic intermediates C and F. It has also been suggested from the experimental NMR work that these anions are stabilised in the active site by the divalent cation and water (39,55). We are currently investigating this possibility.

Table II
Total and Relative Energies of the complexes of intermediates A—F with Mg^{2+} calculated using the STO-3G basis set

Intermediate	Energy*	Relative Energies[†]	
		This Work	Ref. 40
A	−891.951249	0	0
B	−891.643814	193	137
C	−891.609868	214	112
D	−891.129730	515	356
E	−891.933077	+11	−24
F	−891.592188	225	99

*Values in hartree
[†]Values in kcal mol^{-1}

In the light of the energetic results, we have calculated the molecular electrostatic potential (MEP) maps for the substrate (A) and intermediates $B \Rightarrow F$ and also for several of these species complexed to Mg^{2+}. To save space, however, we present and discuss only the most important of these with respect to the inhibitor problem.

The geometry of the model substrate (A) is as expected from elementary considerations of bonding, and is, of course, non planar at the carbon to which the OH group is bonded. The CSC angle of 96° is similar to that observed in CH_3SH (96.5) (56).

The carbonyl and −OH groups lie in a plane, and in this plane, the MEP (Figure 5) are characterised by two deep minima near the $>C=O$ group, and a further minimum near the −OH group. The minimum associated with the −S-atom lies below the plane referred to above. In the plane, the rest of the potential is everywhere positive.

It is, however, apparent that this species has a MEP map whose general features are similar to that of the enediol (E) (Figure 6), the main difference being the distance between the minima (~4A in A; 3.5A in E).

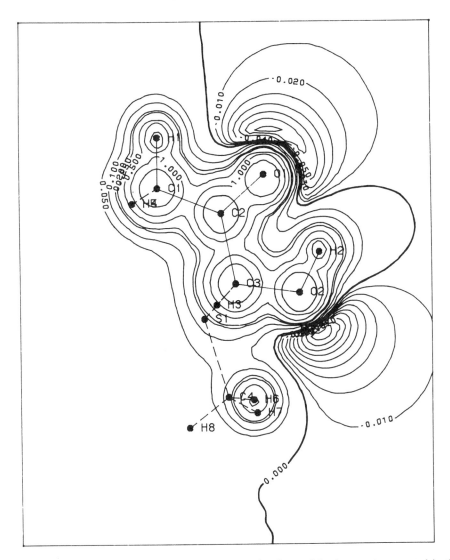

Figure 5. Molecular electrostatic potential (MEP) map for the model substrate A computed in the plane containing the $-CO-$ and $-OH$ groups.

Examination of CPK models re-inforce this view: further, the CH_3 group attached to S (which in glutathione itself is, of course, replaced by part of the tripeptide, with a terminal $-CH_2-$ group) is in a somewhat similar position in the two molecules, and the positive regions of the maps in the vicinity of CH_3S- are similar.

We have also carried out calculations on the substrate A with the actual amino acid residues in glutathione in place of CH_3. These calculations were carried out with the PCILO method (57) and confirm that the conformation of the rest of GSH

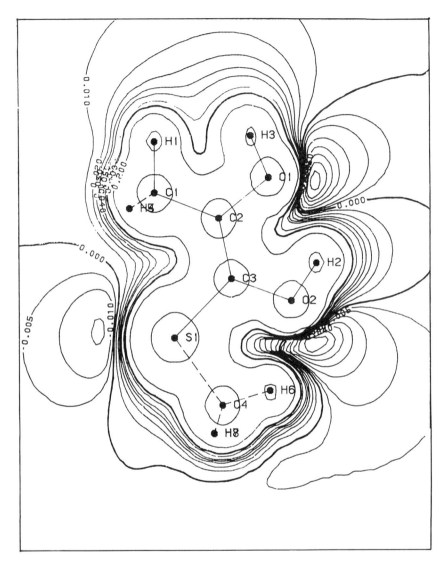

Figure 6. MEP map for the model enediol, computed in the molecular plane.

relative to that of the rest of the molecule is such that the model compound should be a reasonable model for the true substrate.

Therefore, it seems likely that in the active site, a molecule which resembles either the substrate A or part of the enediol (E) might bind to the enzyme.

The other species in Figure 4 are anions, and are thus probably stabilised by the divalent metal cation. The anions have similar MEPs except for more extended minima around the O^-, but the Mg^{2+}/anion complexes have only positive contours.

To save space, these MEP maps are not reproduced here but are available from the authors.

The neutral complex of the enol dianion with Mg^{2+} (Figure 7), has a map which resembles both A and E, except for a larger positive "hill" between the two minima, but there is also a large and broad negative region behind the molecule not found in A or E.

Our tentative conclusion from the preliminary analysis of these MEP are that the shape of the molecules and the maps themselves lend support to the suggestion that the

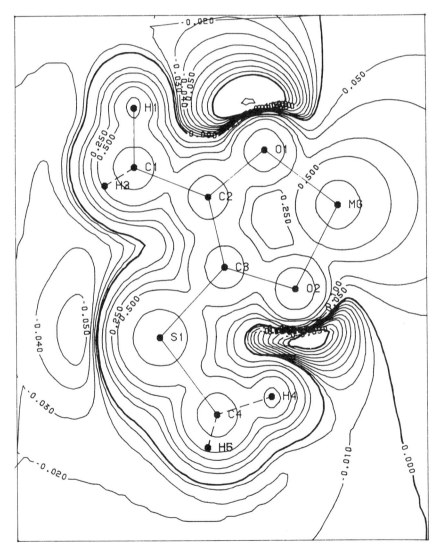

Figure 7. MEP of the complex of the dianion (D) and Mg^{2+}, computed in the molecular plane.

observed inhibitors could function by resembling A or E. We now turn our attention to our further experimental and theoretical work on the inhibitors themselves.

Table III
Inhibition of Red blood cell glyoxalase-I: I_{50} values (mM)

Compound	I_{50}
Esculetin	0.03
4-Methyl-esculetin	0.03
Isoesculetin	0.30
2,3 Dihydroxy-benzoic Acid	0.28
3,4 Dihydroxy-benzoic Acid	0.32
Ascorbic Acid	0.48
2,3 Dihydroxy-pyridine	0.53

Hydroquinones	I_{50}	I_{88}	Flavones[†]	Name	I_{50}
Unsubstituted	4.1	1.5	R_1=OH R_2=OH R_3=OH R_4=H		
Methoxy	1.75	0.73		Myricetin[b]	0.005
2,3 Dimethoxy	2.30	0.80	R_1=OH R_2=H R_3=OH R_4=H		
2,5 Dimethoxy	1.10	0.60		Quercetin[b]	0.009
2,6 Dimethoxy	1.20	0.90	R_1=OH R_2=H R_3=OH R_4=H	Fisetin[c‡]	0.010
Tetramethoxy	1.60	1.35	R_3=OH R_1=H R_2=H R_3=H	3-OH Flavone[d]	0.010
			R_1=H R_2=H R_3=OH R_4=OH	Morin[c]	0.03
Benzoquinones					
Unsubstituted	0.48	0.08	R_1=H R_2=H R_3=OH R_4=H	Kaempferol[c]	0.055
Methoxy	0.48	0.23			
2,3 Dimethoxy	0.39	0.24	R_1=OH R_2=H R_3=O−Rham R_4=H	Quercitrin[c]	0.070
2,5 Dimethoxy*	—	0.10			
2,6 Dimethoxy	0.24	0.01	R_1=H R_2=H R_3=H R_4=H	Apigenin[c]	0.070
Tetramethoxy	0.87	0.20	R_1=OH R_2=H R_2=O−Gluc−Rham R_4=H	Rutin[b]	0.110
			R_1=OH R_2=OH R_3=O-Rham R_4=H	Myricitrin[c]	0.185
			R_1=H R_2=H R_3=H R_4=H	Naringenin[c§]	0.250
			R_1=OH R_2=H R_3=OH R_4=H	Taxifolin[b§]	0.330

*The 2,5 Dimethoxybenzoquinone is very poorly soluble and therefore a I_{88} was made. A comparable figure for the other quinones and hydroquinones is included.
[†]See Figure 11 for designations.
[‡]Fisetin-No hydroxyl at position 5 on flavone ring.
[§]Naringenin, Taxifolin-No double bond between positions 2,3 on flavone ring.

b) Inhibitors of Glyoxalase-I—Experimental Results

In our earlier experimental work (41), the inhibitors studied were chosen on the basis of the similarity of part of the molecule to the enediol structure of intermediate E. Experimental details are given in Ref. 42 and will not be repeated here. The results are quoted in terms of I_{50}, the concentration of inhibitor for 50% inhibition, and are reproduced in Table III.

Several of the compounds in our initial study are also inhibitory to tumour growth in the mouse L1210 leukaemia model. Szent-Györgyi (58) has recently shown that a further series of compounds derived from quinones and hydroquinones have anti-tumour activity in the mouse L1210 system, when added in combination with ascorbic acid. Examination of some of these compounds in our system (Table III) shows them also to be inhibitors of glyoxalase-I, but not particularly good ones. Experiments are underway to test these compounds in conjunction with ascorbic acid, which was found to be a relatively weak inhibitor in our earlier studies.

The latest series of compounds we have tested are flavone derivatives, suggested partly by further work by Douglas and Nadvi (59) on lapachol derivatives, and more directly by experiments by M. A. Johnson (60).

The experimental values of I_{50} for all these compounds are collected together in Table III, where it is apparent that several of the flavone derivatives are more effective inhibitors (61) than those used in our earlier work. Similar results were obtained when the enzyme was yeast glyoxalase-I.

c) Inhibitors of Glyoxalase-I—Electronic Structure and
Molecular Electrostatic Potential Maps

We have computed the wave functions for a variety of the inhibitors described above, and also for several related compounds containing similar functional groups, and used these wave functions to generate the MEP maps.

The observation that the inhibitors must be planar (32) supports the view that one common feature of these inhibitors is the existence of an enediol like moiety in the molecule.

Comparison of the MEP maps of the inhibitors with those of the substrate A and enediol intermediate E leads us to a number of conclusions.

The most effective inhibitor in the series based on the coumarin structure is esculetin (Figure 2), and the MEP map is given in Figure 8. Since coumarin itself is barely inhibitory at all (41), the lactone grouping *per se* is not an important feature as far as inhibition is concerned. This region has three minima in a large region of negative potential extending right around the $-\overset{\text{O}}{\underset{\|}{C}}-O$ grouping.

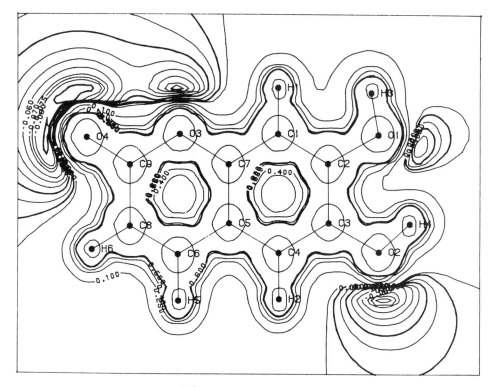

Figure 8. MEP of esculetin.

The minima associated with the two OH-groups lie in a negative potential region similar to that observed with the enediol E, except they are slighty farther apart due to the increased C—C distance in the ring system compared to the isolated double bond. 4-Methyl esculetin is also inhibitory, and the MEP is very similar to that of esculetin, as expected.

It was suggested to us (61) that isoesculetin (Figure 2) might more closely resemble a true enediol and be an effective inhibitor. Examination of its MEP shows similarities to esculetin, but some significant differences (Figure 9).

Although a complete refinement of the geometry of these molecules was too costly, a more recent limited optimisation of the lactone ring shows tht the distance between the minima associated with the two OH groups is slightly smaller (not larger as stated in Ref. 41) in the case of isoesculetin, since one OH group is hydrogen bonded to the carbonyl group. In fact the depth of this OH minimum is reduced as a consequence. However, the orientation of the rest of the molecule with respect to these minima probably results in a steric constraint for this molecule compared to esculetin.

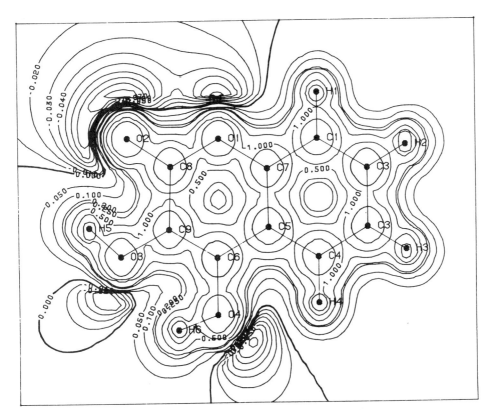

Figure 9. MEP of isoesculetin.

Although squaric acid is an effective inhibitor, the experimental data suggest that its mode of inhibition may be different (41). Furthermore, at physiological pH, it certainly exists as the dianion $C_4O_4^{2-}$, and almost certainly this is complexed with a metal cation. Therefore, we defer discussion of the MEP for this compound, its ions and Mg^{2+} complexes, to a later date.

Maltol was shown to be inhibitory in the work of Ref. 32, and estimated in our system to be about as effective as L-ascorbate. Since this compound is not a diol, at first sight it does not fit into the above picture.

However, examination of CPK models of the substrate, the enediol and the other inhibitors shows that the presence of a $-CH_2-$ grouping near the $-\underset{\underset{O}{\|}}{C}-C(OH)-$ group resembles the arrangement of the $-CH_2-$ group attached to sulphur in the substrate A. Hence this compound may be mimicking the substrate: the MEP in this region support this suggestion (Figure 10).

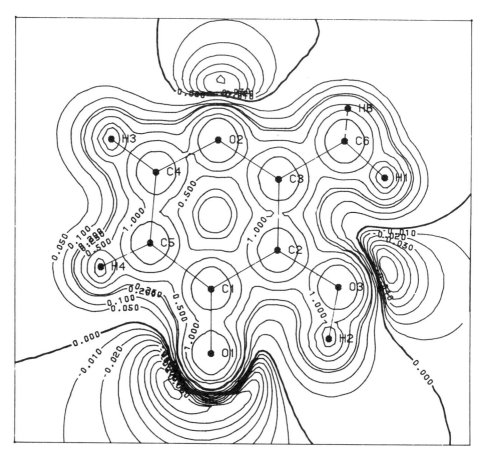

Figure 10. MEP of maltol.

In order to save space, we will not present the maps for the other inhibitors depicted in Figure 3. The general features are as expected for these compounds, but there are complications with some of these species such as the hydroxypyridines, for which different tautomeric forms are possible (62), and the substituted benzoic acids, which are undoubtedly in the form of their ions in solution.

In attempting to correlate the inhibitory activity with quantities derived from the wave function calculations, we have examined a variety of molecular properties, such as the depth of the potential minima, the Mulliken populations, the orbital energies and various other possibilities. However, in view of the sensitivity of such quantities to the basis set, it was considered more likely that some qualitative feature of the MEP might be more useful in this regard, and one such quantity is the distance $R(OH_1:OH_2)$ between the potential minima associated with the two $-OH$ groups. Table IV displays the rough correlation between this distance and the inhibitory behaviour. Although this is an approximate correlation, it does suggest

Table IV

Correlation of Inhibitory effect of various compounds on Glyoxalase-I with the distance $R(OH_1 : OH_2)$

Compound	[mM]*	$R(OH_1 : OH_2)^\dagger$
Esculetin	0.03	3.7
4-Methyl-esculetin	0.03	3.7
Isoesculetin	0.30	3.6
2,3 Dihydroxy-benzoic Acid	0.28	3.9
3,4 Dihydroxy-benzoic Acid	0.32	3.7
Ascorbic Acid	0.48	4.4
2,3 Dihydroxy-pyridine	0.53	4.0

*Concentration for 50% inhibition
†Value in A.

that the effective inhibitors should have values of $R(OH_1:OH_2)$ as small as possible and close to the value in the enediol E.

More important, however, is the suggestion of other inhibitors for testing on the

Table V

Some potential inhibitors of glyoxalase-I?

Name	Formula	$R(OH_1 : OH_2)$
7-methyl esculetin	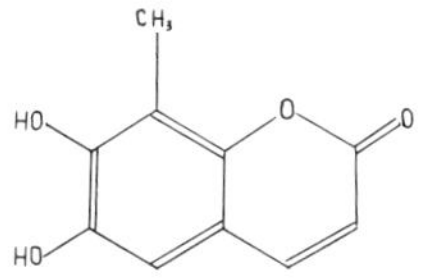	~3.4
2,3 Dihydroxy benzoquinone	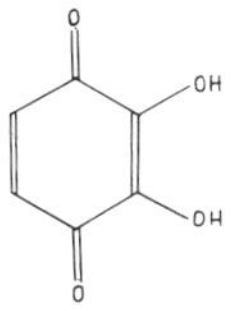	3.8
2,3 Dihydroxy α-pyrone	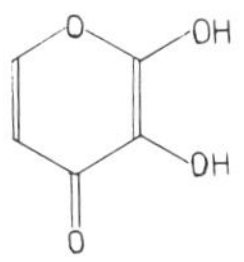	3.5
3,4 Dihydroxy α-pyrone	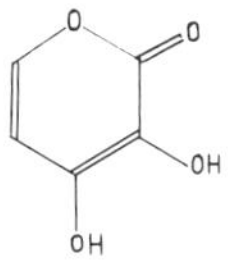	3.6
3,4 Dihydroxy pyridine	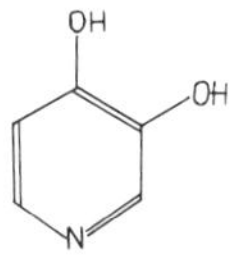	3.9

basis of this correlation. We have examined a variety of compounds containing two OH groups, and distances between the OH minima of the same order of magnitude as found for the known inhibitors (3.4⇒3.9A) were found for the compounds given in Table V which we believe to be potential inhibitors of this enzyme. Undoubtedly, some of these may be difficult to synthesise, or will be unstable, but we certainly expect 7-methyl esculetin to be effective.

Examination of the wave function data and maps for the flavones is not yet complete, but on the basis of the results described above, and the structural features which characterise the flavone inhibitors, we are able to suggest other possible inhibitors which might be tested, and these are depicted in Figure 11.

In the flavone series, these compounds all contain the double bond conjugated with the carbonyl group and also a nearby OH group.

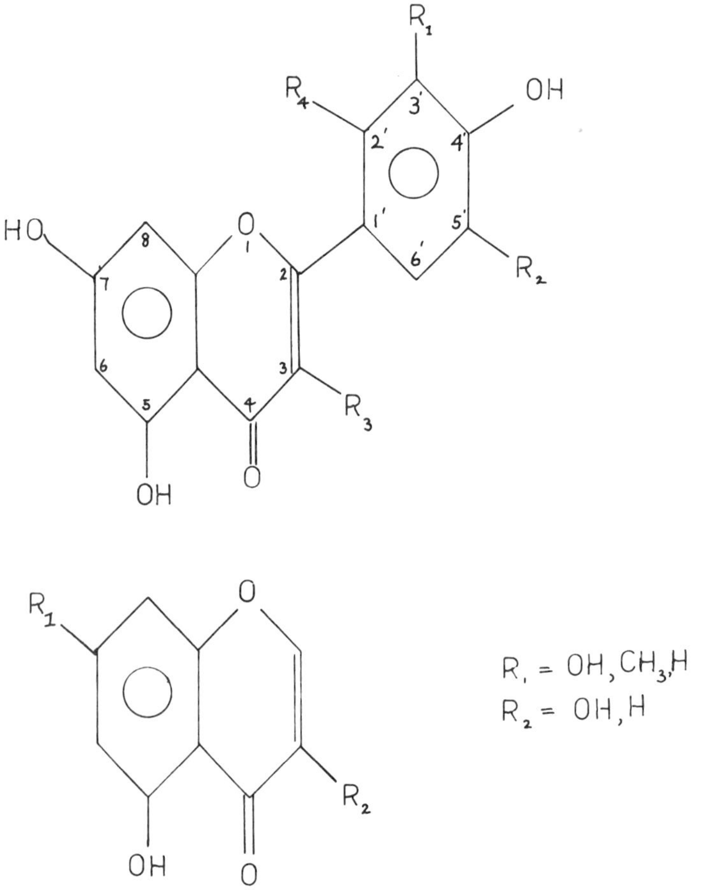

$R_1 = OH, CH_3, H$

$R_2 = OH, H$

Figure 11. Structure of substituted flavones with various R groups as listed in Table III.

We are currently calculating the MEP for the flavone derivatives and testing these compounds in vitro.

Much further work is needed on other related molecules, and also on the structure of the actual transition states themselves, but we believe the time is ripe for the application of quantum chemical techniques to problems in enzyme inhibition, and that ultimately new and more potent inhibitors will be suggested by theoretical calculations.

Acknowledgements

We are indebted to the National Foundation for Cancer Research for continued financial support, and to Professor J. A. Pople for a copy of GAUSSIAN 80. We thank Dr. J. R. Ball and Mr. C. Reynolds for their help with the improvements to GAUSSIAN 80 and for useful discussions. Finally, we thank the University of St. Andrews Computing Laboratory for the large amount of computer time assigned to this project.

References and Footnotes

1. T. I. Kalman, *Drug Development Research, 1,* 311 (1981).
2. W. N. Lipscomb, *Ann. Rev. Biochem., 52,* 17 (1983).
3. M. Dixon and E. C. Webb, in *Enzymes,* 3rd Edition, Longmans, London (1979).
4. C. Walsh, *Enzymatic Reaction Mechanisms,* Freeman, San Francisco (1979).
5. L. Pauling, *Amer. Scientist, 36,* 51 (1948).
6. W. P. Jencks, in *Current Aspects of Biochemical Energetics,* Ed., N. O. Kaplan and E. P. Kennedy, Academic Press, New York, p. 273 (1966).
7. R. Wolfendem, *Acc. Chem. Res., 5,* 10 (1972).
8. G. E. Lienhard, *Science, 180,* 149 (1973).
9. *Transition States of Biochemical Processes,* Ed., R. D. Gandour and R. L Schowen, Plenum Press, New York (1978).
10. G. M. Maggiora and R. E. Christoffersen, Chapter 3 in Ref. 9.
11. P. Pulay, *Mol. Phys., 17,* 197 (1969).
12. P. Pulay, in *Modern Theoretical Chemistry,* Ed., H. F. Schaefer III, Chapter 3, Plenum Press, New York (1977).
13. L. Schäfer, *J. Molec. Struct., Theochem, 100,* 51 (1983).
14. P. R. Andrews, in *Computer Assisted Drug Design,* ACS Symposium Series 112, American Chemical Society, Washington (1979).
15. P. R. Andrews, M. N. Iskander, G. P. Jones and D. A. Winkler, *Int. J. Quantum Chem., QB9,* 345 (1982).
16. A. Szent-Györgyi, L. G. Egyud and J. A. McLaughlin, *Science, 155,* 539 (1967).
17. L. G. Egyud, J. A. McLaughlin and A. Szent-Györgyi, *Proc. Nat. Acad. Sc., U. S., 57,* 1422 (1967).
18. A. Szent-Györgyi, *Science, 161,* 988 (1968).
19. A. Szent-Györgyi, *Electronic Biology and Cancer: a new theory of Cancer,* M. Dekker, New York, (1976).
20. G. Fodor, R. Miyumdar and A. Szent-Györgyi, *Proc. Nat. Acad. Sci., U. S., 75,* 4317 (1978).
21. H. D. Dakin and H. W. Dudley, *J. Biol. Chem., 14,* 423 (1913).
22. C. Neuberg, *Biochem. Z., 51,* 484 (1913).
23. K. Lohmann, *Biochem. Z., 254,* 232 (1932).
24. E. Racker, *J. Biol. Chem., 190,* 685 (1950).

25. A. Szent-Györgyi and L. Egyud, *Proc. Nat. Acad. Sci., U. S., 56,* 203 (1960).
26. R. Vince, S. Daluge and W. B. Wadd, *J. Med. Chem., 14,* 35 (1971).
27. W. L. Kermack and N. A. Matheson, *Biochem. J., 65,* 48 (1957).
28. M. V. Kester, J. A. Reese and S. J. Norton, *J. Med. Chem., 17,* 413 (1974).
29. G. W. Phillips and S. J. Norton, *J. Med. Chem., 18,* 482 (1975).
30. R. Vince, M. Wolf and C. Sanford, *J. Med. Chem., 16,* 951 (1973).
31. M. Ito, K. Okabe and H. Omura, *J. Nutrit. Sci. Vitaminol., 22,* 53 (1976), and references therein.
32. K. T. Douglas and I. N. Nadvi, *F. E. B. S. Lett., 106,* 393 (1979).
33. B. Oray and S. J. Norton, *Biochem. Biophys. Res. Comm., 95,* 624 (1980).
34. B. Mannervik, Chapter 14 in *Enzymatic Basis of Detoxification,* Vol. 2, Academic Press, New York (1980).
35. A. C. Aronsson, E. Marmstal and B. Mannervik, *Biochem. Biophys. Res. Comm., 81,* 1235 (1978).
36. S. S. Hall, A. M. Doweyko and F. Jordan, *J. Am. Chem. Soc., 98,* 7460 (1976).
37. K. A. Davis and G. R. Williams, *Biochim. Biophys. Acta, 113,* 393 (1966).
38. A. C. Aronsson, S. Sellin, G. Tebbelin and B. Mannervik, *Biochem. J., 197,* 67 (1981).
39. S. Sellin, L. E. G. Eriksson, and B. Mannervik, *Biochemistry, 21,* 4850 (1982).
40. R. Lavery and B. Pullman, *Int. J. Quant. Chem., QB6,* 461 (1979).
41. R. B. Brandt, M. E. Brandt, M. E. April and C. Thomson, *Int. J. Quant. Chem., QB9,* 335 (1982).
42. R. B. Brandt, M. E. Brandt, M. E. April and C. Thomson, *Biochemical Medicine, 29,* 385 (1983).
43. C. Thomson and R. B. Brandt, *Int. J. Quantum Chem., QB10,* 357 (1983).
44. P. Politzer and P. Pulay (Eds), *Chemical Applications of Atomic and Molecular Electrostatic Potentials,* Plenum Publishing Co., New York (1981).
45. R. Osman, S. Topiol, H. Weinstein and J. E. Eilers, *Chem Phys. Letters, 73,* 399 (1980).
46. J. S. Binkley, R. A. Whiteside, R. Krishnan, R. Seeger, D. J. DeFrees, H. B. Schlegel, S. Topiol, L. R. Kahn and J. Pople, *GAUSSIAN 80,* Carnegie-Mellon University, Pittsburgh, Penn. (1980).
47. W. J. Hehre, R. F. Stewart and J. A. Pople, *J. Chem. Phys., 54,* 724 (1971).
48. K. Kitaura and K. Morukuma, *Int. J. Quant. Chem., 10,* 325 (1976).
49. J. Tomasi, private communcation.
50. L. Radom, Chapter 8 in *Modern Quantum Chemistry,* Vol. 4, Ed. H. F. Schaefer, III, Plenum Press, New York (1976).
51. J. Chandrasekar, J. G. Andrade and P. von R. Scheyer, *J. Am. Chem. Soc., 103,* 5609 (1981).
52. A. Johansson, P. Kollman, and S. Rothenberg, *Theor. Chim. Acta., 29,* 167 (1973).
53. S. Shinkai, T. Yamashita, Y. Kusano and O. Manabe, *J. Am. Chem. Soc., 103,* 2070 (1981).
54. S. S. Hall, A. M. Dowekyo and F. Jordan, *J. Am. Chem. Soc., 100,* 5934 (1978).
55. S. Sellin, P. R. Rosevear, B. Mannervik and A. S. Mildvan, *J. Biol. Chem., 257,* 10023 (1982).
56. T. Kojima, *J. Phys. Soc. Japan, 15,* 1284 (1960).
57. S. Dinet, J. P. Malrieu, F. Jordan and M. Gilbert, *Chem. Phys. Letters, 2,* 319 (1968).
58. A. Szent-Györgyi, Int. *J. Quantum Chem., QB9,* 27 (1982).
59. K. T. Douglas, I. N. Nadvi and N. Thakar, *IRCS Med, Sci., 10,* 683 (1982).
60. M. A. Johnson (unpublished observations).
61. R. B. Brandt, J. Laux, C. Thomson, and M. A. Johnson, *Int. J. Quantum Chem., QB11,* (1984).
62. G. Fodor (personal communication).
63. G. L. Manna, *J. Mol. Struct., Theochem, 85,* 389 (1981).

Structure & Motion: Membranes, Nucleic Acids & Proteins,
Eds., E. Clementi, G. Corongiu, M. H. Sarma & R. H. Sarma,
ISBN 0-940030-12-8, Adenine Press,   Copyright Adenine Press, 1985.

A Davydov-like Soliton in Acetanilide
Experimental Evidence and Biochemical Considerations

Giorgio Careri
Dipartimento di Fisica
Universita di Roma I
00185 Roma, Italy

Abstract

Acetanilide (ACN) is an appropriate model system to study nonlinear one dimensional hydrogen bonding in proteins. Measurements on crystalline ACN at low temperature show near the Amide I band a new band that cannot be explained by conventional molecular spectroscopy. A soliton model, similar to that proposed by Davydov for alpha-helix, is in agreement with the data. By considering this kind of soliton as a ligand to a polyfunctional macromolecule, the possibility of unidirectional circulation in a Wyman biochemical cycle is discussed.

Introduction

Acetanilide ($CH_3CONHC_6H_5$), or ACN, is an interesting solid at room temperature because two close chains (spines) of nearly planar hydrogen-bonded amide groups run through the crystal, providing an interesting model for an array of hydrogen-bonded amides in one direction. Moreover, the bond distances in ACN are very close to those found in α-helices, where three similar spines are coiled along the helix axis. Since the physical properties of hydrogen-bonded systems are very sensitive to bond distances, we thought ACN would be a useful model system to be used in searching for new physical features of extended poly peptide chains and perhaps also proteins. Therefore, about ten years ago, I reported some details of the I.R. and Raman spectra, showing that a new amide-I band appears at low temperature in crystalline ACN, red-shifted by $15 cm^{-1}$ from the primary amide-I band at 1665 cm^{-1} (1). At that time, we were unable to account for this new band in terms of the exiton model, but no better theory was available. Since then we have reformulated the problem on the basis of a soliton model similar to Davydov's (2) for the α-helix in proteins, and the results have been presented in a recent note (3). We are still studying this problem (4,5) as well as its biological relevance (6). These studies will be reviewed below, after a few observations regarding the possibility of anharmonic coupling in a system of hydrogen-bonded amides, a topic quite appropriate to this conference.

ACN as a physical model system for proteins

Hydrogen bonding is widespread in biomacromolecules, presenting physical features that vary greatly from one case to another. In weak bonds this interaction can be treated as a problem in electrostatics involving a set of fixed and localized interacting charges. On the other hand, in the case of very strong hydrogen bonds, one is faced with a delocalized charge distribution to be treated according to valence theory. Between these two extremes, there are cases where the bonds are of intermediate strength. In these cases, one can model the complex state of affairs by assuming that the local charges depend upon their mutual distances and that these distances in turn depend upon the local charges. Thus in these intermediate cases one can visualize the microscopic source of the anharmonic coupling which gives rise to the non-linear terms in the equations describing the dynamics of the system. This must be the case of the $N-H\cdots O=C$ bond in proteins, for $R(\overline{NO})$ distances close to $2.80A°$. According to a theoretical study by Saito et al. (7), in this case the contribution of the charge transfer structure $\overline{N}:H:O^+$ is about 9.2%. This may be contrasted with the $O-H\cdots O$ bond, where Coulson has evaluated the charge transfer structure at the same distance of $2.80A°$, finding it to be close to 2.8% (8). In addition, it explains why in a variety of situations water can be treated satisfactorily as a system of fixed charges, whereas in the hydrogen-bonded amides present in proteins, the charge transfer induced by the hydrogen bonding changes the charge distribution of the amide group itself.

Since the charge distribution of the amide group is strictly connected with the planarity of this group, the above-mentioned effect must be sensitive to steric constraints and we must anticipate that hydrogen bonding between amides will vary from case to case in globular proteins. In fact, various energy functions have been suggested in conformational analysis, and there is not even a consensus on the analytic representation of this interaction energy function. For all these reasons, crystalline ACN seems the best candidate for investigations of the physical properties of a-helices dependent upon the anharmonic coupling discussed above.

To quantify the influence of the $R(\overline{NO})$ distance on the charge distribution of hydrogen-bonded amides, it is convenient to consider the effect on the frequency v, defining the quantity

$$g = \frac{dv(\text{amide-I mode})}{dR\ (\overline{NO})}$$

In our experiments we have found this quantity to be about $0.62 \cdot 10^{-10}$ newtons, so that it is relevant in the theory of the bond energy transport along one-dimensional molecular amide chains by soliton, as first proposed by Davydov and Kisluskha (2), because subsequent theoretical study showed that there is a definite threshold level for g, below which solitons cannot be expected to form.

Infrared absorption in acetanilide by solitons

In this section I will report the results of experiments that I have performed on ACN and some ACN derivatives in several different physical preparations, in collaboration with Umberto Buontempo, Fabrizio Carta, Fabrizio Galluzzi in Rome, and Enrico Gratton and E. Shyamsunder in Urbana. Some of these data have already been published (3) and a more complete paper is in preparation (4). The most relevant features of these experiments can be summarized as follows:

1) On cooling crystalline ACN samples, a new infrared band appears at $1650 cm^{-1}$, while no other major changes occur in the IR spectrum from 4000 to 800 cm^{-1}. The intensity of this new band steadily increases from room temperature to 70 K, then slowly increases down to 4 K. At $1650 cm^{-1}$ the band is not present in amorphous samples, but after annealing the amorphous material it is fully recovered.

2) We assigned the new band at $1650 cm^{-1}$ to an amide-I component on the following basis: i) N^{15} substitution which induces a small shift on the amide-I at $1665 cm^{-1}$ also shifts the new band by the same amount. ii) Deuterium substitution at the NH position strongly affects both the amide-I and the new band in a complex way. iii) Upon cooling, a parallel decrease of the normal amide-I integrated absorption and increase of the 1650 band integrated absorption is observed. iv) The 1650 band and the amide-I band show the same dichroism over the temperature range investigated.

3) Several other experiments were performed to clarify the nature of the 1650 band. i) The Hydrogen-bonded network of ACN crystals is essential for the appearance of the 1650 band. ii) Raman spectra of ACN show this 1650 band has the same temperature dependence as in the IR. iii) A parallel study of temperature dependence has been carried out on specific heat and volume expansion as detected by x-rays, in order to rule out the occurrence of a polymorphic transition. Such transitions would affect some other IR and Raman absorption bands, a fact which was not observed in the entire spectrum during a large number of runs from 10 K to room temperature. In these experiments, no dependence upon the cooling rate nor hysterisis were ever observed (4).

4) We summarize this brief survey of experimental results by emphasizing that the new band at $1650 cm^{-1}$ is characteristic of the amide group of ACN in crystal form. A detailed analysis by the usual exiton model cannot account for this new band. Having excluded such other conventional explanations as Fermi resonance, we consider the possibility of assigning it to a collective excitation similar to the soliton proposed by Davydov for α-helix in proteins.

Davydov's soliton arises from a cooperative interaction between localized amide-I bond energy and lattice distortion. The bond energy acts through non-linear coupling, as a source of lattice distortion. This lattice distortion reacts, again through non-linear coupling, as a potential well to trap the bond energy and prevent its dispersion via dipole-dipole coupling effects. We followed the same theory with one important difference: for lattice distortion we substituted displacement of the hydrogen-bonded proton. The distinction is vital, because Davydov has shown that photon absorption by his intermolecular vibrational soliton is ruled out by the Frank-Condon principle. Alwyn Scott gives a preliminary theory of our assignment in our paper (3). Here I shall limit myself to outlining the major points and presenting some conclusions. The main idea was that the effect of introducing localized amide-I energy could displace the ground state of the adjacent hydrogen-bonded proton. This displacement of the proton acts to trap the amide-I bond energy and prevent its dispersion via dipole-dipole interaction effects. The combined excitation was proved to be a soliton, and we assigned the binding energy of this soliton to the experimentally observed red shift of $15 cm^{-1}$ from the conventional amide-I band to the unconventional amide-I bond.

For a soliton on a single chain, this simplified theory gives $E = E(O) - 2J - g^4/12f^4m^2J$, where f is the frequency of the NH deformation mode, m is the proton mass, J the first neighbor's coupling energy and E(O) the unperturbed level of the amide-I mode. This implies that $g = 1.9 \cdot 10^{-10}$ newtons, in good agreement with the experimental data quoted above. Moreover, our soliton hypothesis was supported by calculations which showed that the absorption cross section for a soliton should compete with that of an isolated amide group. In the case of a Davydov soliton, absorption of light is impossible because there is not enough time for the necessary lattice distortion to take place while the photon is being absorbed; in our case this does not apply because we require displacement of the hydrogen-bonded proton only. Finally, we calculated the temperature dependence of the intensity of the soliton band, assuming that the effect of temperature was to raise the proton slightly above its ground state, thus detuning the soliton from the incident beam. Of course, this theory holds only on a time scale that is shorter than the lattice relaxation time. On a longer time scale, this soliton should relax into a Davydov soliton by creating lattice phonons.

Alwyn Scott and his co-workers at Los Alamos National Laboratory have developed a more complete theory, which will be reported shortly (5). They start from a Hamiltonian operator which contains three terms. The first contribution represents the Amide-I excitation energy in the field of its first neighbors. The second term is the energy of an unspecified low-frequency vibration, represented by a classical harmonic oscillator. The third term is the interaction energy between amide-I excitation and low-frquency vibration with the non-linear constant g already introduced above. Applying variational methods, dynamic equations are obtained in which one can focus attention on solutions displaying low-frequency oscillators at rest. Such solutions are stationary not only towards low-frequency classical motion but also toward Amide-I excitations, and when $g \neq 0$ both wave-like and localized solutions are allowed. Wave solutions are essentially excitonic solutions, while the self-trapped solutions are the solitons we are looking for. In this theory, the intensity of the soliton line depends primarily upon the inner product of the wave function for low-frequency vibration before soliton absorption and that after absorption. By fitting this theory to the experimental data on the temperature dependence of the intensity of the soliton band, we obtained the value: $130cm^{-1}$ for the low-frequency mode, which is close to some far infrared absorption peaks detected in the laboratory.

In our opinion, the soliton model described here provides an explanation for the unusual behavior of the $1650cm^{-1}$ band of ACN which has been observed in many experiments over the past decade. Although these experiments must be extended to analyze larger single crystals and the theory must be refined by numerical studies of the soliton dynamics, I feel that the experimental and theoretical work reported here points to the existence of what we first called the proton-assisted Davydov soliton (3). This is a new member of the growing family of solitons detected in condensed matter. It is also the first example of a soliton in an organic solid which is a suitable model system for proteins.

Biochemical considerations

Perhaps the most important finding of the work reported above is that a soliton can be created by infrared light on a network of hydrogen-bonded amides. In this section, I shall mention a possible relevance of this fact in enzymology, a problem I am currently studying in collaboration with Jeffries Wyman.

Enzymes are globular proteins and work in a cycle characterized by a turnover number. To understand this general fact, and to explain cooperativity and regulation, an allosteric model was postulated despite the fact that in many cases the system is far from equilibrium. Therefore, Wyman has proposed a "kinetic" variant of the allosteric model, called the Turning Wheel (9). This model will be briefly reviewed in order to show how the uni-directional circulation of an enzyme can be explained, physically, in terms of a soliton trapped in the protein matrix as a ligand, and how this possibility can be tested in the laboratory using infrared radiation.

Consider a macromolecule which exists in several different conformations or states of ligation. Suppose that it acts as an enzyme for one or more of its ligands. Suppose further that it is present in an environment in which its substrates and their products are maintained at constant concentrations, in general not at equilibrium. The system will be described by a set of simultaneous first-order linear differential equations whose solution shows a unique asymptotically stable critical point (10). This means that regardless of the starting point, the system will always settle down to a steady state. At equilibrium, when the product of their constants taken clockwise round any closed path is equal to that taken counter-clockwise, everything will be statistically at rest. In a steady state, on the other hand, there will be a "circulation" of the macromolecule round the elementary closed paths. This will occur when, owing to the fact that the concentrations are maintained by sources and sinks at non-equilibrium values, there is a flow of matter through the system with a corresponding release of free energy and heat. Under these conditions, through the mediation of the polyfunctional enzyme, we have the possibility that one reaction will drive another, even when the two reactions share no common chemical element.

Following current usage (11), we shall apply the term soliton to any "self-trapped" state, namely to an energy packet trapped inside a portion of a protein matrix, until it decays into heat as a result of a strong perturbation from some other portion of the same protein. The question arises whether such a soliton may not provide the mechanism responsible for maintaining the constant uni-directional circulation of the Turning Wheel. Consider the simple case illustrated below and assume that the life of the soliton, which depends upon the state of the macromolecule, is very long in M_2 and very short in M_1. Imagine a soliton S to be formed as a result of the binding of substrate L by one form of the enzyme, namely M_2, and suppose it to find its way, undiminished, in the course of the circulation of the macromolecule, to another form M_1, where it is liquidated in connection with the formation and liberation of the enzymatic product L^1. The difference between the original free energy of the soliton and that involved in the formation and detachment of the product will appear directly as heat. Thus the soliton would provide a physical mechanism for explaining the unidirectional circulation of the Turning Wheel, behaving as a ligand that can be fully absorbed by an ideal sink, the thermal bath. Schematically, this may be written as:

$$M_1 + L^1 + \text{heat} \Leftarrow M_1LS \rightleftharpoons M_2LS \rightleftharpoons M_2 + L$$
$$M_1 \quad \rightleftharpoons M_2$$

Before accepting this proposal, let us use a specific case to consider the plausibility that ligand binding can in fact give rise to a vibrational soliton. To this end, we recall that in many enzymes the negative phosphate moieties of a ligand bind at the N termini of α-helices. Since the first suggestion by Hol et al. (12), we now have an impressive list of cases where the ligand is a co-enzyme, and where the ligand binding site has been well identified by x-ray diffraction. The reason for this has already been provided by Hol et al. (12), who call attention to the fact, first pointed out by Wada (13), that alignment of peptide dipoles parallel to the helix axis gives rise to the macrodipole of considerable strength. The resulting effect is such that a half unit of positive charge becomes located at the N-terminus if the helix, and the binding of one single negatively charged group at that end of the helix involves an attractive energy of about 13 K cal mol^{-1}. This makes the total binding process possible even for a large ligand like a co-enzyme. Hol et al. (12) proposed that this dipole field could assist in the binding of substrates or co-enzymes, and even in accelerating some reactions like the proton transfer from nearby side chains to the N-terminus of the helix. We propose that this dipole field can assist in the creation of a soliton in the α-helix in question (6), because in the transient compression of the helix a variety of distances among the hydrogen-bonded peptide groups will be made available. Therefore, in our opinion, it is likely that in this transient elastic wave a vibrational soliton can be created and localized near the N-terminus.

Let us assume that the creation of this soliton can take place as proposed above, and now let us inquire about its fate in the specific case considered. In practice, the soliton may be expected to decay when the protein matrix undergoes those major conformational changes which are caused by the catalytic events that take place at the active site. As a result, the co-enzymes are destablized and finally move back into the solvent, and the system is ready to repeat an identical cycle.

So far, experimental evidence for such vibrationally excited solitons in proteins is lacking. Nevertheless, the finding that Davydov-like solitons can be created by IR absorptions in ACN (3) suggests that the particular process considered above should be sensitive to the presence of IR-active solitons created on the same protein moiety by means of infrared light. This occurrence should be experimentally detected as a change in the overall turnover of the enzymatic reaction, because the IR-created soliton should increase the concentration of this peculiar ligand. The strong IR absorption by water will make this experiment impossible in an aqueous medium, but recently nearly dry proteins have been shown to be still enzymatically active (14). In a sense, this suggestion naturally follows from the first introduction of the Turning Wheel (9), where a photon was considered as a possible candidate to drive the wheel. However, the decay time of a photon is so short compared with the circulation period of the Turning Wheel that these two processes can hardly be coupled. Only if a photon can create a soliton, and if this soliton decays thanks to the energy-rich conformational changes occurring during the catalytic action, can the soliton decay be triggered by the Turning Wheel itself and the soliton energy become liberated to cause the wheel to turn in one direction.

Although these considerations by no means prove, they nevertheless establish the plausibility of the picture of vibrational energy soliton as a physical agent for the unidirectional circulation of the Turning Wheel. All this seems to follow from the intrinsic structure of the hydrogen-bonded polypeptide chain itself, because of its capability to display different conformations and to bind a vibrational soliton to its matrix.

References and Footnotes

1. G. Careri in *Cooperative Phenomena*, Ed., H. Haken and M. Wagner, Springer Verlag, Berlin, p. 391 (1973).
2. A.S. Davydov and N.I. Kislukha, *Phys. Stat. Sol. (b), 59,* 465 (1973); A.S. Davydov, *J. Theor. Biol., 38,* 559 (1973).
3. G. Careri, U. Buontempo, F. Carta, E. Gratton and A. C. Scott, *Phys. Rev. Lett., 51,* 304 (1983).
4. G. Careri, U. Buontempo, F. Galluzzi, E. Gratton, S. Shyamsunder and A. C. Scott, *Phys. Rev.,* (in press) (1984).
5. J. C. Eilbeck, P.S. Lomdahl and A.C. Scott, *Phys. Rev.,* (in press) (1984).
6. G. Careri and J. Wyman, *Proc. Nat. Acad, Sci. U.S.A. 81,* 4386 (1983).
7. H. Saito, K. Nukada, H. Kato, T. Yomezawa and K. Fukui, *Tetrahedron Lett., 2,* 111 (1965).
8. C.A. Coulson in *Hydrogen Bonding,* Ed., D. Hadzi, Pergamon Press, London, p. 339 (1959).

9. J. Wyman, *Proc. Nat. Acad. Sci. U.S.A., 72,* 3983 (1975).

10. J. Wyman, *Biophys. Chem., 14,* 135 (1981).

11. A.C. Scott, F.Y.F. Chu and D.W. McLaughlin, *Proc. IEEE, 61,* 1443 (1973); R.K. Dodd, J.C. Eilbeck, J.D. Gibbon and H.C. Morris, *Solitons and Non-Linear Wave Equations,* Academic Press, London (1982).

12. W.G.J. Hol, P. T. Van Duijnen and H.J.C. Berendsen, *Nature, 273, 443* (1978).

13. A. Wada, *Adv. Biophys., 9,* 1 (1975).

14. J.A. Rupley, E. Gratton and G. Careri, *Trends in Biochem. Sci., 8,* 18 (1983).

Structure & Motion: Membranes, Nucleic Acids & Proteins,
Eds., E. Clementi, G. Corongiu, M. H. Sarma & R. H. Sarma,
ISBN 0-940030-12-8, Adenine Press, Copyright Adenine Press, 1985.

Further Theoretical Studies of (Nonlinear) Conformational Motions in Double-Helix DNA

J. A. Krumhansl[†*§], G.M. Wysin[†*], D. M. Alexander[*], A. Garcia[*]
Laboratory for Atomic and Solid State Physics
Cornell University
Ithaca, New York 14853 U.S.A.

and

P. S. Lomdahl[‡], Scott P. Layne[‡]
Los Alamos National Laboratory
Los Alamos, New Mexico 87545 U.S.A.

Abstract

We have pursued further the earlier work of Krumhansl and Alexander (KA)[a] in the direction of developing a rationale for the use of a reduced set of variables for describing the large amplitude conformational dynamics and static structure of DNA. Guided by the structure studies of Fratini, et al.[b] the computer experiments of Keepers, et al.[c] NMR data[d], we have used correlation data and other constraints to argue for the use of a set of conformationally significant variables (CVS), of the order of three per base. Equations of motion have been programmed at Los Alamos for segments up to 200 b.p. in length, and can accommodate arbitrary sequences. Using this program for a model dodecamer we can reproduce and generalize Calladine's[e] results somewhat. For longer chains with suitable parameters we can generate ("soliton") junctions between A and B conformations; whether the parameters used are realistic cannot yet be determined. Moreover, the effects of water and counterions must be added. We comment speculatively on implications in cancer research.

Introduction

We are concerned with the static structure or the dynamics of low frequency conformational changes in DNA which can lead to significant topological changes

[†]And Center for Nonlinear Studies, Los Alamos National Laboratory.
[*]Work supported by National Foundation for Cancer Research.
[‡]Work supported by the U.S. Department of Energy.
[§]Brookhaven National Laboratory

[a]J.A. Krumhansl and D. M. Alexander, *"Structure and Dynamics: Nucleic Acid and Proteins,"* Eds. E. Clementi and R. H. Sarma, Adenine Press, NY (1983) p. 61.
[b]A. V. Fratini, M. L. Kopka, H. R. Drew and R. E. Dickerson, *J. Biol. Chem. 257,* 14686 (1982).
[c]J. W. Keepers, P. A. Kollman, P. K. Weiner, T. L. James; *Proc. Nat. Acad. Sci. USA 79,* 5537 (1982) Biophysics.
[d]See T. L. James; *Bull. Magnetic Resonance 4,* 119 (1982-83).
[e]C. R. Calladine; *J. Mol. Biol. 149,* 761 (1981).

(e.g. transitions among A, B, Z structures). Our objective is to develop a descriptive theoretical understanding which can not only provide a general phenomenological framework for understanding that class of biomolecular dynamics, but also can be utilitarian for modeling and exploring specific problems (such as opening, site selective activity, drug intercalation, radiation effects, promoter and repressor sequences, and radiation effects.) In a more general context, though as yet farther from quantitative molecular biology, we have thought for some time that long range effects and solitons might provide a substantially new ingredient to consider in the understanding of oncogenes and sequence recognition in DNA research.

We follow the viewpoint [Lomonossoff, Butler and Klug (1981), (1) Dickerson, et al (1981), (1982), (1982) (2)] that conformational *structural specificity and dynamics* are of significant importance in the function of biopolymers, in addition to chemical *sequence specificity*.

With respect to the structure and dynamics of biopolymers a traditional approach [e.g. CHARMM, Brooks, et al; (3) AMBER, Wiener and Kollman (4)] which has become widespread with the advent of large computational facilities is to simulate the interaction potentials between *all* the atoms in a large model molecule and: (a) attempt to derive the equilibrium structure through energy and energy gradient minimization algorithms, (b) study the dynamical excitations about equilibrium structures, i.e. molecular dynamics. However, even with the largest computers this "first principle" [or "brute force"] approach soon reaches its limitations for molecules containing say more than 100 atoms. Exactly that kind of limitation was found a long time ago in the physics of (condensed) many-body systems. There the solution to this difficulty has been to search (by physical intuition or experimental insight) for a few functionally significant variables (e.g. the "order parameter" introduced by Landau) which relate particularly to a specific class of behaviors which one seeks to describe (e.g. the onset of long range order in a ferromagnet). [Formally, there is a procedure called renormalization group theory which has been developed to average over irrelevant degrees of freedom in order to justify the reduction of a many (ca. 10^{23}) body Hamiltonian to an effective Hamiltonian involving only a small number of order parameters (i.e. significant variables). That, and the simpler method of constraint identification together provide in principle the formal justification of the order parameter idea.]

We follow an analogous philosophy, though so far we have taken a much more pragmatic approach than a totally deductive theoretical procedure to reduce the many body form of molecular dynamics to significantly fewer variables for the specific purpose of studying metastable states and low frequency global vs. bond dynamics. Specifically, we use experimental results to guide a heuristic choice of a few functionally important effective variables—conformationally significant variables (CSV). This approach, as is usually the case in the use of "order parameters" in condensed matter physics, is partly empirical and lies between full blown molecular dynamics and continuum elastic modeling of DNA; it is less general than the former but more general than the latter. In particular, elastic models cannot deal with strongly non-linear excitations whereas we can deal with such (e.g. kinks, nucleation, and solitons) relatively easily, formally or computationally, in terms of reduced effective equations of motion. At the same time the results of static elastic models [Calladine (5) (1982)] may be achieved as a limiting zero frequency case.

In this conference report our aim is simply to outline the main ideas and present initial computer simulation results to illustrate the kind of applications possible. More detailed descriptions of the equations of motion and the parameters therein, as used in the simulations, will be discussed elsewhere.

Method

We outline: (1) The use of experimental data to reduce general equations of motion; (2) The nature of the reduced equations of motion. The approximate molecular equations of motion sought are to be used to describe asymptotic static variations from an ideal Watson-Crick structure, or the possible dynamic excitations from those displaced states which could lead to large conformational changes.

Fratini, et al (6) (1982) conclude: "The main elements of flexibility in double-helical DNA are *sugar puckering, phosphate orientation,* and *propellor twist. . . .*" There have been numerous NMR studies of the motions in DNA. These are reviewed by James (7), and a discussion particularly related to backbone motions has been given by Keepers and James (8). Relaxation rates cover the range from picoseconds to microseconds, but there are particularly important motions in the nanosecond range which cannot be related to helix twisting and bending, tumbling, or the usual small amplitude harmonic motions. It is notable, as we will discuss in further detail in the next section, that the nanosecond conformational motional patterns seems to correlate strongly with the static patterns found by Dickerson, et al.

Qualitatively similar behavior is found in energy minimization computer simulations by Keepers, et al (9). By subjecting a simulated structure to various imposed diagnostic displacements and studying the response of the system to those imposed conditions it is found that only a few of the backbone torsional angles are particularly "responsive" (i.e. mobile) in conformational rearrangements.

With only Dickerson's earlier results in mind Krumhansl and Alexander (7) (1983) had postulated a phenomenological (classical) effective Hamiltonian for use in describing the slow and large conformational motions; the choice of variables represented a qualitative embodiment of the available structural and computer studies. Now, with the above experimental information the rationale for those equations can be made much more systematic as illustrated below in the case of MPD-7. Simply put, in the static and slow motion regime displacement-displacement correlation analyses give defined experimental relationships between atom positions, and as such obviously *provide constraints which must reduce the number of inde-*

pendent degrees of freedom in the equations of motion. In fact, as noted above, the main flexibility of DNA seems to come from only three degrees of motion per base residue.

Thus, in a real sense in the low frequency regime most of the degrees of freedom are either frozen or "slave" to a few significant variables. Because of the non-independence of the variables the choice of responsive coordinates is not unique; in fact, K-A chose sugar puckering, backbone extension and base orientation (twist about the helix axis), again three "free" variables per unit. In view of the high degree of correlation between, for example χ and δ, we may hope that the K-A and Dickerson "free" variables represent alternative but equally acceptable choices. In any case the data of Fratini, et al (6) provides further validation of the K-A equations (10), which remain the same in form except for one additional term which couples stretching and twisting of the helix, and additional terms to simulate cross strand purine interference or the presence of attached complexes.

Rather than lay out the equations in full mathematical detail in this report we outline them functionally:

Hamiltonian = Sum (over base−ribose−phosphate; strand 1,2) of: [(sugar pucker, including possible double well potential) + (backbone stretch; phosphate rotation and base pair separation implied) + (base orientation (twist) rotation about helix axis) + (pucker−stretch coupling) + (pucker−twist coupling) + (stretch−twist coupling) + (steric forces, e.g. Calladine (1982)) + (external forces, e.g. models of intercalants) + (Langevin damping from high frequency modes or temperature bath)].

These equations have been programmed at Los Alamos in general form for oligomers up to 200 base pairs, for either static or dynamic simulations in response to various initial conditions or steric forces, and may describe arbitrary sequences. Temperature effects are present so far only through Langevin damping.

Application and Results

We report initial results of simulations on the dodecamer MPD-7 and on a 50 b.p. model system; the modified K-A model yields the kind of static structural variations along the helix found by Dickerson on MPD-7 and it also simulates a B−A junction on a 50 b.p. model.

(a) MPD-7

Obvious constraints are: Bond lengths constant, bond angles constant. Data shows that several backbone angles are essentially constant; approximately, $\alpha \simeq 65°$, $\beta \simeq 175°$, $\gamma \simeq 175°$, over a wide range of phosphate and ribose configurations. Several variables are strongly correlated: δ and glycosyl χo; δ and backbone angle ζ, ϵ and ζ; rise "h" and global twist "t". A statement that δ varies is now widely accepted as equivalent to sugar pucker variation. The correlation between δ and ζ is equivalent to the conclusion of Keepers and James (8) that the ^{31}P NMR results can be explained by simultaneous pucker and rotational jumps about the C3'$-$O3' bond. It is noteworthy that the experimental constraint relationships found by Dickerson in static X-ray measurements seem to be identical to those holding in the nanosecond range, an important proviso to our use of CSV for both static and dynamic variations.

The above refer to backbone and pucker within one ribose-phosphodiester residue; in addition we introduce an angle ϕ which measures the rotation of this residue and its attached base about the helix axis; finally we introduce the cordinate z which measures displacement of the residue parallel to the axis. These are effectively the K-A CSV. As noted below it may eventually be necessary to introduce a fourth variable to describe the B$-$Z transition.

We imposed cross strand purine$-$purine forces in the same manner as Calladine (5). The computer program allows us also to vary parameters such as torsional stiffness at the ends, and to introduce additional forces (due for example to Br on the MPD-7). Both static and dynamic simulations were carried out. Initially the dodecamer was placed in the ideal Watson-Crick structure; small random initial displacements were applied and the system allowed to run through about 50-100 time units, with a damping constant of 10 inverse time units to simulate viscous or temperature bath effects. Figure 1(a) shows the dynamic history of the interbase twist "DEL-PHI" (similar to Dickerson's t_i) vs. base pair number [time units are the transit time for sound to travel from one base to its neighbor]. Figure 1(b) shows the asymptotic static limit. An attempt to model the Br on cytosine is illustrated. In this run the torsional stiffness at the ends was reduced to half that in the mid-region. The result closely resembles that found by Calladine's rules; this is not surprising because the reduced equations can simulate elastic behavior. These runs took only a few seconds. We are studying the effects of bromine and differences between A$-$A and G$-$G interactions, and eventually will explore the conformation of arbitrary sequences.

(b) 50 b.p. Model

In this series of runs the principal parameters in the Hamiltonian were the same as for the MPD-7 run, but no steric interactions were assumed. The pucker potential (or equivalently for) was a double well with barrier height $= 3K$ cal; one well represents the A-form, the other B, and in the run shown they are taken to have the same (free) energy. This is the condition for B$-$A transition. Figure 2(a) shows the time history of B$-$A domain boundary (i.e. soliton) imposed on the system at t=O (Antiperiodic boundary conditions). The main result of this run is to show that although various transient oscillations are set up the B$-$A interface is asymptotiently stable within our equations of motion *which include damping*. Figure 2(a) depicts the sugar pucker and Figure 2(b) shows how the twist varies with time: It is quite clear that there are two distinctly different motions: (a) a torsional-stretching oscillation of the entire chain (low frequency but dynamic): and (b) a long lived "kink" in which sugar pucker, twist and stretch maintain a very characteristic pattern,

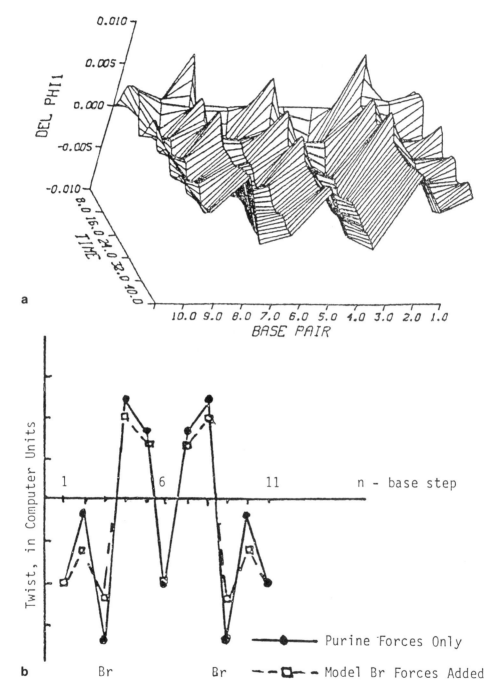

Figure 1. (a) Time evolution of a twist pattern driven by purine—purine forces from a uniform helical twist (DEL-PHI=O) at t=O toward a Calladine-like pattern at 40 time units. (b) The static limit of the variations in twist along the chain. In both cases the model represents MPD-7; first only purine—purine forces were included, then the effect of Br was simulated by adding additional forces.

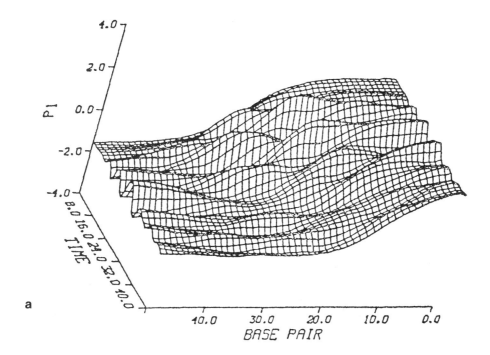

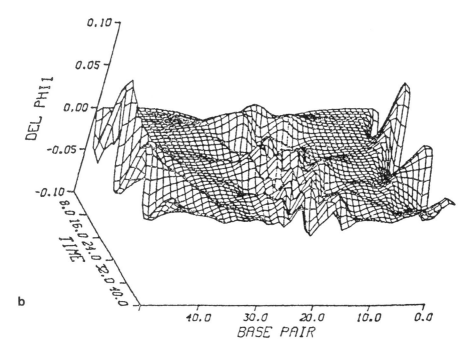

Figure 2. (a) Pucker; and (b) Twist variation along a uniform 50 b.p. model helix due to the introduction of a kink (i.e. B−A boundary) at t=O. Note the oscillatory transients which eventually damp out while the kink remains essentially unaffected.

which is *not removed by damping*. This independence of normal mode damped harmonic-like motions and solitary long-lived excitations is generic in nonlinear soliton bearing systems. There may also be an oscillatory "breathing" component within the kink region.

In either case, the runs took only a small amount of computer time, a few seconds for the dodecamer and no more than a minute for the longer helix, including graphic output.

Discussion

In the most general sense what these considerations mean is that significant features of DNA conformations and their relation to sequence are determined in a systematic way by physical considerations and may be found from suitable equations of motion. Every new structure need not be determined experimentally if this method is successful and basic parameters are determined for reference systems.

It should be clear also that this theoretically approximate method goes considerably beyond elastic models of polynucleotides, yet for the purposes for which it is designed it should be a cheap and quick tool for analyzing and planning experiments. We envisage that a library of effective parameters for the reduced equations might be built up empirically in a systematic joint experimental-analytical (formal theory or simulation) program involving X-ray, NMR, Raman, IR and other studies on prototypical oligomers, enzymes, or other biomolecules.

At the same time the formulation arrived at here has the features of a number of models developed in theoretical condensed matter physics over the past decade, suggesting that a useful interdisciplinary transfer of concepts may take place. For example, a model for the production of Z regions in B due to negative supercoiling or for kinks and bends falls naturally out of an analogy with strain-induced discommensurations in expitaxial layers. Another concept is that of a generalized order parameter which may have several components. It seems to us that many such parallelisms can be found and profitably exploited.

We note briefly some preliminary thoughts on the B−Z transition. The most significant structural feature of Z-DNA is that every other base has been rolled over from the normal anti to syn configurations; the basic unit is a dinucleoside. Earlier in the discussion we noted the possible need for an additional fourth coordinate for each backbone segment. Perhaps this should be the glycosyl angle, and physically the effective potential for that motion might be something like a double well (corresponding to anti or syn). Ordinarily, the hump is too difficult to surmount, but salt and strong (G−C) sequence forces might drive it. If this is the case we have a most intriguing dynamic and static problem: two anharmonic bistable systems, ribose and glycosyl, coupled to backbone twist and helix extension. It is not at all beyond the capability of the present method to model this situation; we plan to do so and see whether a B−Z junction can be found in this manner.

Finally, in view of our sponsorship by the National Foundation for Cancer Research we offer a few speculations about possible implications regarding cancer initiation at the molecular level. Fundamentally, what our work suggests is that in addition to the more conventional modes of atomic vibrations there may be both anharmonically localized static or dynamic conformations, and long range effects due to the motions of these solitary (soliton) conformational excitations. These might, for example, be initiated by a carcinogen at one point on the DNA helix, diffuse or be driven chemically or by coil stress to another region where they might substantially conspire with an oncogenic sequence to direct cancer cell production. Possibly more immediately accessible is the better understanding of the role of anti-tumor drugs. However, much before one pursues such ideas too far we must plod along developing experience on simpler, well defined molecular problems.

Acknowledgements

We wish to acknowledge a number of stimulating discussions in the course of developing these ideas with: C. R. Calladine; N. R. Kallenbach, and A. J. Heeger; A. Klug; A. C. Scott; H. M. Sobell.

Institutional support has been received from the British Science Research Council, the National Foundation for Cancer Research, and Los Alamos National Laboratory, and the hospitality of Brookhaven National Laboratory is to be noted. Work at Brookhaven supported by the Division of Materials Sciences U.S. Department of Energy under Contract No. DE-AC02-76CH00016.

References and Footnotes

1. G. P. Lomonossoff, P. J. G. Butler, A. Klug; *J. Mol. Biol. 149,* 745 (1981).
2. R. E. Dickerson and H. R. Drew; *J. Mol. Biol. 149,* 761 (1981). R. E. Dickerson, H. R. Drew, B. N. Cooper, R. M. Wing, A. V. Fratini and M. L. Kopka; *Science 216,* 475 (1982). R. E. Dickerson; *Scientific American,* p. 94, December (1983).
3. B. R. Brooks, R. E. Bruccoleri, B. D. Olafson, D. J. States, S. Swaminathanand and M. Karplus, *J. Comput. Chem. 4,* 187 (1983).
4. P. K. Wiener and P. A. Kollman, *J. Comput. Chem. 2,* 287 (1983).
5. C. R. Calladine; *J. Mol. Biol. 161,* 343 (1982).
6. A. V. Fratini, M. L. Kopka, H. R. Drew and R. E. Dickerson; *J. Biol. Chem. 257,* 14686 (1982).
7. ref. (d) of Abstract.
8. J. W. Keepers and T. L. James, *J. Am. Chem. Soc. 104,* 929 (1982).
9. ref. (c) of Abstract.
10. J. A. Krumhansl and D. M. Alexander; *"Structure and Dynamics: Nucleic Acid and Proteins,"* Eds. E. Clementi and R. H. Sarma, Adenine Press, NY (1983), p. 61.

Structure & Motion: Membranes, Nucleic Acids & Proteins,
Eds., E. Clementi, G. Corongiu, M. H. Sarma & R. H. Sarma.
ISBN 0-940030-12-8, Adenine Press,   Copyright Adenine Press, 1985.

Fluctuational Motility of DNA

Maxim D. Frank-Kamenetskii
Institute of Molecular Genetics
USSR Academy of Sciences
Moscow 123182 USSR

Abstract

We consider our present-day knowledge of the thermal fluctuations in linear DNA. Fluctuations of these types are considered: bending and torsional motions of the double helix and base-pair openings. The latest experimental and theoretical work has made it possible to estimate quite reliably the bending and torsional rigidities of DNA. As to the fluctuational opening, the current literature is very controversial. The approach based on study of DNA modification by formaldehyde is considered in some detail. Recent studies have shown that allowance for an "outside" reaction, that takes place without base-pair openings, in the our earlier theoretical model of DNA modification by formaldehyde eliminates the last descrepancies between theory and experiment. This testifies to the effect that the base-pair opening probability is correctly predicted by the standard helix-coil transition theory, and that the fraction of open base pairs is as low as 10^{-5}. Critical comments are made concerning the interpretation of the formaldehyde data by McGhee and von Hippel and the hydrogen exchange data by Mandel, Kallenbach and Englander. The highly overestimated value of the fraction of open base pairs claimed by these authors is most probably based on unjustified assumptions.

Introduction

It is an established fact today that, under normal conditions, the structure of DNA is for the most part close to the B form. However, to understand the functioning of DNA in the cell we have to know what states, other than this most probable ground state, are possible. In other words, the problem is formulated in the following way: how easily can a given DNA region occupy an "excited" state other than the regular B form?

It should be emphasized at once that the answer to this question critically depends on the topological state of DNA. Indeed, recent theoretical and experimental studies have demonstrated that the overall pattern of the DNA fluctuational motility changes dramatically when DNA is supercoiled. Namely, some states that are absolutely impossible in linear, nonsupercoiled DNA under normal conditions (such as cruciforms or the Z form) become dominant in regions with special sequences, such as inverted repeats (palindromes) or alternating purine-pyrimidine stretches.

The formation of such noncanonical structures in superhelical DNA is reviewed at length in another chapter by Frank-Kamenetskii and Vologodskii in the volume. Here I consider fluctuational deviations from the ground, B form-like, state in linear DNA.

Three main strategies are used to solve this important but extremely difficult problem. The first strategy is purely theoretical and includes such techniques as quantum chemistry, atom-atom potentials, Monte Carlo simulations, molecular dynamics, etc. The second strategy consists in the study of the transition of DNA or synthetic polynucleotides to a structure other than the B form through a considerable change of the ambient conditions (temperature, ionic strength, solvent, etc., very recently even the topological state of DNA), and the extrapolation of the energy parameters found in these studies to the normal ambient conditions (room temperature, water solution with an ionic strength of about 0.2M, linear DNA, etc.). Finally, the third strategy consists in direct probing of the deviations from the ground state that always take place due to thermal motion.

The three approaches are mutually complementary. Each of them has its own pros and cons and none of them can provide the final solution of the problem of fluctuational motility in DNA. The third approach, however, may bring us nearest to our objective.

In this chapter I discuss the recent advances in the study of thermal fluctuations in linear DNA.

As in any molecular system, one can divide the fluctuations in DNA into two categories, namely small fluctuations near the ground state and large perturbations that correspond to the transient occupation of different local minima. Let us start with the former category.

Bending and Torsional Fluctuations in the Double Helix

Probably the greatest progress has been achieved to date in the study of the bending and torsional fluctuations in the double helix. In these studies DNA has always been modelled as a homogeneous and isotropic elastic rod (1-6). To what extent is this simple model applicable to real DNA? This question is justified since, firstly, conformational analysis shows that the double helix should be very anisotropic in respect of bending in different directions (7) and, secondly, some people believe that the regular double helix is very often (after every hundred base pairs on an average, or even more often) interrupted by a serious structural perturbation (8,9).

The bending and torsional rigidities are usually obtained from and used for long DNA stretches, comprising many helix turns. Such long helices may be sufficiently well described in terms of the homogeneous model with averaged bending and torsional rigidities. The model is probably inadequate for some very special DNA sequences (10) as well as for very short helices.

The second limitation is much more critical, and if such frequent interruptions were really demonstrated it would deeply change the whole concept of the bending and torsional fluctuations in DNA. However, as I am going to argue below when I consider our present knowledge of the large fluctuational perturbations, the existence of such frequent interruptions is questionable, and a hundred thousand base pairs seems a much more reliable estimate of the average length of uninterrupted double helix. Then one should not worry about any interruptions when studying the bending and torsinal fluctuations, and rely on the homogeneous and isotropic elastic rod model, at least as a good zero approximation.

Within the framework of the model, the bending rigidity may be measured in terms of the persistence length value, a. This value determines a number of DNA characteristics, specifically overall dimensions of the polymeric coil of a high-molecular-weight DNA, which can be measured by different hydrodynamic methods. However, the coil dimensions also depend on the excluded volume effects, so that it is very difficult to establish the persistence length from hydrodynamic data.

The study of comparatively short molecules that behave almost as rigid rods is a much more reliable way. While measuring the dependence of some experimental characteristic on the chain length, one can see that this dependence gradually deviates from the theoretical curve predicted for a rigid rod. If a good theory is available predicting the chain length dependence of the measured quantity for the elastic rod model, a comparison of the experimental data with this theory should yield the value of the persistence length a. The advantage of such an approach lies in the fact that only one theoretical parameter (the persistence length a) has to be adjusted to fit theory to experiment.

In the past the main obstacle was the lack of monodisperse preparations of comparatively short (comprising hundreds of base pairs) DNA modules. Since the mid-seventies this is no longer a problem owing to the availability of restriction endonucleases. Kovacic and van Holde (11) were the first to use restriction fragments for determining the persistence length. They used sedimentation data and obtained a value of a about 60 nm for standard ionic conditions.

More recently, Hagerman undertook a comprehensive study of the persistence length problem by looking at the rotational diffusion constant of restriction fragments of different lengths (12). Comparing the experimental results with an accurate theory (13) he obtained the most reliable data to date on the ionic strength dependence of the DNA persistence length. Hagerman's data show that the persistence length strongly depends on ionic strength at very low salt concentrations, but gets saturated at about 50 nm under salt concentration higher than 1-10 mM of sodium. Thus, at physiological levels of the supporting electrolyte the persistence length of DNA is independent of the ionic strength, and is about 50 nm.

This conclusion agrees with the early results of Frisman and co-workers (see (14) and refs therein) obtained from studies of high molecular weight DNA. However, other people have arrived at an entirely different conclusion about both the value of a and its ionic strength dependence (15-18). These data have recently been shown to be the result of an incorrect allowance for various factors (for details see refs. 12, 14, 19, 20). After Manning published his theoretical paper (21) where he predicted a sharp ionic strength dependence of the persistence length of DNA even under physiological conditions, there has been a long controversy on the subject (9,19,22). However, recent theoretical studies have shown that the electrostatic contribution to the a value should be small at physiological ionic strengths (23,24), in accordance with Hagerman's data (12).

Much attention has recently been paid to the torsional rigidity of DNA. A number of different approaches have been applied.

The first approach was used by Vologodskii et al. (5). It was based on experimental data (25-27) concerning the variance of the equilibrium distribution of closed circular DNA molecules over the linking number, Lk, i.e. the value of $<(\Delta Lk)^2>$. This value was shown to consist of two terms (5):

$$<(\Delta Lk)^2> = <(\Delta Tw)^2> + <(Wr)^2> \qquad (1)$$

where Tw and Wr denote the twist and writhe of DNA, and $<\ldots>$ means averaging over the equilibrium distribution.

Vologodskii et al. (5) rigorously calculated the $<(Wr)^2>$ value with the aid of the Monte Carlo method. For the number of base pairs N in DNA larger than 3000 this value proved to be directly proportional to N. The same is true for the experimental value of $<(\Delta Lk)^2>$ and should be true for $<(\Delta Tw)^2>$. On the basis of these calculations and experimental data (25-27) the following equation coupling the persistence length a with the DNA torsional rigidity constant C may be derived:

$$4.7 = 3.6 \ 10^{-19} \ C^{-1} + 153 \ a^{-1} \qquad (2)$$

Here a is measured in nm, and C in erg·cm. The torsional rigidity constant C determines the energy change versus the change of the angle $\Delta\phi$ between adjacent base pairs:

$$E = (C/2h) \ (\Delta\phi)^2 \qquad (3)$$

where h is the distance between adjacent base pairs along the DNA axis (0.34nm), $\Delta\phi$ is measured in radians.

In an earlier study (5) we used the value $a = 57.5$ nm from the data of Kovacic and van Holde (11) and obtained $C = 1.7·10^{-19}$ erg·cm. The value of $a = 50$ nm obtained in the more recent and comprehensive study by Hagerman (12) seems more reliable. By substituting it into Eq. (2) we get $C = 2.2·10^{-19}$ erg·cm.

The second method consists in studying the variance of Lk for small DNA circles for which the variance of Wr should vanish. In this case one directly obtains the value of $<(\Delta Tw)^2>$ (see Eq. (1)). The method was applied in (28-30) to circles comprising as few as 200-300 bp. This yielded the value of $C = (2.4\text{-}2.9) \cdot 10^{-19}$ erg·cm (28-30), though there is some discrepancy between the results obtained by two groups.

The difference between the value obtained in (28-30) and our result for $a = 50$ nm is insignificant and lies within the experimental error of determination of the $<(\Delta Lk)^2>$ quantity. Indeed, the recent data of Horowitz and Wang (30) for the large DNA circle (4362 bp) indicate that the term on the left side of Eq. (2) is 4.3 rather than 4.7. This yields $C = 2.9\ 10^{-19}$ erg·cm. In general, the data of Horowitz and Wang (30) are completely explained in terms of Eq. (1), the data for $<(Wr)^2>$ from (5) supplemented by our recent calculations for small circles, and the values of $a = 50$ nm and $C = 2.9\ 10^{-19}$ erg·cm (to be published).

The third method for determining the DNA torsional rigidity is based on sophisticated measurements of fluorescence depolarization (31,32) or EPR spectra (33) of probes intercalated into the double helix. Here again, there is some controversy. Ironically, both techniques yielded the same value of $C = 1.4 \cdot 10^{-19}$ erg·cm, but according to (32) it is the high salt limit, whereas according to (33), it is the low salt limit. According to (33), C increases with increasing salt concentration, and at 0.2 M Na^+ it is $3.2 \cdot 10^{-19}$ erg·cm. Finally, very recently the value of $C = 2.0 \cdot 10^{-19}$ erg·cm was obtained by the fluorescence depolarization method (34).

The reasons for some of the discrepancies between the results obtained by different methods remain unclear. These discrepancies are not large, and in any case the amplitude of fluctuations in the winding angle between adjacent base is about 4-5°, as was first suggested by Vologodskii et al. (5).

What is known about the relaxation times of the bending and torsional fluctuations? According to the fluorescence, EPR and NMR data, the most rapid fluctuations correspond to the torsional motions and take place on a time scale of 10 ns (32-35). Significant bending motions include translations of large DNA segments, about the persistence length, and as a result they occur on a microseconds scale (4,36).

Fluctuational Openings of Base Pairs

We now turn to fluctuations that correspond to large perturbations of the DNA ground state. Their study presents a challenging problem for both theorists and experimentalists. Theoretical difficulties stem from the obvious fact that the method of expansion over the powers of displacements from the equilibrium position fails in this case. Experimental difficulties stem from the low probability of occupation of these states.

It is because of these difficulties that information about the large fluctuations of the double helix remains controversial in spite of the fact that their existence has been postulated for a long time.

Nevertheless, I believe that we now have enough theoretical and experimental evidence to the effect that the most important of these large perturbations in linear DNA are the fluctuational openings of base pairs, and under normal conditions only single base pairs are open with a probability as low as 10^{-5}.

The principal experimental approach to the problem of base pairs openings consists in studying the accessibility for chemical modification or isotopic exchange of the groups that are buried inside the Watson-Crick double helix, specifically the thymine and guanine imino (N1−H) groups and adenine, cytosine and guanine amino (NH$_2$) groups. First of all, the mechanism and the rate constants on the monomeric level are studied. Then, the DNA or synthetic polynucleotide situation is investigated and is compared with the monomeric situation. This comparison leads to some definite conclusion about the characteristics of the DNA fluctuational motility.

The principal assumption underlying this approach is that the modification is absolutely forbidden in the ground state of the double helix (i.e. in the B form) and proceeds only through fluctuationally occupied "open" states. In many cases this assumption is accepted as "a symbol of faith" and other possibilities are not even discussed. I believe that most controversies accumulated in the field stem from the fact that modification (or exchange) routes other than base-pair opening have been underestimated. Strong evidence of the existence of such an alternative route has recently been obtained in the case of the modification of DNA amino groups by formaldehyde (see Refs 20, 37, and below).

In general, the most comprehensive studies of fluctuational openings in DNA have been based on the formaldehyde method. We begin the discussion of these data by summarizing the results on the reaction between formaldehyde and DNA monomers.

The mechanism and rate constants of the reaction between formaldehyde and nucleotides were most extensively studied by McGhee and von Hippel (38, 39). Formaldehyde reversibly reacts with the imino and amino groups of bases forming hydroxymethyl derivatives.

The reaction with the thymine imino group:

$$N-H + CH_2O = \quad N-CH_2OH$$

is very rapid, highly reversible and strongly pH-dependent (it experiences a ten-fold acceleration when pH is increased by one unit). The latter fact clearly indicates that the formaldehyde molecule attacks the deprotonated imino group. The hydroxyl ions catalyze the forward as well as the reverse reactions of formaldehyde with the thymine imino group without shifting the equilibrium. Thus, in spite of the very sharp pH-dependence of the forward and the reverse rate constants, their ratio, i.e. the equilibrium constant, is strictly pH-dependent. The available data indicate that formaldehyde most probably does not react with the guanine imino group.

The reaction with the amino groups:

$$-N\begin{smallmatrix}H\\\\H\end{smallmatrix} + CH_2O + -N\begin{smallmatrix}CH_2OH\\\\H\end{smallmatrix}$$

proceeds through the tetrahedral intermediate state of the nitrogen atom that entails the formaldehyde attack on the nitrogen lone pair. For the amino groups, especially the cytosine one, the equilibrium is considerably shifted towards hydroxymethylation, though the reaction remains reversible. Both the forward and the reverse rate constants are pH-independent.

When native DNA is dissolved in the formaldehyde solution, it experiences gradual denaturation that is easily detected by measuring the hyperchromicity within the DNA absorption band. Fig. 1 shows the typical experimental kinetic curve of DNA unwinding by formaldehyde measuring at two wavelengths. Note that the term "unwinding" is used throughout the present chapter to name briefly "the formaldehyde-induced helix-coil transition", rather than just a change in the DNA winding angle.

The general explanation of the observed kinetic curve is as follows (40-41). Even well below the DNA melting temperatures, the double helix experiences, due to the thermal motion, fluctuational openings ("breathing") of the base pairs. In the open state, the imino and amino groups of bases become accessible for modification by formaldehyde present in the solution. This modification should resemble the reaction with free nucleotides. However, at any given moment only a tiny fraction of the base pairs is accessible to the reaction. As a result, the overall process of DNA modification is rather slow. Thus, the rate of DNA modification by formaldehyde depends on the fraction of open base pairs, in other words, on the base-pair opening probability.

The process of the base-pair modification is accompanied by the unwinding of the double helix. Indeed, the substitution of the bulky hydroxymethyl group for the thymine imino hydrogen should completely preclude the formation of the Watson-Crick AT pair. Although the amino group modification does not forbid the reformation of Watson-Crick pairs, such pairs are highly weakened (42).

Are fluctuational openings obligatory for the modification of the DNA double helix by formaldehyde? This is the central question, since if the answer were negative for both imino and amino reactions, this would mean that formaldehyde is an inadequate probe for studying the base-pair opening fluctuations. Fortunately, the possibility of modification of imino groups deeply buried inside the double helix without

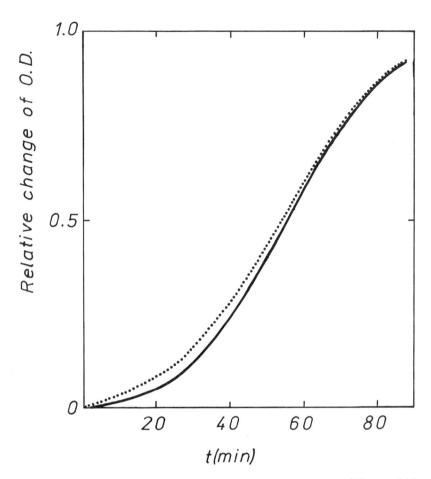

Figure 1. Experimental kinetic curves of the relative change in optical density at 260 nm (solid line) and 285 nm (dotted line) for T7 DNA. The formaldehyde concentration was 5.7M, $[Na^+] = 4\cdot10^{-2}$ M, pH5.5, T = 43°C. The half-unwinding time was 55 min. The data were obtained by V. V. Demidov and L. A. Serdyukova.

base pairs opening can be ruled out. This holds even for the case of hydrogen exchange, and is absolutely obvious for formaldehyde. However, in the case of the amino groups, one can easily imagine the modification without a large perturbation of the DNA structure. Such an "outside" modification of the double helix by formaldehyde has been demonstrated (20,37).

Because of the two types of chemical reactions with quite different quantitative characteristics, the overall process of DNA modification by formaldehyde is rather complicated. So much so that mere experiments could not provide a genuine insight into the process. The problem has been solved after the experimental studies were complemented by theoretical simulation.

Theoretical studies of the DNA modification by formaldehyde proceed from the fact that the rate constants of the imino and amino group modification are much lower than the rate constant of the base-pair opening-closing process. Hence, between two subsequent acts of modification the system is equilibrated in respect of the DNA conformational states. These equilibrium states are calculated on the basis of the standard DNA helix-coil transition theory.

In the most general scheme of calculations proposed by Lukashin et al. (43) the process of DNA modification by formaldehyde is simulated by Monte Carlo method. At any given stage of the simulation the DNA state is calculated from the helix-coil transition theory. Then the site and the moment of modification (or reverse event) is chosen on the basis of the known values of the kinetic constants of reaction with monomers. In an early study (43), we ignored the possibility of the outside reaction of formaldehyde with amino groups or the reformation of base pairs after the amino group modification. These new elements are readily incorporated into the Monte Carlo scheme.

Table I

Characteristics of the kinetic curves of DNA unwinding by formaldehyde calculated for six different models of the process at the following conditions: pH 5.5, T=43°C, formaldehyde concentration 5.7M [Na^+] = 4.10^{-2}M, DNA GC content x_o = 0.48 (corresponding to T7 DNA)

	1	2	3	4	5	6
"outside reaction"	—	—	—	—	+	+
base pairs re-formation after hydroxymethylation of amino groups	—	—	+	+	+	+
reaction with thymine imino group	—	+	—	+	—	+
halfdenaturation time τ (min)	370	140	4500	250	230	90
initial rate of unwinding $v = -(d\vartheta/dt)_{t=o}(min^{-1})$	10^{-4}	10^{-3}	$5 \cdot 10^{-8}$	$5 \cdot 10^{-4}$	$5 \cdot 10^{-6}$	$5 \cdot 10^{-4}$

Table 1 summarizes the results of calculations for the two main characteristics of the kinetic curve of DNA unwinding by formaldehyde (the half-unwinding time and the initial rate of unwinding) when different elements of the process are allowed for. Fig. 2 shows the original kinetic curves for two variants (corresponding to columns 4 and 6).

The principal conclusion is that only our final model of DNA modification by formaldehyde which allows for all the major elements of the process, agrees with the sum total of the experimental data. These major elements are as follow.

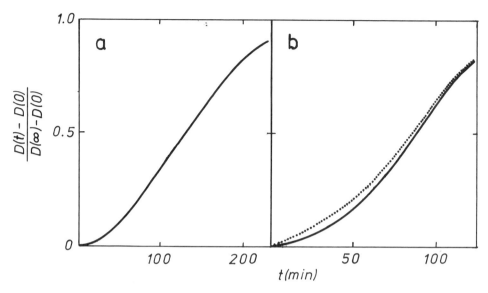

Figure 2. Theoretical kinetic curves of DNA unwinding by formaldehyde, simulating the results of experiments at two wavelengths, 260 nm (solid line) and 285 nm (dotted line). The theoretical parameters corresponded to the conditions listed in the legend to Fig. 1. The theoretical model in (a) corresponds to column 4 (both lines coincided) and in (b) to column 6. The calculations were performed by A. V. Vologodskii and A. V. Lukashin.

i). The base-pair opening probability is calculated on the basis of the standard helix-coil transition theory.

ii). The imino group modification proceeds only through complete base pairs openings.

iii). The modified thymines cannot participate in base pairing.

iv). The amino group modification proceeds through the outside reaction, as well as through the base-pair openings.

v). The nucleotides modified at their amino groups are incorporated into the double helix forming weakened pairs.

It should be emphasized that all parameters entering this final model were determined from independent data. Thus, our theory includes no adjustable parameters.

The model considered by Lukashin et al. (43) corresponds to column 2 of Table 1. One can see that it yields quite good values of τ and ν, that are close to the values for our final model (column 6). However, as was emphasized in (43), the results based on the model of Lukashin et al. differed from the experimental results in two points. First, the experimental kinetic curves has a much more distinct lag phase than the theoretical ones. Secondly, when the pH value was changed from 5.5 to 6.5 the theoretical value of τ increased by a factor of 2.7, whereas the experimental value increased by only a factor of 1.5.

Table 1 shows that when the model of Lukashin et al. is supplemented by allowance for the reformation of the base pairs modified at their amino groups, this gives rise to a considerable discrepancy in respect to the τ value (see column 4). But when we make the next step and allow for outside reaction (column 6), all discrepancies disappear. First, the τ value decreases and comes closer to the experimental value of 55 min (see Fig.1). Secondly, the kinetic curve shows a distinct lag phase, just as on the experimental curve (cf. Figs.1 and 2). Thirdly, when the pH value is changed from 5.5 to 6.5, the τ value increases by a factor of 1.9, which is close to the experimental factor of 1.5. Finally, the theory simulates quite well the difference in the kinetic curves measured at two different wavelengths (cf. Figs 1 and 2). This difference is simulated only when the outside reaction is allowed for.

Thus, allowance for the outside reaction, i.e. the ability of formaldehyde to modify the double helix bypassing the base pair opening route, leads to virtually complete agreement between theory and experiment. Note once again that the theory does not include any adjustable parameters.

The fact that we have succeeded in explaining the DNA modification by formalde-hyde on the basis of the helix-coil transition theory testifies to the effect that this theory provides an adequate picture of the DNA base-pair opening process. The principal feature of this picture is the prediction that, at temperature far below the DNA melting temperature, only single base pairs are open and their opening proba-bility is as low as about 10^{-5} (see refs 6, 20, 40, 41, 43). However, since the final model of DNA modification by formaldehyde proved to be rather complicated, one can wonder to what extend these features are critical for the observed agreement between theory and experiment.

We have specially examined this question and arrived at the conclusion that the base-pair opening probability cannot be markedly increased without ruining the agreement. By contrast, its decrease changes our results only slightly. As is seen from column 5 of Table 1, even a complete block of the modification route passing through base-pair opening does not dramatically affect the calculated kinetic curves.

Thus, our data shows the upper limit of the opening probability to be not larger than 10^{-5}. On the other hand, Chay (44) has performed a comprehensive analysis of the experimental data of Utiyama and Doty (45) corresponding to the alcaline pH region. In contrast with the acidic and neutral regions, in the alkaline region the process of DNA unwinding by formaldehyde completely depends on the reaction with imino groups, so that the outside reaction and the reformation of modified base pairs may be neglected. Chay also arrived at the value of 10^{-5} for the opening probability, and any deviation from this estimate should affect the agreement be-tween theory and experiment.

Concluding our review of the formaldehyde method, we have to mention the paper by McGhee and von Hippel (42) who arrived at entirely different results starting from virtually the same experimental data. Their overestimation (by a factor of 10^{3})

of the opening probability directly stemmed from their unjustified neglect of the reaction with the thymine imino groups. Since the authors did not allow for the outside reaction but allowed for the reformation of base pairs with modified amino groups, their model corresponds to column 3 of Table 1. Comparing this column with column 4, one can see how large the error is. Since McGhee and von Hippel (42) were interested in the initial slope of the kinetic curve, they could be wrong by a factor of ten thousand!

McGhee and von Hippel (42) substantiated their assumption by the fact that in the acidic pH region the kinetics of DNA unwinding weakly depends on pH, whereas the thymine reaction is strongly pH-dependent. However, as we have seen above, our final model successfully simulates this weak pH-dependence, though the thymine reaction plays an important part in the model. As I explained at length elsewhere (6), this apparent pH-independence stems from the fact that, since the thymine reaction is very fast, it reaches equilibrium before any appreciable unwinding takes place. Thus, since under these conditions reaction with thymines alone cannot unwind the double helix, the unwinding process is governed by the equilibrium modification of thymines which is strictly pH-independent, and the overall process virtually does not depend on pH (46).

In summary, our knowledge of the mechanism of DNA modification by formaldehyde has been much expanded in the recent years, so that we are now in a position to explain quantitatively all principal features of the process over a wide range of ambient conditions. These advances have strongly supported our early predictions of the DNA "breathing" based on the helix-coil transition theory (6,20,40,41,43). According to this theory, only single base pairs are open, provided that the temperature is well below the DNA helix-coil transition range. The opening probability for a base pair inside the helix is as low as 10^{-5}, and is about 10^{-1} for a base pair at the edge of the helix. The correlation length is rather small (43). It should be clear from the above discussion that the alternative picture of DNA "breathing" based on the parameters of Gralla and Crothers (47), predicting a much greater opening probability (42,48), is incompatible with the formaldehyde data (49).

The formaldehyde data *per se* give no information about the time scale of the fluctuational base-pair openings. However, such information may be obtained if one combines this data with the data on the characteristic time of the base pair closing reaction. The latter value was reliably estimated at 10^{-6} s in studies of the relaxational kinetics of the helix-coil transition in oligonucleotides (50-52). As it was shown recently (53), a theory of slow relaxational processes in DNA melting that uses this value as the principal parameter, quantitatively explains a number of experimental data on DNA melting under ordinary, quasiequilibrium conditions. Some specific theoretical predictions have recently been experimentally confirmed (54).

Thus, the "lifetime" of a base pair in the open state is about 10^{-6} s. This means that each base pair inside the double helix transiently opens about 10 times in a second. At the helix edge this occurs about ten thousand times more frequently.

Some of these early theoretical predictions (40,41,43) have more recently been confirmed by Early et al. (35,55) with the aid of high-resolution NMR. By studying very short DNA fragments (12-69 bp), they have shown that the opening process for a given base pair does not depend on its neighbors, i.e. only single base pairs can open. Moreover, the rate constant of the opening process proved to be about 10 s^{-1}, in full agreement with the above estimate. Thus, the NMR data provide a completely independent confirmation of our picture of base pair "breathing".

Traditionally, hydrogen exchange has been believed to provide the most adequate data on the DNA "breathing" (56). The obvious advantage of this technique over the formaldehyde method is the simplicity of theoretical analysis stemming from the virtual lack of effect of the isotopic exchange on base-pair stability. Thus, one can obtain the base-pair opening probability simply by dividing the rate constant of exchange for DNA by the rate constants for free nucleotides. However, as with formaldehyde, this is correct only if the chemical route of exchange is the same in both cases. Again as with formaldehyde this is definitely valid only in the case of imino groups. However, the exchange of imino groups is believed to be so fast that it cannot be used for probing the probability of base-pair openings (57,58).

As a result, people use data on the rate of hydrogen exchange of the amino protons to evaluate the base-pair opening probability (57,58). One can anticipate, bearing in mind the formaldehyde story, that in this case the reaction route may be different for monomers and DNA. This is what I am going to argue for.

First of all, let us recollect the route of exchange of the amino protons in free bases. This mechanism, as revealed by Englander and co-workers (57,58), consists in pre-protonation of the bases (adenine and cytosine) at their N1 sites, followed by the removal of the amino proton by the hydroxyl ion. As a result, the overall process proves to be pH-independent because the rate of exchange is proportional to the H^+ as well as the OH^- concentration. Since the exchange of amino protons in polynucleotides (poly U·poly A and poly I·poly C) is also shown to be pH-independent, the authors (57,58) concluded that the exchange in polymers follows the same route as in monomers, namely it pre-requires the N1 protonation. As a result they postulated that the amino exchange requires a base-pair opening event, the same as the imino exchange. This led to a very high opening probability (about $5 \cdot 10^{-2}$) and an unusually large "lifetime" of the open state (about $2 \cdot 10^{-2}$ s). The possible nature of this "open" state remains obscure (58).

This state has clearly nothing in common with the open states predicted by the helix-coil transition theory and revealed by the formaldehyde method. It is difficult to accept the idea that the highly probable and abnormally long-lived "open" state should manifest itself only in the hydrogen exchange experiments. I cannot help feeling that an alternative explanation of the hydrogen exchange data should exist.

Specifically, one can speculate that when the N1 sites are buried inside the double helix, the pre-protonation of other sites, such as N7 of adenine, may affect the amino group in a way similar to the N1 protonation of free bases. Indeed, adenine was shown to be protonated at N7 with a pK of about 0 (59). This takes place even for free bases after the protonation at N1 (pK4). It is clear that the pK value for the N7 protonation should be higher when the N1 protonation is blocked. Nobody knows the magnitude of the effect, but it seems likely that the observed slowing down of the hydrogen exchange for the adenine amino group in poly A·poly U as compared with free adenine may reflect a lower efficiency of the N7 route rather than some mysterious "open" state.

Similarly, in the case of poly I·poly C one can speculate that the N7 site of inosine (guanine without the amino group) is protonated first. This protonation (pK3) shifts the tautameric equilibrium inside the I·C base pair, and the N1 proton of inosine attaches to the N1 site of cytosine. This would ensure the N1 pre-protonation without any structural perturbation of the double helix.

Conclusions

We have considered at length our present-day knowledge of the thermal fluctuations in linear DNA. The picture that emerges from the critical analysis of all available data is as follows. The DNA double helix experiences smooth bending and torsional deformations due to thermal motion, which have been well characterized in the past years. It is because of these movements that the double helix forms a polymer coil in solution, may form a supercoil under topological stress, may be wrapped around the histone core in nucleosomes, etc. The double helix seems to be very resistant to any significant disturbance of its ground state (i.e. the B form). Large perturbations only occur once in every hundred thousand base pairs on an average. They are base-pair openings, which have also been well characterized. There is no conclusive evidence of any other perturbations of the linear double helix under normal conditions.

Acknowledgements

I am grateful to my colleagues who have participated in studies summarized here, and to many people in Moscow and elsewhere who have been discussing with me various aspects of the problem. I thank Professor James Wang for sending me his important paper on the DNA torsional rigidity prior to its publication. I thank Professor Enrico Clementi who suggested that I should write this chapter. Finally, I thank Mr. D. Agrachev for editing my English.

Note added in proof: In their recent papers Schurr with co-workers (Wilcoxon, J. and Schurr, J.M., *Biopolymers, 22,* 2273 (1983), Thomas, J.C. and Schurr, J.M., *Biochemistry, 22,* 6194 (1983), Schurr, J.M., *Biopolymers, 23,* 191 (1984)) have arrived at very similar conclusion about the base-pair opening probability basing on quite different data.

References and Footnotes

1. Schellman, J. A., *Biopolymers 13*, 217 (1974).
2. Le Bret, M., *Biopolymers 17*, 1939 (1978).
3. Benham, C. J., *Biopolymers 18*, 609 (1979).
4. Barkley, M. D. and Zimm, B. H., *J. Chem. Phys. 70*, 2991 (1979).
5. Vologodskii, A. V., Anshelevich, V. V., Lukashin, A. V. and Frank-Kamenetskii, M. D., *Nature 280*, 294 (1979).
6. Frank-Kamenetskii, M. D., *Comm. Mol. Cell. Biophys. 1*, 105 (1981).
7. Zhurkin, V. B., Lysov, Y. P. and Ivanov, V. I., *Nucleic Acids Res. 6*, 1081 (1979).
8. Manning, G. S., in *Structure and Dynamics: Nucleic Acids and Proteins*, Eds. E. Clementi and R. H. Sarma, Adenine Press, New York, *p. 289* (1983).
9. Manning, G. S., *Biopolymers, 22*, 689 (1983).
10. Levene, S. D. and Crothers, D. M., *J. Biomol. Struct. Dynam., 1*, 429 (1983).
11. Kovacic, R. T. and van Holde, K. E., *Biochemistry, 16*, 1490 (1977).
12. Hagerman, P. J., *Biopolymers, 20*, 1503 (1981).
13. Hagerman, P. J. and Zimm, B. H., *Biopolymers, 20*, 1481 (1981).
14. Slonitskii, S. V. and Frisman, E. V., *Molek. Biol., 14*, 496 (1980).
15. Harrington, R. E., *Biopolymers, 17*, 919 (1978).
16. Frontali, C., et al., *Biopolymers, 18*, 1353 (1979).
17. Borochov, N., Eisenberg, H. and Kam, Z., *Biopolymers, 20*, 231 (1981).
18. Kam, Z., Borochov, N. and Eisenberg, H., *Biopolymers, 20*, 2671 (1981).
19. Odijk, T., *Biopolymers, 18*, 3111 (1979).
20. Frank-Kamenetskii, M. D., *Molek. Biol., 17*, 639 (1983).
21. Manning, G. S., *Q. Rev. Biophys., 11*, 179 (1978).
22. Schurr, J. M. and Allison, S. A., *Biopolymers, 20*, 251 (1981).
23. LeBret, M., *J. Chem. Phys., 76*, 6243 (1982).
24. Fixman, M., *J. Chem. Phys., 76*, 6346 (1982).
25. Depew, R. E. and Wang, J. C., *Proc. Natnl. Acad. Sci. U.S.A., 72*, 4275 (1975).
26. Pulleyblank, D. E., Shure, M., Tang, D., Vinograd, J. and Vosberg, H. P., *Proc. Natnl. Acad. Sci., U.S.A., 72*, 4280 (1975).
27. Shure, M., Pulleyblank, D. E. and Vinograd, J., *Nucl. Acids Res., 4*, 1183 (1977).
28. Shore, D. and Baldwin, R. L., *J. Mol. Biol., 170*, 957 (1983).
29. Shore, D. and Baldwin, R. L., *J. Mol. Biol., 170*, 983 (1983).
30. Horowitz, D. S. and Wang, J. C., *J. Mol. Biol., 173*, 75 (1984).
31. Thomas, J. C., Allison, S. A., Appellof, C. J. and Schurr, J. M., *Biophys. Chem., 12*, 177 (1980).
32. Millar, D. P., Robbins, R. J. and Zewail, A. M., *J. Chem. Phys., 76*, 2080 (1982).
33. Hurley, I, Osei-Gymiah, P., Archer, S., Scholes, C. P. and Lerman, L. S., *Biochemistry, 21*, 4999 (1982).
34. Ashikawa, I., Kinosita, K., Ikegami, A., Nishimura, Y., Tsuboi, M., Watanabe, K., Iso, K. and Nakano, T., *Biochemistry, 22*, 6018 (1983).
35. Early, T. A., Kearns, D. R., Hillen, W. and Wells, R. D., *Biochemistry, 20*, 3756 (1981).
36. Opella, S. J., Wise, W. B. and DiVerdi, J. A., *Biochemistry, 20*, 284 (1981).
37. Demidov, V. V., *Dokl. Akad. Nauk SSSR, 251*, 1268 (1980).
38. McGhee, J. D. and von Hippel, P. H., *Biochemistry, 14*, 1281 (1975).
39. McGhee, J. D. and von Hippel, P. H., *Biochemistry, 14*, 1297 (1975).
40. Lazurkin, Y. S., Frank-Kamenetskii, M. D. and Trifonov, E. N., *Biopolymers, 9*, 1253 (1970).
41. Frank-Kamenetskii, M. D. and Lazurkin, Y. S., *Ann. Rev. Biophys. Bioeng. 3*, 127 (1974).
42. McGhee, J. D. and von Hippel, P. H., *Biochemistry 16*, 3276 (1977).
43. Lukashin, A. V., Vologodskii, A. V., Frank-Kamenetskii, M. D. and Lyubchenko, Y. L., *J. Mol. Biol., 108*, 665 (1976)
44. Chay, T. R., *Biopolymers, 18*, 1439 (1979).
45. Utiyama, H. and Doty, P., *Biochemistry, 10*, 1254 (1971).

46. My publication (6) containing this criticism of the McGhee and von Hippel results (42) was supplemented by an intriguing *Editor's Note:* "Dr. P. H. von Hippel has indicated that he and J. D. McGhee will present a short comment from a rather different point of view in a subsequent issue". This "different point of view" has failed to appear in the past three years.

47. Gralla, J. and Crothers, D. M., *J. Mol. Biol., 78,* 301 (1973).

48. Wartell, R. M. and Benight, A. S., *Biopolymers, 21,* 2069 (1982).

49. It should be recalled in this connection that Gralla and Crothers (47) studied RNA rather than DNA (see also discussion in (43), *p. 677*).

50. Craig, M. E., Crothers, D. M. and Doty, P., *J. Mol. Biol., 62,* 383 (1971).

51. Porschke, D. and Eigen, M., *J. Mol. Biol., 62,* 361 (1971).

52. Porschke, D., Uhlenbeck, O.C. and Martin, F. H., *Biopolymers, 12,* 1313 (1973).

53. Anshelevich, V. V., Vologodskii, A. V., Lukashin, A.V. and Frank-Kamenetskii, M. D., *Biopolymers, 23,* 39 (1984).

54. Kozyavkin, S. A. and Lyubchenko, Y. L., *Nucleic Acids Res., 12,* 4339 (1984).

55. Early, T. A., Kearns, D. R., Hillen, W. and Wells, R. D., *Biochemistry, 20,* 3764 (1981).

56. Very often one can encounter in literature a statement of this kind: "According to hydrogen exchange studies of DNA "breathing", about 5% of base pairs are broken at any given moment".

57. Mandal, C., Kallenbach, N. R. and Englander, S.W., *J. Mol. Biol. 135,* 391 (1979).

58. Preiser, R. S., Mandal, C., Englander, S. W., Kallenbach, N. R., Howard, F. B., Frazier, J. and Miles, H. T., in *Biomolecular Stereodynamics 1,* Ed., R. H. Sarma, Adenine Press, New York, p. 405, (1981).

59. Budo, G. and Tomasz, J., *Acta Biochim. Biophys. Acad. Sci. Hung., 9,* 217 (1974).

Structure & Motion: Membranes, Nucleic Acids & Proteins,
Eds., E. Clementi, G. Corongiu, M. H. Sarma & R. H. Sarma,
ISBN 0-940030-12-8, Adenine Press,   Copyright Adenine Press, 1985.

Relation of Macromolecular Structure and Dynamics of DNA to the Mechanisms of Fidelity and Errors of Nucleic Acid Biosynthesis

Poltev, V. I. and Chuprina, V. P.
Institute of Biological Physics
USSR Academy of Sciences
Pushchino, Moscow Region, USSR

Abstract

The relation between properties of DNA double helix as a conformationally flexible structure and its properties as a carrier of genetic information have been considered. The following problems have been discussed: (1) the possible pathways and intermediate mispairs for errors in NA biosynthesis; (2) the role of non-bonded interaction energy calculations in the study of different ways of mispairing; (3) procedure for calculation of the energy of non-bonded interactions in NA fragments; (4) possible conformational changes of the double helix within A- and B-families of NA; (5) conformations of NA fragments containing nucleotide mispairs. Calculations of the energy of non-bonded interactions via atom-atom potential functions contribute much to the understanding of the structure and dynamics of both the regular double helix and fragments containing non Watson-Crick base oppositions. By means of theoretical conformational analysis we have demonstrated that all types of base pairs in usual tautomeric forms could be incorporated into the double helix with no short interatom contact arising and without drastic changes of the sugar-phosphate backbone. All experimental data on spontaneous errors in NA biosynthesis may be explained by formation of such pairs.

Introduction

During the recent decade the idea of DNA as a uniform and static structure is being replaced by a concept of its heterogeneity and flexibility. The properties of DNA as a dynamic structure whose parts can undergo considerable mutual shifts are involved in copying and transcription of the genetic information. These properties should be taken into account when considering the molecular mechanisms ensuring accuracy of the processes of NA biosynthesis. In the present paper we examine the relation between the properties of DNA double helix as a conformationally flexible structure and its properties as a carrier of genetic information.

Almost exclusively complementary (i.e. those forming A:U(T) or G:C pairs) nucleotides are incorporated into the newly synthesized chain in NA biosynthesis. The formation of any other pair is an error leading to a nucleotide replacement:

transistion, if a purine-pyrimidine pair is formed, or transversion, if a purine-purine or pyrimidine-pyrimidine pair is formed. Errors in DNA biosynthesis *in vivo* occur with a probability of 10^{-8}-10^{-11} per base pair per replication (1). Complex enzyme systems ensure extremely high accuracy *in vitro,* also (2). To understand the processes of replication, repair and transcription and to be able to influence these processes (e.g. for chemeotherapy) it is important to elucidate what mechanisms ensure high accuracy of NA biosynthesis, to what extent accuracy is ensured by NA components, what is the role of synthesis enzymes and what are the pathways of infidelity.

Understanding of the mechanism of the processes requires construction of detailed atomic-molecular models of the structures involved and of changes in these structures. Models for such complex systems can be constructed using computer calculations and the experimental data obtained with different methods. NA models, from Watson-Crick double helix (3) to the latest models of polynucleotides in fibres (4) or double helical fragments in crystals (5) are based not only on experimental data but also on theoretical generalizations, energy and structure considerations. These considerations can be taken into account qualitatively, e.g. distances between non-bonded atoms in models must not be smaller than the sum of van-der-Waals radii. A more precise modelling involves calculation and minimization of the structure energy.

Advances in NA synthesis and in experimental technique permit X-ray studies of monocrystals of double-helical NA fragments containing some nucleotide pairs, but the accuracy of X-ray data is insufficient for an independent determination of atom coordinates. Determination of atom coordinates of the sugar-phosphate backbone of a dodecanucleotide became possible in studies of Dickerson et al. (5) due to calculations of the energy of non-bonded interactions. The role of calculations in analysis of supramolecular complexes is no less essential. Modern computers make it possible to calculate the energy of systems containing hundreds of atoms and to search for minima of this energy as a function of many variables. However such calculations are possible only within the limits of classical models, i.e. considering the molecule (or a supramolecular complex) as a system of points interacting according to the laws of classical mechanics. Reliability of such an approach and its correspondence to the real situation requires additional studies. The results of calculations carried out by the method of classical potential functions should be compared with the experimental data and quantum-mechanical calculations. Considerable success has been achieved lately in quantum-mechanical study of nucleic acid fragments. For NA components sufficiently reliable *ab initio* calculations can be performed which help in choosing classic potential functions.

The following problems will be considered in this paper:

— the possible pathways and intermediate mispairs for errors in NA biosynthesis;
— the role of energy calculations of non-bonded interactions in the study of different ways of mispairing;

— procedure for calculation of the energy of non-bonded interactions in NA fragments;

— possible conformational changes of the double helix within A-and B-families of NA;

— conformations of NA fragments containing nucleotide mispairs.

The obtained results suggest that the most probable pathway of point errors in NA biosynthesis is the formation of nucleotide mispairs with bases in normal tautomeric forms. This assumption is in agreement with all the available experimental data. Such nucleotide mispairs can be presumed for each spontaneous nucleotide replacement and for many replacements induced by base analogs.

Properties of Nucleic Acid Components and Pathways of Biosynthesis Errors

Functioning of living systems is possible only within certain limits of error frequency in nucleic acid biosynthesis: a too-high level of errors is incompatible with

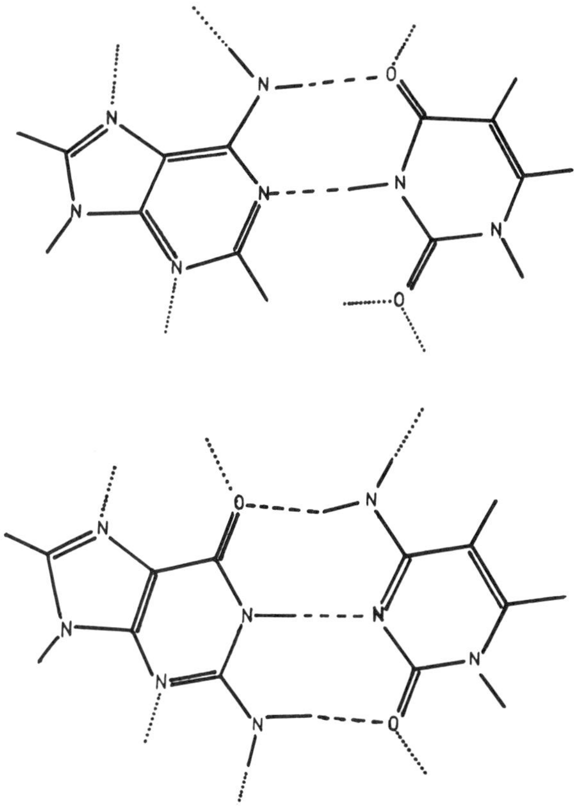

Figure 1. Watson-Crick A:U and G:C pairs. Possible sites of hydrogen bond formation are shown. Transition of labile hydrogens between these sites is possible at tautomerization.

the stability of genetic information in many generations while a too-low level does not promote evolution.

The possibility of errors in biosynthesis is intrinsic to the heredity substance, to the chemical structure of nucleic acids consisting of rigid heterocyclic bases and a flexible sugar-phosphate backbone admitting of conformational rearrangements.

Each nitrogen base has several proton donor and proton acceptor centres (Fig. 1). This allows first, the formation of quite different practically planar hydrogen bonded pairs (as well as triplets and quadruplets) of bases, second, the transition of bases into the ionized state by protonation or proton dissociation and third, the existence of bases as different tautomers, i.e. with a different position of hydrogens at proton acceptor atoms. Therefore three types of mispairs could be represented. First, pairs in which one of the bases is in the ionized state; second, pairs with bases in rare tautomeric forms and third, pairs of neutral bases in normal tautomeric forms.

The possible types of mispairs are exemplified by the G:U pairs in Fig. 2. The pairs with a base in a negatively ionized state (Fig. 2 a,b) have the same mutual position

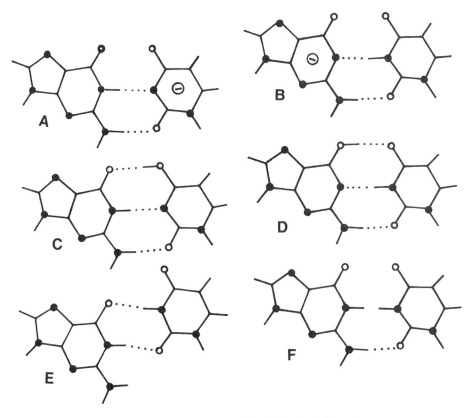

Figure 2. Hypothetical ways of the G:U mispair formation.

of glycosyl bonds as pairs A:T and G:C. Dissociation of a proton from G or T in aqueous solutions at physiological pH results in the appearance of negatively ionized forms constituting about 10^{-2} of the total amount. Such a level of errors is incompatible with life and the available experimental data show that these and other ionized bases seem not to be involved in nucleic acid biosynthesis. This pathway of mispairing suggested more than 20 years ago (6) can be rejected as the main pathway of transitions and transversions based on a series of experimental data. They are the following: first, the polynucleotides consisting of ionized base analogs are not templates during DNA synthesis *in vitro* (7); second, a change of extracellular pH (from 5.5 to 8.5) does not affect frequency of base substitutions (8), and third, a change of pH in a cell-free system (from 6 to 10) does not change misincorporation frequency (9). It can be presumed that some mechanism hinders incorporation of ionized bases into DNA and RNA due to the presence of ionized groups in an appropriate site of the template enzyme and (or) to a change of the conformation of the nucleotide being incorporated when the base is ionized. Therefore only two other types of mispairs, those with a base in a rare tautomeric form and those with usual base tautomers will be discussed further as the two possible mechanisms of infidelity resulting in base substitutions. These two mechanisms do not in principle exclude each other as they could both contribute with different efficiency depending on conditions.

In 1953 Watson and Crick (3) suggested the formation of pairs A:C and G:T with a base in a rare (existing with a probability of $10^{-4} \div 10^{-5}$) tautomeric form as the pathway of transitions (Fig. 2 c,d represents similar G:U pairs). In 1976 Topal and Fresco (10) extended this mechanism to transversions which were assumed to occur through the formation of purine-purine pairs with one base in a rare tautomeric form and the other in *syn* orientation relative to sugar. In two of four pairs suggested one base is in a "twice rare" imino-lactim tautomeric form and the authors of paper (10) do not take into account that the probability of such a form is extremely low. This mechanism presumes that nucleic acids incorporate only bases which form pairs practically coinciding with pairs A:T and G:C by the mutual position of glycosyl bonds. The probability of transitions is equal to that of the existence of rare tautomers and the probability of transversions to the product of probabilities of the existence of rare tautomers and *syn* isomers. The conformation of the sugar-phospate backbone is assumed to be fixed. DNA is considered as a static double-helical structure with all bases in fixed positions. Though some experimental data do not contradict such a mechanism, there are many facts which cannot be explained within its limits. Among these facts are first, nucleotide replacements having pyrimidine-pyrimidine pairs as intermediate stages and the formation of such pairs in biosynthesis *in vitro;* second, the error frequency in some systems is higher than the probability of the rare tautomers (frequencies of different errors for a number of systems are summarized in reviews (11,12) and third, incorporation into DNA and RNA of base analogs with no plausible pairs of this type (e.g. benzimidazol and some alkylated derivatives).

Such a model has been repeatedly subjected to theoretical analysis. Since the NA conformation is assumed to be fixed in this model, it is possible to apply quantum mechanical methods. These methods have been used to estimate the energy of different tautomers and the energy of base pairs involving rare tautomers. Calculations of Rein et al. (13,14) have shown that the formation of purine-purine (*syn*) pairs with one base in a rare tautomeric form is very unfavourable and the probability of such purine-pyrimidine mispairs varies from 10^{-2} to 10^{-11}. These energy estimates show that the formation of pairs intermediate for transitions is plausible (though for some pairs the probability is too low) while the formation of pairs intermediate for transversions is practically impossible according to this mechanism (the probability is 10^{-22} (14)). However, there is genetic and biochemical evidence that frequencies of transitions and transversions differ not so drastically.

Another way of nucleotide mispairing assumes that the bases are in their usual tautomeric forms (13-16). Calculations of the interaction energy of nitrogen bases have shown (15-17) that for each coplanar pair there are energy minima in which the mutual positions of glycosyl bonds differ from that in A:U and G:C pairs by no more than 3 A and $30°$. These minima correspond to the formation of two or one N-H . . . N and (or) N-H . . . O hydrogen bonds. Some pairs have a base in *syn* orientation relative to sugar. Consideration of base pairs in normal tautomeric forms as intermediate steps of spontaneous mutations (15,16) permit qualitative interpretation of all the experimental data on spontaneous mutagenesis and on the errors of *in vitro* nucleic acid biosynthesis.

Fig. 2e represents G:U base pair as an example. This is one of *wobble* pairs suggested by Crick (18) to explain degeneracy of the genetic code. The existence of such a pair has been shown in helical tRNA fragments (19), in double stranded polynucleotide (20) and oligonucleotide (21) complexes.

Mispairs of normal tautomers are characterized by a displacement of bases relative to the position occupied by complementary bases in Watson-Crick pairs, i.e. mutual position of glycosyl bonds is modified. Most of these mispairs are impossible without such a displacement because of reduced interatomic contact arising (e.g. between H (1) of guanine and H (3) of uracil in G:U pair, Fig. 2f). A shift of bases requires distortion of the sugar-phospate backbone. It was not evident that the double helix could incorporate such pairs without a strong energy increase that would make practically impossible the incorporation of wrong nucleotides by this mechanism. Therefore we have examined theoretically the possiblity of formation of such pairs in DNA, have calculated the energy of non-bonded interactions of NA fragments containing these pairs and have searched for low-energy conformations of these fragments. We have shown plausibility of incorporation of pairs of all types into the double helix. Therefore the mechanism of infidelity in biosynthesis through the formation of wrong base pairs in normal tautomeric forms (13-16) became more plausible than the model involving rare tautomers (10). The possibility of incorporation of purine-pyrimidine, purine-purine and pyrimidine-pyrimidine mispairs has been demonstrated by us (22-24); the incorporation of G:T, A *syn* : G and A:C pairs was considered later by Rein et al. (14) and in this paper we shall give the results for each pair. Before presenting the results of calculations for fragments with mispairs we shall discuss the method for calculating the energy of non-bonded interactions using classical potential functions and consider conformations of a regular double helix.

Potential Functions

To examine conformations of NA fragments we need a simple and convenient method for calculating the energy of non-bonded interactions. The method should permit calculation, if possible, by similar formulae, of both intramolecular interactions in nucleic acids and NA interactions with water molecules to simulate NA fragment behaviour in solutions. Simplicity is required as the energy of even a minimal repetitive fragment of a regular polynucleotide is a function of many conformational variables. The energy minima can be found and conformational possibilities can be studied only after having calculated the functions in tens and hundreds thousands of points in the space of conformational variables. Therefore even the most powerful modern computers cannot perform such calculations when sufficiently strict quantum mechanical methods are used. The only way of calculating such systems is to apply classical potential functions which consider the interaction between the molecules (or their fragments) as a sum of pairwise atom-atom interactions. This method will be denoted AAPF, the method of atom-atom potential functions. Each interaction depends on the properties of atoms and the interatomic distance. These interactions are usually calculated as a sum of electrostatic (proportional to $1/r$ where r is the interatomic distance) and van-der-Waals terms. The attraction term is assumed to be proportional to r^{-6} and the repulsion term to r^{-12} or r^{-9} or exponentially decreasing with the distance. Calculations of intramolecular interactions also take into account energy changes connected with the rotation around the bonds (torsion potentials) and distortion of bond angles and bond lengths (if they are not assumed to be constant). For simple fragments, e.g. base pairs, studies of the interaction energy including the search for minima can be performed by more complicated though also classical methods. These methods (25) take into account not only charges but also dipoles and quadrupoles and not only on atoms but also on middles of real and virtual bonds. However, complication of the method and an increase of the computer time hinder such calculations for NA fragments containing several nucleotide pairs or for simulation of systems containing a NA fragment and tens of water molecules.

To enable investigation of interactions in complex systems, the AAPF method should reproduce experimentally determined characteristics of NA components. This suggests the way of determining AAPF: to choose them from analysis of experimental data for related compounds, e.g. crystal structures and sublimation heats. However an independent determination of AAPF form and coefficients for all types of atoms represents a too complicated mathematical task, the experimental data are insufficient and the available data are not always determined precisely enough. Therefore the form of the energy dependence on the interatomic distance is *a priori* presumed. But even having chosen a definite form of potential functions it is difficult to find all the parameters of potentials from the experimental data without additional assumptions because there is not the only strictly definite solution of the system of equations for determining the parameters. The determined parameters of the potentials can considerably change with a change of the assumed magnitude of an experimental parameter (e.g. sublimation heat) within the limits of the experimental error. On the other hand it is impossible to theoretically deduce potential functions from a quantum mechanical determination of the dependence of interaction energy of atoms on distance. At the same time, quantum mechanical calculations may be of use in determination of the AAPF. AAPF parameters can be chosen so that when the molecules (or parts of a molecule) are in certain mutual positions AAPF reproduce the energy of intermolecular (or intramolecular) interactions calculated with the help of strictest quantum mechanical methods. Then the energy values at every mutual position of the molecules can be found using AAPF without quantum mechanical calculations. Selection of the parameters by this procedure is easier than the choice from experimental data as energy values (not values depending on energy) in definite points are used and a sufficient number of points can be chosen. This way of choosing the parameters of classical potential functions suggested by Clementi et al. more than ten years ago is now being widely used for many classes of molecules. AAPF parameters for calculating intramolecular interactions in NA (26) and interactions of NA with water molecules (27) obtained in such a way gave interesting results including those on NA hydration (28).

At the same time, AAPF parameters obtained only on the basis of quantum mechanical results may have, in our opinion, several shortcomings. These shortcomings are due first, to the limited precision of quantum mechanical calculations. For example, many *ab initio* calculations do not take into account correlation energy correction, i.e. dispersion interactions are ignored. The assumption on invariability of bond angles and bond lengths can distort the dependence of the intramolecular interaction energy on the dihedral angle. Second, the parameters of potentials depend on the choice of points in which energy values have been calculated quantum mechanically. A systematic absence of any type of mutual positions of molecules can distort the pattern of intermolecular interactions obtained with the use of chosen AAPF parameters. Another shortcoming of AAPF obtained in this way is due to the fact that intermolecular interactions are pair-additive only approximately.

We think that potential functions for calculating non-bonded NA interactions should be based both on quantum mechanical results and considerations and the experimental data. The scheme of choice and application of potential functions (29,30) can be represented in the following way (Fig. 3). Approximate AAPF are based on theoretical consideration and experimental data. Quantum mechanical considerations help to choose the shape of the functional dependence. The used dependence of energy on distance is as follows

$$U(r_{ij}) = \frac{e_i \cdot e_j}{\epsilon \cdot r_{ij}} - A_{ij} r_{ij}^{-6} + B_{ij} r_{ij}^{-12} \qquad (1)$$

where r_{ij} is the interatomic distance, e_i and e_j are charges of atoms, ϵ is the effective

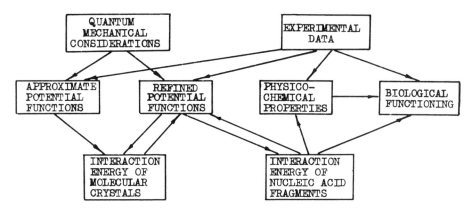

Figure 3. Scheme of choice and application of atom-atom potential functions.

dielectric constant, A_{ij} and B_{ij} are the parameters depending on the type of atom, its valent state and (in some cases) neighboring atoms. A_{ij} coefficients are calculated by approximate London formula. Effective charges on atoms are evaluatd by semi-empirical quantum chemical methods. The experimental data used at the first stage are those on dipole moments and refractions of molecules and van-der-Waals radii of atoms. These potential functions are used to calculate the energy of intermolecular interactions in crystals. Having compared calculation results with the experimental data on structures and sublimation heats of crystals we change the AAPF parameters. The modified potential functions are used to calculate crystal energies. This process is repeated several times so as to choose AAPF which reproduce most adequately the data on crystals of related compounds. Some parameters slightly changed to facilitate agreement with the experimental data on conformations of simple NA fragments (sugars, nucleosides).

When choosing AAPF describing interaction of NA with water molecules (31-33) we took into account experimental values of the interaction energy of water molecules with bases and the distances between water molecules and the atoms of bases in crystallohydrates. The results of calculations with the chosen AAPF have been compared with those obtained by other e.g. quantum-mechanical methods. Sometimes this comparison suggested that some parameters should be modified. Then simple systems were recalculated and the results were compared with the experimental data. Only a part of experimental data on simple systems (crystals, associates) has been used for choosing AAPF parameters. Another part has been used to evaluate reliability and precision of the obtained results.

We have recently refined AAPF for the interaction of nitrogen bases (17) and NA with water molecules and between water molecules (33). We assumed that the interaction energy of atoms capable of forming a hydrogen bond (hydrogen and acceptor) can be formulated as

$$U(r_{ij}) = \frac{e_i \cdot e_j}{r_{ij}} - A_{ij}^{(10)} r_{ij}^{-10} + B_{ij}^{(10)} r_{ij}^{-12} \qquad (2)$$

Such a dependence has been used in AAPF for conformational analysis of protein fragments (34) and allows one to obtain more linear H-bonds and, in some cases, to improve agreement of calculation results with the experimental data. This refinement is not quite essential for calculations discussed in next sections.

AAPF chosen in this way may be used to calculate the energy of non-bonded interactions in nucleic acid fragments and to find minima of this energy as well as for calculation of systems consisting of NA fragments and dozens of water molecules. A comparison of these calculations with physico-chemical, biochemical and genetic experimental data suggests conclusions on the structure and properties of NA ensuring its functioning as the heredity material.

Conformational Flexibility of the DNA Double Helix

Polymorphism of the DNA double helix and transitions between different forms are the two facts well established at the present time. We can consider several classes of conformations (4), within each of them there is a continuum of rather similar forms. The transition from one class to another is connected with a considerable change of the conformation of the sugar-phosphate backbone (transition of at least one dihedral angle into another range) and in many cases, energy barrier should be overcome. Conformations pertaining to different classes can pay a different biological role. The existence of several classes of double helices was predicted from calculations of energy of non-bonded interactions in 1975 (35); such calculations have been used recently to estimate differences of energies between conformations pertaining to different classes (36). In solution the double helix undergoes permanent conformational changes, both rather slow (due to transitions between classes) and quicker ones (due to small movements of separate atomic groups). Here we shall be concerned with possible conformational changes within A-and B-conformations of NA. After the studies on the structure of monocrystals of NA double helical fragments (5), the dependence of the conformation of double helical fragment on nucleotide sequence seems quite natural. Depending on the position of the nucleotide in the fragment, each conformational parameter can assume different values. Calculations of the energy of non-bonded interactions have shown (37) that in a homopolynucleotide there can be considerable changes of conformational parameters, displacements of atomic groups which require an energy increase of no more than 1-2 kcal/mole. In the multidimensional space of conformational variables there are extended regions of minimal energy values (valleys) corresponding to A- and B-families of NA conformations. We have recently refined (38) these calculations using new potential functions and considering energy as a function of more variables than earlier (37). This permitted evaluation of dimensions of the valleys, i.e. the limits of changes of conformational parameters at a certain increase of the energy of non-bonded interactions. We were primarily interested in possible changes of the mutual position of bases. Therefore as independent conformational variables in energy calculations we chose Arnott parameters (39) determining the position of bases in the double helix, the parameters determining conformations of sugar rings (2 dihedral and 2 bond angles per sugar of each chain) and glycosyl dihedral angles χ. The other dihedral angles and coordinates of all atoms of the fragment of the regular helix were found from these variables assuming bond lengths and bond angles to be constant (besides angles of the sugar ring since their fixation would mean rigidity of the ring conformation). Plotting conformational maps and isoenergetic levels of the double helical polynucleotide fragment in the space of these variables is very time-consuming. The exact shape of isoenergetic levels depends on the potential functions used and the number of independent variables. Therefore we have not carried out the whole bulk of this work; we have calculated values of the minimum energy in several rows of points corresponding to gradual change of one of conformational parameters and performed minimization by other parameters. We have most extensively studied changes of helical rotation angle τ for conformations of the B-family and distances d between neighbouring base pairs along the helix axis for the A-family. Based on these calculations we have plotted the dependences of the minimum energy of non-bonded interactions on conformational parameters and the dependences of conformational parameters for low-energy conformations (projections of the valley bottom on the planes including different pairs of conformational variables). Some of the dependences are presented in Fig. 4. It should be noted that the position of minima strongly depends on the account of charges on phosphate groups. The results presented in Fig. 4 correspond to completely ionized phosphate groups with the total charge on each nucleotide equal to -1, the interaction between two sugar-phosphate chains being neglected; when phosphate groups are completely neutralized the minimum energy values are shifted to smaller (down to $36°$) τ for the B-family and to smaller d for the A-family practically without a change of the shapes and dimensions of the valleys and relations between conformational parameters in the points corresponding to each valley bottom.

It is important to note that rather large changes of conformational parameters describing the mutual position of base pairs are possible at small deviations of the energy from the optimum. Even when the energy change is 0.3 kcal/mole ($RT/2$ for room temperature) τ can vary in the range of 2.4°, $d - 0.2$ A, and $D - 0.8$ A. This explains considerable thermal fluctuations of the structure (40).

Dihedral angles of the sugar-phosphate backbone for conformations approaching the valley bottom and differing from the optimum by no more than 2 kcal/mole lie in rather narrow regions whose dimensions for the B-family are no more than 20° (this region is presented in Table II together with such regions for unregular fragments). These regions lie entirely within wider regions corresponding to possible values of these angles in crystals of NA fragments. It should be noted that conformational parameters corresponding to low-energy B-conformations (38) are correlated with each other. These correlations have been observed later for angles in dodecanucleotide crystal (5), e.g. between χ and δ, δ and τ, δ and ζ. We have not considered conformations with essentially different angles of the two chains, but the fact that pyrimidine nucleotides prefer smaller χ than purine ones was also observed in our calculations of the regular polynucleotide poly-dA:poly-dU.

In crystals of double-helical fragments having a B-like conformation (5) ranges for some dihedral angles are considerably wider than it follows from theoretical conformational analysis. This seems to be mainly due to the regularity distortion of the double helix. A shift of conformations of some sugar rings of dodecanucleotide (5) into the energy-unfavourable O'-*endo* region encountered extremely rarely in monomer crystals seems to be due to application of potential functions which do not reproduce the energy barrier between C(2')-*endo* and C(3')-*endo* conformations (41).

Models of double-helical structures constructed by Arnott et al. in 1969-1975 from X-ray data on fibers (39,42-44) can have the energy of non-bonded interactions differing considerably (by tens of kcal/mol) from the optimum for the corresponding family. However, as seen from Fig. 4, the points in the space of Arnott's parameters corresponding to these models are located along the valley bottom (Fig. 4). The only exception is the angle θ_2 (*twist*) in the A-family of conformations. According to our data, all low-energy conformations of the A-family have the same sign of θ_2 as B-family conformations and when θ_2 approaches the values assumed in (39,42,43) the energy increases by several kcal/mol. Now, after Calladine's analysis (45) and studies on crystals of double helical fragments (5) this seems natural. But when we obtained such a result from our calculations (38) there were no other indications on the necessity to change the sign of the angle between bases in the pair for the A-family. A systematic study of low-energy conformations of the A-family permitted us to predict conformations having large values of the angle θ_2 between bases in the pair (38). These regions are outlined by dashed line in Fig. 4 and one of these conformations is represented in Fig. 5. They have an energy lower than "ordinary" A-conformations owing to negative charges on phosphate groups, but even at complete neutralization of these charges the most favourable of such conformations has the energy only 0.5 kcal/mol. higher than the conformation with the same d and a "moderate" θ_2.

The new refined model of the DNA A-form constructed from X-ray data on fibers (46) is rather similar to the conformation having the miniumum energy at $d = 2.56$ A and $\tau = 32.70°$ (Table I). However this is not the case for the new model of the DNA B-form ($d=3.38$ A, $\tau=36°$) having D approaching zero and the dihedral angle ζ equal to $203°$, i.e. lying in the *trans*-range (46,47). Our calculations, as well as those performed by other authors, give for low-energy B-DNA conformations the angles of $240-270°$ and D about 1A. Our calculations (Table I) show that a shift of angle ζ to trans range and energy minimization (d, τ and C(2')-*endo* sugar puckering are fixed) leads to a decrease of D to negative values and a shift of other variables to the values of the X-ray model (46,47). As a result, the energy of non-bonded interactions considerably increases. This is mainly due to an increase of the energy of the sugar-phosphate backbone (van-der-Walls interactions and torsion energy in

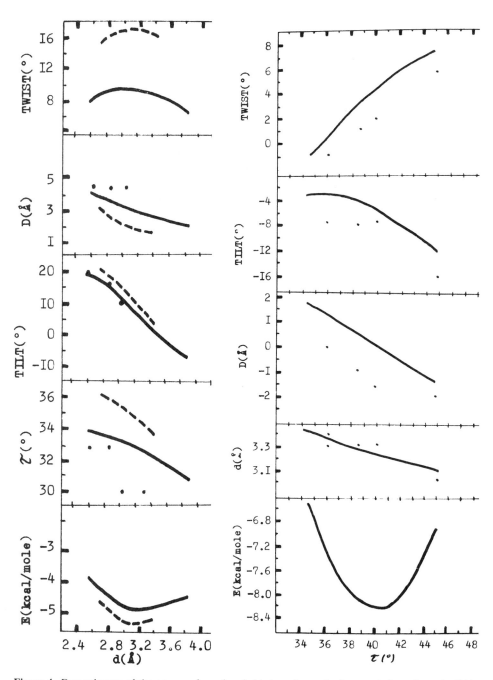

Figure 4. Dependences of the energy of non-bonded interactions of a fragment of regular poly d(A): poly d(U) polynucleotide on *d* for the A-family (left) and on *τ* for the B-family (right) and projections of the bottoms of two valleys of minimal energy values. Points correspond to the models constructed from X-ray data (39,42-44).

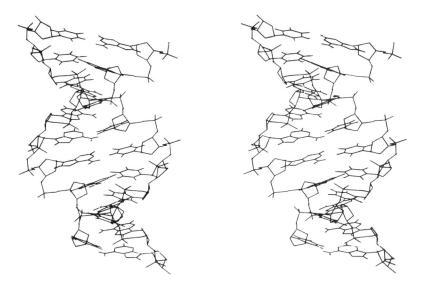

Figure 5. Computer stereo drawing of one of low-energy A-conformations with a large (34.6°) angle between bases in pairs, d=3.06 A, τ=35.1°, D=1.9 A, $TILT$=14°.

Table I
Conformational Parameters of DNA Models Constructed from X-ray Data (1 and 3)
and of Some Calculated Conformations of poly d(A) : poly d(U)

No	D	$TILT$	$TWIST$	χ	α	β	γ	δ	ϵ	ζ	Reference or E
			A-conformations, $d = 2.56$ A, $\tau = 32.7°$								
1	4.8	22	3.0	198	310	172	42	79	215	282	(46)
2	4.4	19	7.8	198	282	179	60	78	203	295	(38)
			B-conformations, $d = 3.38$ A, $\tau = 36°$								
3	−0.1	2	6.6	256	327	138	33	142	219	203	(46)
4	1.27	−3.8	0.1	240	290	181	59	137	184	251	−11.3
5	−0.1*	2*	6.6*	256*	336	167	31	141	190	236	4.4
6	−0.4	2.8	4.1	255	336	153	26	141	198	223*	12.8
7	−0.9	2.5	3.9	257	351	139	23	140	198	210*	26.7
8	1.1	2.7	−0.8	246	292	189	53	139	179	255	−9.8

Note: $d, \tau, D, TILT, TWIST$ are Arnott parameters (39); dihedral angles are designated as in Ref. (4,5). E is the non-bonded interaction energy (kcal/mol of nucleotide pair), phosphate charges are completely neutralized.
*nonvaried parameters; standard C(2′)-endo sugar puckering was fixed in conformations 5-8.

approximately equal portions). It should be also noted that we have obtained structures which are similar to this model and have almost no too short interatomic contacts, though they are far from the optimum by energy (Table I). Since calculations performed by different workers according to different procedures as well as X-ray studies on crystals of double-helical fragments (5) give ζ lying in another range than in the model constructed from X-ray data on fibres (46,47), the model seems to need further refinement.

Incorporation of Incorrect Pairs Into the Double Helix and the Mechanisms of Errors in NA Biosynthesis

When conformational flexibility of the double helix consisting of Watson-Crick nucleotide pairs had been demonstrated, it became quite natural to propose that pairs with a changed geometry could be incorporated into the double helix. Here we shall show that all spontaneous nucleotide replacements can be explained by the formation of incorrect base pairs in normal tautomeric forms but with a changed, as compared to A:T and G:C pairs, mutual position. To be incorporated into the double helix in biosynthesis, the incorrect pair should not change much the conformation of the double helix. It can be assumed that enzyme systems performing template-directed biosynthesis of NA would impede incorporation of an incorrect nucleotide if this is connected with a considerable change of the sugar-phosphate backbone and the transition of dihedral angles into other ranges. To prevent the conformation of the sugar-phosphate backbone from considerable changing upon formation of a mispair, the mutual position of glycosyl bonds in these pairs should not differ much from that in A:T and G:C pairs. For all base pairs there are minima of base interaction energy satisfying this requirement. For some pairs such minima correspond to the formation of two H-bonds of the type N-H . . . O and (or) N-H . . . N. For other pairs the position of bases in the minima is stabilized by one H-bond of the type N-H . . . O or N-H . . . N and one reduced C-H . . . O or C-H . . . N contact which can be considered as an additional weak H-bond. Mutual positions of bases in the examined pairs are represented in Fig. 6. These pairs are sufficient

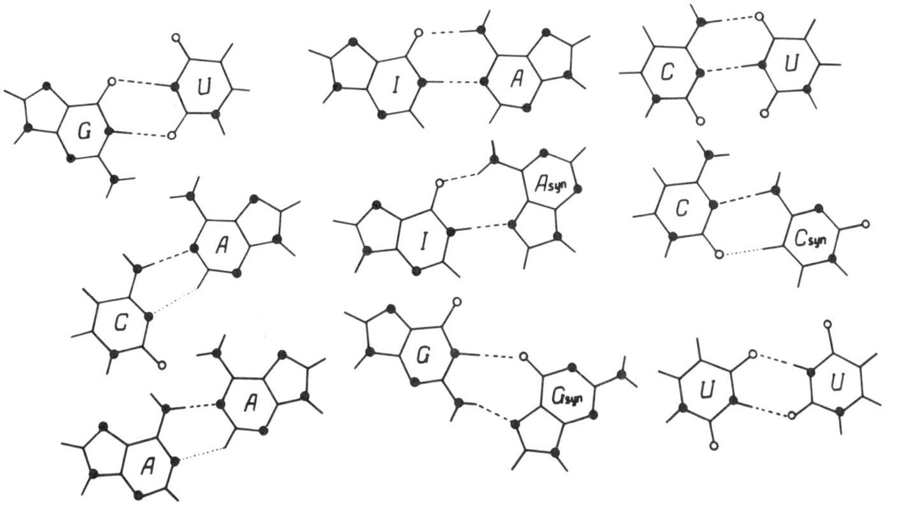

Figure 6. Mutual positions of bases in mispairs corresponding to energy minima of intermolecular interactions.

for explaining all possible nucleotide replacements, all spontaneous mispairing in NA biosynthesis. For convenience, in some pairs we have replaced thymine by uracyl and guanine by hypoxantine which practically does not affect the mutual position of bases in the minima. For each pair we have considered only one configuration, the most suitable, at first glance, for incorporation into the double helix consisting mainly of pairs A:T and G:C. Only for one pair G:A (I:A) we made an exception having considered two configurations, one of which corresponds to the *anti* conformation of both nucleotides and the other has A in *syn* conformation. Between these two configurations it is difficult to choose *a priori* the most suitable pair for incorporation. For some other pairs two or more configurations are also

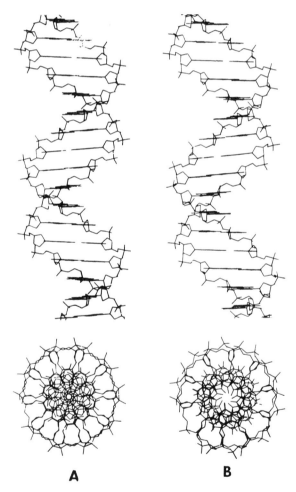

A   B

Figure 7. Projections on the plane of the helix axis (top) and on the plane perpendicular to the helix axis (bottom) of low-energy B-conformations of regular polynucleotides poly d(A) : poly d(U) (a), poly d(U) : poly d(G) (b), poly d(I) : poly d(A) (c) and poly d(U) : poly d(U) (d). All conformations were determined with the assumption that base pairs are planar.

possible. They correspond to the energy minima of base interaction and the position of their glycosyl bond is rather close to that in A:T and G:C pairs (some of these pairs are given in Ref. 15-17).

We started studies on conformations of NA fragments containing nucleotide mispairs by demonstrating that regular double helices can be formed which consist entirely of purine-pyrimidine, purine-purine and pyrimidine-pyrimidine mispairs. We have shown that double-stranded poly d(G) : poly d(U), poly d(A) : poly d(I) and poly d(U) : poly d(U) complexes can assume conformations similar to A- and B-conformations of NA. Fig. 7 represents low-energy B-like structures of each complex. It is seen in the figure that conformations of the sugar-phosphate backbone for each structure and low-energy poly d(A) : poly d(U) structure are rather similar.

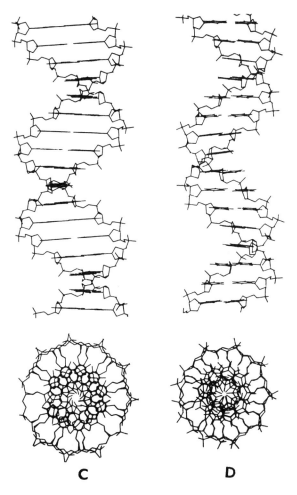

C   D

Figure 7 continued from previous page.

Fig. 8 represents the low-energy A-family conformation of the complex poly d(U) :
: poly d(U). For each complex, the same as for poly d(A) : poly d(U), there are
regions of minimum energy values, the valleys, corresponding to A- and B-families
of conformations. The shape of the valleys in the space of conformational variables
is similar to the valleys of regular polynucleotides consisting of canonical pairs. In
low-energy conformations bases of the opposite chains form pairs with two practically linear H-bonds and all dihedral angles of the sugar-phosphate backbone lie in
the same regions as for the corresponding families of poly d(A) : poly d(U)
conformation. Though other conformations of complexes are not excluded in solutions and fibres, our aim has been to demonstrate plausibility of conformations
similar to A- and B-forms of NA for double-helical polynucleotides having the
distance between C(1') atoms of the pair about 2 A larger or 2 A smaller than A:T
and G:C pairs or a 2 A shift of one base along the dyad axis. Each dihedral angle of
the sugar-phosphate backbone of these polynucleotides differs from the corresponding angles of low-energy conformations of the same family of poly d(A) :
: poly d(U) by no more than 20°. This small difference gave an additional indication on the possibility of mispair incorporation into the double helix consisting of
Watson-Crick pairs and at this incorporation the dihedral angles should not considerably exceed the limits characteristic of A- and B-families of conformations.

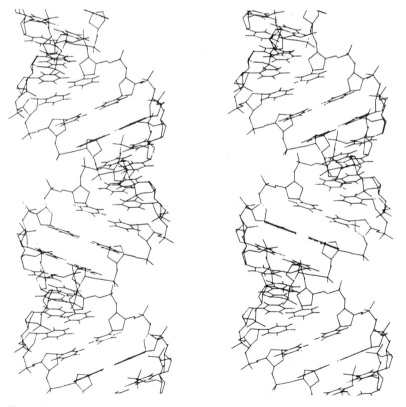

Figure 8. Computer drawing of low-energy A conformation of poly d(U) : poly d(U).

Then we studied mispair incorporation into the double helix consisting of A:U pairs. We have shown recently (22-24) that purine-pyrimidine (G:U, G:T), purine-purine (I:A, G:A, I:A *syn*, G:A *syn*) and pyrimidine-pyrimidine (C:U, C:T) pairs with two H-bonds can be incorporated into the double helix both in A- and B-conformations. Distortion of regularity of the double helix caused by incorporation of these pairs does not result in the appearance of reduced interatomic contacts and the angles of the sugar-phosphate backbone remain within the same limits as for double helices consisting of A:T and G:C pairs. This means that the sugar-phosphate backbone possesses a sufficient flexibility to allow incorporation into the double helix of nucleotide pairs having the dimensions of about 2 A larger (I:A pair) and 2 A smaller (U:C pair) than Watson-Crick pairs as well as incorporation of pairs in which one base is shifted relative to the other by about 2 A along the dyad axis (G:U pair) and pairs with one nucleotide in *syn* conformation. Almost all distortions of the backbone occur in the part connecting the mispair with the neighboring correct pairs.

Here we present the results of the study of double helical fragments consisting of A:U pairs and containing all possible base oppositions. For each X:Y pair represented in Fig. 6 we considered two fragments dApdApdX : dYpdUpdU and dXpdApdA : dUpdUpdY containing the incorrect pair on one or another side of two successive correct pairs. One of these fragments results from incorporation of the incorrect base X during synthesis on the Y-containing template and the other from incorporation of the incorrect base Y during synthesis on the X-containing template. At synthesis on one template one fragment corresponds to incorporation of an incorrect base and the second one to elongation of synthesis after this incorporation. When searching for low-energy conformations we considered energy of the fragment as a function of conformational variables determining the position of bases in planar pairs and of pairs relative to the helix axis, helical parameters, sugar ring conformations and glycosyl dihedral angles χ. The search was performed by gradual shifting of bases in mispairs towards the position corresponding to the minimum of interaction energy of two isolated bases (these positions are represented in Fig. 6) and energy was minimized by other conformational variables. When bases achieved this or a similar position, minimization was performed by all independent variables. Conformations of all fragments (except one fragment with a U:U pair) are characterized by deviations of the sugar-phosphate backbone angles from their values for low-energy poly d(A) : poly d(U) conformations by no more than 30° and an increase of the interaction energy in the sugar-phosphate backbone (including torsional energy but disregarding the electrostatic contribution) by no more than 3 kcal/mol. Calculations show that a shift of bases relative to each other along the dyad axis as compared to Watson-Crick pairs leads to a smaller increase of the sugar-phosphate backbone energy than a shift in the perpendicular direction. For low-energy conformations obtained (except for conformations with pairs G:G, U:U and C:U) the interaction energy of bases in a mispair is only 0.1 ÷ 0.7 kcal/mol higher than the minimum, the mutual position of bases in these pairs is close to the position in the energy minimum for isolated pairs. For the G:G pair there is a rather wide region of minimal values of the interaction energy of bases, i.e. there can be shifts of one base relative to another by $1 \div 2$ A along the direction approximately perpendicular to the H-bonds when the energy increases by only $1 \div 2$ kcal/mol. This seems to cause a shift of the mutual position of bases in low-energy fragment conformations as compared to the isolated pair. For one of the fragments with the U:U pair the energy-minimum conformation corresponds to the mutual position of bases in mispair approaching the minimum for the isolated pair and for the other fragment the energy minimum corresponds to the glycosyl bond position close to that in Watson-Crick pairs. If the mutual position of wrong bases in this fragment is maintained in the energy minimum for the pair, the fragment conformation will differ from low-energy poly d(A) : poly d(U) conformations by $50°$ for dihedral angles and by about 8 kcal/mol for the interaction energy in the sugar-phosphate backbone. Energy minima for the fragments with the C:U pair correspond to the distance between bases in the pair 0.5 A larger than in the energy minimum for the isolated pair. These conformations have the energy $0.5 \div 1.0$ kcal/mol lower than low-energy conformations corresponding to the energy minimum of cytosine-uracyl interaction. Our preliminary data show that if simplifying assumptions (planarity of pairs, fixation of many bond angles) are not used, the mutual position of bases in low-energy conformations of fragments with U:U and C:U pairs approaches the mutual position of bases in isolated pairs.

For low-energy conformations of fragments dApdApdX : dYpdUpdU and dXpdApdA : dUpdUpdY the difference in conformational parameters of the nucleoside X:Y pair is a few degrees by the angle variables and some tenths of A by the shift variables. Then the conformational parameters of nucleoside mispairs of the two fragments were made equal by shifting them gradually towards each other and minimizing by other variables at each step. The energy of non-bonded interactions

Table II

Ranges of the Angles of the Sugar-Phosphate Backbone and of the Angle of Sugar Pseudorotation (P) in Conformations dApdApdXpdApdA : dUpdUpdYpdUpdU and Poly d(A) : Poly d(U) from Calculations and in Crystals of Nucleic Acid Double-Helical Fragments from X-Ray Data

X:Y	α	β	γ	δ	ϵ	ζ	P
G:U	$287 \div 293$	$165 \div 192$	$57 \div 63$	$140 \div 141$	$177 \div 205$	$228 \div 260$	$155 \div 166$
C:A	$284 \div 295$	$159 \div 190$	$52 \div 61$	$138 \div 141$	$178 \div 221$	$214 \div 265$	$151 \div 155$
A:I	$280 \div 294$	$158 \div 200$	$53 \div 65$	$133 \div 144$	$168 \div 213$	$225 \div 272$	$145 \div 161$
A syn : I	$286 \div 298$	$179 \div 190$	$51 \div 65$	$133 \div 143$	$177 \div 186$	$245 \div 258$	$145 \div 159$
A:A	$274 \div 298$	$151 \div 204$	$59 \div 83$	$135 \div 145$	$165 \div 215$	$225 \div 275$	$146 \div 162$
G:G	$271 \div 291$	$163 \div 191$	$60 \div 81$	$131 \div 145$	$162 \div 210$	$215 \div 281$	$140 \div 163$
C:U	$284 \div 289$	$150 \div 200$	$53 \div 66$	$137 \div 144$	$168 \div 222$	$220 \div 271$	$150 \div 159$
C:C	$283 \div 305$	$157 \div 192$	$45 \div 68$	$135 \div 142$	$174 \div 213$	$224 \div 265$	$148 \div 156$
U:U	$254 \div 296$	$142 \div 205$	$49 \div 82$	$135 \div 143$	$166 \div 242$	$218 \div 275$	$147 \div 158$
A:U (38)	$286 \div 293$	$169 \div 186$	$55 \div 73$	$137 \div 147$	$181 \div 187$	$239 \div 253$	$144 \div 166$
crystals (5)	$278 \div 309$	$139 \div 190$	$40 \div 66$	$70 \div 160$	$170 \div 260$	$150 \div 274$	

Note: The presented considerable deviations of angles at incorporation of the U:U pair are caused by a distortion of the sugar-phophate backbone in the three-pair fragment whose low-energy conformation has no U:U pair with H-bonds.

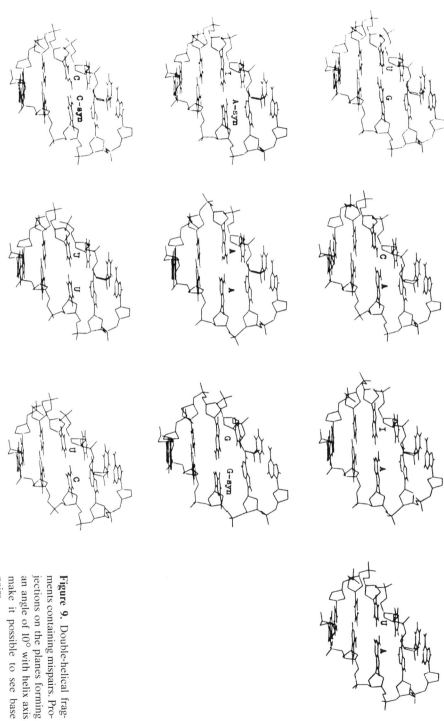

Figure 9. Double-helical fragments containing mispairs. Projections on the planes forming an angle of 10° with helix axis make it possible to see base pairs.

of each fragment somewhat (by no more than 4 kcal/mol) increases, but reduced interatomic contacts do not appear. Thus we have found sterically allowed conformations of fragments dApdApdXpdApdA : dUpdUpdYpdUpdU containg a X:Y mispair incorporated between correct pairs. Mutual positions of bases in mispairs correspond to those in Fig. 6. Table II summarizes ranges for dihedral angles of B-like conformations of such fragments containing X:Y pairs represented in Fig. 6. It should be noted that for each pair we have found only one low-energy conformation, one point within the extended region of minimum energy values. As in the case of regular double-stranded polynucleotide, the fragments considered have an extended region of minimum energy values, i.e. a valley corresponding to the B-family of conformations. Dihedral angles of all conformations obtained are within the limits or do not differ much from the angles chracteristic of DNA double-helical fragments in crystals. For the glycosyl dihedral angle χ there are two regions, *anti* and *syn*.

We shall not consider here in detail energetics of mispair incorporation into the double helix, we shall only note some qualitative regularities. Calculations show that incorporation of some pairs (e.g. U:G, I:A *syn*) leads to a change of the non-bonded interaction energy of fragments by no more than 3 kcal/mol as compared to the energy of a fragment consisting of A:U pairs only. This provides the error rate of $10^{-2} \div 10^{-3}$ which gives the possibility to explain high frequency of biosynthesis errors in some systems, the fact inexplicable from the point of view of rare tautomers. Much smaller error frequencies in template synthesis are considered to be due to the action of enzymes which increase accuracy. Perhaps this increase is achieved through interaction of the enzyme recognizing site with structural invariants of complementary nucleotide pairs (48). These invariants could be the position and conformation of the sugar-phosphate backbone and N(3) atoms of purines and O(2) of pyrimidines in complementary pairs. According to our calculations, the formation of most mispairs is accomplished by changes of the sugar-phosphate backbone conformation which are not larger than for polynucleotides consisting of Watson-Crick pairs. Therefore the interaction with only sugar-phosphate backbone seems to be insufficient to clearly distinguish between correct and incorrect pairs. It can be assumed that interaction of the enzyme recognizing site with nitrogen bases contributes to the high accuracy of biosynthesis.

Our calculations show that the G:U pair is the most easily incorporated into the double helix. Incorporation of this pair leads to a small increase of energy and the smallest (of all mispairs) change of dihedral angles of the sugar-phosphate backbone relative to poly d(A)·poly d(U) (see Table II). This result can explain a higher frequency of transitions than transversions (49,50) and the data that transitions mainly occur through the formation of G:T pairs (50-53). Much more rare is the formation of the A:C pair, the other pair through which transitions can occur (49-53), because of a larger energy increase and a larger change of the sugar-phosphate backbone angles at its incorporation into the double helix than at G:T incorporation. Incorporation of any purine-purine pair leads to a greater change of the sugar-phosphate backbone conformation than for the G:U pair. The less probable, according to our calculations, is incorporation of pyrimidine-pyrimidine pairs which leads to an increase of energy by $8 \div 10$ kcal/mol. If geometry of the U:U and C:U pairs in double-helical fragments corresponds to the minima of interaction energy in isolated pairs, these fragments have dihedral angles of the sugar-phosphate backbone differing the most considerably from the corresponding angles of fragments consisting of Watson-Crick pairs (Table 2). These results suggest an explanation of the experimental data (49-53) that transversion occurs mainly through the formation of purine-purine pairs. The formation of pyrimidine-pyrimidine pairs occurs extremely rarely in DNA biosynthesis (49-53) and is not revealed in many experiments since frequencies of the corresponding nucleotide replacements are lower than the method sensitivity.

We hope that calculations similar to those described but without simplifying assumptions (minimum fragment dimensions, planarity of pairs, rigidity of most bond angles) and a detailed analysis of these calculations taking into account the role of enzymes increasing accuracy will permit quantitative comparison of theoretical predictions with the experiment and contribute to refinement of the model of point mutations. Such calculations are in progress in our laboratory.

Discussion and Concluding Remarks

Representation of the double helix as a universal and static structure was in agreement with the mechanism of biosynthesis errors involving mispairs whose incorporation does not require changes of its conformation. Such pairs could be represented if one base is in a rare tautomeric form. The possibility of considerable changes of the double helix conformation permits the assumption that the formation of incorrect pairs with a changed geometry as compared to Watson-Crick pairs, is possible in NA biosynthesis as a rare event. By this mechanism of mispairing and taking into account the enzymatic increase of accuracy it is possible to explain all regularities of point errors in NA biosynthesis in vitro and in vivo. Relative frequencies of purine-pyrimidine, purine-purine and pyrimidine-pyrimidine mispairs are in correlation with our results on incorporation of mispairs with bases in normal tautomeric forms into the double helix. Our data suggest an explanation of the results which cannot be explained by the mechanism involving base pair with rare tautomers. It is known (11,12) that in some enzymatic systems biosynthesis errors occur at a higher frequency than that of rare tautomers. A high level of non-complementary nucleotide incorporation is observed in non-enzymatic template synthesis (54). In the latter case the error frequency of $10^{-2} \div 10^{-3}$ in synthesis on poly C at pH 7.9 cannot be explained by either pairs with rare tautomers or pairs with ionized forms.

Though the formation of incorrect pairs with rare forms of bases is sometimes considered as a mechanism of biosynthesis errors, all experimental data proposed as a proof of such mechanism can be explained without the assumption on formation of such mispairs. This pertains to biochemical, genetic and spectral data. NMR studies on tRNA (55) and spectral studies on melting of polynucleotide complexes containing A:C and I:U oppositions (56) have been interpreted using the assumption on rare tautomers but have not provided clear evidence of the existence of such pairs. The data on mutagenic activity of 5-haloiduracyls and 2-aminopurine cannot be considered as a proof of the formation of pairs with rare tautomers either. Numerous data indicate that mutagenesis caused by haloiduracyls is not related directly to the increase of mispairing (57). Moreover, no increase of mispairing has been revealed (58) in RNA-polymerase reaction of *in vitro* synthesis on poly d(A-BrU) template as compared to poly d(A-T). Errors induced by 2-aminopurine (AP) can be explained by the formation of AP:U and AP:C pairs in normal tautomeric form and the formation of pairs with a rare AP tautomer contradicts the data on frequencies of base incorporation *in vitro* (59). Compounds whose tautomers exist in comparable concentrations, e.g. hydroxycytidine and metoxyadenine can be incorporated into the polynucleotide chain as two tautomers (60-62).

Calculations of the energy of non-bonded interactions contribute much to the understanding of the mechanism of double helix conformational flexibility. These calculations were used to demonstrate both variety of classes of double-helical structures having quite different conformations of the sugar-phosphate backbone but retaining Watson-Crick base pairs (35,36) and continuum of forms within a class, the possiblity of considerable continuous change of conformation without changes of energy more than 1-2 kcal/mol (37,38). Only with the help of theoretical conformational analysis it became possible to demonstrate that all base oppositions can be incorporated into the double helix without considerable changes of the conformation of the sugar-phosphate backbone. Feasibility of such incorporation was not evident when we started calculations. The G:U pair was at that time the only mispair whose existence in the t-RNA double-helical fragment (the dimensions of this fragment is smaller than the helix turn) was experimentally demonstrated (19-21). Conformational parameters of this fragment have been reported e.g. in Ref. (19) and it has been shown that this is a *wobble* pair and not a pair with a rare tautomer. These and other data on G:U pair pertained to polyribonucleotides, i.e. to double helices in A-conformation. It could not be excluded that incorporation of even this pair (or G:T pair) into the double helix existing in B-conformation would be accompanied by such an increase of energy that local opening of the helix would be more favorable. There was no data on either plausibility of incorporation of other mispairs into the double helix or the conformation of double helical fragments containing these mispairs.

NMR studies on the conformation of a double-helical complex of deoxydodecanucleotides containing G:T oppositions (63) were carried out after our theoretical investigations of fragments containing mispairs (22-24). NMR data corroborate the formation of the G:T wobble pair and relatively small changes of the double helix conformation limited to the nearest environment of the mispair. Conformation of the double helical fragment containing a G:U pair is dicussed in Ref. 64 in relation to NMR data. Quite recently a complex of dodecanucleotides containing G:A oppositions has been studied (65). Although conformational parameters cannot be determined from NMR data only, it has been shown (65) that a G:A pair similar to pair I:A (Fig. 6) is formed. We hope that new experimental data on mispairs in double helices will follow.

The conclusion that the sugar-phosphate backbone of the double helix permits rather large local changes of the NA conformation owing to a considerable shift of the mutual position of bases is important for dynamics of nucleic acids. A shift of one base relative to another by 2 A along the line connecting C(1′) atoms of the Watson-Crick nucleotide pair, by 3 A along the dyad axis of this pair and shifts in both these mutually perpendicular directions are possible without the formation of reduced interatomic contacts in sugar-phosphate backbone. These shifts could be achieved by rather small distortions of the helix involving only the nearest and next-to-nearest neighbors of the wrong nucleotide pair. These distortions are connected with the deviation of dihedral angles by no more than 30° (for the fragment with the U:U pair up to 50°) from their values for low-energy poly d(A):poly d(U) conformations. Calculations performed permit us to estimate which modifications of bases are possible with the double helix remaining intact and which ones require its break-down. Since pairs with 2-aminopurine, 2,6-diaminopurine (66), with some alkylated purines and pyrimidines (67,68) differ from A:T and G:C pairs by dimensions and mutual position of bases no more than the mispairs considered their formation in template biosynthesis is quite plausible.

In conclusion we would like to note once again that mispairs in normal tautomeric forms considered here are sufficient to explain the whole bulk of experimental data on spontaneous point errors in NA biosynthesis. This does not however exclude the possibility that rare tautomers participate in such errors but pairs with rare tautomers only cannot explain all the experimental data.

References and Footnotes

1. Drake, J. W., *Nature 221* 1132-1133 (1969).
2. Kunkel, T. A., Meyer, R. R. and Loeb, L. A., *Proc. Nat. Acad. Sci. U.S.A.*, 76 6331-6335 (1979).
3. Watson, J. D. and Crick, F. H. C., *Nature 171*, 963-964 (1953).
4. Arnott, S. and Chandrasekaran, R., *Biomolecular Sterodynamics*, ed. R. H. Sarma, Adenine Pres., New York, *1* 99-122 (1981).
5. See e.g. Dickerson, R. E., Kopka, M. L. and Drew, H. R., in *Structure and Dynamics: Nucleic Acids and Proteins*, eds. E. Clementi and R. H. Sarma, Adenine Press, New York, pp. 149-179 (1983) and references therein.
6. Lowley, P. D. and Brookes, P., *Nature 192*, 1081-1082 (1961).
7. Richardson, C. C., Schildkraut, C. L. and Kornberg, A., *Cold Spring Harbor Symp. Quant. Biol. 28*, 9-18 (1963).
8. Apple, N. L. and Drake, J. W., *Mut. Res. 20*, 271-273 (1973).
9. Battula, N. and Loeb, L. S., *J. Biol. Chem. 250*, 4405-4409 (1975).
10. Topal, M. D. and Fresco, J. R., *Nature 263*, 285-289 (1976).
11. Bernardi, F. and Ninio, J., *Biochimie 60*, 1083-1095 (1978).
12. Loeb, L. A. and Kunkel, T. A., *Ann. Rev. Biochem. 52*, 429-457.
13. Garduno, R., Rein, R., Egan, J. T., Coeckelenbergh, Y. and Mac Elroy, R. D., *Int. J. Quant. Chem.: Quant. Biol. Symp. 4*, 197-204 (1977).

14. Rein, R., Shibata, M., Garduno-Juarez, R. and Kieber-Emmons, T., *Structure and Dynamics: Nucleic Acids and Proteins,* eds. E. Clementi and R. H. Sarma, Adenine Press, New York, 269-288 (1983).
15. Poltev, V. I. and Bruskov, V. I., *Mol. Biol. (USSR) 11,* 661-670 (1977).
16. Poltev, V. I. and Bruskov, V. I., *J. Theor. Biol. 70,* 69-83 (1978).
17. Poltev, V. I. and Shulyupina, N. V., *Mol. Biol.* (In Press) (USSR) (1984).
18. Crick, F. H. C., *J. Mol. Biol. 19,* 548-555 (1966).
19. Quigley, G. J., Seeman, N. C., Wang, A. H.-J., Suddath, F. L., and Rich, A., *Nucl. Acid. Res. 2,* 2329-2341 (1975).
20. Ackermann, Th., Gramlich, V., Klump, H., Knable, Th., Schmidt, E. D., Seliger, H. and Stulz, J., *Biophys. Chem. 10,* 231-238 (1979).
21. Romaniuk, P.J., Hughes, D. W., Gregoire, R. J., Bell, R. A., and Neilson, T., *Biochemistry 18,* 5109-5116 (1979).
22. Poltev, V. I., Chuprina, V. P., Shulyupina, N. V. and Bruskov, V. I., *Studia Bioph. 87,* 247-248 (1982).
23. Poltev, V. I., Chuprina, V. P., *Incorporation of Incorrect Nucleotide Pairs into the DNA Double Helix and Molecular Mechanisms Mutagenesis.* Preprint, Pushchino (1982).
24. Chuprina, V. P., Poltev, V. I., *Nucl. Acid. Res. 11,* 5205-5222 (1983).
25. Langlet, J., Claverie, P., Caron, F., and Boeuve, J. C., *Int. J. Quant.Chem. 19,* 299-338 (1981).
26. Matsuka, O., Tosi, C. and Clementi, E., *Biopolymers 17,* 33-50 (1978).
27. Scordamaglia, R., Cavallone, F. and Clementi, E., *J. Amer. Chem. Soc. 99,* 5554-5550 (1977).
28. Clementi, E., in *Structure and Dynamics: Nucleic Acids and Proteins,* eds. E. Clementi and R. H. Sarma, Adenine Press, New York, pp. 321-364 (1983).
29. Poltev, V. I., *Int. J. Quant. Chem. 16,* 9863-9868 (1979).
30. Zhurkin, V. B., Poltev, V. I. and Florentiev, V. L., *Mol. Biol. (USSR) 14,* 1116-1130 (1980).
31. Poltev, V. I., Shulyupina, N. V., D'yakonova, L. P. and Malenkov, G. G., *Studia Bioph. 84,* 187-194 (1981).
32. Poltev, V. I., Danilov, V. I., Sharafutdinov, M. R., Shvartsman, A. Z., Shulyupina, N. V. and Malenkov. G. G., *Studia Bioph. 91,* 37-43 (1982).
33. Poltev, V. I., Grokhlina and T., Malenkov, G. G., *J. Biomol. Structure Dynam.* (submitted for publication).
34. McGuire, R. F., Momany, F. A., and Scheraga, H. A., *J. Phys. Chem. 76,* 375-393 (1972).
35. Zhurkin, V. B., Lysov, Yu. P. and Ivanov, V. I., *FEBS Lett. 59,* 44-47 (1975).
36. Keepers, J. W., Kollman, P.A., Weiner, P. K., James, T. L., *Proc. Nat. Acad. Sci. U.S.A. 79,* 5537-5541 (1982).
37. Khutorsky, V. E. and Poltev, V. I., *Nature, 264,* 483-484 (1976).
38. Chuprina, V. P., Khutorsky, V. E. and Poltev, V. I., *Studia Bioph. 85,* 81-88 (1981).
39. Arnott, S., *Progr. Bioph. Mol. Biol. 21,* 265-319 (1980).
40. See also Zhurkin, V. B., Lysov, Yu. P., Florentiev, V. L. and Ivanov, V. I., *Nucl. Acid. Res. 10,* 1811-1830 (1982).
41. See in this connection also Olson, W. K. and Sussman, J. L., *J. Amer. Chem. Soc. 104,* 270-278 (1982); Zhurkin, V. B., *Mol. Biol. (USSR) 17,* 622-638 (1983).
42. Arnott, S. and Hukis, D. W. L., *Biochem. Biophys. Res. Commun. 47,* 1504-1510 (1972).
43. Arnott, S. and Hukis, D. W. L., *J. Mol. Biol. 81,* 93-105 (1973).
44. Arnott, S. and Selsing, E., *J. Mol. Biol. 98,* 265-269 (1975).
45. Calladine, C. R., *J. Mol. Biol. 161,* 343-352 (1982).
46. Chandrasekaran, R. and Arnott, S., personal communication (1982); the dihedral angles of the B-form differ slightly from listed in Arnott, S., Chandrasekaran, R., Birdsale, D. L, Leslie, A.G.W. and Ratliff, R. L., *Nature 283,* 743-745 (1980).
47. Coordinates of B-DNA with reference to S. Arnott and R. Chandrasekaran, personal communications are listed in Appendix to the paper Giessner-Prettre, C., Prado, F. R., Pullman, B., Kan, L.-S., Kast, J. R. and P.O.P.Ts'o, *Computer Programs in Biomedicine, 13,* 167-184 (1981). These coordinates correspond to conformational parameters close to those from ref. (46), besides $D = -0.77$ A (not -0.1 A).
48. Bruskov, V. I. and Poltev, V. I., *J. Theor. Biol. 78,* 29-41 (1979).
49. Sankoff, D., Cedergren, R.J., and Lapalme, G., *J. Mol. Evol. 7,* 133-149 (1976).
50. Sinha, N.K., and Haimes, M.D., *J. Biol. Chem. 256,* 10671-10683 (1981).

51. Fersht, A.R. and Knill-Jones, J.W., *J. Mol. Biol. 165,* 633-654 (1983).

52. Fersht, A.R., Shi, J.P. and Tsui, W.C., *J. Mol. Biol 165* 655-667 (1983).

53. Grosse, F., Krauss, G., Knill-Jones, J.W. and Fersht, A.R., *EMBO J. 2,* 1515-1519 (1983).

54. Inoue, T. and Orgel, L.E. *J. Mol. Biol. 162,* 201-217 (1982).

55. Rueterjans, H. Kann, E., Hull, W.E. and Liubach, H.H. *Nucl. Acid Res. 10,* 7027-7039 (1982).

56. Fresco, J.R. Broitman, S. and Lane, A.E., *ICN-UCLA Symposia on Molecular and Cellular Biology 19,* 753-768 (1980).

57. See e.g. Rydberg, B., *Mol. Gen. Genet. 152,* 19-28 (1977).

58. Bick, M.D., *Nucl. Acid Res. 2,* 1513-1519 (1975).

59. Goodman, M.F., Watanabe, S.M., and Branscomb, E.W. in *Molecular and Cellular Mechanisms of Mutagenesis,* eds. J. F. Lemont and W. M. Generoso, N.Y.—London, Plenum Press, pp. 213-229, (1982).

60. Budowsky, E.I., Sverdlov, E.D., and Spasokukotskaya, *Biochim. Biophys. Acta 287,* 195-210 (1972).

61. Singer, B., and Spengler, S., *FEBS Lett. 139,* 69-71 (1982).

62. Sledziewska, E. and Janion, C., *Mut. Res. 70,* 11-16 (1980).

63. Patel, D.J., Kozlowski, S.A., Marky, L.A., Rice, J.A., Broka, C. Dallas, J., Itakura, K. and Breslauer, K.J., *Biochemistry 21,* 437-444 (1982).

64. Wagner, B.J., Dhingra, M.M., Sarma, M.H. and Sarma, R., in *Structure and Dynamics: Nucleic Acids and Proteins,* eds. E. Clementi and R. H. Sarma, Adenine Press, New York, pp. 197-208 (1983).

65. Kan, L.-S. Chandrasekaran, S., Pulford, S.M., and Miller, P.S., *Proc. Nat. Acad. Sci. USA 80,* 263-4265 (1983).

66. Poltev, V.I., Shulyupina, N.V. and Bruskov, V.I., *Mol. Biol. (USSR) 13,* 822-827 (1979).

67. Poltev, V.I., Shulyupina, N. V. and Bruskov, V.I., *Studia Bioph. 79,* 45-46 (1980).

68. Poltev, V.I., Shulyupina, N.V., and Bruskov, V.I., *Mol. Biol. (USSR) 15,* 1286-1294 (1981).

Structure & Motion: Membranes, Nucleic Acids & Proteins,
Eds., E. Clementi, G. Corongiu, M. H. Sarma & R. H. Sarma.
ISBN 0-940030-12-8, Adenine Press.   Copyright Adenine Press, 1985.

The Binding of Netropsin to Double-Helical B-DNA of Sequence C−G−C−G−A−A−T−T−^{Br}C−G−C−G: Single Crystal X-Ray Structure Analysis

Mary L. Kopka, Philip Pjura, Chun Yoon, David Goodsell and Richard E. Dickerson

Molecular Biology Institute, and Institute of Geophysics and
Planetary Physics
University of California, Los Angeles
Los Angeles, CA 90024, U.S.A.

Abstract

A single crystal x-ray structure analysis has been carried out of the 1:1 complex of the antitumor antibiotic *netropsin* with the double-helical B-DNA dodecamer of sequence C−G−C−G−A−A−T−T−^{Br}C−G−C−G. The complex has been refined at 2.2 A resolution to a residual error of 26.0% for all reflections, or 23.0% for two-sigma data. A single netropsin molecule sits in the minor groove in the A−A−T−T center of the double helix, equidistant from the two walls of the groove. The three amide NH of the drug molecule point toward the floor of the groove. Each of the two pyrrole rings is within 3° of being parallel to its own region of the minor groove. As a consequence, the two rings make an angle of 33° to one another. To a first approximation the netropsin molecule simply replaces the ordered spine of hydration that helps to stabilize the B form of the helix in the absence of drug molecules. Each of the two outermost amide NH makes one hydrogen bond to a thymine O2, but the central amide NH is prevented from coming close enough for a standard hydrogen bond by steric replusions between base pairs and the two flanking netropsin pyrroles. The A·T base specificity of netropsin arises from the fact that binding within the minor groove leaves no room for the N2 amine group of a guanine. Both of the charged ends of the drug molecule lie in the bottom of the groove, rather than being associated with particular phosphates. The propionamidinium end forms a normal hydrogen bond to an adenine N3, but the guanidinium end is more loosely associated. Binding of netropsin neither unwinds nor elongates the double helix, but it does force open the minor groove at the binding site. The physiological properties of netropsin as an antitumor antibiotic presumably arise from the fact that the drug displaces the spine of hydration in B-DNA, locking the two strands of the double helix together and preventing either replication or transcription.

Introduction

Netropsin is an antiviral, antitumor antibiotic which, although too toxic for routine clinical use, has received intensive study by chemists and molecular biologists because of its unusual DNA-binding properties. Unlike many other drugs that bind to DNA, it does so without intercalating between base pairs. It also demonstrates one of the most pronounced specificities of any DNA-binding drug, interacting only with A·T base pairs in a double helix of the B type. We have cocrystallized netropsin with the B-DNA dodecamer $C-G-C-G-A-A-T-T-^{Br}C-G-C-G$, have located the drug molecule by difference map methods, and refined the structure of the netropsin-DNA complex. These results demonstrate unambiguously the mode of binding and the reasons for netropsin's observed specificity.

The structure of netropsin in shown in Figure 1. It was first isolated from *Streptomyces netropsis* in 1951 (1,2), and has exhibited activity against many strains of bacteria and viruses, including some that produce tumors in mammals (1,3,4). Its toxicity in mammalian cells prevents widespread medical use, but its unusual DNA-binding properties have made it the object of intensive laboratory study. At present it, and its close relative distamycin, are the paradigms of non-intercalative, groove-binding antibiotics.

Netropsin binds only to double-stranded DNA in the B form. It shows little or no affinity for single-stranded DNA or RNA, or for double-stranded RNA or DNA/RNA hybrids (5-8). This suggests strongly that it cannot bind to the A form of the double

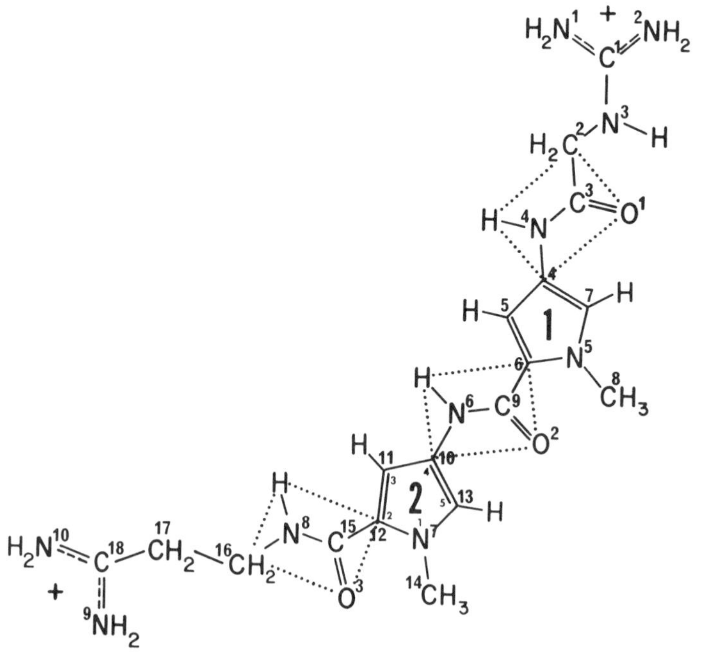

Figure 1. Structure of the netropsin molecule, with numbering as used in this and other papers. The guanidinium group at upper right is considered the beginning of the molecule, and is followed in succession by a planar amide, the first pyrrole ring, another amide, the second pyrrole, amide, and a terminal propionamidinium group. Carbon, nitrogen and oxygen atoms are numbered separately but sequentially from the guanidinium end, as in the x-ray structure analysis of free netropsin (31). Numbering 1-5 of atoms in each pyrrole ring is as in the nmr studies (19).

helix. It also fails to bind to poly(dC-d^{Br8}G) under conditions where the latter adopts the Z form, implying that it cannot bind to the left-handed Z-DNA double helix (9). The binding of netropsin itself favors A-to-B and Z-to-B helix transitions in DNA (10-12). Binding also is diminished if the B form is driven to the A by ethanol or other means (13,14).

The antibiotic and antiviral properties of netropsin are thought to arise because, by binding tightly to DNA, it blocks both replication and transcription. The interaction is directly between drug and DNA itself, and not to the DNA or RNA polymerase enzymes (7,8,15,16). No major structural alterations in the double helix occur when netropsin binds (17,18). The Watson-Crick base pairs remain intact (19), any helical unwinding is less than 3° per bound drug molecule (6), and no change in overall tilt of base pairs results.

Binding of netropsin to DNA involves only the minor groove. Drug binding shields only the minor groove of DNA from subsequent methylation (20), and partial blockage of the major groove with 5-bromouracil or 5-bromocytosine has no effect on binding (7,8). One important contribution to the energy of binding is electrostatic; the positively charged ends of the drug molecule are necessary for binding but are not the source of base specificity (6,7,21,22). Hydrogen bonding also plays a significant role. Nuclear Overhauser studies of netropsin bound to the same C−G−C−G−A−A−T−T−C−G−C−G sequence used in this study suggest that three hydrogen bonds per drug molecule are formed between N−H of netropsin amides and the N3 and/or O2 of adenines and thymines within the minor groove of the DNA (19).

Interactions between netropsin and DNA are extraordinarily specific. The drug attaches only to A·T or I·C base pairs, never to G·C (5,6,7,23,24). Since G·C and I·C base pairs differ only in the presence or absence of a −NH$_2$ group at the C2 position of the purine ring within the minor groove, it has been proposed that the inability of netropsin to bind to G·C base pairs must arise from steric hindrance by this −NH$_2$ group (7). Alternation of A·T and T·A base pairs seems to be less favorable for netropsin binding than are uninterrupted runs of poly(dA)·poly(dT) (16,25-27). Clusters of three or four successive A·T base pairs are most conducive to binding (7,27). Kinetic studies indicate a minimal binding site consisting of 5 or 6 base pairs (6,7,28), and methidiumpropyl-EDTA·Fe(II) footprinting reveals that a single netropsin molecule protects at least 4 base pairs from cleavage (24). Raman spectroscopy (29) and nuclear magnetic resonance (19,30) furnish important details about the orientation of the netropsin molecule within the minor groove: the three amide N−H are pointed toward the bottom of the groove where they are potentially involvable in hydrogen bonds with DNA. The pyrrole H3 atoms (for numbering see Figure 1) lie roughly 2.5A away from adenine H2 atoms, and the pyrrole methyl groups and H5 atoms extend out of the groove away from the base pairs.

The x-ray crystal structure analysis of netropsin alone (31) indicates that the free drug molecule adopts a crescent shape as in Figure 1, with the three amide N−H groups on the concave edge of the crescent involved in hydrogen bonds to water molecules. The molecule as a whole is only approximately flat; the two pyrrole rings actually are inclined to one another by a 20° rotation about the long axis of the drug molecule. Berman *et al.* (31) have proposed that netropsin would bind asymmetrically to poly(dA)·poly(dT) along one side of the minor groove, making hydrogen bonds to O2 atoms of three successive thymines along one strand of the double helix, but not to the N3 atoms of adenines along the other strand. Against this background of information about netropsin and its interaction with DNA, the x-ray analysis of the complex was begun.

Materials and Methods

For the crystal structure analysis of the netropsin-DNA complex, crystals were grown at 4°C from a solution initially:

 0.30 mM in double-helical DNA dodecamer
 0.60 mM in netropsin, or a 2:1 drug-to-double helix ratio.
  (Similar diffraction pattern intensity changes were observed with 1:1 and 1:2 ratios).
 0.30 mM in spermine hydrochloride
 6.30 mM in magnesium acetate
 10% MPD (2-methyl-2,4-pentanediol) by volume

Crystals were grown by vapor diffusion against reservoir solutions of higher MPD concentration, and first appeared around 32.5% MPD.

In crystal habit, space group ($P2_12_12_1$) and unit cell dimensions (24.27 A x 39.62 A x 63.57 A), the netropsin-DNA crystals closely resembled both the native dodecamer (32) and its variant with 5-bromocytosine at the ninth position along each strand, designated as MPD7 (33). Crystal intensity data were collected on an automatic diffractometer to a resolution of 2.21 A. Attempts were made to extend this resolution to 1.9 A, but at that limit less than 10% of the possible reflections had intensities greater than two sigma. The crystals were maintained at a temperature between −1°C and +5°C in a cold stream during data collection, to avoid the rapid deterioration in x-ray pattern that accompanies warming.

After standard instrumental corrections and scaling, the agreement index between two data sets, $R' = \Sigma(F_1 - F_2)^2/\Sigma F_1^2$ was 52% between the netropsin-DNA complex and the native DNA structure, in which the helix has a 19° axial bend, but was only 15% between netropsin-DNA and the MPD7 structure, in which the helix axis is unbent. Hence it was concluded that the netropsin molecule was bound to a straight helix very much like the MPD7 structure (33). The observed 15% change is reasonable for a situation in which one ordered molecule made up of C, N, O and H atoms replaces another highly ordered structure in the minor groove, the spine of hydration consisting of a chain of water molecules (34, 35).

The final refined DNA coordinates from the MDP7 structure were used as a starting point for restrained Jack-Levitt least squares refinement against the x-ray data from the netropsin-DNA cocrystals. The initial value of the standard crystallographic R factor, or the residual error between observed experimental structure factors, F_o, and those calculated from the assumed model, F_c,

$$R = \Sigma(F_o - F_c)/\Sigma F_o$$

was 45.8%. This beginning DNA structure from the MPD7 analysis, before any refinement against the netropsin data, will be termed the R46 coordinate set.

Nineteen cycles of restrained least squares refinement reduced the residual error to 34.8%. At this point we began cautiously adding solvent molecules where indicated by the difference electron density maps, deliberately avoiding the region of the minor groove where a continuous crescent of electron density could be seen. We were reasonably confident that this represented a netropsin molecule, but decided to make the maximum possible improvement in phase by adding solvent peaks and adjusting the DNA structure, before attempting any interpretation of the netropsin image. This "tuning up" process was brought to a close with 26 solvent molecules in place and with a residual error of R = 32.6%. Hence this DNA structure after preliminary refinement but before addition of the netropsin molecule will be termed the R33 coordinate set.

At this point the crescent of density extending down the minor groove was examined to see whether it could be interpreted sensibly as an image of an ordered netropsin molecule. Examination of the electron density map in the conventional manner, as a set of stacked plexiglas sections on a light box, proved to be difficult and inefficient. Instead, the three-dimensional difference map was displayed on an Evans and Sutherland Picture System II graphics terminal, using the BILDER programs written by Robert Diamond and modified by Robert Ladner. This approach turned out to be both easy and rapid.

The difference map image of the netropsin molecule at 2.2 Å resolution is shown in stereo in Figure 2, along with the inferred molecular skeleton that was built inside it. The central amide-pyrrole-amide-pyrrole-amide region was added first. A unique orientation for the molecule was established by protruberances fitting the C=O and pyrrole $-CH_3$ groups. The two "fishtail" ends of the electron density then were fitted easily by guanidinium and amidinium groups.

It is important to recognize that the difference electron density seen in Figure 2 contains no prior assumptions about the structure, or even the existence, of the netropsin molecule. The map is calculated by using as Fourier coefficients the quantities $(F_o - F_c)$, where the F_o come only from the x-ray data and the F_c derive only from the starting model: DNA double helix plus a few solvent molecules. Figure 2 represents the first stage of the analysis that yields direct information

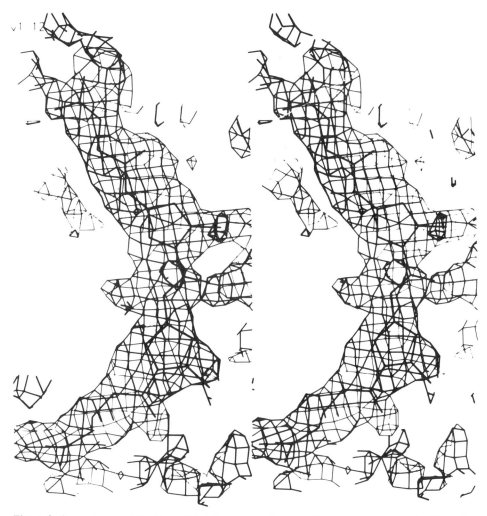

Figure 2. Stereo image of the $(F_o - F_c)$ difference electron density map, as photographed from the screen of an Evans and Sutherland graphics terminal. F_o = Netropsin-DNA cocrystal x-ray data. F_c = Refined DNA structure alone, with addition only of a few major groove water molecules. What remains in such a difference map should be those portions of the true structure that have not yet been included in the F_c model. The netropsin image is contoured in a three-dimensional "chickenwire" cage style, and the molecular skeleton of netropsin has been built inside. The orientation of the drug molècule is as in Figure 1, with the guanidinium at top. Note the fitting of the two pyrrole rings into broad, flat regions of density, the projections to the right for carbonyls and ring methyls, and the fitting of the "fishtail" guanidinium and amidinium ends at top and bottom. The floor of the minor groove of DNA is at left, on the concave side of the netropsin molecule.

about the netropsin molecule, and the last stage that is not influenced by prior assumptions about its structure.

After incorporation of the entire netropsin molecule into the structure, restrained Jack-Levitt refinement was continued, alternating with examination of difference electron density maps to correct errors of position and to search for further solvent peaks. The current stage of refinement has a residual error of R = 26.0% for all data or 23.0% for two-sigma data. In subsequent discussions this is termed the R26 structure. Both the original x-ray intensity data and a full set of atomic coordinates shortly will be deposited with the Brookhaven Protein Data Bank for general distribution.

Results and Discussion: Binding of Netropsin Within the Minor Groove

A single netropsin molecule binds within the minor groove in the central A−A−T−T region of double-helical B-DNA of sequence C−G−C−G−A−A−T−T−C−G−C−G, in the manner shown in Figure 3. The binding appears to be ordered; that is, one does not find a mixture of netropsin molecules with guanidinium ends pointing respectively up and down as viewed in Figure 3. Furthermore, the drug molecule does not attach to one side of the minor groove or the other; it is inserted squarely in center of the groove. This situation becomes obvious when one realizes that the minor groove of the parent B-DNA dodecamer is narrower in the A·T-rich center than in the G·C-rich ends (33). The groove in the free DNA structure is only 3.2 to 4.0 A wide in this region (separation of closest phosphorus atoms across the groove, less 5.8 A for the combined van der Waals radii of two phosphate groups). Hence there is barely room for the insertion of flat organic ring of van der Waals thickness 3.5 A, and no possibility that the netropsin molecule could lie asymmetrically against one side of the minor groove or the other.

The two pyrrole rings of the netropsin molecule are not coplanar, but are skewed relative to one another as was observed in the x-ray structure analysis of the free drug molecule (31). But the dihedral angle between the two pyrrole rings in the DNA complex is 33°, rather than the 20° observed in the isolated molecule. This arises because each pyrrole ring individually is slipped into the minor groove nearly parallel to the walls of the groove in its own locality. The twist of the groove of the DNA accentuates an already existing steric tendency of the netropsin molecule.

The adaptation of the netropsin molecule to the geometry of the DNA is not perfect, however. Careful analysis of the normal vectors of the two pyrrole rings relative to local helix geometry reveals that each pyrrole ring is inclined 3° away from absolute parallelism to the walls of the minor groove, in a direction that would be expected if each pyrrole ring exerted a drag on the other through the central amide linkage because of a natural stiffness of the netropsin molecule. Hence the binding of netropsin to DNA can be described as an induced fit, in the Koshland enzyme-substrate sense, but an incomplete fit, overcoming resistance from the "substrate" molecule.

The tightness of the fitting of the netropsin molecule within the minor groove can be seen from the interatomic distances between netropsin atoms and the O4' atoms of backbone deoxyribose rings, diagrammed in Figure 4. The 28 distances less than 4.0 A have a mean value of 3.61 A, almost exactly the expected van der Waals nonbonded packing distance, and a standard deviation of only 0.26 A.

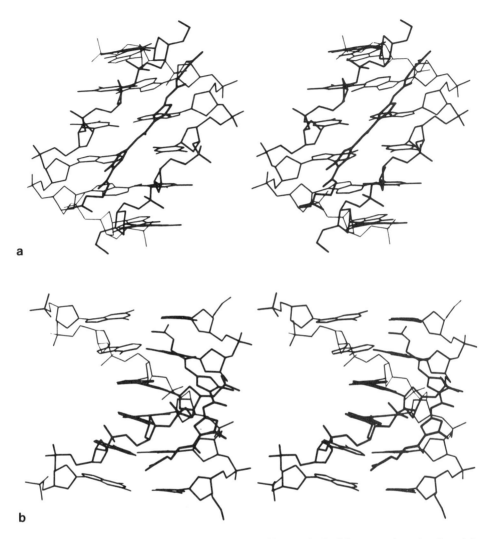

Figure 3. Stereo drawings, photographed from the graphics terminal, of the netropsin molecule and the central six base pairs of the DNA helix. (a) Symmetrical view into the center of the minor groove. Base pair G4·C21 and the guanidinium end of the drug molecule are at top, followed by A5·T20, A6·T19, T7·A18, T8·A17, and finally C9·G16 and the amidinium end. Note that netropsin sits in the center of the groove, and that its two pyrrole rings are each oriented so as to sit perpendicularly within its local region of the groove. Hence the two pyrrole rings are not themselves coplanar. (b) Side view, rotated by 90° about a vertical axis from the previous illustration. The minor groove now opens to the right. Note the pronounced base pair propeller twist, especially visible in the central two base pairs. (c) View directly down the minor groove from the top of the molecule as drawn in the two previous figures. Note the way in which the flat netropsin molecule extends down the very center of the groove. If space-filling representation had been used rather than a stick skeleton, it would be obvious that the minor groove is barely wide enough to accommodate insertion of a flat organic ring. As an incidental observation, note also how the bases along each separate strand of B-DNA are stacked in columns like a pile of coins, almost as though the other strand and its base stack were not present. The two stacks of bases then are intertwined around one another to build a helix. This stacking is responsible for the propeller twist observed in individual base pairs.

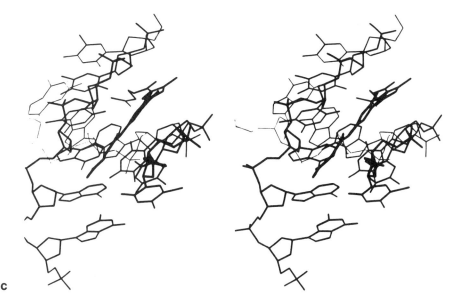

c

Figure 3 continued from previous page.

No investigators prior to this study have suggested that the minor groove of B-DNA was so narrow that it could accommodate only one tightly-packed organic ring, and this narrow groove indeed was one of the unexpected features of the original C−G−C−G−A−A−T−T−C−G−C−G structure analysis. The narrowness of the minor groove of B-DNA in A·T regions as compared wth G·C regions probably is attributable to the greater propeller twist of A·T base pairs, as schematized in Figure 9 of Reference 33. This increased propeller twist, in turn, has two probable causes: the presence of only two hydrogen bonds between bases rather than three, exerting a lesser flattening effect on the base pair, and the absence of guanine-guanine repulsions between adjacent base pairs arising from steric clash of N2 amine groups (36,37). The fundamental reason why propeller twist should occur at all in double-helical A or B-DNA is especially well illustrated in Figure 3c. The bases along each individual strand of the double helix are stacked upon one another like a stack of coins, a situation that helps stabilize that strand. The two stacks of base planes then are twisted around one another to form a double helix, and the sugar-phosphate backbones that connect them almost seem like an afterthought. Although the two bases of a pair are hydrogen-bonded, the primary interaction of a given base seems to be with its neighbors up and down the same chain, rather than with its hydrogen-bonded partner on the other strand. The twisting of stacks of bases around one another inevitably means that the bases of one pair cannot be coplanar, and the observed positive propeller twist is the consequence. Rather than asking why A·T base pairs have such a large propeller twist, one should ask why G·C pairs are so flattened, and the probable reasons are suggested above.

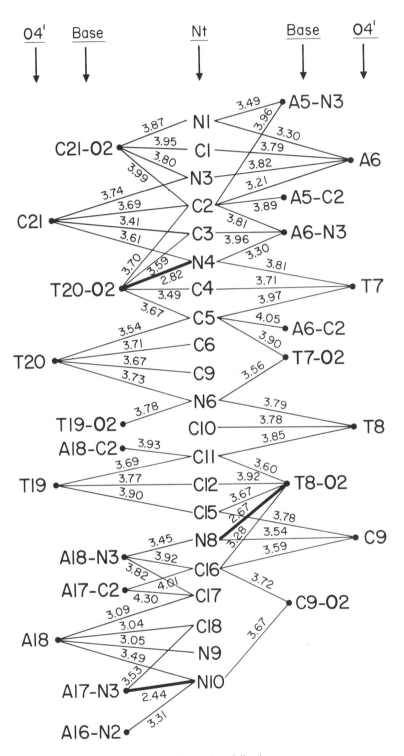

Caption for Figure 4 on following page.

Interactions of Netropsin With the DNA Molecule

The three amide N—H groups of the netropsin molecule, numbered N4, N6 and N8 in the Berman-Neidel numbering scheme of Figure 1, all point symmetrically inward toward the bottom of the minor groove. As Figure 5 indicates, N4 is positioned as though it were a bridge between the O2 of thymine 20 and the N3 of adenine 6, N6 similarly sits between the O2 atoms of thymines 19 and 7, and N8 is a bridge between the N3 of adenine 18 and the O2 of thymine 8. These are exactly the N and O atoms that are bridged by the first layer of water molecules in the spine of hydration in the parent DNA helix (34,35) an ordered water structure that is believed to make a major contribution to the stability of the B helix under conditions of high water activity. The netropsin backbone itself then plays the role of the second hydration shell of the spine, knitting the first shell into a continuous chain.

Closer examination of interatomic distances, however, reveals that not all of the interactions drawn in Figure 5 can be conventional hydrogen bonds; the distances from amide N to DNA N or O atoms is too great. From the numerical values depicted in Figure 4 it is seen that only three are short enough to be true hydrogen bonds: netropsin N4 to thymine 20, 2.82 A; netropsin N8 to thymine 8, 2.67 A; and the amidinium end of the netropsin molecule to the N3 of adenine 17, 2.44 A. The five dotted interactions in Figure 5 all are 3.3 A or longer, whereas the normal N—O or N—N spacing in a sterically unconstrained hydrogen bond is around 2.8 A. A hydrogen bond essentially is a local electrostatic interaction between a hydrogen atom bearing a small positive charge because of its attachment to a more electrophilic atom, and another electrophilic atom such as N or O bearing a partial negative charge. Hence the interactions diagrammed in Figure 5 by dotted lines, although longer than would be expected for free, unconstrained hydrogen bonds, probably have some of the character of normal hydrogen bonds, and contribute both stabilization energy and specificity in a comparable manner.

The reason why these dotted bonds in Figure 5 are so long is that tight contacts between netropsin and DNA at the bottom of the minor groove prevent closer approach. Pyrrole ring atom C5 in the first ring and C11 in the second (the carbons bearing protons "H3" in the nmr notation) are in tight van der Waals contact with adenine 6 and thymines 7 and 20 in the first instance, and adenine 18 and thymine 8 in the second. These and other close contacts prevent the central amide N—H from moving close enough to the DNA to make hydrogen bonds of normal length. But to

---

Figure 4. Short contacts between netropsin (central vertical column) and N or O atoms of DNA (strand 1 at right, strand 2 at left). The drug molecule is packed securely between the two walls of the minor groove, as reflected by distances to sugar O4′ atoms. Distances to several base atoms also are indicated: thymine O2 and adenine N3 and C2. At the present incomplete stage of refinement, these distances must be regarded as indicative but provisional. Relative trends certainly are correct, but some values such as the very short 2.44 A at the bottom almost certainly will shift. The proximity of adenine H2 protons to the H3 atoms on pyrrole ring carbons (C5 and C11 in the overall numbering) was predicted from nuclear Overhauser studies (19).

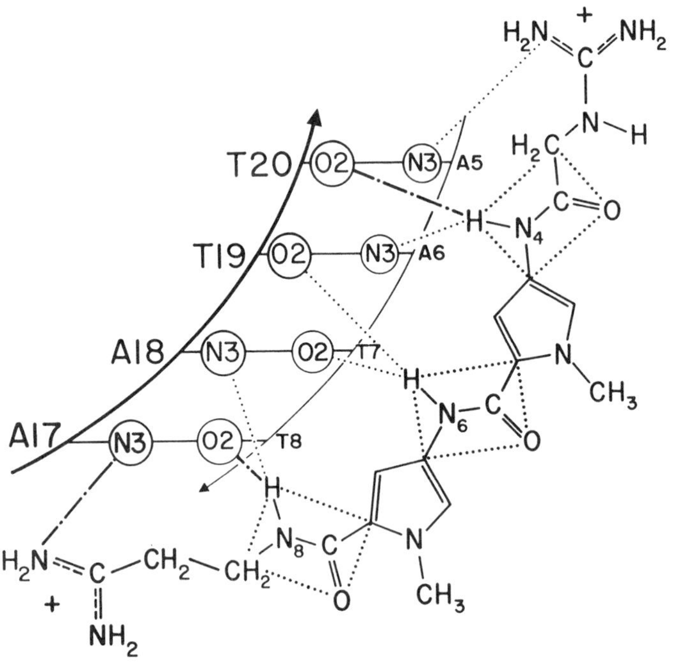

Figure 5. Schematic diagram of the interaction of netropsin with adenine N3 and thymine O2 atoms on the floor of the minor groove. Only the three dot-dashed bonds are short enough to be true hydrogen bonds, with lengths: amidinium to A7−N3: 2.44 A; NH to T8−O2: 2.67 A; and NH to T20−O2: 2.82 A. Dotted lines represent looser associations of 3.3 A or longer. The central amide NH is prevented from approaching close enough for a normal hydrogen bond by van der Waals contacts between the DNA and hydrogen atoms at the C3 positions of the flanking pyrrole rings.

either side, the outer two netropsin amides are flanked only on one side by a pyrrole ring, and on the other by a flexible chain. These outer N−H each can move close enough to form one hydrogen bond of normal length, although not a linear one. The close interaction in each case is to a thymine O2, but the stabilizing influence of the interaction with the adjacent adenine N3 should not be overlooked. Of the two positively charged ends of the netropsin molecule, the amidinium moves close enough to form a hydrogen bond to the N3 of adenine 17, but the guanidinium dips less deeply into the minor groove. This difference in behavior may be attributable to the different geometry of the chains from the outermost amide N−H to the end of the molecule: a flexible hydrocarbon chain in the case of the amidinium, but a more restricted −CO−CH$_2$NH− for the guanidinium end.

It is interesting that the narrowness of the minor groove in regions of A·T base pairs induces the netropsin molecule to behave in a manner predicted by Pullman from considerations of electrostatic potential around the DNA double helix (38-40). The two cationic ends of the drug molecule are not associated with one or another of the phosphate groups along the rim of the groove; instead they occupy positions of potential minima at the bottom center of the groove. As Zakrzewska *et al.* (40)

comment, neither the cationic ends nor the intermediate hydrogen bonds are absolutely required for preferential binding of the drug to A·T regions of the minor groove. The drug SN 18071, with charged ends but lacking hydrogen bonding possibilities in the middle, binds in a manner comparable to that of netropsin. And deletion of the cationic ends in a netropsin analogue still leads to a preference, although weaker, for binding to A·T (8).

The Origin of Base Specificity in Netropsin

The requirement of A·T base pairs for binding of netropsin, and the weakening of binding by intrusive G·C base pairs (27), now can be seen to have exactly the explanation predicted earlier (7): binding of the drug within the minor groove leaves no room for the N2 amine group of a guanine. The close contacts between netropsin backbone atoms and C2 atoms of the four adenines are listed in Table I; they average 3.97 A in length. If one adds a hypothetical $-NH_2$ to each, then the mean distance from N to the same netropsin atoms, also listed in Table I, is only 3.05 A, less than van der Waals contact distance. These adenine C2 to netropsin distances also confirm the findings from nuclear magnetic resonance studies that the protons on adenine C2 and the nearest netropsin backbone atoms should be around 2.5 A apart (19). The requirement of at least four successive A·T base pairs for tight binding arises from the length of the netropsin molecule, which can cover four base pairs and extend out toward the neighboring base pairs at either end (Figure 3). As Patel has noted, binding of netropsin to the A−A−T−T center of C−G−C−G−A−A−T−T−C−G−C−G also perturbs the G·C base pairs to either side (41). This may be, as had been suggested, a consequence of cooperative drug-induced conformation changes in the double helix, but it could also be only a case of simple overlap of the ends of the netropsin molecule into the G·C regions.

Table I

Close Approach of Netropsin to Adenine Rings

Base	Netropsin Atom	Distance to adenine atoms: C2	"Amine N"
A5	C2	3.89A	3.10A
A6	C5	4.05	3.13
A18	C11	3.93	2.98
A17	C16	4.01	3.00
Mean (Std.dev.)		3.97(0.07)	3.05(0.07)

"Amine N" is the nitrogen of a hypothetical amine group attached to the adenine C2 as in guanine.

This x-ray structure analysis suggests no obvious reason why the binding of netropsin to poly(dA-dT)·poly(dA-dT) should be disfavored relative to binding to poly(dA)·poly(dT), as has been reported from circular dichroism studies (16,25-27). From Figure 5 one might expect that any sequence of the form A−x−y−T, where x and y are either A·T or T·A pairs, would bind netropsin equally well. The most likely

explanation is that the alternating poly(dA-dT) copolymer adopts a helical conformation slightly different from that of standard B-DNA, as has been suggested on
several occasions (42-44), and that netropsin binds less well to this form and must
induce a transition to the standard B structure upon binding.

Effect of Netropsin Binding on B Helix Geometry

A detailed tabulation of local helix parameters is given in the Appendix for the
refined R26 netropsin-DNA structure. Mean values of helix parameters are compared in Table II at three stages of refinement: the starting DNA coordinates from
the MPD7 structure (R46), the DNA structure after preliminary refinement but
before addition of the netropsin molecule (R33), and the current stage of refinement of DNA plus netropsin (R26). Contrary to what had been predicted in some
earlier work, binding of netropsin to the DNA molecule neither winds nor unwinds
the helix, and does not elongate it. The helix with one netropsin molecule bound
per twelve base pairs still has 10.0 base pairs per turn, and a spacing of 3.34 A per
step along the helix axis. The mean inclination of base pairs to the helix axis
changes by less than one standard deviation based on the variation within an
uncomplexed DNA structure. The change in displacement, or distance of a base
pair from the helix axis, is comparably trivial. Changes in propeller twist and main
chain torsion angle δ, which measures sugar puckering, all are far smaller than the
variation observed within one helix itself. It is safe to conclude that the binding of
netropsin introduces no gross or overall changes in the structure of the DNA
double helix.

Table II
Mean Helical Parameters of DNA

Stage of refinement:	R46	R33	R26
t_g = Helical twist angle ($^\circ$)	36.0(4.6)	36.1(3.8)	36.0(3.6)
n = Base pairs per turn	10.0	10.0	10.0
h_g = Rise per base pair (A)	3.37(0.52)	3.34(0.49)	3.34(0.55)
Base pair inclination ($^\circ$)	1.8(5.2)	−2.7(6.1)	−2.4(5.4)
Base pair displacement (A)	−0.36(0.58)	−0.53(0.87)	−0.46(0.84)
Propeller twist ($^\circ$)	17.7(6.4)	17.4(5.3)	15.9(5.0)
Torsion angle δ ($^\circ$)			
Purines	131.2(25.0)	124.2(18.9)	124.2(17.6)
Pyrimidines	108.3(25.7)	107.8(20.0)	105.8(20.7)

Values given are mean values, with standard deviations in parentheses.
Inclination = Angle between vector from C6 of pyrimidine to C8 of purine, and a plane normal to the
  helix axis.
Displacement = Distance from helix axis to the C6-C8 vector, in projection onto a plane normal to the
  helix axis.
Propeller twist = Dihedral angle between bases of one pair, rotated around their C6-C8 vector as an axis.
Torsion angle δ (C5′-C4′-C3′-O3′) is a measure of sugar puckering.
For further definitions, see references 33, 37 and 45.

If one now compares individual values of helix parameters at the three stages of refinement, some changes are visible, but few that indicate systematic alterations of helix structure. (R26 values are in the Appendix; R46 values are listed as the MPD7 structure in the Appendix to reference 45; R33 values are available from the authors upon request.) Most of the changes observed between R46 and R26 already are visible in the partially refined R33 coordinates—that is, *the information needed to produce the observed shifts in DNA structure was present in the x-ray data alone, without the necessity of adding a model netropsin structure.*

The most striking changes produced by binding the netropsin molecule are in the width of the minor groove (Table III and Figure 6). Insertion of netropsin into the groove appears to require it to open somewhat, even to produce such tight contacts as listed in Figure 4. At the guanidinium end the groove widens by approximately one-half Angstrom, but the tighter interactions at the amidinium end require an opening of 1.5 to 2.0 A. This wedging open of the minor groove reinforces the statement that there is no way in which a netropsin molecule could lie asymmetrically against one side of the minor groove, or that two netropsin molecules could be accommodated in tandem in the same region of the groove.

Table III
Minor Groove Widths in Netropsin-DNA Complex

		P-P Distances less 5.8 A			
Strand 1	Strand 2	Run R46	Run R33	Run R26	R26-R46
P5	P24	5.63	5.11	5.02	−0.61
P6	P23	4.56	5.14	5.11	+0.55
P7	P22	3.84	4.50	4.45	+0.61
P8	P21	3.67	4.14	3.84	+0.17
P9	P20	4.02	4.48	4.43	+0.41
P10	P19	3.17	5.05	5.20	+2.03
P11	P18	5.19	6.56	6.92	+1.73
P12	P17	7.02	8.05	8.06	+1.04

Effective minor groove widths have been obtained from closest phosphorus-phosphorus separations by subtracting 5.8 A to represent two phosphate group radii.

Other standard local helix parameters show less perturbation from binding of the netropsin molecule, as evidenced by Figures 7 and 8. The same base sequence-dependent variation in helical twist angle is observed in the netropsin-DNA complex as in the DNA alone (Figure 7a), and the changes induced by netropsin are considerably smaller than the variation within the parent DNA helix. The spacing from one base pair to the next along the helix axis shows little alteration upon netropsin binding (Figure 7b). The roll angle between base pairs becomes generally more positive in the netrospin-binding center of the molecule (Figure 7c). This indicates an opening of angles between base pairs toward the minor groove, and

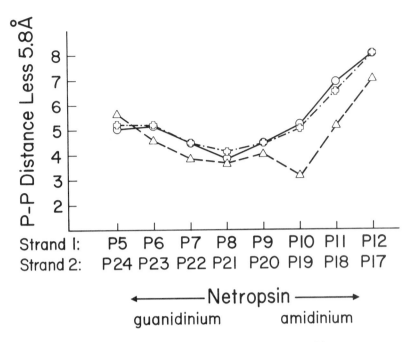

Figure 6. Minor groove widths down the twelve base pair double helix. Distances are measured between closest phosphorus atoms across the groove, and have been diminished by 5.8 A to allow for the van der Waals radii of the phosphate groups. Dashed line: initial MPD7 helix, or coordinate set R46. Dotted line: intermediate set R33, just prior to addition of the netropsin molecule to the analysis. Solid line: refined set R26 containing DNA, netropsin, and solvent molecules. Note that the minor groove has been opened up by one to two Angstroms on the end where the amidinium binds. Note also that this opening up occurred during refinement of the DNA against the netropsin-DNA x-ray data, *before* inclusion of the netropsin molecule in the refinement process. Hence the shifts must be a consequence of the x-ray data, and not of the particular model used for refinement.

recurrence of this behavior at several successive steps would require bending of the overall helix axis.

Binding of netropsin has little effect on propeller twist of base pairs (Figure 8a), but does shift them relative to the helix axis (Figure 8b). They move ca. 0.4 A toward the netropsin molecule at its center, and a slightly greater distance away from it at its ends.

As with the parent DNA dodecamer helix, main chain and glycosyl torsion angles are relatively insensitive measures of helix deformation. A χ/δ correlation plot for the netropsin-DNA complex shows the same overall behavior as with the DNA alone: linear correlation of χ and δ, a preference on the part of purines for higher angles, and anticorrelation between the purine and pyrimidine of each base pair.

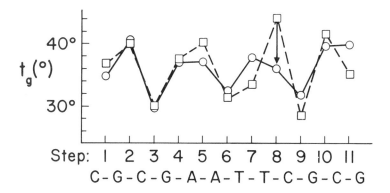

a

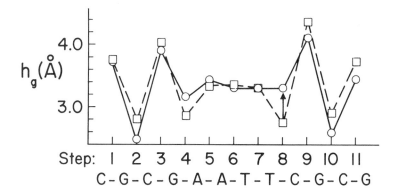

b

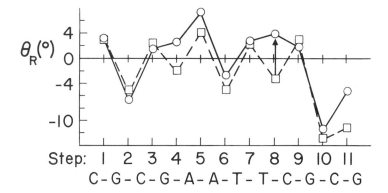

c

Figure 7. Behavior of various DNA helix parameters in the presence and absence of netropsin. Base sequence is indicted beneath each plot. Step *n* is from base pair *n* to pair *n* + 1. Dashed lines: DNA alone (MPD7 structure or coordinate set R46). Solid lines: DNA-netropsin complex (coordinate set R26). (a) Global helix twist angle, t_g. (b) Rise per base pair step, h_g. (c) Base pair roll angle, Θ_R. Arrows indicate pronounced changes at step 8.

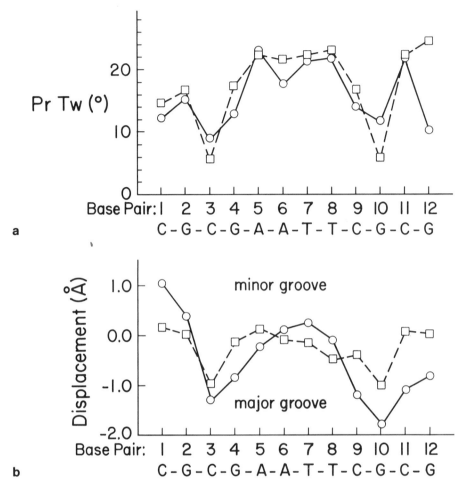

a

b

Figure 8. Behavior of base pair propeller twist and displacement in the presence and absence of netropsin. Dashed lines: DNA alone. Solid lines: DNA-netropsin complex. (a) Propeller twist. (b) Displacement, or distance of the purine C8 to pyrimidine C6 long axis of the base pair from the helix axis. Positive displacement means a shift of the base pair in the direction of the minor groove.

Physiological Effects of Netropsin Binding

The netropsin molecule appears to be admirably structured to bind tightly within the minor groove of the B form of the DNA double helix, as long as that groove is not blocked by N2 amine groups on the purines. It displaces the ordered spine of water molecules and mimics it by its interactions with adenine N3 and thymine O2 atoms on base edges. Breslauer and coworkers (46) have measured the thermodynamic parameters of this displacement reaction:

$$\text{DNA·}n\text{H}_2\text{O} + \text{Netropsin} \Rightarrow \text{DNA·Netropsin} + n\,\text{H}_2\text{O}$$

and find the following values at 25°C:

$$\Delta S^\circ \; = \; +10.3 \; \text{cal/deg mol}$$

$$\Delta H^\circ \; = \; -9.2 \; \text{kcal/mol}$$

$$\Delta G^\circ \; = \; -12.3 \; \text{kcal/mol}$$

The enthalpy drop is reasonable for the replacement of hydrogen bonds to water molecules by a combination of electrostatic and hydrogen bonds to a netropsin molecule. The rise in entropy upon binding also is easily accounted for in terms of the increased disorder resulting from the displacement of water molecules. Netropsin in a sense is like a spine of hydration, but one whose components are permanently held together by covalent bonds. The tightness of its binding to double-helical B-DNA means that it cannot be removed in order to permit the strands to come apart, either for replication of DNA, or for transcription to messenger RNA. The genetic record is frozen in place, and the cell perishes.

Acknowledgements

This work was performed with the support of NIH grants GM-30543 and GM-31299, and NSF grant PCM82-02775. One of the authors (DG) is the holder of a predoctoral traineeship under USPHS National Research Service Award GM-07104. This is publication No. 2547 from the Institute of Geophysics and Planetary Physics.

References and Footnotes

1. Finlay, A. C., Hochstein, F. A., Sobin, B. A. and Murphy, F. X. (1951). *JACS 73:* 341-343.
2. Thrum, H. (1959). *Naturwissenschaften 46:* 87.
3. Schabel, F. M. Jr., Laster, W. R., Jr., Brockman, R. W. and Skipper, H. E. (1953). *Proc. Soc. Exp. Biol. Med. 83:* 1-3.
4. Hahn, F. E. (1975). In *"Antibiotics"* (J. W. Corcoran and F. E. Hahn, Eds.), Vol. III, pp. 79-100, Springer Verlag, New York.
5. Zimmer, C., Reinert, K. E., Luck, G., Wähnert, U., Löber, G. and Thrum, H. (1971). *J. Mol. Biol. 58:* 329-348.
6. Luck, G., Triebel, H., Waring, M. and Zimmer, C. (1974). *Nucl. Acids Res. 1:* 503-530.
7. Wartell, R. M., Larson, J. E. and Wells, R. E. (1974). *J. Biol. Chem. 249:* 6719-6731.
8. Zimmer, C. (1975). *Prog. Nucl. Acid Res. Mol. Biol. 15:* 285-318.
9. Lafer, E. M., Möller, A., Nordheim, A., Stollar, B. D. and Rich, A. (1981). *Proc. Natl. Acad. Sci. USA 78:* 3546-3550.
10. Ivanov, V. I., Minchenkova, L. E., Minyat, E. E., Frank-Kamenetskii, M. D., and Schyolkina, A. K. (1974). *J. Mol. Biol. 87:* 817-833.
11. Minchenkova, L. E. and Zimmer, C. (1980). *Biopolymers 19:* 823-831.
12. Zimmer, C., Marck, C. and Guschelbauer, W. (1983). *FEBS Lett. 154:* 156-160.
13. Zasedatelev, A. S., Gursky, G. V., Zimmer, C. and Thrum, H. (1974). *Mol. Biol. Reports 1:* 337-342.
14. Luck, G. and Zimmer, C. (1973). *Studia Biophysica 40:* 9-12.
15. Zimmer, C., Puschendorf, B., Grunicke, H., Chandra, P. and Venner, H. (1971). *Eur. J. Biochem. 21:* 269-278.
16. Wähnert, U., Zimmer, C., Luck, G. and Pitra, C. (1975). *Nucl. Acids Res. 2:* 391-404.
17. Sutherland, J. C., Duval, J. F. and Griffin, K. P. (1978). *Biochemistry 17:* 5088-5091.

18. Zimmer, C., Marck, C., Schneider, C., Thiele, D., Luck, G. and Guschlbauer, W. (1980). *Biochim. Biophys. Acta 607:* 232-246.
19. Patel, D. J. (1982). *Proc. Natl. Acad. Sci. USA 79:* 6424-6428.
20. Kolchinskii, A. M., Mirzabekov, A. D., Zasedatelev, A. S., Gurskii, G. V., Grokhovskii, S. L., Zhuze, A. L., and Gottikh, B. P. (1975). *Mol. Biol. (USSR) 9:* 14-20.
21. Zimmer, C., Luck, G. and Thrum, H. (1970). *Studia Biophysica 24/25:* 311-317.
22. Zimmer, C., Luck, G., Thrum, H. and Pitra, C. (1972). *Eur. J. Biochem. 26:* 81-89.
23. Reinert, K. E. and Thrum, H. (1970). *Studia Biophysica 24/25:* 319-325.
24. Van Dyke, M. W., Hertzberg, R. P. and Dervan, P. B. (1982). *Proc. Natl. Acad. Sci. USA 79:* 5470-5474.
25. Zimmer, C., Luck, G. and Fric, I. (1976). *Nucl. Acids Res. 3:* 1521-1532.
26. Gursky, G. V., Tumanyan, V. G., Zasedatelev, A. S., Zhuze, A. L., Grokhovsky, S. L. and Gottikh, B. P. (1977). In *"Nucleic Acid-Protein Recognition"* (H. J. Vogel, Ed.), Academic Press, New York, pp. 189-217.
27. Zimmer, C., Marck, C., Schneider, C. and Guschlbauer, W. (1979). *Nucl. Acids Res. 6:* 2831-2837.
28. Patel, D. J. and Canuel, L. L. (1977). *Proc. Natl. Acad. Sci. USA 74:* 5207-5211.
29. Martin, J. C., Wartell, R. M. and O'Shea, D. C. (1978). *Proc. Natl. Acad. Sci. USA 75:* 5483-5487.
30. Patel, D. J. (1979). *Eur. J. Biochem. 99:* 369-378.
31. Berman, H. M., Neidle, S., Zimmer, C. and Thrum, H. (1979). *Biochim. Biophys. Acta 561:* 124-131.
32. Drew, H. R., Wing, R. M., Takano, T., Broka, C., Tanaka, S., Itakura, K. and Dickerson, R. E. (1981). *Proc. Natl. Acad. Sci. 78:* 2179-2183.
33. Fratini, A. V., Kopka, M. L., Drew, H. R. and Dickerson, R. E. (1982). *J. Biol. Chem. 257:* 14686-14707.
34. Drew, H. R. and Dickerson, R. E. (1981). *J. Mol. Biol. 151:* 535-556.
35. Kopka, M. L., Fratini, A. V., Drew, H. R. and Dickerson, R. E. (1983). *J. Mol. Biol. 163:* 129-146.
36. Calladine, C. R. (1982). *J. Mol. Biol. 161:* 343-352.
37. Dickerson, R. E. (1983). *J. Mol. Biol. 166:* 419-441.
38. Pullman, B. and Pullman, A. (1981). *Studia Biophysica 86:* 95-102.
39. Lavery, R. and Pullman, B. (1981). *Int. J. Quant. Chem. 20:* 259-272.
40. Zakrzewska, K., Lavery, R. and Pullman, B. (1983). *Nucl. Acids Res. 11:* 8825-8839.
41. Patel, D. J., Kozlowski, S. A., Rice, J. A., Broka, C. and Itakura, K. (1981). *Proc. Nat. Acad. Sci. USA 78:* 7281-7284.
42. Scheffler, I. E., Elson, E. L., and Baldwin, R. L. (1968). *J. Mol. Biol. 36:* 291-304.
43. Klug, A., Jack, A., Viswamitra, M. A., Kennard, O., Shakked, Z. and Steitz, T. A. (1979). *J. Mol. Biol. 131:* 669-680.
44. Patel, D. J., Kozlowski, S. A., Suggs, J. W. and Cox, S. D. (1981). *Proc. Natl. Acad. Sci. 78:* 4063-4067.
45. McPherson, A. and Jurnak, F., eds. (1984). "Structural Biology," Wylie, New York.
46. Marky, L. A., Blumenfeld, K. S. and Breslauer, K. J. (1983). *Nucl. Acids Res. 11:* 2857-2870.

APPENDIX. Helix Parameters for the Netropsin-DNA Complex

Tables A1 through A3 list important helical parameters for the complex of netropsin with C−G−C−G−A−A−T−T−^{Br}C−G−C−G for refined coordinate set R26, or the stage with 26.0% residual error. These were calculated using the HELIX program written by John Rosenberg, and the ROLL, CYLIN and TORAN programs by R. E. Dickerson. In the Appendix to reference 45 may be found identical tables for four single-crystalline A-helical oligomers, five B-DNA structures, and two of Z-DNA. Brief definitions of selected helix parameters are given below: for more details and analytical expressions, see references 33, 37 and 45.

In Table A1, roll is the orientation of a base pair around its long axis, defined as the line connecting atom C6 of a pyrimidine to C8 of a purine. Tilt is the orientation

Table A1

A1. ROLL AND TILT PARAMETERS FOR NETROPSIN-DNA COMPLEX

STRAND 1 ROLL AND TILT ANGLES					STRAND 2 ROLL AND TILT ANGLES			
PHI/R	PHI/T	THET/R	THET/T		PHI/R	PHI/T	THET/R	THET/T
-1.28	2.24	2.39	6.16		10.31	3.80	5.75	-4.52
-1.86	7.74	-6.23	2.31		13.28	6.32	-11.68	2.38
-13.29	4.39	1.38	13.59		-6.13	11.59	5.67	-4.07
-16.36	10.90	-0.63	6.11		-4.32	4.88	0.86	-11.39
-22.52	4.40	10.82	1.84		-1.98	-9.24	3.86	2.10
-10.62	-4.50	-0.48	2.16		6.92	-5.48	3.84	-2.04
-7.59	-7.63	0.32	3.05		13.17	-1.58	5.38	-2.69
-2.09	-7.68	1.42	2.51		17.55	5.95	0.04	-3.76
3.57	-4.60	3.36	11.41		11.48	11.52	3.64	-19.41
5.38	9.72	-19.96	-4.64		13.06	-1.54	-10.39	-1.95
-19.73	-0.05	9.59	10.44		1.59	2.20	-16.14	6.90
-9.90	1.16	0.00	0.00		-17.69	3.76	0.00	0.00

BEST PLANE THROUGH BOTH BASES

PHI/R	PHI/T	THET/R	THET/T	INCLIN	PR TW	BUCKLE	SLIDE	DISP	SLIP	C6/C8
4.81	5.34	3.46	-1.01	-5.48	12.13	-4.83	-0.43	1.04	0.62	9.67
4.58	7.19	-6.71	3.53	-7.56	15.18	-0.87	-0.85	0.38	1.07	9.68
-8.95	9.29	1.50	1.98	-9.28	8.88	-2.27	-0.57	-1.28	1.10	9.71
-11.46	6.18	2.27	-0.49	-6.65	13.02	0.15	-0.83	0.60	9.89	
-10.54	-1.75	7.30	-1.10	1.56	22.69	-1.84	0.18	-0.22	-0.07	9.87
-0.16	-6.56	-2.81	-0.35	6.30	17.64	2.47	0.39	0.12	-0.07	9.74
0.86	-6.81	2.64	3.02	5.89	21.26	2.37	-0.09	0.25	-0.02	9.92
6.31	-1.14	3.97	0.39	0.85	21.90	-0.07	-0.61	-0.10	0.32	10.00
9.50	4.27	1.87	-2.20	-4.94	14.05	-2.84	-0.97	-1.21	0.40	9.71
8.43	6.93	-11.61	-2.59	-6.90	11.69	-5.41	-0.96	-1.81	0.13	9.84
-8.38	4.49	-5.37	0.38	-4.76	21.98	-7.08	-0.59	-1.08	-0.32	9.64
-14.68	-3.01	0.00	0.00	2.06	10.22	8.91	0.00	-0.81	-0.44	9.77

NOTE: ANGLES ARE CALCULATED FROM TOP TO BOTTOM OF STRAND 1, AND SIGNS OF
ANGLES ALSO ARE CALCULATED WITH RESPECT TO STRAND 1. TO EXAMINE INDIVIDUAL
STRAND 2 BASES WITH RESPECT TO THEIR OWN STRAND, REVERSE THE SIGNS OF
PHI/R AND THETA/T FROM THOSE GIVEN HERE

about the short axis of the base pair, at right angles to the C6-C8 line. Angles Φ_R and Φ_T describe the roll and tilt of individual base pairs, whereas Θ_R and Θ_T measure changes in roll and tilt from one base pair to the next. Θ_R is positive if the angle between base pairs opens toward the minor groove, and negative if the angle opens toward the major groove. Other sign and zero conventions are as described in 33 and 45. Inclination is the angle that the C6-C8 line of a base pair makes with a plane normal to the helix axis, and this is nearly identical to $-\Phi_T$. Propeller twist is the dihedral angle between bases in a pair, viewed along the long C6—C8 axis, with positive angles for clockwise rotation of the nearer base of the pair. Buckle is the dihedral angle between bases, measured around the minor axis of the pair, after propeller twist has been rotated back to zero. Displacement is the distance from the helix axis to the C6—C8 line, viewed in projection down the helix axis. For definitions of slide and slip, see reference 45.

In the upper half of Table A2 (R, ϕ, Z) are the cylindrical coordinates of phosphorus atoms. Distance d is the P—P separation along one chain in A, q is the projec-

Table A2

A2.   CYLINDRICAL PARAMETERS FOR NETROPSIN-DNA COMPLEX

PHOSPHATE BACKBONE TABLE, 5" TO 3" DIRECTION IN EACH STRAND

R	PHI	Z	D	Q	H	PI
STRAND	1					
10.63	-25.22	30.14	6.82	5.19	-4.41	-40.35
10.73	-53.37	25.72	6.47	5.03	-4.07	-38.96
9.57	-81.30	21.66	7.01	6.70	-2.04	-16.93
9.05	-123.39	19.62	6.89	6.76	-1.37	-11.49
9.04	-167.25	18.24	6.71	5.99	-3.01	-26.70
9.64	155.51	15.23	6.68	5.79	-3.34	-29.95
9.62	120.50	11.89	6.73	4.92	-4.60	-43.06
9.95	91.45	7.30	6.45	4.81	-4.30	-41.78
9.43	62.84	3.00	6.57	4.93	-4.35	-41.40
8.80	31.69	-1.35	6.61	5.85	-3.07	-27.66
8.52	-7.77	-4.42	0.00	0.00	0.00	0.00
STRAND	2					
9.65	-175.70	-2.68	6.91	6.07	3.30	28.54
9.40	-138.60	0.62	6.39	5.19	3.73	35.69
8.67	-105.56	4.35	7.02	6.45	2.77	23.25
8.49	-61.42	7.12	7.00	6.76	1.80	14.91
9.32	-17.10	8.92	6.78	6.20	2.74	23.87
9.61	21.11	11.66	6.53	5.39	3.68	34.29
9.60	53.74	15.34	7.04	5.38	4.55	40.23
9.01	87.14	19.89	6.31	4.99	3.87	37.81
8.66	119.84	23.76	6.69	5.33	4.04	37.18
7.70	157.24	27.80	6.78	6.67	1.22	10.37
9.22	-157.38	29.02	0.00	0.00	0.00	0.00

ROTATION AND RISE TABLE, 5" TO 3" DIRECTION

S5"	S3"	R(P)	T(C1")	TG	H(C1")	HG	SLIDE	DISP	SLIP	C1/C1
STRAND	1									
0.00	-7.80	0.00	-26.84	34.78	-3.868	3.695	1.08	3.51	0.48	10.60
-19.04	-9.10	-28.15	-28.28	40.59	-2.835	2.518	1.33	2.81	1.20	10.65
-19.18	-8.75	-27.93	-34.40	29.64	-3.724	3.974	0.69	1.21	1.04	10.37
-25.65	-16.44	-42.09	-44.84	37.02	-2.143	3.020	2.13	1.68	0.73	10.46
-28.40	-15.45	-43.85	-41.52	36.94	-2.945	3.480	1.98	2.21	-0.13	10.41
-26.06	-11.19	-37.25	-32.01	32.19	-3.160	3.168	2.25	2.64	0.01	9.82
-20.82	-14.19	-35.00	-33.12	37.93	-3.740	3.286	1.85	2.71	-0.10	10.41
-18.94	-10.12	-29.05	-25.77	36.03	-4.084	3.296	1.32	2.35	0.28	10.42
-15.65	-12.95	-28.60	-25.41	31.67	-4.465	4.242	0.36	1.26	0.24	10.55
-12.46	-18.69	-31.15	-47.30	39.56	-1.919	2.466	1.26	0.64	0.19	10.65
-28.61	-10.85	-39.46	-43.42	39.92	-3.713	3.561	1.01	1.26	-0.51	10.83
-32.74	0.00	0.00	0.00	0.00	0.000	0.000	0.00	1.59	-0.47	10.73
STRAND	2									
0.00	11.48	0.00	36.61	39.92	3.410	3.561	0.00	0.00	0.00	0.00
25.13	11.97	37.10	34.72	39.56	3.013	2.466	0.00	0.00	0.00	0.00
22.75	10.29	33.04	38.65	31.67	4.019	4.242	0.00	0.00	0.00	0.00
28.36	15.78	44.14	47.39	36.03	2.509	3.296	0.00	0.00	0.00	0.00
31.61	12.71	44.32	39.56	37.93	2.831	3.286	0.00	0.00	0.00	0.00
26.86	11.36	38.21	33.38	32.19	3.176	3.168	0.00	0.00	0.00	0.00
22.03	10.60	32.62	31.20	36.94	4.014	3.480	0.00	0.00	0.00	0.00
20.60	12.81	33.41	34.97	37.02	3.896	3.020	0.00	0.00	0.00	0.00
22.16	10.54	32.69	25.44	29.64	4.223	3.974	0.00	0.00	0.00	0.00
14.90	22.51	37.41	58.58	40.59	2.202	2.518	0.00	0.00	0.00	0.00
36.08	9.30	45.38	36.54	34.78	3.522	3.695	0.00	0.00	0.00	0.00
27.24	0.00	0.00	0.00	0.00	0.000	0.000	0.00	0.00	0.00	0.00

D, Q, TG AND HG ARE MAGNITUDES.   ALL OTHER DERIVED QUANTITIES HAVE
PROPER SIGNS--OPPOSITE FOR DESCENDING AND ASCENDING STRANDS.

tion of that distance on a plane normal to the helix axis, h is the rise between phosphates or the projection of d onto the helix axis, and π is the local pitch angle: $\sin \pi = h/d$.

In the lower half of Table A2 the global twist angle, t_g, is the change in orientation of C1'-C1' vectors of two adjacent base pairs, viewed in projection down the helix axis. The global rise per residue, h_g, is the mean of the rise from one C1' atom to the next along the helix axis, on the two sides of the base pair. For definitions of other parameters, see reference 33 and 45.

Table A3

A3. TORSION ANGLES FOR NETROPSIN-DNA COMPLEX

STRAND 1

ALPHA	BETA	GAMMA	DELTA	EPSILON	ZETA	CHI	DEL
0.0	0.0	116.6	121.1	-158.4	-104.1	-112.1	24.3
-70.2	171.2	61.6	119.9	-164.0	-117.0	-104.4	-14.4
-60.1	143.9	62.0	70.9	-173.9	-86.7	-138.9	-67.0
-39.0	163.1	55.2	142.7	173.3	-107.0	-101.2	36.2
-65.1	-169.0	60.2	131.0	165.0	-80.3	-101.9	7.7
-85.4	-161.9	63.2	122.2	-173.9	-109.9	-105.7	19.7
-35.7	150.5	49.5	116.1	-152.3	-110.0	-119.6	-15.3
-71.5	157.5	69.2	92.0	-176.6	-84.7	-133.6	-44.5
-56.0	169.2	56.1	108.6	-157.7	-125.3	-124.0	-14.7
-28.2	144.2	40.4	138.4	-126.5	-179.0	-86.0	53.3
-59.4	123.1	53.3	78.3	-164.3	-70.3	-144.0	-46.5
-77.9	-177.7	62.1	83.7	0.0	0.0	-132.9	-47.1

STRAND 2

ALPHA	BETA	GAMMA	DELTA	EPSILON	ZETA	CHI	MEAN
0.0	0.0	69.3	130.8	-96.4	162.0	-105.7	107.2
-52.1	134.0	37.3	124.8	-170.8	-106.7	-117.9	101.5
-64.8	159.0	60.6	85.2	-169.3	-93.6	-140.0	111.8
-37.3	159.4	56.1	123.2	166.2	-95.1	-99.4	115.9
-70.6	-172.1	67.9	136.5	-179.7	-105.6	-99.8	114.2
-35.0	169.4	38.1	131.5	174.7	-99.0	-101.7	123.8
-50.8	168.1	52.0	102.5	-174.3	-87.5	-121.3	112.4
-65.9	172.5	62.6	123.3	-177.9	-105.0	-124.3	127.2
-57.7	174.5	58.0	106.5	-159.3	-107.3	-118.7	124.6
-32.1	147.8	39.2	138.0	-92.8	152.2	-95.6	104.5
-72.9	142.3	45.6	134.3	-144.8	-104.3	-110.1	127.1
-59.7	158.2	40.2	96.9	0.0	0.0	-108.5	109.0

In Table A3, torsion angles are named by the current IUB/IUPAC convention:

$$P \xrightarrow{\alpha} O5' \xrightarrow{\beta} C5' \xrightarrow{\gamma} C4' \xrightarrow{\delta} C3 \xrightarrow{\epsilon} O3' \xrightarrow{\zeta} P$$

Glycosyl torsion angles χ are defined by:

$$O4' \text{———} C1' \xrightarrow{\chi} N1 \text{———} C2 \qquad \text{for pyrimidines}$$

$$O4 \text{———} C1' \xrightarrow{\chi} N9 \text{———} C4 \qquad \text{for purines}$$

In every case, positive torsion angles result from clockwise rotation of the more distant bond.

Structure & Motion: Membranes, Nucleic Acids & Proteins,
Eds., E. Clementi, G. Corongiu, M. H. Sarma & R. H. Sarma,
ISBN 0-940030-12-8, Adenine Press, Copyright Adenine Press, 1985.

Sequential Resonance Assignments in the ¹H NMR Spectrum of a lac Operator Fragment by Two-Dimensional NOE Spectroscopy at 500 MHz

R. M. Scheek, R. Boelens, N. Russo and R. Kaptein

Department of Physical Chemistry, University of Groningen
Groningen, The Netherlands

Abstract

Sequential assignments were obtained for the non-exchangeable proton resonances in the 500 MHz NMR spectrum of a lac operator fragment, consisting of the synthetic strands d(GGAATTGTGAGCGG) and d(CCGCTCACAATTCC). Two-dimensional NMR spectra are presented that contain all information necessary to assign the adenine H8 and H2, guanine H8, cytosine H6 and H5, thymine H6 and 5-methyl and the deoxyrobose H1′, H2′, H2″, H3′ and H4′ resonances in this fourteen basepairs DNA fragment. These assignments are required for the interpretation of two-dimensional NMR data on this DNA fragment and its complex with the lac repressor headpiece in terms of the molecular structure and dynamics in solution.

Introduction

Currently there is great interest in the mechanism by which proteins recognize specific base sequences in nucleic acids, because this process plays a central role in the regulation of gene expression (1). The most extensively studied gene-control system is undoubtedly that of the lac operon of Escherichia coli. Binding of the lac repressor to a short DNA segment (the lac operator) blocks the transcription of structural genes in this operon. This situation can be reversed by the binding of an inducer to the lac repressor, which abolishes the extremely high affinity of the repressor for its operator site on the DNA (2,3).

We use high-resolution ¹H NMR to study this lac repressor-lac operator interaction at a molecular level. This approach has benefited greatly from a number of important recent developments. First, the large-scale synthesis of well-defined DNA fragments, comprising complete operator sequences, is now within the reach of an increasing number of research groups. Second, the enormous problem of resonance assignments in the ¹H NMR spectra of the molecules of interest can be tackled nowadays by the application of two-dimensional NMR techniques (4-6). The application of these techniques to the lac repressor headpiece (the N-terminal DNA-binding domain of the lac repressor consisting of 51 amino acid residues) has yielded the majority of ^{1}H resonance assignments (7) and many details on second-ary and tertiary structure (8). Similarly, the two-dimensional approach has proven successful in the analysis of DNA ^{1}H NMR spectra (9-11). A sequential assignment procedure for the non-exchangeable proton resonances was developed indepen-dently in our laboratory and by the group of Reid (12-14). Since then this strategy has been applied by several other groups and has become the method of choice for assigning the NMR resonances of the non-exchangeable protons in oligonucleotides (15-20). Here we wish to present some new features of the proposed strategy and report on its application to a fragment of the lac operator consisting of the comple-mentary strands d(GGAATTGTGAGCGG) and d(CCGCTCACAATTCC). This 14 base pairs fragment comprises the stronger-binding half of the lac operator as was shown in elegant experiments by Sadler et al. (21). It was utilised previously in titration studies with the lac repressor headpiece (22). At that time only the low-field imino-proton region of the DNA spectrum was assigned by a combination of melting studies (23) and one-dimensional NOE measurements (22). This lack of resonance assignments has impeded the complete analysis of NMR data on the complex of the lac repressor headpiece and the 14 base pairs lac operator fragment (24). The present study shows that all non-exchangeable proton resona..ces of the lac operator fragment can be assigned, except the heavily-overlapping H5′ and H5″ resonances.

Methods and Materials

The complementary strands d(GGAATTGTGAGCGG) and d(CCGCTCACAA−TTCC) were synthesized as described elsewhere (25,26). Solutions were 6 mM duplex in 50 mM KPO$_4$, 0.2 M KCl, 0.02% azide, pH 6.5 (meter reading) in ^{2}H$_2$O. Spectra were recorded at 35 °C. Two-dimensional NOE spectra were recorded at 500 MHz on a Bruker WM 500 spectrometer, interfaced to an ASPECT 2000 computer. A (90°-t_1-90°-t_m-90°-Acq)$_n$ pulse scheme was employed (5). 512 free induction decays of 2048 data points each were acquired. The phase-cycling scheme proposed by States et al. (27) was used and 32 experiments were performed for each t_1 value, using a 1s relaxation delay. The carrier was positioned outside the spectrum, so that only real t_1 data had to be collected.

For quantitative measurement of NOE cross peak intensities the above pulse scheme was modified according to (28). Four experiments, with (average) mixing times of 25, 50, 75, and 100 ms were interweaved. While incrementing t_1 with 0.1 ms (corresponding to a spectral width of 5000 Hz), the mixing time was incremented with 0.01 ms. This has the effect that cross peaks arising from zero-quantum coher-ences are separated from NOE cross peaks, allowing a reliable estimation of the NOE cross peak intensities, even between J-coupled spins.

The time-domain data were processed on a Cyber 170/760 mainframe computer, using a software package that will be described elsewhere. Free induction decays and interferograms were weighed with 45°-60° shifted sine-bell functions before zero-filling, Fourier transformation and phase correction. The frequency domain consisted of 1024×1024 data points.

Results and Discussion

Figure 1 shows a two-dimensional NOE spectrum of the 14 base pairs lac operator fragment. A contour plot of part of the same spectrum is shown in figure 2. Macura and Ernst (5) have shown that in this type of two-dimensional NMR spectroscopy cross peaks arise when nuclei are close enough in space so that magnetisation can be transferred among them via cross relaxation. The rate of cross relaxation falls rapidly with increasing inter-nuclear distances: under the conditions of the experiment shown in figure 1 direct magnetisation transfer is detectable for inter-nuclear distances up to about 0.4 nm (see below). Magnetisation may be transferred over longer distances if an effective pathway exists for limited spin diffusion (29). Thus, as shown in figure 2, magnetisation labelled with the frequency of any of the deoxyribose protons during t_1 is detectable in the aromatic region of the spectrum during t_2. In a regular B-DNA conformation (30) the shortest distances between sugar and aromatic protons are those between a H2′ and the B proton (purine H8 or pyrimidine H6) of the same nucleotide (0.21 nm) and those between a H2″ and the

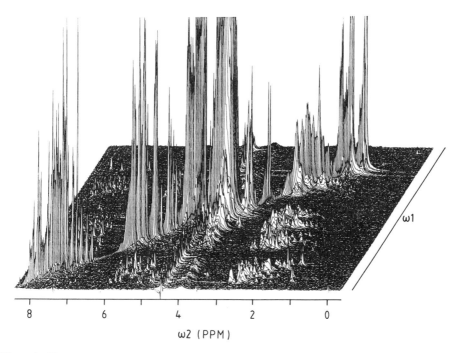

Figure 1. The two-dimensional NOE spectrum of the lac operator DNA fragment recorded at 500 MHz. A 200 ms mixing time was used. Further details were given in the Methods section.

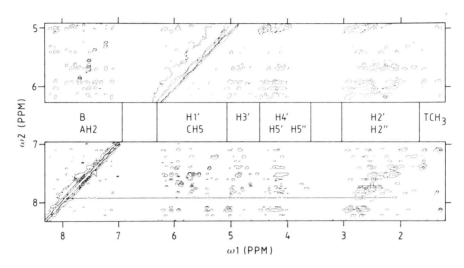

Figure 2. Contour plot of part of the spectrum shown in figure 1. The cross peaks involving aromatic and deoxyribose H1′ resonances are shown in the lower and upper parts of this figure, respectively. The drawn line connects all cross peaks involving the H8 resonance of guanine 10.

B proton of the 3′-neighbouring nucleotide (0.23 nm). It can be anticipated that a significant part of the magnetisation transfer from other sugar protons to the aromatic ring protons occurs via one of these pathways. Thus, apart from the relatively strong B−H2′(i,i) intra-nucleotide cross peaks (31) (see figure 2) we find B−H2″(i,i), B−H3′(i,i), and even B−H4′(i,i), B−H5′(i,i) and B−H5″(i,i) cross peaks. Similarly, apart from the relatively strong B−H2″(i,i−1) inter-nucleotide cross peaks, we find B−H2′(i,i−1), B−H1′(i,i−1), B−H3′(i,i−1) and some B−H4′(i,i−1) cross peaks. As will be shown below, the observation of such indirect contacts greatly facilitates the assignment of the deoxyribose proton resonances.

Sequential assignments of B, H1′, H2′ and H2″ resonances.

The systematic occurrence of both intra- and inter-nucleotide cross peaks of the type B−H2′, B−H2″ and B−H1′ (see figure 3 and 4) makes it possible to apply a sequential assignment strategy (6,12-14). Here such a strategy amounts to the tracing of sequential proton-proton contacts that define a continuous cross-relaxation network throughout a particular strand of the DNA. This is illustrated in figure 3 and 4 for both strands of the 14 base pairs DNA fragment. For each strand three non-interrupted cross-relaxation networks can be traced, starting from the 5′-terminal B resonance via either H1′ resonances (region a) or H2′ and H2″ resonances (region b) to the 3′ terminal nucleotide. The three networks within a certain strand are related, not only through the common B(i) resonance frequencies, but also through pairs of strong H1′−H2′,H2″(i,i) cross peaks in region c, which correlate the H1′ and H2′, H2″ resonances of nucleotide(i). In this manner we can assign a sequence number to each of the B, H1′, H2′ and H2″ resonances in a strand. In

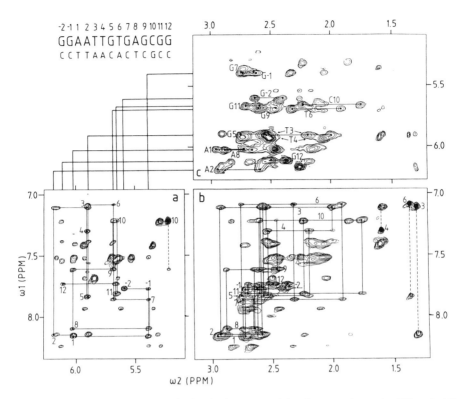

Figure 3. The cross-relaxation networks that lead to sequential assignments for purine H8, pyrimidine H6 and deoxyribose H1' (region a), H2' and H2'' (region b) resonances are indicated with drawn lines, connecting intra- and inter-nucleotide cross peaks involving the B, H1', H2' and H2'' resonances of the top strand d(GGAATTGTGAGCGG). Dashed lines connect cross peaks involving cytosine H5 and thymine 5-methyl resonances. Intra-nucleotide cross peaks are indicated with a dot. See text for the explanation of the sequential assignment strategy.

addition the cytosine H6 and thymine H6 resonances can be discriminated from the remaining B-resonances, because the CH6 resonances are characterized by strong CH6−CH5(i,i) cross peaks in region a, while the TH6 resonances show strong TH6−TCH$_3$(i,i) cross peaks in region b. This extra information is necessary to correlate in a unique manner a sequence of B resonances that follows from the spectrum with the known sequence of nucleotides in one of the strands of the duplex. As a result individual assignments are obtained at this stage for all AH8, GH8, CH6, CH5, TH6, TCH$_3$ and H1' resonances, and pairwise assignments for the H2' and H2'' resonances (see table I).

Discrimination between H2' and H2'' resonances.

We used a quantitative analysis of cross-peak intensities in region b (B−H2',H2''(i,i)) and region c (H1'−H2',H2''(i,i)) to obtain stereospecific assignments of the C2'

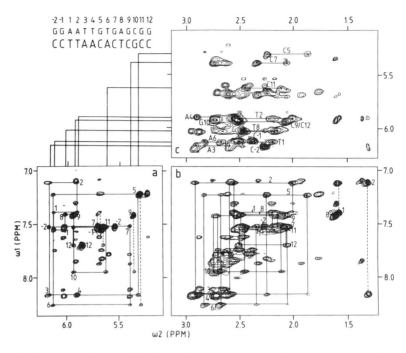

Figure 4. See legends to figure 3. This figure illustrates the sequential assignment procedure for the resonances of the bottom strand d(CCGCTCACAATTCC).

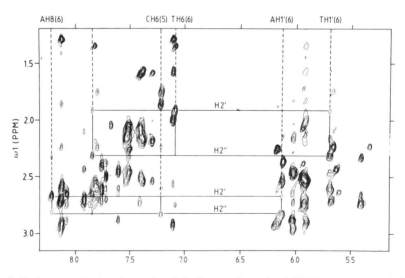

Figure 5. Regions b and c (see figures 3 and 4) of a two-dimensional NOE spectrum recorded with a short mixing time, which was incremented systematically in an accordion-like manner as described in the Methods section. The average mixing time was 25 ms. Cross peaks involving H2′ and H2″ resonances can be discriminated on the basis of their relative intensities, as is shown for the nucleotides A(6) and T(6). Further details were given in the Methods section.

Table I
Chemical shifts (in ppm relative to DSS) of proton resonances in the duplex of d(GGAATTGTGAGCGG) and d(CCGCTCACAATTCC). Numbering of the base pairs is according to the convention proposed by Ogata and Gilbert (35).

nucleotide	AH8 GH8	AH2 CH5 TCH$_3$	H1'	H2'	H2"	H3'	H4'
G-2	7.77		5.61	2.40	2.64	4.80	4.18
G-1	7.77		5.41	2.63	2.73	4.98	4.32
A1	8.15	7.32	6.03	2.72	2.96	5.09	4.48
A2	8.14	7.63	6.19	2.59	2.94	5.03	4.49
T3	7.11	1.29	5.92	2.01	2.55	4.85	4.22
T4	7.30	1.59	5.92	2.20	2.54	4.92	4.23
G5	7.83		5.92	2.58	2.75	4.97	4.39
T6	7.09	1.35	5.69	1.92	2.32	4.84	4.16
G7	7.85		5.41	2.65	2.75	4.98	4.31
A8	8.09	7.60	6.04	2.65	2.88	5.05	4.42
G9	7.61		5.70	2.46	2.61	4.95	4.34
C10	7.22	5.24	5.66	1.76	2.24	4.77	4.11
G11	7.81		5.68	2.61	2.71	4.95	4.32
G12	7.73		6.12	2.51	2.37	4.64	4.21
C-2	7.52	5.55	6.17	2.26	2.26	4.54	4.01
C-1	7.56	5.69	6.02	2.12	2.47	4.85	4.20
T1	7.38	1.56	6.13	2.18	2.55	4.90	4.25
T2	7.11	1.30	5.90	2.00	2.54	4.82	4.17
A3	8.14	7.55	6.16	2.58	2.91	5.01	4.46
A4	8.14	6.99	5.91	2.70	2.91	5.02	4.36
C5	7.22	5.31	5.33	1.87	2.24	4.76	4.09
A6	8.22	7.58	6.13	2.67	2.83	5.01	4.39
C7	7.51	5.71	5.41	2.06	2.34	4.83	4.11
T8	7.41	1.60	6.01	2.07	2.48	4.86	4.18
C9	7.41	5.33	5.93	2.05	2.52	4.71	4.25
G10	7.93		5.93	2.71	2.75	5.01	4.40
C11	7.52	5.66	5.63	2.11	2.44	4.86	4.15
C12	7.69	5.86	5.94	2.06	2.51	4.66	4.11

methylene proton resonances (12; see also (14) for an alternative approach). A series of two-dimensional NOE spectra was recorded with mixing times of 25, 50, 75 and 100 ms. Cross peaks arising from zero-quantum coherences were separated from NOE cross peaks by a systematic incrementation of the mixing time as described in the methods section (28). Figure 5 shows the regions b and c of the spectrum that was recorded with a 25 ms mixing time. This mixing time is short enough to ensure that the intensities of cross peaks are a direct reflection of their initial build-up rates. Thus the strongest cross peak of a H1'−H2',H2"(i,i) pair in region c is assigned to the H1'−H2"(i,i) contact (0.24 nm vs. 0.30 nm in regular B-DNA), while the strongest peak of a B−H2',H2"(i,i) pair in region b is assigned to the B−H2'(i,i) contact (0.21 nm vs. 0.36 nm). Similarly, the strongest peak of a B−H2',H2"(i,i−1) pair in region b is assigned to the B−H2"(i,i−1) contact (0.23 nm vs. 0.37 nm). This is shown for the nucleotides A6 and T6 in figure 5. Table I presents the complete results of this analysis.

Assignment of the deoxyribose H3' and H4' resonances.

The mixing time used for the experiment shown in figure 1 (200 ms) is long enough so that intra- and inter-nucleotide contacts of the type B−H3' and B−H4' are observed, mainly due to indirect magnetisation transfer (spin diffusion). This is shown in figure 6. Also shown are the intra-nucleotide H1'−H3' and H1'−H4' cross peaks. From the known assignments of the B and H1' resonances those of the

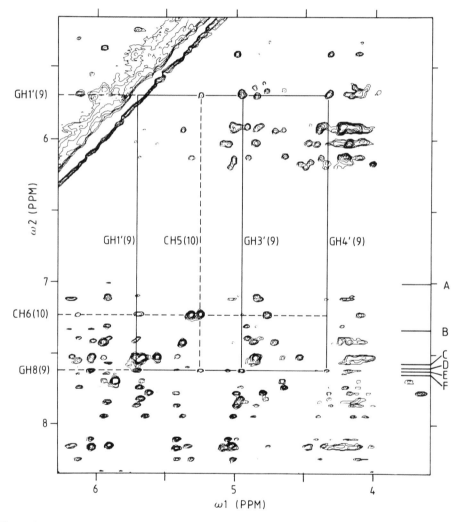

Figure 6. Intra-nucleotide cross peaks that lead to the assignments of the H3' and H4' resonances in the two-dimensional NOE spectrum of figure 1. The B−H3', B−H4', H1'−H3' and H1'−H4' cross peaks of guanine(9) are indicated with drawn lines. Dashed lines indicate the contacts between the H5 of cytosine(10) and the H8 and H1' of guanine(9). At the frequencies indicated with A to F weak contacts are visible between adenine H2 and anomeric proton resonances between 5 and 6 ppm (see Table II).

H3′ and H4′ resonances follow directly, as is shown in figure 6 for the nucleotide G9. Ambiguities may arise when B−H5′,H5″ and H1′−H5′,H5″ contacts show up next to the B−H4′ and H1′−H4′ contacts. The criterium we used to assign a cross peak to a H1′−H4′(i,i) contact is that it should also be present in a spectrum recorded with a 50 ms mixing time (not shown). This is possible because the H1′−H4′ distances are short enough (0.34 nm in regular B-DNA) to observe the corresponding cross peaks before indirect magnetisation transfer occurs to a significant extent. The resulting assignments are listed in table I.

Assignment of adenine H2 resonances

A powerful method exists for the assignment of adenine H2 resonances of double-stranded oligonucleotides in ^1H_2O solutions (32,33). It makes use of the strong nuclear Overhauser effect that can be observed between the hydrogen-bonded imino proton and the adenine H2 of the same AT base pair. Provided that the imino-proton resonance can be assigned to a specific AT base pair, the AH2 assignment follows directly. However, in ^2H_2O solutions the labile imino protons are rapidly replaced by deuterons, prohibiting this approach. Still, as will be shown below, two-dimensional NOE spectra of oligonucleotides in ^2H_2O contain all the information necessary to arrive at specific assignments for the adenine H2 resonances.

The strongest NOE cross peaks involving AH2 resonances are of the type AH2−AH2(i,i+1). Thus, in the two-dimensional NOE spectrum recorded with a 200 ms mixing time two cross peaks (see table II) define the two pairs of adjacent adenine nucleotides that occur in the 14 base pairs DNA fragment. Apart from these, no non-exchangeable protons occur in regular B-DNA within a sphere of 0.4 nm around an adenine H2. However, we do observe additional cross peaks in region a of the spectrum that are of the type AH2−H1′ (see figure 6; the six AH2 resonances frequencies are indicated with A-F). There are three anomeric protons in regular B-DNA that occur between 0.4 and 0.5 nm from a given adenine H2. These are the H1′ of the same nucleotide, that of the 3′-neighbouring nucleotide, and that of the nucleotide, base paired to the 5′-neighbouring nucleotide. Indeed, as summarised in table II, each AH2−H1′ cross peak observed could be indentified as

Table II
Adenine H2 resonance assignments; chemical shifts in ppm relative to DSS.

resonance (AH2)	cross peak (AH2−AH2)	cross peak (AH2−H1′)	cross peak (AH2−H1′)	assignment (AH2)
A 6.99	7.55 (C)	6.16 (A3)	5.91 (A4/G5)	A4
B 7.32	7.63 (F)	6.19 (A2)	6.03 (A1/C−1)	A1
C 7.55	6.99 (A)	6.15 (A3)	5.91 (T2/T4)	A3
D 7.58		6.13 (A6)	5.33 (C5)	A6
E 7.60			5.41 (C7)	A8
F 7.63	7.32 (B)	6.13 (T1)	5.92 (T3)	A2

one of these contacts. This leads to the unambiguous assignment of the H2 resonances of A1, A6, A8 and A2, and pairwise assignments of the AH2(3) and AH2(4)
resonances. It is interesting to note that the observed AH2—H1′ contacts are not
mediated by indirect magnetisation transfer, because no other cross peaks involving AH2 resonances than the ones discussed here are detectable in the spectrum.

The adenine H2 assignments presented in table I and II are consistent with those
obtained from 2D NOE experiments performed on the same DNA fragment in ^1H_2O
(34). From these experiments also the remaining ambiguity about the AH2(3) and
AH2(4) resonance assignments could be resolved.

Concluding remarks

We wish to discuss here the remaining inter-nucleotide cross peaks that we found
useful in the process of tracing the cross-relaxation networks shown in figures 3 and
4. In these figures dashed lines were used to indicate the cross peaks involving CH5
and TCH_3 resonances. Apart from the strong intra-nucleotide CH6—CH5(i,i) and
TH6—TCH_3(i,i) cross peaks all inter-nucleotide cross peaks of the type CH6—B(i,i−i)
and TCH_3—B(i,i−i) are indicated, which define alternative connections between
pyrimidines and their 5′-neighbouring nucleotides. Furthermore, figure 6 shows
one example of a CH5—H1′(i,i−i) contact, all of which could be identified in the
spectrum. Likewise, all CH5—H2′,H2″(i,i−i) cross peaks, as well as the analogous
TCH_3—H1′(i,i−i) and TCH_3—H2′,H2″(i,i−i) contacts were identified, except in
some cases, where spectral overlap hinders their observation.

Finally we mention the cross peaks of the type B—B(i,i+1) in the aromatic region
of the spectrum (see figure 2). Indeed, most of the inter-nucleotide 'steps' in the
cross-relaxation networks of figure 3 and 4 could be checked for consistency with
this region of the spectrum.

Current work in our laboratory is devoted to the quantitative analysis of two-
dimensional NOE data on oligonucleotides and proteins. This will allow us to
exploit the full potential of high-resolution NMR as a technique for the study of
macromolecular structure and dynamics in solution.

Acknowledgements

This work was supported by the Netherlands Foundation for Chemical Research
(SON) with financial aid from the Netherlands Organization for the Advancement
of Pure Research (ZWO). We thank P.A.W. van Dael, Dr. C.A.G. Haasnoot and K.
Dijkstra for their expert technical assistence and stimulating discussions.

References and Footnotes

1. F. Jacob and J. Monod, *J. Mol. Biol. 73*, 139-152 (1961).
2. S. Bourgeois and M. Pfahl, *Adv. Protein Chem. 30*, 1-99 (1976).

3. J. H. Miller and W. Reznikoff, *The Operon,* Cold Spring Harbor Laboratory, Cold Spring Harbor (1978).
4. W. P. Aue, E. Bartholdi and R. R. Ernst, *J. Chem. Phys. 64,* 2229-2246 (1976).
5. S. Macura and R. R. Ernst, *Mol. Phys. 41,* 95-117 (1980).
6. K. Wüthrich, G. Wilder, G. Wagner and W. Braun, *J. Mol. Biol. 155,* 311-319 (1982).
7. E. R. P. Zuiderweg, R. Kaptein and K. Wüthrich, *Eur. J. Biochem. 137,* 279-292 (1983).
8. E.R.P. Zuiderweg, R. Kaptein and K. Wüthrich, *Proc. Natl. Acad. Sci. USA, 80,* 5837-5841 (1983).
9. J. Feigon, J.M. Wright, W.A. Denny, W. Leupin and D.R. Kearns, *Cold Spring Harbor Symp. Quant. Biol. 47,* 207 (1982).
10. J. Feigon, J.M. Wright, W. Leupin, W.A. Denny and D.R. Kearns, *J. Am. Chem. Soc. 104,* 5540 (1982).
11. A Pardi, R. Walker, H. Rapoport, G. Wider and K. Wüthrich, *J. Am. Chem. Soc. 105,* 1652-1653 (1983).
12. R.M. Scheek, N. Russo, R. Boelens and R. Kaptein, *J. Am. Chem. Soc. 105,* 2914-2916 (1983).
13. R.M. Scheek, R. Boelens, N. Russo, J.H. van Boom and R. Kaptein, *Biochemistry 23,* 1371-1376 (1984).
14. D.R. Hare, D.E. Wemmer, S.-H. Chou, G. Drobny and B.R. Reid, *J. Mol. Biol. 171,* 319-336 (1983).
15. D. Frechet, D.M. Cheng, L.-S. Kan and P.O.P. Ts'o, *Biochemistry 22,* 5194-5200 (1983).
16. J. Feigon, W. Leupin, W.A. Denny and D.R. Kearns, *Biochemistry 22,* 5943-5951 (1983).
17. J.-R. Mellema, C.A.G. Haasnoot, G.A. van der Marel, G. Wille, C.A.A. van Boeckel, J.H. van Boom and C. Altona, *Nucleic Acids Res.* 5717-5738 (1983).
18. S.C. Brown, K. Mullis, C. Levenson and R.H. Shafer, *Biochemistry 23,* 403-408 (1984).
19. M.R. Sanderson, J.-R. Mellema, G.A. van der Marel, G. Wille, J.H. van Boom and C. Altona, *Nucleic Acids Res. 11,* 3333-3346 (1983).
20. M.A. Weiss, D.J. Patel, R.T. Sauer and M. Karplus, *Proc. Natl. Acad. Sci. USA 81,* 130-134 (1984).
21. J.R. Sadler, H. Sasmor and J.L. Betz, *Proc. Natl. Acad. Sci. USA 80,* 6785-6789 (1983).
22. R.M. Scheek, E.R.P. Zuiderweg, K.J.M. Klappe, J.H. van Boom, R. Kaptein, H. Rüterjans and K. Beyreuther, *Biochemistry 22,* 228-235 (1983).
23. E.R.P. Zuiderweg, R.M. Scheek, G. Veeneman, J.H. van Boom, R. Kaptein, H. Rüterjans and K. Beyreuther, *Nucleic Acids Res. 9,* 6553-6569 (1981).
24. R. Kaptein, R.M. Scheek, E.R.P. Zuiderweg, R. Boelens, K.J.M. Klappe, J.H. van Boom, H. Rüterjans and K. Beyreuther in *Structure and Dynamics: Nucleic Acids and Proteins,* Eds., E. Clementi and R.H. Sarma, Adenine Press, New York, p. 209-225 (1983).
25. J.H. van Boom, P.H.J. Burgers and P.H. van Deursen, *Tetrahedron Lett.* 869 (1976).
26. J.F.M. de Rooij, G. Wille-Hazeleger, P.H. van Deursen, J. Serdijn and J.H. van Boom, *Recl. Trav. Chim. Pays. Bas 98,* 537 (1979).
27. D.J. States, R.A. Haberkorn and D.J. Ruben, *J. Magn. Reson. 48,* 286-292 (1982).
28. S. Macura, K. Wüthrich and R.R. Ernst, *J. Magn. Reson. 46,* 269-282 (1982).
29. A. Kalk and H.J.C. Berendsen, *J. Magn. Reson. 24,* 343-366 (1976).
30. S. Arnott and D.W.L. Hukins, *Biochem. Biophys. Res. Commun. 47,* 1504-1509 (1972).
31. In our notation for cross peaks the numbering of nucleotides starts at the 5'-terminus. Thus a B-H2"(i,i-l) cross peak is between the B resonance of nucleotide(i) and the H2" resonance of its 5'-neighboring nucleotide(i-l).
32. A.G. Redfield, S. Roy, V. Sanches, J. Tropp and N. Figueroa, in *Biomolecular Stereodynamics I,* Ed. R.H. Sarma, Academic Press, New York, p. 195-208 (1981).
33. D.J. Patel, S.A. Kozlowski, L.A. Marky, C. Broka, J.A. Rice, K. Hakura and K.J. Breslauer, *Biochemistry 21,* 428-436 (1982).
34. R. Boelens, R.M. Scheek, K. Dijkstra and R. Kaptein, *J. Magn. Reson.* (1984) in press.
35. R.T. Ogata and W. Gilbert, *J. Mol. Biol. 132,* 709-728 (1979).

Structure & Motion: Membranes, Nucleic Acids & Proteins,
Eds., E. Clementi, G. Corongiu, M. H. Sarma & R. H. Sarma,
ISBN 0-940030-12-8, Adenine Press,   Copyright Adenine Press, 1985.

Flexibility and Base Composition Dependence of DNA Conformation In Solution From Laser Raman Scattering

Warner L. Peticolas and Gerald A. Thomas

Department of Chemistry and the Institute of Molecular Biology
University of Oregon, Eugene, Oregon 97403

Abstract

Techniques are described for taking and interpreting the Raman spectra of nucleic acids in terms of their conformation and fluctuations in their conformation. From the Raman spectra of DNA in the A, B, C, and Z forms, Raman marker bands which depend upon conformation may be assigned. The conformation of DNA is shown to be dependent on base sequence, solvent, and temperature. In particular, tetramers of deoxynucleotides containing only guanine and cytosine are shown to form B-genus double helical duplex structures but only d(CGCG) undergoes the transition to the Z form. PolydA·polydT is shown to contain an equal amount of the two ring puckers C3'-endo and C2'-endo. The conformation of polyd(dG−dC) is shown to change markedly upon interaction with cis and trans dichloro-diaminoplatinum. In particular it is shown that both cis and trans platinum have a marked effect on the B to Z transition which affects both the nature of the resultant Z form and the cooperativity of the transition.

Introduction and Review

Our only knowledge of the atomic coordinates of nucleic acids comes from an analysis of X-ray diffraction patterns obtained from nucleic acid samples in the crystalline or fibrous state. However it has been known since the earliest days of X-ray diffraction studies on fibers that DNA can exist in more than one conformation depending upon such factors as the degree of hydration, the fiber salt concentration, etc. All of the early X-ray work indicated that under conditions of high humidity, fibers of RNA-like polymers existed in the A form while fibers of DNA existed in the B form. Early work in our laboratory showed that correlations existed between the X-ray determined conformation and the Raman spectra of fibers of deoxyribonucleic acids under conditions of varying humidity (1-3). In particular it was possible to show the existence of several Raman marker bands which appeared to be indicative of the A, B, and C forms of DNA (1-3). Normal mode calculations showed that these Raman bands were due to differences in the backbone geometry which is usually expressed in terms of the furanose ring pucker being C3'-endo in the A form and C2'-endo (or the closely related C3'exo) in the B form (4-5). The convention of defining the quantity of the A and B form in terms of the intensity of the Raman marker bands has been followed by most of the workers in the field (6-8) see reference 8 for review. However these assignments have been recently revised to be more precise (9-10) and this convention will be followed here.

There are two Raman bands at least one of which is generally observed in ordered nucleic acid structures. A sharp intense band at 805-814 cm^{-1} which lies close to the strong purine vibrational band at 800 cm^{-1} is now assigned to a conformationally dependent vibration of a 5′—phosphate—furanose ester or diester with a single negative charge i.e., $3′-R-O-P(O_2-)-O-R′-5′$ where R and R′ are furanose rings. R is attached through the 3′ position while R′ is attached through the 5′ position. Experiments on ordered gels of 5′—GMP and 3′—GMP show that R may be replaced by a hydrogen but R′ must be a furanose ring (11-12). This marker band is observed when the furanose ring attached at the 5′ position is in a conformation such that the torsional angle defined by the C5′—C4′—C3′—O3′ bonds are in the range of 80-95°. See Table 1. Since this is the range usually associated with the C3′-endo furanose ring pucker, we may use the intensity of this band to estimate the fraction of the furanose rings in the C3′ endo form. This assumes that there is a unique relation between this backbone torsional angle and the sequential torsional angles around the furanose ring. This is a reasonable assumption if the maximum amplitude of these sequential torsional amplitudes is a constant from oligomer to oligomer but as the last column in Table 1 shows this is not usually the case. See reference 9 and 10 for a more detailed discussion of this point. Nevertheless we will ignore this complication and assume when all of the furanose rings possess the C3′-endo ring pucker that the intensity of this 814-cm^{-1} band has a maximum value of 1.65 (3-6). This value is determined by measurement of the ratio of the peak height of the 814 cm^{-1} band to the height of the band at 1100 cm^{-1} due to the phosphate ion symmetric stretching vibration (See Ref. 6 for details on base-line conventions.). This latter band has been found to be rather insensitive to conformational changes and thus may be used as an internal standard (3,6,13).

For deoxyribonucleotides whose torsional angle about the C4′—C3′ furanose bond lies in the range 151-156° the band at 814 cm^{-1} is absent and a broad weak band at 833-835 cm^{-1} is observed (1-3). As will be discussed below, this band appears to be quite intense in pTpT crystals where both furanose ring puckers have been shown to be C2′-endo. The intensity of this band may be used to estimate the amount of C2′-endo ring pucker in oligomers and polymers of deoxynucleotide. It has not yet been observed in the Raman spectrum of a mononucleotide. The assignment of these two marker bands is known from normal mode calculations (4,5). In this paper the intensity of these bands will be used for conformational analysis.

Table 1 summarizes our findings concerning the relation between the marker band frequencies and intensities and the nucleic acid helical geometry. Not only are the frequencies of the Raman marker bands listed but also their intensity which is given as the ratio of the marker band peak height to the height of the conformationally intensitive band at 1100 cm^{-1}. It may be seen that crystalline samples of GpC and

Table I

Correlation of Raman Marker Band Frequencies in cm^{-1}
with the C5$'$−C4$'$−−C3$'$−O3$'$ Torsional Angle in Nucleotides

Molecule or Polymer	Sample Form	Dihedral Angle Degrees	Ring Pucker endo	Marker Band cm^{-1}	Intensity Ratio	Max Amp Tau Degrees
GpC	cryst	89	C3$'$	813	1.6	42
UpA	cryst	81.5	C3$'$	802	1.5	33
3$'$−GMP	gel	—	C3$'$	none	none	—
5$'$−GMP	gel	—	C3$'$	814	1.7	—
A-RNA	fiber	95	C3$'$	814	1.65	—
A-DNA	fiber	83	C3$'$	807	1.65	—
pTpT	cryst	157	C2$'$	833	2.3	29
B-DNA	fiber	156	C2$'$	835	0.33	—

a. For detailed references to x-ray and Raman Data see references 9, 10.

UpA as well as fiber samples of RNA polymers, low humidity (ie A-genus) DNA polymers and gels of 5$'$-GMP all show the characteristic C3$'$-endo Raman marker band. In every example but one (UpA) the frequency falls in the range 807-824 cm^{-1}. UpA is a bit of an exception as it has a conformation in which the bases are not arranged in a helical configuration although both sugars are in a C3$'$-endo form (14). The somewhat low frequency of this vibration which is found at 802 cm^{-1} in UpA may be due to the non-helical conformation of this dinucleotide in the crystal. However in every case the intensity ratio is found to be close to the canonical value of 1.6. From this fact it has been concluded that the RNA polymers including biologically active RNA molecules contain ribose rings which are held in a rather rigid inflexible position such as that found in the crystalline state of dinucleotides (9,10). On the other hand the C2$'$ marker band found at 833-835 cm^{-1} is only found in B-genus DNA and crystals of pTpT where both furanose rings are known to possess the C2$'$-endo pucker. However the extraordinary large intensity of this band in the pTpT crystal (See table 1) where the furanose ring is rigid and the rather weak intensity of this band in fibers of B-genus DNA has lead to the conclusion that DNA must be much more flexible than RNA (9,10). (The 833 cm^{-1} band which is strong in pTpT crystals is nonexistent in pTpT solutions showing that little if any of this specific conformation exists in the dinucleotide in solution.) The small intensity of the 835 cm^{-1} marker band which is observed in fibers and solutions of DNA is believed to be due to the existence of a large number of furanose ring conformations which are sufficiently different from the C2$'$ pucker that they do not contribute to the intensity of the Raman band in ordinary DNA (9,10).

Recently various oligonucleotides have been synthesized, crystallized and their conformation in the crystalline state determined by X-ray diffraction techniques. This work may be divided into three classes, the work on oligomers of alternating dA-dT (15), the work on oligomers of alternating dC-dG (16,17) and a paper on the crystal structure of dCdCdGdG in which the first cytosine has an iodine substituted in the C5 position (18). In all cases conformations which are unusual for deoxynucleic acids have been found.

The tetramer pdAdTdAdT in the crystalline state was found to form a right handed duplex structure which exhibits an alternating C3'-endo−C2'-endo furanose ring pucker (15). The dCdCdGdG tetramer was found to belong to the A-genus conformation (18). Crystals of oligomers of the form (dCdG)$_N$ have been found to form a selfcomplementary duplex structure called the left handed Z helix (16,17).

The biological significance of these unusual conformations is the subject of much discussion. It is usually assumed that these unusual conformations do not support biological functions directly but serve as structural control elements for processes involved in genetic expression. Thus if certain repeating sequences are used as control devices, it should be possible to determine some relation between certain specific sequences of the bases and the conformation. Using Raman scattering we have determined the conformation of a number of deoxynucleic acid polymers and oligomers. In addition we have examined the effect of the complexation of poly(dG-dC) double helix under physiological salt conditions with cis- and trans-diaminodichloro platinum compounds. Circular dichroism spectra were also taken on some of these materials. In each case we have looked for evidence of unusual conformations induced either by complexation or by environmental effects in polymers and oligomers of specific base sequences.

Conformational Changes in DNA which Result from Changes in Solvent Environment

In view of the recent X-ray work on crystals of alternating d(A−T) oligomers which has shown that alternating ring puckers can exist in such oligomers it is of interest to examine double helical polymers of both alternating d(A−T)·d(A−T) as well as poly(dA)·poly(dT) by Raman Spectroscopy in order to see if evidence can be found for the presence of both ring puckers in these double helices in solution. Figure 1 shows the Raman spectrum of poly(dA)·poly(dT) under conditions of 0.020M NaCl, 0.020M sodium cacodylate buffer at pH=7.0 and various temperatures. This figure is of interest- because both the C3'-endo and C2'-endo marker bands are plainly evident in the Raman spectrum taken at the lower temperature. This indicates that both ring puckers are present in this material at low temperature although at higher temperatures the Raman band due to the C3'-endo ring pucker tends to broaden indicating some loss of rigidity and some change in conformation at higher temperatures. This change in conformation occurs well below the melting point of the double stranded helix. Table 1 shows that for rigid ring puckers the intensity of the Raman marker band at 814 cm^{-1} (C3'-endo) and that at 833 cm^{-1} (C2'-endo) have approximately the same intensity as measured by the ratio of the peak height of the marker band to the band at 1100 cm^{-1}. This ratio is approximately 2. The fact that the marker band at 814 cm^{-1} and 833 cm^{-1} have the same intensity at low temperatures in polyd(A)·polyd(T) indicates that this deoxy-double helix is unusually rigid and that both conformers are present in approximately equal amount. It is thus reasonable to assume from the Raman data alone that one chain is in the C3'

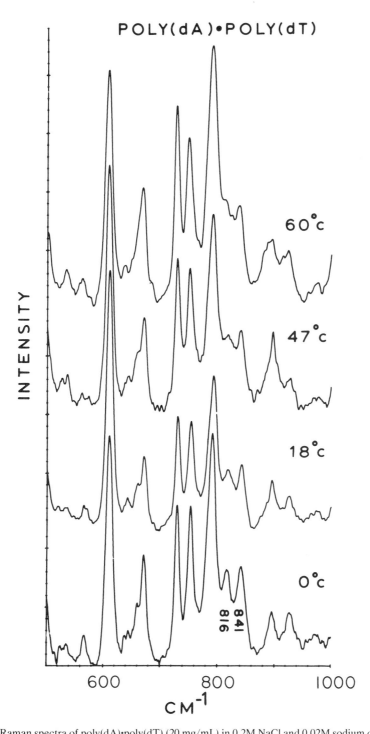

Figure 1. Raman spectra of poly(dA)•poly(dT) (20 mg/mL) in 0.2M NaCl and 0.02M sodium cacodylate at pH 7.0 obtained with 300 mW of 514.5-nm light.

ring pucker and one is in the C2′ ring pucker. This conclusion has been verified by recent X-ray analysis of fibers of these materials in which it was shown that the poly(A) strand has C3′-endo ring pucker and the poly(T) strand has C2′-endo ring pucker (19).

Measurements on double helical complexes of the alternating poly(dA−dT)· poly(dA−dT) copolymer in dilute salt solution at pH=7.0 also show the existence of both C3′ and C2′ marker bands (9,10). Since again these bands have approximately equal intensity, we may conclude that the C3′-endo ring pucker can be associated with the adenine base and the C2′ ring pucker can be associated with the thymine base since this is the conformation which has been found for the pdAdTdAdT tetramer in the crystal (15). (As we will discuss below in the Z-DNA the C3′-endo ring pucker is associated with the guanine residue and the C2′-endo ring pucker is associated with the cytosine. In the Raman spectrum of Z-DNA both marker bands have been shown to be present (20). Thus in every case where both Raman marker bands are present it appears the the purine base is attached to the deoxyribose ring in the C3′-endo position while the pyrimidine base is attached to the deoxyribose ring in the C2′-endo position.)

Having shown that A−T double helices can have a B-form with mixed ring pucker, it is of interest to examine the conformation of deoxyoligomers containing guanine and cytosine under a variety of conditions to see what conformational states may exist in these materials in solution. We have investigated the dependence of conformation on specific sequences of self-complementary tetranucleotides containing only guanine and cytosine.

Figure 2 shows Raman spectra of poly(dG−dC)·poly(dG−dC) taken in the B form in solution at low salt concentration and in the Z form in high salt concentration. These spectra were taken in order to obtain very high signal to noise spectra with which to be able to compare d(C−G) polymers and oligomers which may be only partially in the B form, i.e. in some sort of intermediate form. Table 2 lists the guanine and cytosine frequencies which characterize the B-Z transition.

In order to see which sequences of guanine and cytosine can support the B-Z transition, highly purified samples of the self-complementary tetramers, 5′−d(CpGpCpG)−3′, 5′−d(GpCpGpC)−3′, 5′−d(CpCpGpG)−3′ and 5′−d(GpGpCpC)−3′ were purchased from P.L. Biochemicals with complete analytical data. These materials were dissolved to a concentration of 0.02M in nucleotides base in 0.5M NaCl and cooled to 2°C. Raman spectra were first taken at 2°C and 50°C in 0.5M NaCl pH=7, and it was determined that each tetramer formed a duplex with the characteristic C2′-endo marker band at 835 cm^{-1}. The temperature was raised to 50°C and for each tetramer the B-type duplex was observed to melt with low cooperativity to a disordered form over a wide temperature range of 10°C-30°C. The salt concentration was then increased to 4M and finally to saturation.

At the concentration of 0.02M mononucleotide which is quite high compared to

Table II

Raman bands of the nucleic acid components which are sensitive to conformation

Frequency (cm^{-1})	Assignment
1100	PO2 symmetric stretch
805-815	DNA or RNA chain mode for furanose-phosphate ester linkage in C3'-endo conformation. (A-DNA)
835	Weak but prominent Raman mode for DNA with C2'-endo (B-DNA)
870-880	Similar to mode above but for DNA in C-form.
682	Guanine ring breathing for C2'-endo-anti (found in B-DNA)
665	Same as above but for C3' endo anti (found in A-DNA)
625	Same as above but for C3'-endo syn (found in Z-DNA)
1260	weak cytosine band in B DNA
1265	strong cytosine band in Z DNA
1318	moderate guanine band in B DNA
1318	very strong cytosine band in Z DNA
1334	moderate guanine band in B DNA
1355	moderate guanine band in Z DNA
1362	moderate guanine band in B DNA
1418	weak guanine band in Z DNA
1420	moderate guanine band in B DNA
1426	weak guanine band in Z DNA

optical concentrations, all four oligomers were observed to form a duplex structure of the B-type in low salt solutions and to form a disordered non-duplex structure above 40°C. However only one of the tetradeoxynucleotides gave evidence of going into the Z form at 4M or even saturated salt solution. Figure 3a shows the Raman spectra of d(CpGpCpG). The middle spectrum in figure 3a shows the spectrum of this tetramer in a 0.5M NaCl solution recorded from a sample maintained at 2.2°C. This spectrum is quite similar to the spectrum observed for poly(dG−dC) at an NaCl concentration of 0.02M. The band at moderate strength at 830 cm^{-1} is readily identifiable as being due to the C2'-endo furanose ring conformation, and it has the intensity found in ordinary DNA. Further evidence that the tetramer has formed a double stranded duplex structure is obtained form the melting behavior. The top spectrum in figure 3a shows the spectrum of the melted duplex structure. As expected elevation of the temperature to 50°C causes marked changes in the spectrum of the tetramer due to disruption of base stacking and base pairing interactions. The band at 830 cm^{-1} is no longer visible due to the flexible nature of the melted DNA backbone (9,10). Intensity reductions at 681 cm^{-1} and 1415 cm^{-1} may also be noted as has been reported in previous work on DNA melting by

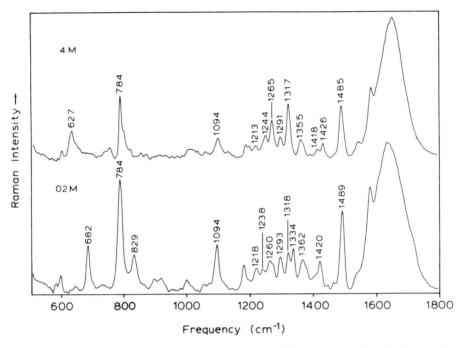

Figure 2. Raman spectra of native poly(dG−dC) recorded at NaCl concentrations of 0.02 and 4M. Raman spectra taken with 300 mW of an Argon laser.

Erfurth and the present author (22). The intensity of the bands at 1237 cm^{-1} and 1319 cm^{-1} increases markedly upon melting. The increases of the band at 1319 cm^{-1} is at present not assigned but the increase of the band at 1237 cm^{-1} is due to the unstacking of cytosine (23,24).

At high salt concentration this tetramer exhibits a Raman spectrum characteristic of the Z-form. The bottom spectrum in figure 3a is that of this tetramer in 4M NaCl solution recorded at 2°C. As in the case of the double helical polyd(G−C) in high salt solution we find a reduction in the intensity of the Raman bands at 680 cm^{-1}, 830 cm^{-1}, and 1415 cm^{-1}. A strong new band appears at 621 cm^{-1}. Intensity increases are observed in the Raman bands at 1264 cm^{-1} and 1319 cm^{-1}. The 1319 cm^{-1} band is a replacement of the two bands at 1317 cm^{-1} and 1333 cm^{-1} which are observed in the low salt form. The band at 1363cm^{-1} in the low salt tetramer spectrum shifts to 1356 cm^{-1} in the high salt form. In total these changes are essentially the same as those found in the spectrum of polyd(G−C) in going from the low salt B form to the high salt Z form. However the changes are not quite so pronounced and the band at 680 cm^{-1} is still apparent. Thus we must conclude that under conditions of high salt this tetramer is a duplex structure with a predominant but not necessarily complete left-handed Z conformation. Figure 3b shows an equivalent set of spectra to those in figure 3a but for the tetramer d(GpCpGpC). This tetramer does not go into the Z-form and indeed the 680 cm^{-1} band seems to be even stronger in the high salt than in the low salt form. Since the Raman spectra of the other three tetramers studied show no change with increasing salt concentration up to saturation it is evident that they do not undergo a B-Z transition. It is therefore well established that the specific sequence d(CpGpCpG) is the only self complementary tetramer involving cytosine and guanine which will support the Z form.

Since both d(GpCpGpC) and d(CpGpCpG) can be considered as a repeating segment of polyd(G−C)·polyd(G−C) it is of interest that the former retains the B form up to high salt concentrations while the latter changes into the Z form. The

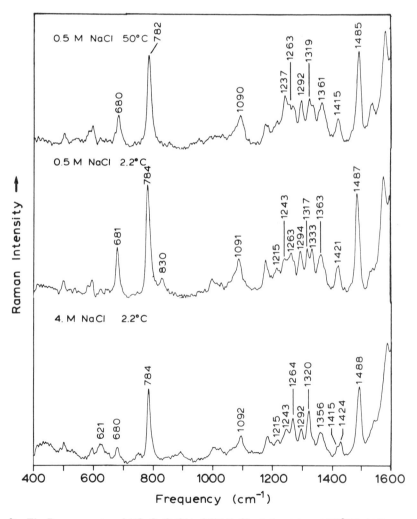

Figure 3a. The Raman spectra of d(CpGpCpG) in 0.5M NaCl solution taken at 2.2°C (middle spectrum); at 50°C (top spectrum), and in 4M NaCl solution at 2.2°C (bottom spectrum). The center spectrum is the one belonging to the B genus with which the other two should be compared. The top spectrum is the melted or disordered form while the bottom spectrum is of the high salt form belonging to the Z-genus.

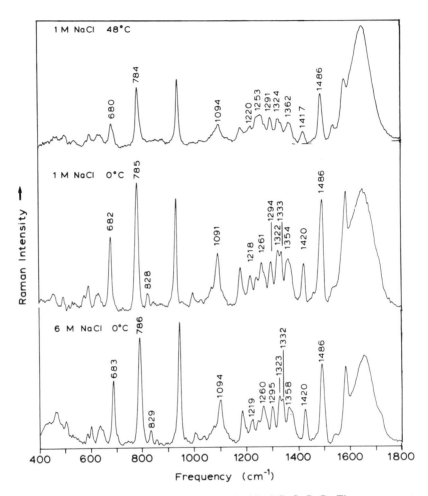

Figure 3b. The Raman spectrum of the deoxytetranucleotide d(GpCpGpC). The center spectrum is that of the B-form at 0°C, the top spectrum is that of the melted B-form at 48°C while the bottom spectrum is of the tetramer in 4M to saturated NaCl solution.

retention of the B form by d(GpCpGpC) in solutions of very high salt concentration is probably due in part to both a sequence dependent enhanced stabilization of this B form for this tetramer as well as a sequence dependent destabilization of the corresponding left-handed Z form. It is of interest to consider just what molecular interactions might account for the conformational differences in these two tetramers with alternating cytosine and guanine residues. Some clues may be obtained from a careful study of the models showing the structure of Z-DNA.

In high molecular weight Z-DNA composed of alternating cytosine and guanine residues the cytosine ring at the five prime position stacks closely to the corresponding cytosine ring of the opposite strand. This means that even numbered oligomers with sequences of alternating cytosine and guanine residues which start from the

5-prime end with cytosine will have one more such interstrand stacking interaction than those starting from the 5-prime position with guanine. Thus d(CpGpCpG) contains two such interstrand cytosine-cytosine stacking interactions while d(GpCpGpC) contains only one such stacking interaction if both tetramers are constrained to the Z-type conformation. Secondly there is a strong interaction in the Z-type helix involving the oxygen of the furanose ring associated with the cytosine in the 5-prime direction and the corresponding guanine ring in the 3-prime direction. It is reasonable to suppose that it is the furanose oxygen-guanine ring interaction that is responsible for the remarkable changes in the frequency of the ring vibrations of guanine in Z-DNA. Examination of models of the Z-DNA structure shows that this interaction would occur with all four guanine residues in the tetramer d(CpGpCpG) but that it occurs with only two or at most three guanine residues in the d(GpCpGpC) sequence. Finally in the Z-type DNA there is an interaction along the same chain proceeding from the 5-prime to the 3-prime direction in which the guanine ring on each chain stacks with the succeeding cytosine ring on the same strand. However there are only two of these interactions (one in each strand) in the Z form of the tetramer d(CpGpCpG) while there are four such interactions in a hypothetical Z form of d(GpCpGpC). Thus it may be that this third interaction tends to destabilize the Z-form.

As a result of these three distinct interactions one can distinguish between short alternating purine−pyrimidine and pyrimidine−purine sequences. It seems reasonable to expect that these factors will be more prominent in oligomers of short length or in short segments of defined alternating sequence along a normal DNA chain. As a consequence one would expect the difference between alternating pyrimidine−purine and purine−pyrimidine sequence to diminish in sequences of increasing length.

Conformational Changes Induced by Platinum Binding

The drug cis-dichlorodiamine platinum(II) is a therapeutically useful agent in the treatment of tumors. It has been demonstrated that cis-Pt binds to DNA and several lines of evidence suggest that this binding is related to the anti-tumor activity of this compound (for review, see 25,26). On the other hand, the geometrically related isomer trans-dichlorodiamine Platinum(II), trans-Pt, also binds strongly to DNA, particularly at the guanine sites but it shows no measureable anti-tumor activity (25,26). In view of the fact that these reagents attack primarily the guanine residue, it is of interest to see what they do to the conformation of polyd(G−C)•polyd(G−C). In particular we are interested to know if treatment with either δ the cis- or trans-Pt would result in the formation of unusual conformational states. In particular we would be interested to learn the effect of the Pt-drugs on the B-Z transition. Hopefully in studying these transitions we will learn something about the biological role played by conformational changes in DNA.

Previous work by other investigators has shown that the polyd(G−C)•polyd(G−C) treated with cis- or trans-Pt show CD spectra which could be indicative of left-handed helical structures in high (4M) salt solutions (27,28). However although the

CD spectrum of the trans-Pt treated and native polymer are quite similar, the cis is quite different and it is a matter of considerable subjectivity as to whether either or both of the Pt-treated polymers can be considered to belong to the Z family in 4-6M salt concentration.

In order to ascertain the effects of cis- and trans-Pt drug binding upon the conformation of polyd(G−C)·polyd(G−C) during its salt induced transition from right to left handed helix, the conformation of trans-Pt treated, cis-Pt treated, and native poly(dG−dC) was monitored by circular dichroism (CD) and Raman spectroscopy at a number of salt concentrations between 0.02 and 6M. All of our experiments were done at a ratio r of Pt to total base of 0.1. The r-value of 0.1 was chosen as a compromise between a low value which would have some relevance to the therapeutic effect of the Pt-compounds and a larger value which would cause more pronounced changes in the spectra.

Materials and Methods

Samples of polyd(G−C) were prepared from material obtained from P.L. Biochemicals lot 692, 217190, or 676-7. For optical studies a stock solution was prepared by dissolving 10 or 20 OD units of the polymer in 0.3 to 0.5 ml of doubly distilled water. This solution was then used to prepare a stock solution with a polymer concentration of about 1.2×10^{-4}M in mononucleotide. The concentration was chosen to give a maximum absorbance of 1 at a pathlength of 10 mm. Three equal volumes of 3mL were removed from the stock solution and two of the volumes were treated with an appropriate platinum drug. One sample was left untreated and used as a control. The cis and trans dichlorodiamine platinum were obtained from Alfa Chemicals. The platinum compounds were added to the polymer solutions in the form of concentrated solutions. The base concentration was determined by UV absorption and the platinum compounds added so to obtain one Pt for each 10 bases. To assure that all of the platinum had reacted, the unbound platinum content in the solution was determined by spectrophotometric analysis by tin chloride complexation. It was determined that less than 2% of the Pt compounds remained unbound in the polymer solutions. Raman Samples were prepared in a similar manner except the final polymer concentration was calculated to be approximately 2.5×10^{-2}M in mononucleotide.

Discussion

Since differences have been found in the CD spectra of the Pt treated polymers it is of interest to see how the CD spectrum develops with increasing salt concentration. Figure 4 shows the CD spectrum of native polyd(G−C)·polyd(G−C) at 0.02M and 4M NaCl. This figure shows the characteristic change in CD in going from the right handed B form to the left handed Z form, and it will be used as the standard with which to compare the Pt-treated polymers. The observed CD spectrum of the native polymer under low salt conditions is characterized by a broad positive lobe with a double maximum at 287 nm and 275 nm followed by a large negative lobe at

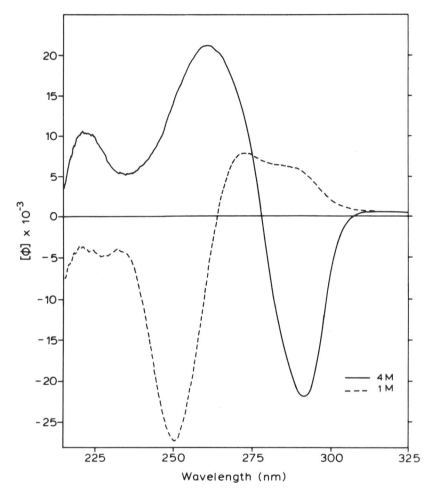

Figure 4. Circular dichroism spectra of native poly(dG−dC) recorded at NaCl concentrations of 0.025M and 4.0M. The solid line is the spectrum at 4.0M.

251 nm. The corresponding low salt CD spectrum of the trans-Pt treated polymer shown in figure 5 is almost identical with the exception of a minor loss in intensity at the second positive maximum at 275 nm. The sequential addition of NaCl causes an inversion in the circular dichroism spectrum of the trans-Pt treated polymer. As the NaCl concentration is increased the intensity of the positive lobe at 287 nm is reduced resulting in the generation of a negative band at 291 nm. Simultaneously the negative band at 252 nm diminishes resulting in the formation of a positive lobe at 268 nm. This inversion is similar to the well-known change in the circular dichroism spectrum of native poly(dG−dC) at elevated NaCl concentrations which is characterized by a strong negative band at 291 nm and a positive band at 261 nm. The inversion in the circular dichroism spectrum of the native polymer is associ-ated with a transformation from a right to a left handed helical secondary structure

(29,30). An analysis of the molar ellipticities of the trans-Pt treated polymer at high salt shows that it has a lower absolute value than the native polymer. This suggests that the extent of Z-conversion may be less in the trans-Pt treated polymer than in the native polymer or that the trans-Pt in some way reduces the intensity of the CD spectrum.

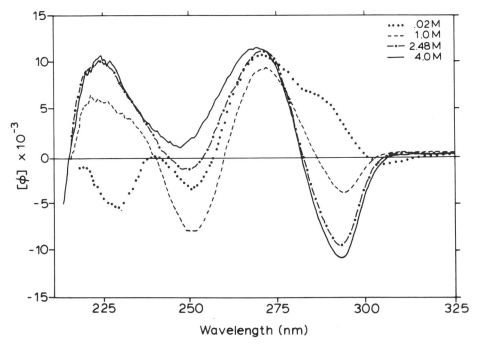

Figure 5. Circular dichroism spectra of poly(dG−dC) treated with trans dichlorodiamine Platinum at a drug to base ratio of 0.1 recorded at various NaCl concentrations.

Malfoy et al. (31) show a CD spectrum for the trans-Pt treated polyd(G−C)· polyd(G−C) in high salt which is similar to ours but concluded that this polymer when modified by the trans-Pt does not adopt the Z form. Since the CD spectra are similar but not identical, this point is hard to resolve from only the CD spectra although it may be greatly clarified by using the Raman spectra.

The Raman spectra of the trans-Pt treated polymer recorded at three salt concentrations has been reproduced in figure 6. The Raman spectra of the native and trans-Pt treated polymers in 0.2M and 1M NaCl solutions are very similar which suggest a close correspondence between the secondary structures of these two polymers in salt solutions of low and moderate concentrations. Of particular note is the moderately strong band at 830 cm^{-1} which occurs in the trans-Pt treated polymer at low salt concentration. As we have seen this band is a specific marker band

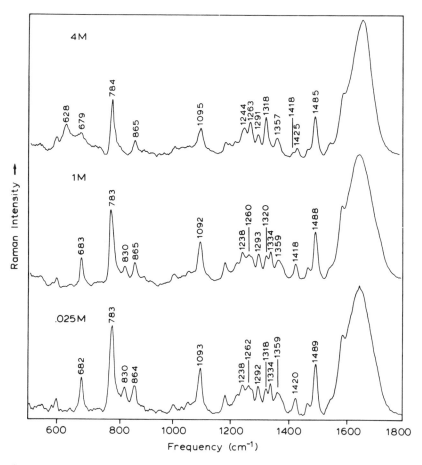

Figure 6. Raman spectra of poly(dG—dC) treated with trans dichlorodiamine platinum at a drug to base ratio of 0.1 at three salt concentrations, 0.025M, 1.0M and 4.0M.

for the C2'-endo ring pucker, and as a consequence we can assign the secondary structure at low salt concentration of the trans-Pt treated polymer to a B-genus helical form. Though the peak positions observed in the low salt spectrum of the trans-Pt treated and native polymers correspond to one another almost exactly, subtle intensity variations are observed between the two spectra. The most pronounced difference consists of a small instensity increase in the bands at 1240 cm⁻¹ and 1262 cm⁻¹. Bands found in the Raman spectra of nucleic acids between 1200 and 1800 cm⁻¹ are generally assigned to in-plane vibrations of the nucleic acid bases. Variations in the intensity of these bands are usually associated with alternations in base stacking and hydrogen bonding. The small intensity differences observed between the trans-Pt treated and native polymers may be associated with subtle base stacking modifications due to the interaction of the platinum drug. The intensity decline at 275 nm observed in the CD spectrum at the trans-Pt treated polymer in low salt concentrations may in part be due to these minor modifications in base stacking.

Elevation of the NaCl concentration of the trans-Pt treated polymer to 1M produced only minor alternations in the Raman spectra of this polymer. Further addition of NaCl to 4M produced pronounced changes in the Raman spectra of the trans-Pt treated polymer similar to those found in the native polymer. Significant intensity increases are observed in the spectral regions located at about 630, 1240, 1260, and 1320 cm^{-1}. Intensity reductions are evident at 1418 cm^{-1} and 830 cm^{-1}. At 4M NaCl the 830 cm^{-1} band appears to have vanished and a shoulder at 810 cm^{-1} is now visible. In the native polymer at 4M salt concentration both peaks are visible indicating the presence of both C3'- and C2'-endo ring pucker. In the Raman spectrum at 4M of the trans-Pt treated polymer these bands are only suggested as possible shoulders. In all of the Raman spectra of the Pt treated material, we find a band of unknown origin at 865 cm^{-1}. This band does not exist in either GMP or polyC treated with the Pt compounds (32,23). Peaks in the base vibrational region frm 1200 to 1800 cm^{-1} are located at 1244, 1266, 1318, 1355, and 1424, in comparison to the low salt spectrum in which Raman bands are observed at 1238, 1262, 1318, 1334, 1354, and 1414 cm^{-1}. The band observed at 682 cm^{-1} in the low salt spectrum is almost completely absent in the high salt spectrum of the trans-Pt treated polymer but a small peak at 679 cm^{-1} does remain. Thus although there are some differences in the details of its high salt spectrum we must conclude that both CD and Raman spectra clearly indicate that increasing the salt concentration in solutions of the trans-treated polymer to 4M induces a transition to a conformation which belongs to the left-handed Z-genus.

The influence of the cis-Pt upon the conformation of poly(dG−dC) is quite different from that of the trans isomer and the differences begin to appear even at very low salt concentrations. The distinctive influence of the cis-Pt drug can be seen by an examination of figure 7 which contains a collection of CD spectra of poly(dG−dC) treated with cis-Pt diamine at a maximum drug to base ratio of 0.1. The initial low salt spectrum observed for the cis-Pt treated polymer is quite different from that of the trans-Pt treated and native polymer spectra. The cis- polymer spectrum is characterized by a broad maximum at 270 nm with a positive shoulder at 282 nm. A small negative band at 305 nm is also present along with a larger negative band at 250 nm. The magnitude of the 270 nm band is larger in the cis-Pt treated polymer than in the native polymer. The negativity of the 250 nm band is much less relative to the native polymer. The intensity variations of the 270 nm and 250 nm bands in the CD of the cis-Pt treated polymer may reflect modifications in the secondary structure of the polymer such as base twisting which do not affect the force field of the guanine bases and thus do not change the Raman spectrum of the Z-type guanine marker bands.

Modifications in secondary structure are also evident in the Raman spectrum of the cis-Pt treated polymer at 0.025M NaCl which is displayed in figure 8. One of the most notable differences is an apparent reduction in the intensity of the conformationally sensitive furanose−phosphate backbone band at 830 cm^{-1} in the Raman spectrum of the cis-Pt treated polymer. The decline in the 830 cm^{-1} band reflects a reduction in the population of furanose rings with the C2' endo ring pucker. Several

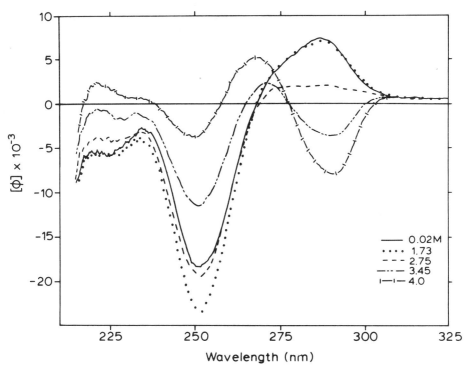

Figure 7. Circular dichroism spectra of poly(dG−dC) treated with cis dichlorodiamine platinum at a drug to base ratio of 0.1 recorded at various NaCl concentrations.

other spectral alternations are observable in the base vibrational region from 1200 to 1800 cm^{-1}. Minor intensity increases are observed for the cis-Pt treated polymer at 1262 and 1318 cm^{-1}. A more prominent intensity increase is observed for the band at 1241 cm^{-1}. The intensity increase at 1241 cm^{-1} somewhat parallels the intensity increase observed for the trans-Pt treated polymer but is appreciably larger in magnitude in the cis-Pt case. The band at 1241 cm^{-1} arises primarily from contributions from the cytosine ring. The intensity of this cytosine mode is particularly sensitive to alterations in base stacking and hydrogen bonding (23). One may infer from the behavior of other cytosine containing compounds that the increase at 1241 cm^{-1} in the cis-Pt treated polymer is evidence of alterations in base stacking due to the presence of the platinum drug.

An increase in the NaCl concentration results in a progressive noncooperative alteration in the conformational state of the cis-Pt treated polymer (see figures 7 and 8). In terms of the CD data, as the NaCl concentration is increased the intensity of the shoulder at 283 nm decreases, finally yielding to the formation of a negative band centered at 295nm. This change results in an apparent increase in the negative band at 305 nm. The minimum at 228 nm becomes less negative yielding to the formation of a positive band at 225 nm. The negativity of the 250 nm band decreases as the salt concentration is increased. At an NaCl concentration of 1M the circular dichroism spectrum of the cis-Pt treated polymer is reminiscent of the

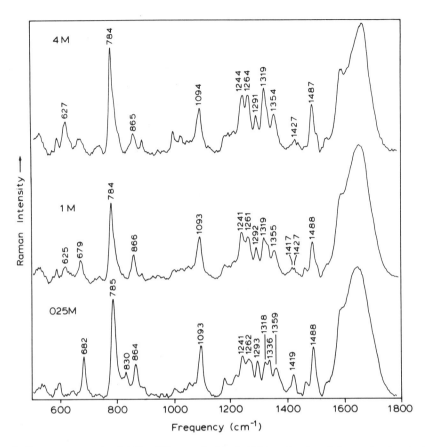

Figure 8. Raman spectra of poly(dG−dC) treated with cis dichlorodiamine platinum at a drug to base ratio of 0.1 recorded at selected NaCl concentrations.

inverted spectrum observed for the trans-Pt treated polymer at 4M NaCl. Furthermore the Raman spectrum at 1M (center spectrum, figure 8) already shows evidence of Z-helix formation. There is a band at 625 cm^{-1} starting to grow, and the bands at 1319, 1241, and 1261 cm^{-1} are starting to resemble those characteristic of the Raman spectrum of Z-DNA. The bands continue to change in the same direction as the salt is increased in concentration. The spectrum of the cis-Pt treated polymer clearly shows many of the characteristic bands of the Z-DNA. This point will be discussed in more detail below.

A plot of the molar ellipticity at 290 nm as a function of NaCl concentration for cis-Pt treated, trans-Pt treated and native polymers is contained in figure 9a. From an examination of this plot it is evident that the maximum variation in the ellipticity at 290 nm for both the trans-Pt treated and native polymers occurs over a narrow range in salt concentration. The transition in both polymers appears to be cooperative. A plot of the percent transition as a function of NaCl concentration is

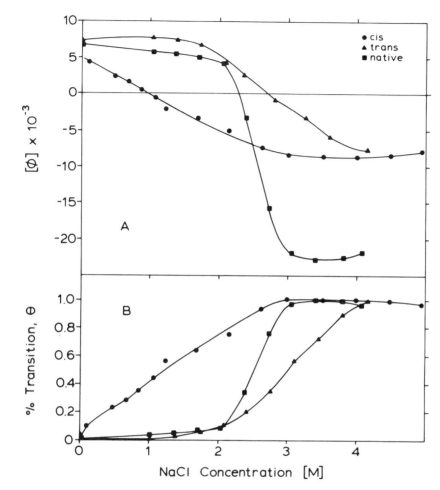

Figure 9. a) Plot of the molar ellipticity at 290 nm as a function of NaCl concentrations for the native, trans treated and cis treated polymers. b.) Plot of the degree of transitions as a function of NaCl concentration for the native, trans treated and cis treated polymers.

presented in figure 9b. The transformation from a right-handed to a left-handed form in the case of the native polymer is quite sharp and highly cooperative. The transition midpoint occurs at 2.5M NaCl for the untreated polymer. In the case of the trans-Pt treated polymer the percent transition curve is less sharp indicating that the transition is broader and less cooperative. The midpoint of transition for the trans-Pt treated polymer occurs at about 3.1M NaCl. The cis treated polymer appears to show no cooperativity—the transition begins directly at low concentrations and continues without a break in slope.

The cooperativity differences between the untreated and treated polymers can be placed upon a more quantitative foundation by an analysis of the percent transition data in terms of a Hill plot. A plot of the log of percent transition divided by 1 minus percent transition versus sodium chloride concentration is linear over the 10% to 90% transition range for all three polymers. From this plot Hill coefficients of 13.3, 6.7, and 1.8 are obtained for the native, the trans-Pt treated, and the cis-Pt treated polymers, respectively. A Hill coefficient of 12 to 18 has been reported by Sande and Jovin (33) for the magnesium induced B to Z transition in poly(dG−dC). Based upon the differences in the Hill coefficients for the polymers, one can conclude that the transition cooperativity is reduced by a factor of 2 for the trans-Pt treated polymer, and a factor of almost 8 for the cis-Pt treated polymer. Thus not only does the cis-Pt drug drastically reduce the cooperativity of the right to left-handed transition but it also shifts the midpoint of the B-Z transition to lower salt concentrations. In contrast the trans-Pt treated polymer shifts the midpoint of the B-Z transition to higher salt concentrations.

Figure 11 shows a comparison of the relative intensities of selected Raman bands for the trans-Pt treated, the cis-Pt treated and the native polymers as a function of

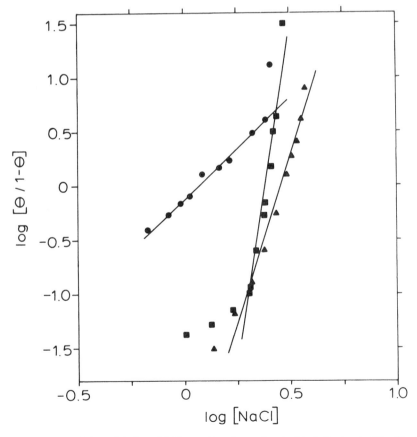

Figure 10. Hill plot of the salt induced B-Z transition for native double helical polyd(G−C) and the same material treated with cis- and trans-Pt with an r value of 0.1. The squares are for the native polymer while the circles and triangles are for cis and trans treated polymers.

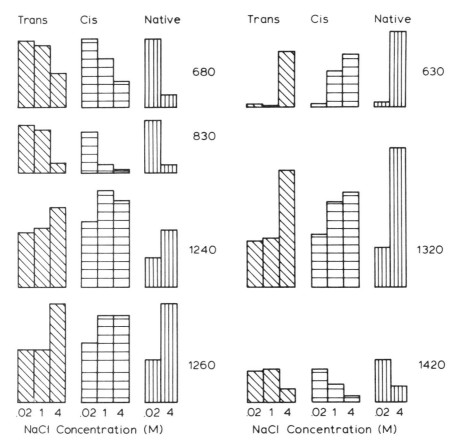

Figure 11. Comparison of the relative intensities of selected Raman bands for the trans treated, cis treated, and native polymer as a function of NaCl concentration.

NaCl concentration. It is evident from this figure and the CD data that at high salt concentrations all three of the polymers go into a conformation which belongs to the Z family. The conclusion that all three polymers go into a left-handed conformation comes from a simple analysis of the inversion of the lobes of the CD spectrum as the salt concentration is increased. Furthermore in view of the similarity of the Raman spectra of all three polymers at high salt concentration we must conclude that the guanine residues are under the influence of the same forces as they are in Z-DNA since they give the same characteristic frequency change.

The question naturally arises as to the reason for the rather substantial differences between the CD spectrum of the low salt form of the cis-Pt treated and the CD spectra of the other two polymers. We would like to suggest that since the Raman spectra are so similar in all three polymers the secondary structure of all three polymers is essentially the same as that of the native polyd(G$-$C)·polyd(G$-$C) both in low and high salt solution, insofar as the forces on the guanine ring are concerned. However it is quite possible that the cis-Pt compound tilts the bases upon reaction so that the interaction between the transition dipole moments is altered.

This could give rise to the observed drastic change in the CD spectrum without changing the Raman spectrum substantially. This tilting could also be responsible for the disruption of the backbone chain order which results in the anomalously small intensities of the backbone marker bands of the cis-Pt treated polymer when compared with the native polymer both in low and high salt concentrations. It may be that the therapeutic anti-tumor activity of the cis-Pt compound lies in its ability to modify the secondary structure of the DNA in a very subtle way in regions of high quanine content thus preventing it from performing its normal function. This would not necessarily have anything to do with the B-Z transition because as we have seen even the low salt CD spectrum of the cis-treated polymer is substantially disturbed while the Raman spectrum is only slightly changed. However by following both the CD and Raman spectra as a function of the NaCl concentration it has been possible to show consistent differences in conformation between these three polymers at each different salt concentration. But by far the greatest difference is found between the cis-platinum treated polymer and the other two polymers whose conformational behavior appears to be quite similar at all salt concentrations. Thus this may be the first clear example of a measurement that shows a distinct qualitative and quantitative difference between the behavior of the cis-Pt and trans-Pt compounds.

Acknowledgement

Support of this work by National Science Foundation Grant 76-82222 and National Institutes of Health Grant GM-15547 is gratefully acknowledged.

References and Footnotes

1. Erfurth, S. C., Kiser, E. J. and Peticolas, W. L., *Proc. Natl. Acad. Sci. U.S.A., 69,* 38 (1972).
2. Erfurth, S. C., Bond, P. J. and Peticolas, W. L., *Biopolymers, 14,* 1245 (1975).
3. Brown, K. B., Kiser, E. J. and Peticolas, W. L., *Biopolymers, 11,* 1855 (1972).
4. Brown, E. B. and Peticolas, W. L., *Biopolymers, 14,* 1259 (1975).
5. Lu, K. C., Prohofsky, E. W. and Van Zandt, L. L., *Biopolymers, 16,* 2491 (1977).
6. Thomas, Jr., G. J. and Hartman, K. A., *Biochim. Biophys. Acta, 312,* 311 (1973).
7. Goodwin, D. C. and Brahms, J., *Nucleic Acid Res., 5,* 835 (1978).
8. Peticolas, W. L. and Tsuboi, N., *Infrared and Raman Spectroscopy of Biological Molecules,* p. 153 (1979).
9. Thomas, G. A. and Peticolas, W. L., *J. Am. Chem. Soc., 105,* 986 (1983).
10. Thomas, G. A. and Peticolas, W. L., *J. Am. Chem. Soc., 105,* 983 (1983).
11. Small, E. W. and Peticolas, W. L., *Biopolymers, 10,* 69 (1971).
12. Small, E. W. and Peticolas, W. L., *Biopolymers, 10,* 1377 (1971).
13. Tsuboi, M., Takahashi, S., Muraishi, S., Kajiura, T. and Nishimura, S., *Science, 174,* 1142 (1971).
14. Sussman, J. L., Seeman, N. C., Kimm, S. H. and Berman, H. M., *J. Mol. Biol, 66,* 403 (1972).
15. Viswamitra, M. A., Kennard, O., Shakked, Z., Jones, P. G., Sheldrick, G. M., Salisburg, S. and Fauello, L., *Nature, 273,* 687 (1978).
16. Wang, Andrew H. J., Quigley, G. J., Kolpak, F. J., Crawford, J. L., van Boom, J. H., vand der Marel, G., and Rich, A., *Nature, 282,* 680 (1979).

17. Drew, H. R. and Dickerson, R. E., *J. Mol. Biol., 152,* 723 (1981).
18. Conner, B. N. Takano, T., Tanaka, S., Itakura, K. and Dickerson, R. E., *Nature, 295,* 294 (1982).
19. Arnott, S., Chandrasekaran, R., Banerjee, R. H. and Walker, J. K., *J. Biomol. Struc. and Dyn., 1, 437* (1983).
20. Thamann, T. J., Lord, R. C., Wang, A. H. T. and Rich, A., *Nucleic Acid Res., 9,* 5443 (1981).
21. Nishimura, Y., Tsuboi, M., Nakano, T., Higuchi, S., Sato, T. S., Uesugi, S., Ohtsuka, E. and Ikehara, M., *Nucleic Acids Res., 11,* 1579 (1983).
22. Erfurth, S. C. and Peticolas, W. L., *Biopolymers, 14,* 247 (1975).
23. Strommen, D. P. and Peticolas, W. L., *Biopolymers, 21,* 69 (1982).
24. Chou, C. H. and Thomas, Jr., G. J., *Biopolymers, 16,* 765 (1977).
25. Kleinwachter, V., *Studia Biophysica, 73,* 1 (1978).
26. Roberts, J. J. and Thomson, A. J., *Progress in Nucleic Acids Research and Molecular Biology, 22,* 71 (1979).
27. Malfoy, B., Hartmann, B. and Leng, Marc, *Nucleic Acids Res., 21,* 5659 (1981).
28. Ushay, H. M., Santella, R. M., Caradonna, J. P., Grunberger, D. and Lippard, S. J., *Nucleic Acids Res., 11,* 3573 (1982).
29. Pohl, F. M. and Jovin, T. M., *J. Mol. Biol., 67,* 375 (1972).
30. Pohl, F. M., Jovin, T. M., Baehr, W. and Holbrook, J. J., *PNAS, 69,* 3805 (1972).
31. Pohl, F. M., *Nature, 260,* 365 (1976).
32. Chu, Y. H., Mansy, S., Duncan, R. E. and Tobias, R. S., *J. Am. Chem. Soc., 100,* 593 (1978).
33. Van de Sande, Johan H. and Jovin, Thomas M., *The EMBO Journal, 1,* 115 (1982).

Structure & Motion: Membranes, Nucleic Acids & Proteins,
Eds., E. Clementi, G. Corongiu, M. H. Sarma & R. H. Sarma,
ISBN 0-940030-12-8, Adenine Press,   Copyright Adenine Press, 1985.

FT-IR Spectroscopic Evidence of C2'-*endo, anti,* C3'-*endo, anti* Sugar Ring Pucker in 5'-GMP and 5'-IMP Nucleotides and their Metal-Adducts

Theophile Theophanides* and H.A. Tajmir-Riahi
Department of Chemistry, Université de Montréal
C.P. 6210, Succ. A, Montreal, Quebec, Canada H3C 3V1

Abstract

The C2'-*endo, anti* to C3'-*endo, anti* transition of guanosine-5'-monophosphate disodium salt (5'-GMPNa$_2$) and inosine-5'-monophosphate disodium salt (5'-IMPNa$_2$) and sodium salt (5'-IMPHNa) and their metal adducts Cd(GMP)•8H$_2$O, Co(GMP)•8H$_2$O, Ni(GMP)•8H$_2$O, Cu$_3$(GMP)$_3$•8H$_2$O, Ca(IMP)•6.5H$_2$O, Ba(IMP)•6H$_2$O, Ni(IMP)•7H$_2$O, Co(IMP)•7H$_2$O, [Cu(IMP)]$_n$ and [Zn(IMP)]$_n$ was followed by Fourier Transform Infrared Spectroscopy (FT-IR) in the spectral region 1000-600 cm^{-1}. Both nucleotides are in the C2'-*endo, anti* conformation as sodium salts in the solid state. The metal adducts are known from X-ray diffraction to be in the C3'-*endo, anti* conformation. The FT-IR spectra of the above metal complexes have been recorded in the solid state and the transition from C2'-*endo, anti* to C3'-*endo, anti* was characterized by a few marker bands for the two sugar puckers. The spectra of these metal adducts show changes in the guanine breathing vibration at 688 cm^{-1} and in the sugar-phosphate bands at 935 cm^{-1} and 820 cm^{-1} which decrease in intensity and disappear, while new bands appear at about 678 cm^{-1}, 665 cm^{-1} and 805 cm^{-1} when the C2'-*endo, anti* changes into C3'-*endo, anti* conformation.

Comparison of the FT-IR spectra of several metal adducts reveals these characteristic bands which are sensitive to the presence of metal ions that may induce conformations such as A, B and Z-DNA.

Introduction

The X-ray crystal structures of several metal-nucleotide complexes have shown (1) that the C2'-*endo, anti* conformation of the sugar pucker changes into C3'-*endo, anti* deoxyribose conformation when the nucleotide forms metal adducts and the metal is fixed at the N7 position of the guanine in general (2). The C2'-*endo* into C3'-*endo* transition has also been observed in the B⇒Z transition by X-ray crystallographic diffraction studies of DNA oligomers (3). The change of B-DNA to another form was first suggested from solution circular dichroism spectra (4). It was shown that poly(dG-dC)•poly(dG-dC) undergoes a conformational transition in high-salt NaCl or MgCl$_2$ solutions (4). Raman spectra of this polymer showed that the high-

521

salt form is the double helical Z-DNA and the low-salt form is the B-DNA (5). The metal ions by coordinating at specific sites of the bases, the phosphate oxygen atoms or the sugar hydroxyls can induce the above B$\Rightarrow$Z transition in the polymer poly(dG-dC)·poly(dG-dC). It is known that in solution there is an equilibrium between the C2'-*endo, anti* and C3'-*endo, anti* pucker in the deoxyribose (6-7). This equilibrium is influenced significantly by metal ions and their concentration (6-7). The nature of the metal, the charge it carries and its size do influence substantially the conformation and structure of the nucleotide by direct or indirect interaction through the water molecules of coordination with the basic sites of the nucleotide. These interactions produce significant chemical modifications in nucleotides or in DNA. Recently, it was shown (8,9) by Raman spectroscopy that in nucleic acids the band at 835 cm^{-1} due to a ribose ring vibration and that at 682 cm^{-1}, the ring breathing mode may be considered as marker bands for the C2'-*endo* conformation, while the band at 805-816 cm^{-1} is a marker band for C3'-*endo* conformation.

In the present work the metal-nucleotide complexes of 5'-GMP and 5'-IMP have been studied by FT-IR spectroscopy. The crystal structures of the metal complexes were known from X-ray diffraction analysis (1,2). The infrared spectra were taken in the 1000-600 cm^{-1} region of the spectrum in search of marker bands for the different sugar puckerings. The metal ions stabilize complexes in which the sugar pucker of the deoxyribose is C3'-*endo, anti*. It seems that the driving of a right-handed helix to a stable left-handed helix by metal ions is an intramolecular interaction with a conformational pathway involving the coordination of a hydrated metal ion to a specific base site or to the phosphate or sugar moiety which perturbs the C2'-*endo, anti*⇋C3'-*endo, anti* equilibrium, as well as the *syn*⇋*anti* and gg⇋gt, tg equilibria (gg: gauche-gauche and gt: gauche-*trans*). Theophanides and Polissiou (6) and Polissiou *et al.* (6,7) showed by ^{1}H NMR spectroscopy that 5'-GMPNa$_2$ and its magnesium and platinum adducts do influence the above equilibria. The tendency found is that the metal ions increase the C3'-*endo* sugar pucker and the gg conformation when a positive charge is fixed at the N7 site of the guanine (6,7). However, when the negative charge carrying group PtCl$_3^-$ is fixed at the N7 site of guanine the gt conformation was favoured (7), most probably because the negative charge of this group (PtCl$_3^-$) would repel the negatively charged oxygen atoms of the phosphate group (7). It was observed that when a positively charged group (H$^+$, CH$_3^+$, Mg^{+2} and other positively charged metal cations) were fixed at N7, the gg conformation was predominent and increased almost to 100% (6).

In this study we present experimental observations on the C2'-*endo*, C3'-*endo* conformations by using FT-IR spectroscopy. There are marker bands here too, whose peak positions and intensities indicate the presence or absence or the coexistence of both C2'-*endo* and C3'-*endo* sugar puckers. A comparison of the infrared spectra of 5'-GMPNa$_2$ and 5'-IMPNa$_2$ disodium salts with those of the other transition and post-transition metal adducts indicates the extent of the changes in the pucker of the guanine furanose ring from C2'-*endo* to C3'-*endo* conformation judged from marker bands involving motions of atoms of the sugar-phosphate moiety.

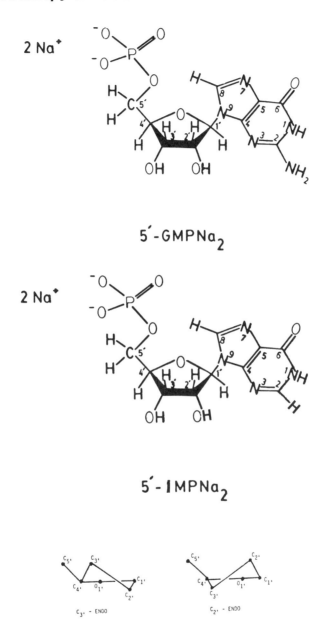

5′-GMPNa$_2$

5′-IMPNa$_2$

Materials and Methods

Materials. Mononucleotides in acid and sodium salt forms were purchased from Sigma Chemical Company. All other chemicals were reagent grade and were used as supplied.

Preparation of metal-nucleotide complexes. All the metal-GMP and metal-IMP complexes were prepared and recrystallized as described in the literature (10-19).

Physical measurements. Infrared spectra were recorded on a DIGILAB FTS-15C/D Fourier Transform Infrared Interferometer equipped with HgCdTe detector (Infrared Associate, New Brunswick, NJ), a KBr beam splitter and a Globar source. The spectra were obtained as KBr pellets with a resolution of 4 to 2 cm^{-1}.

Interaction of Metal Ions with Nucleotides

The metal binding effects on the structural parameters and conformations of nucleic acid constituents can be probed from molecular structural information on metal-nucleotide complexes. The metal ion-nucleotide complexes could be used as models of metal nucleic acid interactions and the binding mode of the antitumor agent *cis*-platinum complex with nucleic acids (20). The DNA binding metal can be a specific binding to bases at particular sites of the bases, purines or pyrimidines and may recognize base sequence in the double helix. In the cation-nucleotide interaction of the nucleotides, guanosine-5'-monophosphate(5'-GMP) and inosine-5'-monophosphate(5'-IMP (see Scheme) the various metals bind to the N7 site of guanine and in some cases to the phosphate group (PO_3^{-2}) and/or to O2' and O3' of the ribose ring. Crystal structure data confirm the above statements (1). The conformational data on the sugar pucker C2'-*endo*, C3'-*endo*, *syn, anti* and gg, gt, tg conformations are also available (1). These data show that the metal interaction with the above nucleotides leads to the C3'-*endo* conformation when the metal is linked at the N7 site of guanine and/or at the phosphate group directly or indirectly through outer waters of hydration coordinated to the metal.

Infrared Spectra of 5'-GMP-Metal Complexes

X-ray structural analysis has shown (10) that the sugar in 5'-GMPNa$_2$·7H$_2$O has the C2'-*endo, anti* conformation and the gg rotation about C4'-C5' bond, while in Cd(GMP)·8H$_2$O it has the C3'-*endo, anti* and gg conformation and the Cd atom is bound to N7 (11). In the crystal structure of the polymeric Cu$_3$(GMP)$_3$·8H$_2$O (13) two of the GMP anions have the C3'-*endo, anti*, gg conformation, while the third GMP anion has the C2'-*endo, anti* gg conformation, the copper being bound to the N7 and the phosphate group. The FT-IR spectra of 5'-GMPNa$_2$·7H$_2$O shows the marker bands at 936 cm^{-1}, 821 cm^{-1} and at 688 cm^{-1}. The first two bands are related to sugar phosphate vibrations and that at 688 cm^{-1} to the guanine ring breathing mode (see Fig. 1 and Table 1). In the spectra of the complex Cd(GMP)·8H$_2$O the band at 936 cm^{-1} is not observed and that at 821 cm^{-1} lost its intensity and at the same time a strong band appeared at 799 cm^{-1}. The crystal structures of Co(GMP)·8H$_2$O and Ni(GMP)·8H$_2$O have been published (12) and in both cases the metal is N7 bound, but the sugar pucker was not reported. The infrared spectra of these complexes and their similarity with the Cd(GMP)·8H$_2$O spectra allows us to assign a C3'-*endo* sugar pucker to the above Co(II)- and Ni(II)-GMP complexes.

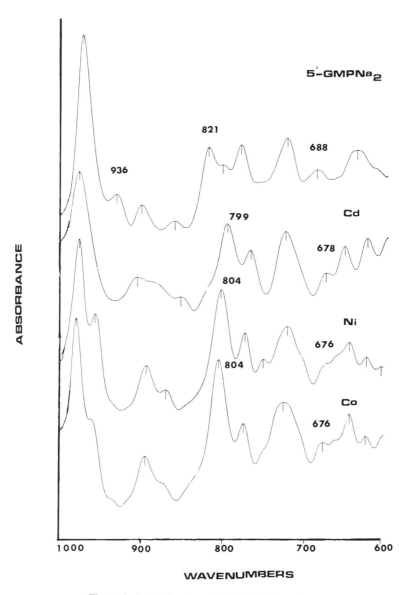

Figure 1. FT-IR Spectra of Metal-GMP Complexes.

Infrared Spectra of 5'-IMP-Metal Complexes

X-ray structural analysis showed that the sugar pucker in 5'-IMPHNa•6H$_2$O (18), 5'-IMPNa$_2$•7H$_2$O and Ba(IMP)•6H$_2$O (17) is C2'-*endo, anti,* gg. In the structurally known Co(IMP)•7H$_2$O and Ni(IMP)•7H$_2$O compounds (19) the ribose is in the C3'-*endo, anti,* gg conformation. The infrared spectra of Na$_2$-IMP, NaH-IMP and

Table I

FT-IR Absorption Frequencies (cm^{-1}) for 5'-GMPNa$_2$ and its Metal Adducts in the Region 1000-600 cm^{-1} with Possible Band Assignments and the Sugar-Phosphate Conformations.

5'-GMPNa$_2$·7H$_2$O	Cd(GMP)·8H$_2$O	Ni(GMP)·8H$_2$O	Co(GMP)·8H$_2$O	Assignments	Sugar-Conformations
982 vs	980 vs	980vs	980 vs	PO_3^{-2} sym. stretching	
936 m	912 m	895 m	895 m	sugar-phosphate	C2'-*endo, anti* (gg)
905 m	880 sh	—	—	sugar stretching	
864 m	858 m	870 m	870 sh	sugar-phosphate	C2'-*endo, anti* (gg)
821 s	822 sh	—	—	sugar phosphate	C3'-*endo, anti* (gg)
804 w	799 vs	804 vs	804 vs	P–O stretching	
782 s	771 m	774 m	774 m	base	
—	—	751 sh	750 sh	base	
725 s	728 s	722 s	725 s	ring breathing	C2'-*endo, anti* (gg)
688 m	679 m	678 sh	676 m	modes	C3'-*endo, anti* (gg)
—	—	665 w	664 vw	δ H$_2$O	
656 vw	654 m	644 m	642 m	base	
632 m	629 m	622 m	622 m		
610 sh	—	613 w	—		

Ba-IMP compounds, show the marker bands, at 937 cm^{-1}, 939 cm^{-1} and 934 cm^{-1}, 826 cm^{-1}, 824 cm^{-1} and 827 cm^{-1}, respectively assigned to the sugar vibrations and the bands at 691 cm^{-1}, 693 cm^{-1} and 695 cm^{-1}, respectively assigned to the breathing mode of the purine ring. These bands are characteristic (diagnostic) of C2′-*endo* conformation. In the spectra of the transition metal adducts, Co(II)-and Ni(II)-IMP

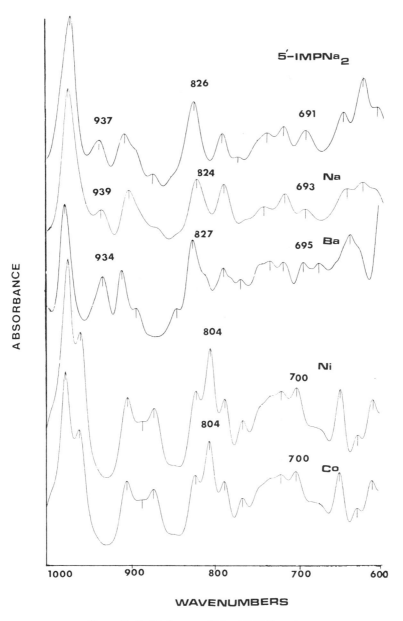

Figure 2. FT-IR Spectra of Metal-IMP Complexes.

Table II

FT-IR Absorption Frequencies (cm^{-1}) for 5'-IMPNa$_2$ and its Metal Adducts in the Region 1000-600 cm^{-1} with Possible Band Assignments and the Sugar Phosphate Conformations.

5'-IMPNa$_2$•7H$_2$O	5'-IMPHNa•6H$_2$O	Ba(IMP)•6H$_2$O	Ni(IMP)•7H$_2$O	Co(IMP)•7H$_2$O	Possible Assignments	Sugar-Phosphate Conformations
977 vs	976 vs	978 vs	976 vs	976 vs	PO$_3^{-2}$ symmetric stretching	
—	—	—	950 s	960 s		
937 m	939 m	934 m	—	—	sugar-phosphate	C2'-endo, anti (gg)
907 m	905 m	911 m	903 m	903 m		
—	—	—	—	—		
872 w	868 w	871 m	871 m	971 m	sugar-stretching	
580 sh	850 sh	846 sh	850 vw	850 vw		
826 s	824 s	827 s	820 m	820 m	sugar-phosphate	C2'-endo, anti (gg)
—	—	—	804 s	804 s	sugar-phosphate	C3'-endo, anti (gg)
791 m	791 m	790 m	786 m	786 m	P—O stretching	
771 w	770 vw	770 w	764 m	764 m		
737 m	743 m	734 m	740 m	741 w	base	
718 m	718 m	719 m	718 m	718 m		
691 m	693 m	695 m	700 m	700 m	ring-breathing mode	
—	—	—	665 w	665 w		
648 m	643 m	639 m	648 m	648 m	δH$_2$O	C2'-endo, anti (gg)
625 s	624 m	620 sh	625 m	625 m	base	C3'-endo, anti (gg)

the bands near 937 cm^{-1} disappear and the new marker band for C3'-*endo* is shown at 804 cm^{-1} and the ring breathing mode showed a splitting and shifting at 700 cm^{-1} and 665 cm^{-1} (see Fig. 2 and Table II). Furthermore, in the spectra of the polymeric compounds [Cu(IMP)]$_n$ (14) and [Zn(IMP)]$_n$ (15) the coexistence of both the C2'-*endo* and C3'-*endo* sugar puckers was shown from the spectra.

Concluding Remarks

The A, B and Z-DNA forms have been differentiated by Raman spectroscopy from characteristic bands in the above region (21,22). Peticolas and collaborators showed that some base and backbone vibrations are sensitive to conformational structures by comparing the Raman spectra of DNA fibers from known X-ray diffraction analysis to be in the A-type (C3'-*endo*) or B-type (C2'-*endo*) conformations (23,24). It was found that a B-type DNA is characterized by a band at 835 cm^{-1} which has been related to a deoxyribose-phosphodiester vibration. In an A-type DNA this band disappears and a large peak at 805-816 cm^{-1} is observed (25). The induction of B-DNA to a left-handed Z-DNA in poly (dG-dC)•poly(dG-dC) can be followed by Raman spectroscopy (5,25). The 682 cm^{-1} band related to the guanine ring breathing mode disappears in B to Z transition and a new band at 626 cm^{-1} is observed (26). In addition, the 831 cm^{-1} backbone band disappears and a new band at 750 cm^{-1} is present in the spectrum of Z-DNA, which may be diagnostic of C3'-*endo*, *syn*-guanosine found in Z-DNA (3,25). Furthermore, the intensities of the bands at 1265 cm^{-1} and 1317 cm^{-1} are considerably increased in the spectra of Z-DNA (25). In the FT-IR spectra of metal-nucleotide complexes we have similar characteristic changes due to the transition of C2'-*endo* to C3'-*endo* conformation in the spectra of the solid metal complexes intensity changes have also been observed when the sugar pucker changed from C2'-*endo* to C3'-*endo*. In the cases where there is one conformer only the marker bands for that conformer are observed. It is thus suggested from the present metal-nucleotide infrared spectra that the metal may induce the B to Z transition by coordination of the bulky metal hydrated cation (M(H$_2$O)$_5^{n+}$) to the N7 site of the guanine base and to the phosphate group by its charge. This may be the driving force which produces the C3'-*endo* sugar pucker. The C2'-*endo* to C3'-*endo* transition in the metal-adducts occurs without involving the *anti-syn* rotation about the guanine C1'-N bond as is shown from the X-ray diffraction analysis (1). This rotation takes place in the poly(dG-dC)•poly(dG-dC) when it passes from B-DNA to Z-DNA. However, for the metal-nucleotide complexes we have only one nucleotide and the purine is not compelled to change its orientation from *anti* to *syn* as in the case of Z-DNA (3), which has a rigid structure.

Acknowledgments

This research was supported by a grant from the NSERC of Canada.

References and Footnotes

1. K. Aoki, *J. Cryst. Soc. Japan 23,* 309 (1981).

2. R. W. Gellert and R. Bau, *"X-Ray Structural Studies of Metal-Nucleoside and Metal-Nucleotide Complexes"* in *"Metal Ions in Biological Systems"*, Vol. II, Ed. H. Siegel, M. Dekker, Inc. New York and Basel, p. 1 (1979).

3. A.H.-J. Want, G.T. Quigley, F.J. Kolpak, J.L. Crawford, J.H. van Boom, G. Van der Marel and A. Rich, *Nature 282*, 680 (1979).

4. F.M. Pohl and T.M. Jovin, *J. Mol. Biol. 67*, 375 (1972).

5. T. Thaman, R.C. Lord, A.H.-J. Wang and A. Rich, *Nucleic Acids Res. 9*, 5443 (1981).

6. T. Theophanides and M. Polissiou, *Inorg. Chim. Acta 56*, L1 (1981); *ibid.* M. Polissiou and T. Theophanides, *"Biomolecular Stereodynamics"*, Vol. II, Ed. R.H. Sarma, Adenine Press, New York, p. 487 (1981).

7. M. Polissiou, M.T. Phan Viet, M. St-Jacques and T. Theophanides, *Can. J. Chem. 59*, 3297 (1981).

8. G.A. Thomas and W.L. Peticolas, *J. Am. Chem. Soc. 105*, 986 (1983); *ibid. 105*, 993 (1983).

9. Y. Nishimura, M. Tsuboi, T. Nakamo, S. Higuchi, T. Sato, T. Shida, S. Uesugi, E. Ohtsuka and M. Ikehara, Nucleic Acid Res. *11*, 1579 (1983).

10. S.K. Katti, T.P. Seshadri and M.A. Viswamitra, *Acta Crystallogr. B37*, 1825 (1981).

11. K. Aoki, *Acta Crystallogr. B32*, 1454 (1974).

12. P. de Meester, D.M.L. Goodgame, T.J. Jones and A.C. Skapski, *C. R. Acad. Sci. Paris 22*, 667 (1974); P. de Meester, D.M.L. Goodgame, A.C. Skapski and B.T. Smith, *Biochim. Biophys. Acta 340*, 113 (1974).

13. K. Aoki, G.R. Clark, and J.D. Orbell, *Acta Crystallogr. B34*, 2119 (1978).

14. G.R. Clark, J.D. Orbell and J.M. Waters, *Biochim. Biophys. Acta 562*, 361 (1979).

15. P. De Meester, D.M.L. Goodgame, T.J. Jones and A.C. Skapski, *Biochim. Biophys. Acta 353*, 392 (1974).

16. E.A. Brown and C.E. Bugg, *Acta Cryst. B36*, 2597 (1980).

17. N. Nagashima and Y. Iitaka, *Acta Cryst. B24*, 1136 (1968).

18. S.T. Rao and M. Sundaralingam, *J. Amer. Chem. Soc. 91*, 1210 (1969).

19. K. Aoki, *Bull. Chem. Japan 48*, 1260 (1975)

20. A.W. Prestoyko, S.T. Crook and S.K. Carter, *"Cis Platinum"*, Eds. Acad. Press, New York, (1980).

21. S.C. Erfurth, E.J. Kiser and W.L. Peticolas, *Proc. Nat. Acad. Sci. 69*, 938 (1972).

22. S.C. Erfurth, P. Bond and W.L. Peticolas, *Biopolymers 14*, 1245 (1975).

23. K.C. Lu, E.W. Prohofsky and L.L. van Zandt, *Biopolymers 21*, 449 (1982).

24. S.C. Erfurth and W.L. Peticolas, *Biopolymers 14*, 247 (1975).

25. R.M. Wartel, J.T. Harell, W. Zacharias and R.D. Wells, *J. of Biomolecular Structure and Dynamics*, Vol. I, p. 83 (1983).

26. J.C. Martin and R.M. Wartel, *Biopolymers 21*, 499 (1982).

Structure & Motion: Membranes, Nucleic Acids & Proteins,
Eds., E. Clementi, G. Corongiu, M. H. Sarma & R. H. Sarma,
ISBN 0-940030-12-8, Adenine Press,   Copyright Adenine Press, 1985.

Lattice Modes, Soft Modes and Local Modes in Double Helical DNA

S. M. Lindsay and J. Powell

Department of Physics
Arizona State University
Tempe, Arizona 85287

and

E. W. Prohofsky and K. V. Devi-Prasad

Department of Physics
Purdue University
West Layfayette, Indiana 47907

Abstract

We outline the lattice dynamics analysis of double helical DNA and compare it to other methods. Dispersion curves are presented for all modes up to 100 cm^{-1} as well as a table of inner products between several low frequency modes so as to facilitate assignment of their characters.

We discuss mode softening as the cause of conformational change and show how Raman data support both the phonon description of DNA excitations and the soft mode theory of the A to B conformation change. The former is based on the existence of a quasi-momentum selection rule for Raman scattering from low frequency backbone excitations and the latter on the association of mode softening with conformation change in samples which undergo the A to B transition at different degrees of hydration.

We have located the lowest lying Raman bands in A- and B-DNA. In A-DNA the lowest lying depolarized mode is at about 15 cm^{-1} with a weak fully polarized mode near 12 cm^{-1}. The B-DNA spectra are dominated by a 12 cm^{-1} band which is probably fully polarized. These mode frequencies and selection rules are in agreement with the lattice dynamics analysis.

The acoustic modes show some evidence of mode softening as the Brillouin linewidths undergo a maximum in the vicinity of the A to B transition, however these data are extremely sensitive to laser heating effects (in contrast to the low lying Raman bands) and this behavior suggests either some unusual dynamics of the water of hydration or a complex fiber morphology.

We end by reviewing the evidence for microwave resonances associated with local modes at a terminus.

531

Introduction

We are pursuing a coordinated program of experiment and theory with the goal of discovering the motions of the DNA double helix at the level of atomic detail. We hope, of course, that such knowledge will explain the biological function of the molecule, eventually leading to a detailed understanding of its interaction with proteins. The first task is to understand the dynamics of the double helix alone. The complexity of the material places severe constraints on both theory and experiment. Nonetheless this work appears promising in several respects:

—Evidence is accumulating for a rather simple mechanism for conformational changes (the so-called 'soft mode').

—Theory has guided experiment in locating the lowest lying vibrational modes with quite remarkable *quantitative* agreement

—There is evidence for resonant local modes (such as base destacking at a terminus).

In this article we shall outline the lattice dynamical theory, contrasting it with other approaches and clarifying its strengths and weaknesses. Our goal here is to sketch the main ideas for the reader who is not familiar with solid state physics. We discuss soft mode mediated conformation change in a similar vein. We describe how inelastic light scattering may be used as a probe of some of the predicted modes and survey some recent discoveries (the details of which we leave to the regular refereed literature). Both theory and experiment are developing apace, so we end by anticipating some future work.

DNA Dynamics—some rough estimates.

We can get a crude idea of some important frequencies by imagining the cooperative motions of DNA as involving one or more base pairs vibrating against a single covalent or Van der Waals bond with spring constants of perhaps mdyne/A (covalent) to μdyne/A (VdW). The mass of a single base pair would vibrate with frequencies of a few cm^{-1} while vibrations involving a significant fraction of the mass in a persistence length might be as low in frequency as 0.1 cm^{-1}. Motions as slow as this may be over-damped in solution but it is not at all clear that simple hydrodynamic considerations apply.

Lattice Dynamics and the "infinite" chain

Of course the motion of any part of the double helix is much more complicated than outlined above—displacing any one atom from equilibrium puts forces on all the others, so it would appear that calculating the motions of an infinite chain is itself an infinite problem. That this is not so is illustrated in Figure 1 where we show a very simple polymer made up of an infinite chain composed of atom A and atom

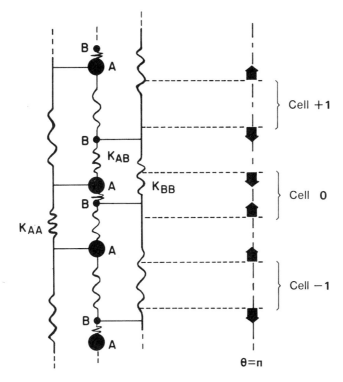

Figure 1. Illustrating how translational (or helical) symmetry can be exploited to make the dynamics of an infinite chain a finite problem. In the polymer sketched (above left) the symmetry cell contains just two atoms (A and B) connected with three types of spring—only a few cells are shown but the vertical dotted lines are meant to imply a continuation to plus and minus infinity. Clearly all the vibrational properties of the system are determined by the two masses, m_A and m_B, and the three spring constants, k_{AA}, k_{BB} and k_{AB} given just the relative cell to cell displacement (i.e. phase) so that the degree of stretching of the intercell springs can be calculated. The diagram shows a mode with a cell to cell phase shift of π radians, so that the atomic displacements change sign from cell to cell. This is demonstrated by the arrows on the right which show the maximum displacements from the equilibrium positions (dotted horizontal lines) of the atoms.

B connected up with springs. The spring k_{AB} connects atom A to atom B (this assembly representing the basic repeat unit or symmetry cell of the polymer) while k_{AA}, k_{BB} and also k_{AB} connect neighboring cells (we could put in longer range forces, but that would complicate our figure). Now clearly the whole ploymer can be made up by repeating this cell, so that once we know the three k's and the masses m_A and m_B we know almost all we need to. The actual displacements along the chain must be just the possible displacements of any one cell. The extra piece of information that incorporation into an infinite chain demands is the relative phase of the displacement from cell to cell. For example, at one extreme the cell on each side of cell zero could be moving in antiphase (as, by symmetry, their neighbors would, and so *ad infinitum*) and in this case the full effects of k_{AA} and k_{BB} are felt in stiffening the vibration of A against B. This is the situation illustrated in Figure 1.

At the other extreme neighboring A's and B's could beat *in phase* (so the whole lattice of A's moves against the whole lattice of B's) and in this case k_{AA} and k_{BB} do not contribute at all. The calculation proceeds by using appropriate symmetry coordin-

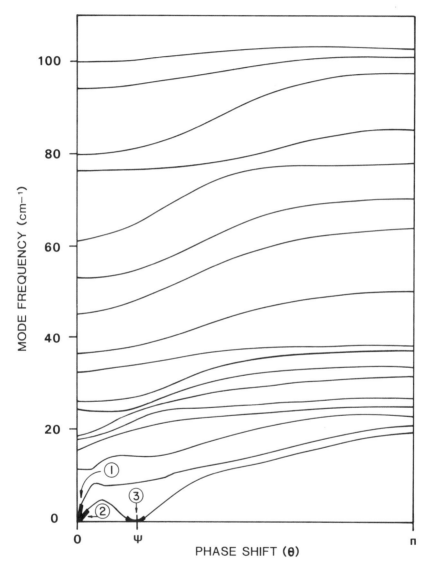

Figure 2. Calculated dispersion curves for A-poly(dG).poly(dC). Mode frequencies (in wavenumbers) are plotted versus cell to cell phase shift, θ up to the largest meaningful value of π. The parts of the acoustic branches containing the modes which go to zero frequency are outlined heavily and labeled as follows: 1) Longitudinal (z) compressional acoustic mode. 2) Torsional acoustic mode. 3) Bending modes (both x and y). Note that the location of the bending mode minimum is a consequence of the helical coordinates used in the calculation (1). This minimum occurs at a phase shift equal to the helix pitch ψ (32.7°). Branches are show for the first 15 optic modes also. Solid curves are used for clarity, but the calculations were actually limited to 20 phase points across the zone.

ates (helical in the case of DNA) so that the equations of motion are uncoupled, requiring only a choice of relative cell to cell phase, θ. In practice the calculation is carried out for as many phases as required to obtain smooth curves of mode frequencies (for example) versus phase shift from cell to cell. Such a set of curves (dispersion curves) is shown for A-form poly(dG)•poly(dC) in Figure 2. The two lowest branches incorporate the 4 acoustic modes (longitudinal compression, X and Y bending and torsion) and the next 15 branches are the optic modes up to 100 cm^{-1}. (This terminology and the mathematical formalism are discussed in our earlier contribution to this series (1)—note that the schematic dispersion curves in that article are wrongly labeled as having 3N-3 optic branches (N is the number of atoms in a symmetry cell). There are four acoustic modes as just listed, so, of course, 3N-4 optic modes). For each point on such dispersion curves one also obtains a full set of displacements for all the atoms in the cell (the *eigenvectors* or displacement patterns).

What goes into a lattice dynamical calculation

Any calculation on complex systems is easy prey for the dreaded GIGO syndrome (Garbage In = Garbage Out) so let us lay out the input clearly;

— Helical symmetry is assumed. Thus the calculations are carried out for an infinite homopolymer in the first place. This is not a serious limitation, for as described elsewhere (1,2) local sites of interest can be embedded into an infinite homopolyer chain using the Green's function method.

— The use of spring constants implies harmonic oscillator behavior, while certain bonds (such as hydrogen bonds) are clearly little like Hookean springs, having a force "constant" that changes quite rapidly with amplitude. In problems such as melting this nonlinear response is crucial and a so-called "self-consistent phonon" approach has been developed to describe large amplitude motions (3).

— The atomic positions and interatomic spring constants are crucial, but we would like to draw a distinction between the difficulties associated with the high and low frequency regimes. The force constants used to describe bond bending, stretching and torsion are fairly well established and refined to include a large body of high frequency Raman and infrared data (4). Thus while many factors may conspire against the idealized harmonic picture (local structure may vary from that determined from fiber studies (5), anharmonicities may contribute and motion of water molecules as well as counter ions must be important) the broad details must reflect an average over all these complications because of the experimental refinement of force constants. For example, when refinement changes the force constant for phosphorous—oxygen stretching in the backbone, this does not imply that the original force constant was wrong in representing the 'bare' P—O motion, rather that a simple harmonic oscillator with a different (refined) force constant gives a frequency of motion, amplitude and character of motion that represents a reasonable average over the much more complicated motion involving water and counter ions around the P—O bonds. If water and counter ions were to play a passive role in the dynamics, then such an approach may be reasonable at all frequencies, complicated only by the lack of detailed spectral information for the low frequency modes which are most sensitive to the long range nonbonded potentials. In fact we believe that the most serious problems lie with treatment of the water and counter ions, and we return to this point in a later section.

Comparison with other approaches to DNA dynamics

Many models have been built for some special aspect of the dynamics, such as the elastic filament approach used in the analysis of torsional fluctuations (6,7) or a sugar-pucker model for nonlinear excitations (8). However, the only other approach which incorporates atomic details is molecular dynamics simulation in which a complete structure is allowed to move under the action of the various interatomic *forces* according to Newton's laws of motion. This should be contrasted with the lattice dynamical approach where the structure is fixed and the motion *about equilibrium* is calculated from the *force "constants"*. The molecular dynamics approach is in principle more realistic than lattice dynamics, however it suffers some limitations at present. The most obvious are, of course, the size of the system and the real time for which calculations may carried out in a reasonable computing time. Levitt has examined 12 and 24 bp double helices over 90 pSec (9) while Tidor *et al.* (10) have carried out simulations for several hexamers over 60 pSec. The dynamics of such fragments are representative of the long chain to only a limited extent (for example see the discussion of motion in a terminal region in Reference 2) while both the limited size and the time scales exclude the lowest lying modes. A less obvious difficulty is associated with the structural freedom that gives the techniques value in the first place; errors in potentials may express themselves as changes in the structure. A gross manifestation of this phenomenon is discussed by Levitt (9) but less dramatic structural fluctuations might arise from this source. In some respects this makes the problem of dealing with the nonbonded interactions more severe for these simulations than it is for the lattice dynamics analysis at present. On the other hand (as pointed out by Levitt (9)) molecular dynamics holds out the prospect of a more realistic treatment of the surrounding water and counter ions when calculations of the required size become possible. However it should be pointed out that existing experimental observations of low lying modes on long helices show quasi-momentum and polarization selection rules which are consistent with lattice mode analysis. This indicates that phonon modes do exist as will be discussed later.

The nonbonded interactions and the water problem

Both lattice dynamics and molecular dynamics assume pair-wise Coulomb-like interactions between charged atoms, paying scant regard to the intervening highly polarizable cloud of water and counter ions, save in as much as parameters (such as charges and ranges) are adjusted in some way. Now the problems associated with an explicit treatment of the water are formidable; induced dipoles react back on the original charge distributions which modifies the induced dipoles—all over finite timescales during which DNA, water and counter ions are executing thermal motion! Finney has discussed some of these problems in structure calculations (11). Some measure of the difficulties associated with such calculations can be gained from the history of the lattice dynamics of the humble alkali halide crystals where the problem of dynamic polarizibilities occupied theorists for decades. Nonetheless the hope is that, for many frequencies of motion, the details of the water dynamics are unimportant. Thus, in these lattice dynamics calculations the force constant connecting charged atom i with charged atom j is obtained from

$$F_{ij} = a \frac{q_i q_j}{\sqrt{\epsilon_i \epsilon_j} \; r_{ij}^3} \qquad (1)$$

where q_i and q_j are the magnitudes of the charges on the atoms and r_{ij} the distance between them. ϵ_i and ϵ_j are the dielectric constants pertinent to atoms i and j. For simplicity, only two dielectric constants are used; 3.6 for atoms in the bases and 81 for atoms in the ribose-phosphate backbone. a is a phenomenological parameter which is adjusted to fit the observed compressional sound speed. Equation 1 corresponds to a long range Coulomb interaction and this would seem to be a questionable assumption. However reducing the range of the nonbonded interactions results in unrealistic short range bond strengths as discussed elsewhere (12). The essential point is that the Raman bands and the sound speed data *cannot be fitted consistently without such long range interactions.* This is not meant to imply that there necessarily are charges which behave as though they are nearly unscreened; rather that the overall effect of the intervening water and counter ions is best incorporated into this 'dry' model in this way.

In the absence of any special water and/or counter ion dynamics, the scheme outlined above may be a successful 'mean field' treatment of the water. Indeed, the lowest lying Raman bands seem to be predicted with remarkable accuracy by this model as we shall show. On the other hand, there are some peculiar effects of very small laser powers on sound speeds, perhaps a hint of a complicated role for the water. Very recent microwave absorption studies suggest that acoustic modes may not be damped even in dilute solutions (13) in contrast to the results of calculations in which the water is taken to behave hydrodynamically right into the grooves of the double helix (14). Furthermore Corongiu and Clementi have discovered an instability in the static structure (as determined by Monte Carlo simulations) of water hydrating the double helix at degrees of hydration which are in the neighborhood of the A to B transition (15). While we believe that there is now considerable experimental evidence for a soft mode picture of the A to B conformation shift, this need not exclude a special role for the water in the hybrid dynamical system of DNA plus water and counter ions.

The soft mode as a mechanism for conformation change and the A to B transition in DNA.

Soft modes—vibrational modes whose frequency falls to zero—have been observed to mediate a large number of displacive phase transitions. These are changes in structure which involve the cooperative rearrangment of the atoms within each symmetry cell. The change may occur as a function of temperature (or some other parameter like pH or degree of hydration) abruptly, in which case the transition is said to be first order, or smoothly, in which case the transition is said to be second order. If a simple displacement of some atoms in the cell is all that is required to bring about the phase change, then a description of the transition is as follows; as the critical temperature (or critical parameter value) is approached, the effective force constants for a particular optic mode tend to zero. This could be, for example, because long and short range contributions tend to cancel at the critical point. The optic mode involved would have a displacement pattern such that each atom moved to the required position in the new phase. Acoustic modes could drive such phase changes in special circumstances—for example if the new phase had a unit cell related to the old phase by a simple shear distortion. In the simplest picture the force constant for the soft mode goes to zero (so the mode frequency goes to zero) driving the amplitude of the mode ever larger (since we want 1/2 (force constant) times (square of amplitude) to stay constant at kT). Of course at some point anharmonicities intervene (as they must for a new phase to become stable) to prevent such singular behavior (16). The progress of the phase change is measured by an *order parameter.* In the simplest case the order parameter is just the eigenvector of the soft mode; that is, the multicomponent vector formed by taking the difference of all the atomic positions in the two phases is identical to the vector which describes the displacement pattern for the soft mode. In this regard the A to B transition in DNA is much simpler than the (probably more important) B to Z transition which involves a major reconstruction of the unit cell and almost certainly involves intermediate structures (17).

The lowest lying optic mode was found to soften as a function of very small changes in the parameter describing the internal dielectric constant (18)—indeed the observation was made compelling because this softening does not occur in poly(dA)·poly(dT), a polymer that does not undergo the B to A transition. The mode that is believed to be the driving force for the conformation change is the mode at 12 cm^{-1} at $\theta=0$ in Figure 2. It does indeed have the right character to be the soft mode. This is demonstrated by forming the inner product of the various low lying modes near $\theta=0$ with the normalized order parameter for the transition (which is formed by subtracting the A and B coordinate set and normalizing the resulting vector). A value of 0 for this inner product means no coincidence with the order parameter. A value of one means perfect alignment. The $\theta=0$ lowest optic mode yields a value near 0.9 (0.84 when squared—see Table I). All the other modes yield values near 0 (as one would expect from the orthogonality of the normal modes). This is remarkably good evidence that the geometry of the mode is as required for a soft mode/order parameter. These inner products are listed in Table I where we also give values for the inner products of various low lying modes at $\theta=0$

Table I

Characteristics of the lowest lying modes in A- poly(dG)·poly(dC)

The top row lists the frequencies of the seven lowest lying $\theta = 0$ modes in A-poly(dG)·poly(dC) while the eight lowest $\theta = \psi$ mode frequencies are listed down the left hand column. The square of the inner product of th mode eigenvector with the normalized order parameter for the A to B transition is show in parenthesis below each frequency. Most of the character of the order parameter appears in the $\theta = 0$ 11.48cm^{-1} mode. The remainder of the table shows the squares of the inner products taken between the $\theta = 0$ and the $\theta = \psi$ modes. This is useful in picking modes which contain much of the character of a simple mode. For example, much of the character of the torsional acoustic mode (lowest $\theta = 0$ mode) is contained in the 8.49 cm^{-1} mode at $\theta = \psi$ because the square of the appropriate inner product is 0.38.

$\theta=0$ FREQUENCIES IN cm^{-1}

	0.00 (0.00)	0.00 (0.00)	11.48 (0.84)	15.65 (0.01)	17.78 (0.02)	18.76 (0.00)	24.17 (0.01)
0.00 (0.26)	0.15	0.06	0.24	0.25	0.13	0.01	0.06
8.49 (0.14)	0.38	0.00	0.20	0.18	0.12	0.04	0.01
14.11 (0.00)	0.17	0.29	0.01	0.03	0.22	0.20	0.02
20.04 (0.17)	0.03	0.06	0.14	0.20	0.17	0.26	0.05
22.46 (0.02)	0.08	0.23	0.03	0.04	0.13	0.24	0.10
23.86 (0.02)	0.07	0.00	0.02	0.08	0.06	0.08	0.56
24.90 (0.10)	0.03	0.15	0.16	0.16	0.04	0.06	0.03
27.01 (0.15)	0.01	0.10	0.12	0.01	0.05	0.02	0.10

$\theta=\psi$ FREQUENCIES IN cm^{-1}

and $\theta = \psi$. They are of value in characterizing the various modes. We have focused on the $\theta=0$ modes because of the requirement that the phase change correspond to a uniform translation from cell to cell. In fact, there is a small change in pitch between A and B DNA (32.7° and 36° respectively), so the mode must soften a little away from $\theta=0$, meaning that it will try to cross the acoustic branches on its way down (see Figure 2). Thus the effects of mode softening may be visible in the acoustic spectrum also.

We should mention the application of a nonlinear approach to this transition by Krumhansl and Alexander (8). Formally, of course, nonlinear parts of the potential become important, even dominant, at some point in the course of the transition (16), however Krumhansl and Alexander (8) choose sugar pucker as the controlling factor in the A to B transition whereas our calculations show that the nonbonded water and ion mediated interactions dominate. The soft mode picture seems to be holding up remarkably well so far, having the particular advantage of allowing quantitative predictions to be made and compared to observations. Most importantly from the biological standpoint, the soft mode picture allows one to describe the factors that bring about the onset of conformation change—in this case changes in the water and ion mediated nonbonded interactions. Just how far one must delve into the complications of nonlinearity remains to be decided by experiment.

Inelastic light scattering as a probe of dynamics

We choose inelastic light scattering because of its ability to probe the very low frequency modes that we believe to be important. In an inelastic light scattering experiment a quantum of light (photon) scatters from a material, simultaneously creating or destroying a vibrational quantum (phonon). Energy conservation requires that the scattering photon be up or down shifted in frequency by an amount equal to the phonon frequency. A schematic layout of an apparatus for such experiments is shown as Figure 5 in our earlier contribution to this series (1). When the phonon involved is a high frequency (generally localized) vibration the process is called Raman scattering, and when the process involves the lower frequency (extended) sound waves the process is called Brillouin scattering. The distinction is historical, based largely on the different experimental procedures. In fact, the "Raman bands" in DNA are so low in frequency that even this rather arbitrary distinction is invalid, and we use the two terms interchangeably. The physical process is the same, however, and (despite having begun with the quantum terminology) we continue with a simple classical description. The incident light field ($E = E_o\, e^{i\omega t}$ say) induces an oscillating dipole moment M which, to a first approximation, oscillates at the frequency of the incident field ($M = M_o\, e^{i\omega t}$), re-radiating the light at the incident frequency (elastic scattering). If, however, some part of the electric susceptibility, χ, depends on atomic displacements, Q, then the induced dipole moment is

$$M = \chi E + \frac{\partial \chi}{\partial Q} Q E + \text{.......} \qquad (2)$$

where the dots remind us that there are many more complex ways in which the susceptibility may depend on displacements. If Q is taken to be some normal mode of the system then it varies harmonically as $Q = Q_o\, e^{i\Omega t}$ so clearly the scattered field is modulated at the mode frequency, Ω, giving rise to peaks at $\omega \pm \Omega$ in the scattered intensity. Equation 2 illustrates another powerful aspect of Raman scattering, which is the existence of selection rules related to the symmetries of the normal modes, Q. The full set of the components of the susceptibility derivatives, $\frac{\partial \chi}{\partial Q}$, must reflect the point symmetry of the molecule, so for a given Q and orientation of the incident field only certain components of the induced dipole are allowed. For helical polymers, modes at $\theta = 0$ (see Figure 2) scatter an incident field along the helix axis (taken as the z axis) in the same polarization, i.e. with the induced dipole also along z .A shorthand notation is to say these modes scatter zz.

Modes at $\theta = \psi$ (the helix pitch angle) scatter xz and yz while modes at $\theta = 2\psi$ scatter xx, xy and yy.

The periodicity of the system imposes another selection rule in addition to the point symmetry constraints just discussed. To calculate the scattered intensity one must sum the fields from each induced dipole—if the sample is a liquid or glass, then one gets some average over all orientations, except at very long wavelengths where vibrations are coherent independent of the atomic structure. However when the dipoles are embedded in a periodic lattice, the scattered fields can cancel in certain directions, while reinforcing each other in directions where the incident and scattered fields are phase matched to the vibrational wavelength (this is just $2\pi/\theta$ times the lattice spacing, or rise per base pair in this case). Thus, if k_i is the wave vector of the incident field and k_s the wave vector of the scattered field then this condition is

$$k_i - k_s = q \qquad (3)$$

where q is the wavevector of the phonon. This is analagous to conservation of momentum for real particles and is called conservation of quasi-momentum. Of course DNA is periodic in one dimension only, so this amounts to the requirement that q (as defined above) lie along z.

We have said nothing of the experimental techniques. Actually, the low frequencies involved require higher resolutions than are available with conventional methods for separating small shifts in scattered laser light, so a special instrument (the "Multipass Vernier Tandem Interferometer", MVTI) was developed for this work. It is described in some detail elsewhere (19). While we can now carry out experiments at the required resolution, the measurements are far from routine. The signals are extremely weak. This is because the count rate in the minimum resolved frequency interval varies inversely with resolution so, given that the acceptance angle must also be reduced with increasing resolution, one loses signal as the square of the increased resolution obtained with the MVTI. There is also more than an additional order of magnitude loss resulting from the small aperature of the MVTI when compared with a grating monochromator. Thus some of our high resolution Raman spectra require many *days* of signal averaging. When possible we work with a conventional Raman monochromator (a Spex 1402). Laser illumination perturbs the samples, so we must work with very low (mW) laser powers.

The lowest lying DNA backbone modes are phonons

The applicability of the lattice dynamics approximation to natural DNA is a matter for experimental test. While all vibrational excitations are 'phonons' in the sense of being quantized, what is important is that they also be periodic along the chain, i.e. that they carry quasi-momentum as assumed in separating the equations of motion. At this point we should clear up a misconception; the presence of Brillouin scattering (scattering from sound waves) is often taken as evidence for phonon-like excitations. However Brillouin scattering is also seen in glasses and liquids because the long range (elastic) forces that give the waves their coherence (and thus quasi-momentum) are not sensitive to interatomic detail, just macroscopic averages. One needs to look for quasi-momentum conservation in modes that work against interatomic springs—that is the Raman bands. The lower lying bands make particularly sensitive probes as the longer range forces begin to be felt (so that the hallmark of a 'glassy' Raman spectrum is an excess *low frequency* Raman scattering due to breakdown of quasi-momentum conservation).

Urabe and Tominaga have reported a q vector selection rule for a band found near 23 cm^{-1} in A-DNA—observation of the band requiring that q have a component along z (20). We have repeated this work using some highly crystalline film kindly given to us by Alan Rupprecht of the University of Stockholm. This material has a strong line near 30 cm^{-1} in B-DNA and 35 cm^{-1} in A-DNA. It is not subject to such a constraint on the orientation of the q vector and therefore serves as an internal standard with which the scattering intensity may be monitored. We have constructed an extremely accurate rotary holder for these films. It allows rotation of the film about the point at which the light scattering occurs with no detectable loss of signal due to change in scattering geometry (the whole system being hermetically sealed over a salt solution appropriate to maintaining the desired relative humidity). With a 90° scattering geometry in the horizontal plane, and the film in the vertical plane with z bisecting the incident and scattering beams, q lies in the film in the horizontal plane. We can change the orientation of q with respect to z (otherwise leaving the scattering geometry undisturbed simply by rotating the film. Figure 3 shows spectra obtained with q both along and perpendicular to z for B- (upper curves) and A-DNA (lower curves). (Both spectra are obtained at 66% r.h., the A-DNA being 1% NaCl salted, the B-DNA 4% LiCl salted.) The A spectrum contains, in addition to the 35 cm^{-1} mode, a 26 cm^{-1} band which we believe to be the mode reported at 23 cm^{-1} by Urabe and Tominaga (the difference being due to salt). This band is a backbone mode (20) and these spectra show the need for q to lie along z in a dramatic manner, coming and going as the film is rotated, being almost absent from the spectrum when q is perpendicular to z. These spectra demostrate that vibrational coherence exists even *in a natural DNA* (calf thymus).

Mode softening accompanies the conformation changes

Urabe and Tominaga have also studied the 23 cm^{-1} band as the r.h. is varied, finding that the band softened to below 15 cm^{-1} as the material went to B form (21). If the lattice dynamics calculations are correct, then this mode is not the principal soft mode, but is related to it in character, so that some softening is expected. The mode polarization is xz (20) indicating that it belongs to the $\theta = \psi$ part of the dispersion curves (Figure 2). In fact it can be seen that there are many modes in the vicinity of 20 to 26 cm^{-1} at this point in the zone, and in natural DNA one expects to see a single inhomogeneously broadened band. An exact description of the character of these modes requires careful examination of the calculated eigenvectors, however one can get some qualitative information from the dispersion curves by

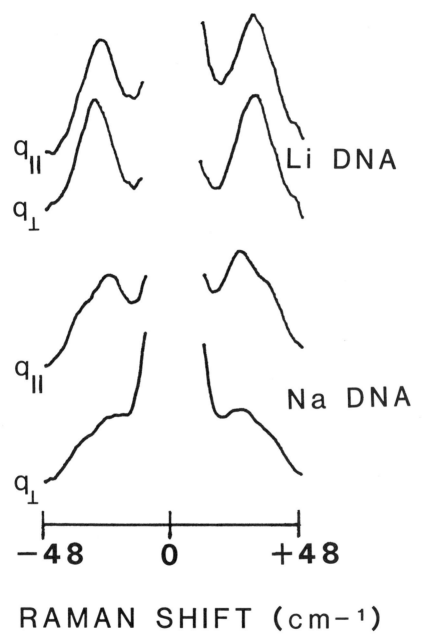

Figure 3. Showing quasi-momentum conservation for a low lying backbone mode at 26 cm^{-1} in A-DNA. The upper traces are for B-DNA films with the q vector parallel (top trace) and perpendicular (next trace) to the fiber axis. The spectra are approximately normalized to the high frequency Raman band intensities and show the relative insensitivity of a band at 30 cm^{-1} to alignment of the q vector. This band appears at about 35cm^{-1} in A-DNA (lower traces). The 26 cm^{-1} component of the A-DNA spectrum disappears as the q vector is rotated so as to be perpendicular to the fiber axis (lowest trace). This demonstrates a q vector selection rule for phonons propagating along the fiber axis.

following a mode of given character, bearing in mind that branch repulsion occurs where there would otherwise be a crossover. Thus the $\theta = 0$ lowest optic mode is deflected from its original path as it approaches crossover with the upcoming longitudinal compressional branch (Figure 2) only to be deflected down again as it approaches crossover with the branch that starts at 16 cm^{-1} (at $\theta = 0$). At $\theta = \psi$ the branch lies in the path of the obvious extention of the torsional acoustic branch to this point in the zone, and does indeed have much torsional character (see Table I) Thus the 23 (at 75% r.h.) cm^{-1} mode possesses some of the character of the soft mode. This is confirmed by the value of the square of the inner product of this mode with the $\theta = 0$ soft mode (0.16—see Table I).

We have repeated Urabe and Tominaga's measurements with two different samples. They are the fibers we have described before (1) which are salted at 1% excess NaCl, undergoing the A to B transition in the classic manner between 75% and 92% r.h. (as verified with fiber diffraction and high frequency Raman spectra) and Rupprecht's wet spun films which do not yield B diffraction patterns at high humidities (22) despite being salted at the same level. In fact, we find that these materials do become B at very high relative humidities (98% +) losing crystallinity as they do so. (We believe the conformation to be pinned by the extensive crystal structure in these materials). Data for the "26/23 cm^{-1}" mode frequency is summarized in Figure 4 on which we have included the data of Urabe and Tominaga (21). These workers did not add salt to their DNA so it does not undergo the A to B transition in the usual manner, though Raman spectra show that it is B by 98% r.h. The data in Figure 4 are converted to absolute water contents (the conversion is based on accurate data for *fibers,* less accurate data indicates that films behave similarly but there may be some small (10%) difference between fibers and films). The established points of the A to B transitions are marked with the arrows (labeled with the symbols which identify them with a given sample). The mode is no longer resolved after the transition, so the 'lines to guide the eye' are continued downward merely to indicate that the endpoint is an unknown low value. It can be seen that both the samples salted at 1% NaCl soften together, parting company as the transition is suppressed in the wet spun film. Softening is delayed, however, in the unsalted material, setting in only as the transition is approached. Thus the common feature in all these samples is that the softening is not associated just with a change in water content, but *specifically with the conformational shift.*

Location of the lowest lying optic modes

The validity of the foregoing discussion rests on the mode assignment. We need to establish the entire low lying Raman spectrum and compare it with the calculated dispersion curves before such analysis can be taken seriously. Though far from complete, we have made a start on this. Progress is slowed not only by the difficulties we have described, but also because we cannot signal average for long periods to improve signal to noise in the spectra. For reasons we do not understand, the samples can suffer damage on prolonged (3 to 4 days) exposure to low power (5 to 15 mW) laser beams. This damage can be visible to the naked eye—we guess that it

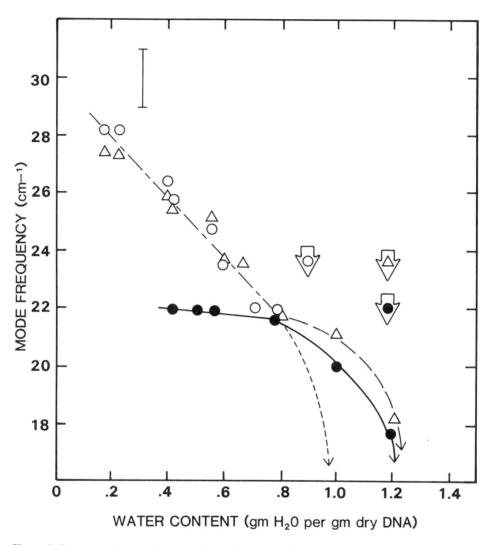

Figure 4. Frequency of a low lying xz polarized Raman band in several samples as a function of their degree of hydration. ○ 1% NaCl salted fibers which undergo the A to B transition at about 0.9 gm H_2O per gm dry DNA (as marked by the arrow containing the appropriate symbol). △ 1% NaCl salted wet spun films in which the transition is delayed to 1.2 gm H_2O per gm dry DNA (as marked by the labeled arrow). ● Unsalted fibers (from the data of Urabe and Tominaga (21) which also do not undergo the transition until 1.2 gm H_2O per gm dry DNA (also marked by an arrow). The lines are guides to the eye— their main function being to illustrate that as the conformation change is undergone, the mode drops to become a unresolved shoulder. The line of short dashes connects the 1% salted-fiber data, the line of long dashes connects the 1% salted-film data and the solid line connects the unsalted fiber data. For clarity we only show one representative error bar in the top left corner.

is caused by some concentration or depletion of salt due to small temperature gradients set up at the point of illumination. We reject spectra when such damage is found (though the usual symptom is a loss of signal). At present we have to work with the 5145 A laser line which is accompanied by a plasma line at 11.5 cm^{-1} on the Stokes side (on our machine we scan into the Stokes direction, so any spurious signal from heating of the photomultiplier also accumulates there). As a final caution we should point out that the transmission of the MVTI is not uniform (in a predictable and measured way which is recorded with each spectrum). With these many qualifications, we turn to the measured spectra.

Spectra for A-DNA (1% NaCl, 75% r.h.) are shown in the top part of Figure 5. In this case the geometry is nearly backscattering, but with the sample tilted so that a component of q lies along z. The upper trace shows the spectrum recorded for horizontal (H) incident polarization with both H and V (vertical) collected polarizations (the z axis is horizontal). Thus this spectrum contains both xz and zz contributions. The spectra with resolved polarizations are affected by losses introduced by the polarizers, but appear to indicate that the 14.5 cm^{-1} peak is xz polarized, while there is a weak shoulder at about 12 cm^{-1} with zz polarization. (A further word of caution; the optical quality of even the best samples does not permit a good assigment of mode polarizations because of scrambling by inhomogeneities.)

This is in remarkable accord with the calculated dynamics. Figure 2 indicates a zone center ($\theta = 0$) mode at 12 cm^{-1}. This branch continues out to a $\theta=\psi$ mode (xz polarization) of longitudinal compressional character at 14.5 cm^{-1}. Note that the branch on which this mode lies is very flat near $\theta = \psi$ which indicates a high density of states (that is, the number of modes per unit quasi-momentum with the same frequency is high) and would indicate a relatively pronounced feature in the Raman spectrum. We do not see a feature near 8 cm^{-1} from the longitudinal acoustic branch at $\theta = \psi$. At this point the mode would have torsional character (the square of the inner product with the torsional acoustic mode is 0.38—see Table I) and we expect it to be suppressed in ordered fibers, although the signal to noise ratio of these spectra does not allow a definite conclusion to be reached on this point.

Up to date calculations for B-DNA are not yet available. Experimentally we find the lowest lying mode of zz polarization to be close to 12 cm^{-1} in B-DNA samples also (see Figure 5).

Acoustic modes, soft modes and water

We would expect to find an interaction between the acoustic modes and a softening optic mode as it attempted to cross the acoustic branches. Such interactions are well documented in solid state physics—as the optic mode approaches the zone center acoustic mode frequencies, the modes couple, providing a loss mechanism for the acoustic modes (often generating a central peak in the spectrum). We have reported Brillouin spectra for A- and B-DNA elsewhere (23), and have now used these methods to examine the acoustic spectra across the entire range of humidities. We find that the compressional sound speeds are completely dominated by the degree of hydration, falling from nearly 4 km/s in dry DNA to about 1.7 km/s in

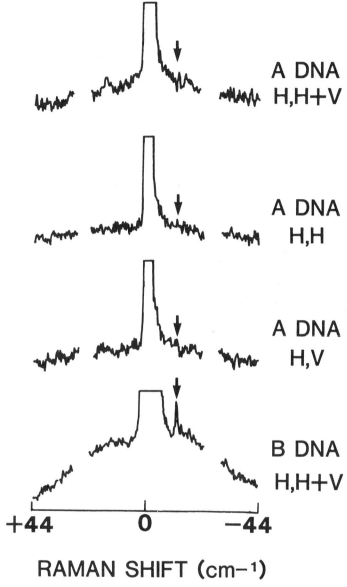

Figure 5. High resolution Raman spectra showing low lying bands in A- and B-DNA. A laser plasma line (11.5 cm^{-1} Stokes shift) obscures some of each spectrum, and is marked with an arrow. The signal to noise is limited b the deterioration of the sample on long exposures and is poorer in the polarization resolved spectra because of losses caused by the polarizer. The gaps are where instrumental ghosts have been removed (19) (transmission of the MVTI falls off somewhat beyond these ghosts). From top to bottom these spectra are: 1) A-DNA (1% NaCl, 75% r.h.), horizontal polarization incident with both polarizations collected. The fiber axis is horizontal and inclined so that there is a component of q along the fiber axis. The 23 cm^{-1} band is under the ghost. 2) As above but with just horizontal polarization collected. 3) As above but with just vertical polarization collected. 4) B-DNA (4% LiCl, 66% r.h.) in the same geometry as the top A-DNA spectrum. Polarization resolved spectra indicate that the broad 12cm^{-1} feature is probably zz polarized.

DNA containing 1.5 gm H_2O per gm dry DNA. This fall-off does not appear to depend on conformation (Li salted DNA behaving similarly) and becomes less dramatic at the point that the fibers start to swell. There is, however some evidence of an association between a maximum in the Brillouin linewidths and the A to B transition, exactly as one might expect if the optic mode softened to very low frequencies. We would, however, like to end this section on a cautionary note. The Brillouin linewidths are also affected by quite small changes in the incident laser powers (as are the Brillouin shifts, as we describe in detail for B-DNA in Reference 23). We do not understand the origin of this behavior, except that since the sound speeds in "humidified" fibers at high power correspond to the sound speeds in dry fibers at low laser powers it is natural to suppose that the (slight) heating on laser exposure (23) dries out the sample. This should mean that similar effects should be seen in the Raman spectra where the 26 cm^{-1} band is a sensitive marker of water content (Figure 4). We took A form fibers at 75% r.h. (because they show a distinct 23 cm^{-1} band which goes to 28 cm^{-1} at 23% r.h.) and measured the Raman shifts as a function of laser power. We did the same for the Brillouin shifts (the effect is less dramatic than that reported for B-fibers because they start out drier). A typical set of spectra (from the same fiber) are shown in Figure 6. The Brillouin lines shift out by about 25% on going from 6mW to 60mW (also narrowing considerably). *The Raman band does not shift detectably.* (The same appears to be true for the lower lying bands described in the last section.)

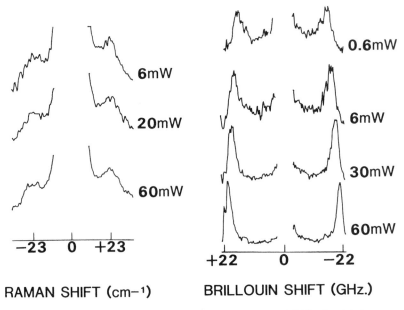

Figure 6. Illustrating the different sensitivity of the Raman (left) and Brillouin (right) spectra to laser heating. The sample is an A-DNA fiber at 75% r.h. The Brillouin shift increases by nearly 25% on going from 6 to 60 mW incident laser power (note the narrowing of the Brillouin peaks too). The 23 cm^{-1} Raman band is not noticeably affected over the same power range.

One can only conclude the (moderate) laser heating does not move the water close to the double helix that determines its conformation (and low lying Raman spectrum). However one must then conclude that the elastic properties of these materials *are dominated by water quite remote from the helix backbone.* Examination of the Brillouin spectra of a wide range of samples points to a most complex behavior. Some features (such as the elastic anisotropy of highly ordered material) argue for a dominance of local (i.e. double helical) properties in determining the long range interactions. Others such as the behavior just discussed lead to the conclusion that some complex superstructure dominates. In this regard it is interesting to note that as fibers continue to swell, the speed of sound *does not* continue to fall to the value appropriate for water, but locks up at a value distinctly above it (24). Giodarno *et al.* report that even dilute DNA solutions are thixotropic (25) (that is they form weakly structured networks over long periods), and the strange time dependent laser diffraction effects we have observed (23) would be consistent with such structures so it may be the case that these data are a consequence of a complex fiber morphology. Nonetheless, a model for the nonbonded interactions fitted to some of these data appears to have predicted the location of the lowest lying Raman bands with uncanny accuracy.

Local modes

We ended our previous contribution to this series with a description of a microwave mode we associated with a 'defect resonance' at a terminus. This work is now described in detail elsewhere (26) and we have yet to carry out the experiments that could confirm the association of this mode with local motion in the terminal region. The notion is of such importance, however, that we would like to recall the main points here. Defect resonances, or virtual local modes (as we shall henceforth call them to avoid misunderstandings about "*defects* in the DNA") arise when motion at a special site (terminus, replicating fork, special sequence, protein binding site, etc.) couples into the lattice modes of the infinite chain. In this case the modes can be pumped as amplitude from lattice modes of the right symmetry piles up at certain frequencies. The increase of amplitude can be enormous and very sharp (2). At the moment we can point to nothing more than the remarkable coincidence of predicted and observed frequencies, awaiting the experimental data for a final verdict. The list of 'remarkable coincidences' is, however, getting long enough for us to end this contribution in anticipating some future developments.

Future developments

Many experiments, obvious corollaries of the work outlined here, are in hand at the time of writing and it is always dangerous in such situations to assume that the present happy state of agreement between theory and experiment will persist! However we would like to underscore the importance of the lattice dynamics analysis if, indeed, its validity continues to be substantiated by this and other experimental programs. In our view, the most exciting prospects lie mainly with the local mode calculations referred to in the last section. This is because, whatever the fascination of the 'average' double helix (and there is a lot of fascination in the dynamics of a complex polyelectrolyte), the biological business of DNA is conducted at special sites. As we said at the beginning of this article, a particular power of the lattice dynamics approach is the detailed treatment of dynamics at special sites that is made possible through the Green's function method (1,2). We would hope that continued experimental support for the unperturbed dynamics would justify the (difficult and time consuming) examination of protein binding sites such as operators. At a more immediate level the nature of the base destacking mode, combined with the newly developed methods for incorporating anharmonicities (3) suggests mechanisms for controlled melting of the DNA by unwinding proteins, for example. The fact that such mechanisms may be constructed pays testimony to the power of lattice dynamics as a tool for unravelling an enormously complex dynamical problem from a theoretical standpoint. It remains for experiment to show to what extent that simplification reflects reality.

Acknowledgements

We are grateful to Chris DeMarco, Martin Pokorny and Mary Hakim for help in the laboratory. This work was supported in part by the NSF biophysics program through grant PCM 8215433.

References and Footnotes

1. Lindsay, S. M. and Powell, J. "Light Scattering Studies of the Lattice Vibrations of DNA" in *Structure and Dynamics: Nucleic Acids and Proteins,* Eds., Clementi, E. and Sarma, R. H., Adenine Press, N.Y., pp. 241-259 (1983).
2. Putnam, B. F., Van Zandt, L. L., Prohofsky, E. W. and Mei, W. N., "Resonant and Localized Breathing Modes in Terminal Regions of the DNA Double Helix", *Biophys. J. 35,* 271-287 (1981).
3. Gao, Y., Devi-Prasad, K. V. and Prohofsky, E. W., "A Selfconsistent Phonon Theory of the Melting of the DNA Double Helix" (In Press), *J. Chem Phys.* (1984).
4. Eyster, J. M. and Prohofsky, E. W., "Lattice Vibrational Modes of Poly(rU) and Poly(rA)", *Biopolymers 13,* 2505-2526 (1974); Eyster, J. M. and Prohofsky, E. W., "Lattice Vibrational Modes of Poly(rU).Poly(rA). A Coupled Helical Approach", *Biopolymers, 13,* 2527-2543 (1974).
5. Dickerson, R. E., Drew, H. R., Conner, B. N., Wing, R. M., Fratini, A. V. and Kopka, M. L., "The Anatomy of A-, B-, and Z-DNA", *Science, 216,* 475-485 (1982).
6. Barkley, M. D. and Zimm, B. H. "Theory of Twisting and Bending of Chain Macromolecules: Analysis of the Flourescence Depolarization of DNA", *J. Chem. Phys., 70,* 2991-3007 (1979).
7. Allison, S. A. and Schurr, J. M, "Torsion Dynamics and Depolarization of Flourescence of Linear Macromolecules, I. Theory and Application to DNA", *Chem. Phys., 41,* 35-39 (1979).
8. Krumhansl, J. A. and Alexander, D. M., "Nonlinear Dynamics and Conformational Excitations in Biomolecular Materials" in *Structure and Dynamics: Nucleic Acids and Proteins,* Eds., Clementi, E. and Sarma, R. H., Adenine Press, N.Y., 61-80 (1983).
9. Levitt, M., "Computer Simulation of DNA Double-helix Dynamics", *Cold Spring Harbour Symposium on Quantitative Biology, 46A,* 251-262 (1982).
10. Tidor, B., Irikura, K. I., Brooks, B. R. and Karplus, M., "Dynamics of DNA Oligomers", *J. Biomolecular Structure and Dynamics, 1,* 231-252 (1983).
11. Finney, J. L. and Goodfellow, M., "Cooperative Effects in Aqueous Biomolecular Systems and Processes" in *Structure and Dynamics: Nucleic Acids and Proteins,* Eds., Clementi, R. and Sarma, R. H. Adenine Press, N.Y., 81-94 (1983).

12. Mei, W. N., Kohli, M., Prohofsky, E. W. and Van Zandt, L. L., "Acoustic Modes and Non-Bonded Interactions of the Double Helix", *Biopolymers, 20,* 833-852 (1981).

13. Davis, C. C., Personal Communication.

14. Van Zandt, L. L., "Damping of DNA Vibration Modes by Viscous Solvents," J. Quantum Chem., *Quantum Biology Symposium 8,* 271-276 (1981).

15. Corongiu, G. and Clementi, E., "Simulations of the Solvent for Macromolecules. II. Structure of Water Solvating Na$^+$ B DNA at 300K and a Model for Conformational Transitions Induced by Solvent Variations", *Biopolymers, 20,* 2427-2483 (1981).

16. Muller, K. A., "Anharmonic Properties Near Structural Phase Transitions" in *Nonlinear Phenomena at Phase Transistions and Instabilities* Ed., Risio, T. , Plenum, N.Y., pp. 1-34 (1982).

17. Wartell, R. M., Harrell, J. T., Zacharias, W. and Wells, R. D., "Raman Spectroscopy Study of the B-Z Transition in (dG-dC)$_n$·(dG-dC)$_n$ and a DNA Restriction Fragment", *J. Biomolecular Structure and Dynamics, 1,* 83-96 (1983).

18. Eyster, J. M. and Prohofsky, E. W., "On the B to A Conformation Change of the Double Helix", *Biopolymers, 16,* 965-982 (1977).

19. Lindsay, S. M., Anderson, M. W. and Sandercock, J. R.,"Construction and Alignment of a High Performance Multipass Vernier Tandem Fabry-Perot Interferometer", *Rev. Sci. Instrum., 52,* 1478-1486 (1981).

20. Urabe, H. and Tominaga, Y., "Experimental Evidence of Collective Vibrations in DNA Double Helix (Raman Spectroscopy)", *J. Chem. Phys., 78,* 5937-5939 (1983).

21. Urabe, H. and Tominaga, Y., "Low-Lying Collective Modes of DNA Double Helix by Raman Spectroscopy", *Biopolymers, 21,* 2477-2481 (1982).

22. Rupprecht, A. and Forslind, A., "Variation of Electrolyte Content in Wet Spun Lithium and Sodium DNA", *Biochimica et Biophysica Acta, 204,* 304-316 (1970).

23. Hakim, M. B., Lindsay, S. M. and Powell, J., "The Speed of Sound in DNA", *Biopolymers, 23,* 1185-1192 (1984).

24. Hakim, M. B., M.S. Thesis, Arizona State University, unpublished (1983).

25. Giordano, R., Mallamace, F., Micali, N,. Wanderlingh, F., Baldini, G. and Doglia, S., "Light Scattering and Structure in a Deoxyribonucleic Acid Solution", *Phys. Rev. A, 28,* 3581-3588 (1983).

26. Lindsay, S. M. and Powell, J., "Possible Observation of a Defect Resonance in DNA", *Biopolymers, 22,* 2045-2060 (1983).

Structure & Motion: Membranes, Nucleic Acids & Proteins,
Eds., E. Clementi, G. Corongiu, M. H. Sarma & R. H. Sarma,
ISBN 0-940030-12-8, Adenine Press,   Copyright Adenine Press, 1985.

Some New Results on the Electronic Structure of DNA and a New Possible Long Range Mechanism of Chemical Carcinogenesis

Janos J. Ladik
Chair for Theoretical Chemistry at the Friedrich-Alexander-University
Erlangen-Nurnberg, Egerlandstrasse 3,
D-8520 Erlangen, West Germany

and

Laboratory of the National Foundation for Cancer Research
at the Chair for Theoretical Chemistry of the University Erlangen-Nurnberg,
D-8520 Erlangen, West Germany

Abstract

On the basis of detailed energy band structure calculations on DNA possible conduction mechanisms in its periodic part are discussed. To estimate the gap between valence and conduction bands of periodic nucleotide base stacks, recent results with a double zeta basis taking into account also about 50% of the correlation are reviewed. An estimation of the value of the gap (about 6 eV) raises the question whether periodic DNA stacks can be intrinsic conductors due to the internal charge transfer.

To consider the effect of disorder on the electronic structure of DNA the results obtained with the help of several techniques are reviewed. On the basis of this, some conclusions about the mechanism of charge transport in disordered DNA are drawn.

Finally, it is shown how the binding of a bulky carcinogen to DNA can generate a conformational soliton. This can serve as a very effective mechanism for long range effects of carcinogens. The mathematical details are reviewed.

Introduction

The determination of the electronic structure of DNA presents a formidable problem. Though with the development of the theoretical methods (both in the sense of physical theory and of numerical analysis) and with the aid of always faster computers large steps have been made forward, there remains a long way to go until we can say that we have really a sound quantum theoretical description of DNA. Recently large scale minimal basis ab initio calculations have been performed on periodic DNA models including not only the base stacks (1,2,3) and polybase-pairs (consisting of A−T or G−C as unit cells) (2) but also on polyhomo-nucleotides containing a whole cytidine (1) or an adenilic acid and thymidine nucleotide (2), respectively, as unit cell. In a first model calculation (3) the effect of the water structure (being determined by a Monte Carlo simulation technique) (4) around the cytosine stack on its band structure has been assessed.

The first steps for the treatment of aperiodicity in DNA from a solid state physical point of view have been done also (5). The proper treatment, however, of this extremely important problem is of course far from being completed.

Recently for a cytosine stack also a double ζ band structure calculation has been performed. In the future one has to execute also at least for some of the other periodic DNA models better basis set calculations and to be able to estimate the basis set effects of the rest. Further one has to treat correlation in those periodic DNA models for which better basis sets were used at least in the level of the second order of the Moller-Plesset perturbation theory (6) applied for infinite systems (7,8). In this respect one should emphasize that the treatment of correlation in extended systems is by no means solved and new methods (like the use of the localized orbitals in the framework of coupled cluster theory (9,10) are under development. Further at the solution of the correlation problem an improved version of the electronic polaron model (tested until now in the cases of polyacetylene (11), polyethylene (12) and a cytosine stack (13)) has to be used to obtain not only a better total energy per unit cell, but also corrected energy bands and a reduced gap.

The methods for the treatment of correlation have to be combined with the methods developed for taking into account aperiodicity to describe more or less correctly the electronic structure of DNA (14). These calculations have to take into account of course also the effect of the water and ion structure around DNA and also the effective field caused by the protein chain around the DNA double helix. (In a pilot calculation the interaction between a homopolynucleotide and polyglycine chain in different relative positions has been already computed (15). In larger perspective all these calculations have to be performed for all the major conformations of DNA (at least for A, B and Z DNA, respectively).

Only after reaching such a level of sophistication one can expect to be able to describe in a reliable way the different properties (like electronic and phonon spectra, transport properties, energy transfer mechanisms) of DNA which determine final end its biological functions.

Periodic DNA Models

Ab initio SCF LCAO crystal orbital (CO) (16) calculations (in which the quasi momentum k is defined on the combined symmetry operation (17) translation + rotation which allows to reduce the unit cells from a turn of the helix to a single chemical unit like A or A−T etc.) of the four nucleotide base stacks (1,2,3) using different minimal basis sets (19,20), the poly (A−T) and poly (G−C) systems (2) modelling double stranded DNA and of the homopolynucleotides polycytidine (11), polyadenilic acid and polythymidine (2) have been performed.

In these *ab initio* CO calculations one solves the generalized eigenvalue problem

$$F(k) \, C_i(k) = \epsilon_i(k) \, S(k) \, C_i(k) \tag{1}$$

of the Hermitian complex matrices

$$F(k) = \sum_{q=-N}^{N} e^{ikqa} \, F(q) \tag{2}$$

and

$$S(k) = \sum_{q=-N}^{N} e^{ikqa} \, S(q) \tag{3}$$

at different values of the crystal momentum k. In the expressions (2) and (3) (which one obtains by taking advantage of the translational (periodic) symmetry (16)) a is the elementary translation, 2N+1 is the number of unit cells in the quasi-one-dimensional chain and the matrices $F(q)$ and $S(q)$ are m × m matrices, if m denotes the number of basis functions in the unit cell. Further the elements of the matrices $S(q)$ and $F(q)$ are defined (16) as:

$$[S(q)]_{r,s} = \langle \chi_r^o | \chi_s^q \rangle \tag{4}$$

$$[F(q)]_{r,s} = \langle \chi_r^o | \hat{F} | \chi_s^q \rangle = \langle \chi_r^o | \hat{H}^N | \chi_s^q \rangle + \sum_{u,v} \sum_{q_1,q_2} P(q_1 - q_2)_{u,v}$$

$$(\langle \chi_r^o(r_1) \chi_u^{q_1}(r_2) | \frac{1}{r_{12}} | \chi_s^q(r_1) \chi_v^{q_2}(r_2) \rangle - 1/2 \langle \rangle_{exch.}) \tag{5}$$

In these formulae the subscripts denote the basis function and the superscripts are cells indices, that is $\chi_s^q = \chi(r\text{-}R_q\text{-}r_s)$ is the AO centered on the atom s_A to which the basis function χ_s belongs ($s \epsilon s_A$) in the q-th cell. H^N is the one-electron part of the Fock operator of the chain (containing the kinetic energy term and the interaction of the electron with all the nuclei of the chain). Finally, the generalized charge-bond order matrix elements are given by the equation (16)

$$P(q_1 - q_2)_{u,v} = (a/2\pi) \int_{-\pi/a}^{\pi/a} C^*(k)_u C(k)_v \, e^{ik(q_1 - q_2)a} dk \tag{6}$$

As we can see the two-electron part of the Fock matrix (Equ.-s) (5) and (6) makes Equ. (1) non-linear in the same way as it happens in the simpler atomic or molecular problem. The only difference here is that to be able to perform the numerical integration in Equ. (6) one has to solve Equ. (1) at each iteration step at different values of k (according to our experiences using 9 different k-values makes the results in this sense consistent.

A more serious problem is presented by the question, until how many neighbors one has to take into account in the sums (2) and (3) to obtain reliable results. The answer is that one has to study the dependence of the different types of integrals in (4) and (5) on the distances between the centers which occur in them to obtain satisfactory results. This procedure (in which one has to be very cautious to preserve also electrical neutrality) means that one cannot cut off all integrals at the same intercenter distances and if one has unit cells with more than one atom one cannot cut off the integrals at all according to the number of neighboring cells. (for more details see (18)).

If one takes good care for a balanced cut off of the different integrals the SCF procedure (if no covergence difficulties occur) needs about 8-12 steps to obtain a consistency of 10^{-4} in all the charge-bond order matrix elements. It should be pointed out, however, that in the polymer calculations one encounters more frequently convergence difficulties (especially if a better basis is used and the number of neighbors taken into account is not sufficiently large) than in molecular calculations. One has to apply in such cases different procedures and tricks to bring the system into convergence. The details of these cannot be described, however, here.

After successfully soving eq. (1) for a linear chain the eigenvalues $\epsilon_i(k)$ provide the band structure. The indices i of $\epsilon_i(k)$ define the bands and the k-values give the levels within the band (if $N \Rightarrow \infty$ or N is large enough that the level spacing $2\pi/aN < k_B T$, value k_B is the Boltzmann constant distribution we can consider the levels within the band continuous. Finally, the vectors $C_i(k)$ give the coefficients of the generalized Bloch function

$$\psi_i(k,\vec{r}) = \sum_{q=-N}^{N} e^{ikqa} \sum_{r=1}^{m} c_{i,r}(k) \chi_r^q \tag{7}$$

Applying the geometry of B-DNA determined by X-ray diffraction (21) these calculations have resulted in band structures which show the following general features:

1) The conduction and valence bands of all the investigated nucleotide base stacks have widths between 0.3 eV and 0.8 eV. These widths lie in the same range as those of the valence band of a TTF stack (0.3 eV) and of the conduction band of a TCNQ stack (1.2 eV) as our previous ab initio Hartree Fock calculations have shown it (22). Therefore, one would expect that by appropriate doping (which is by no means a trivial task) double stranded DNA B with periodic sequences A−T or G−C, respectively, could be made well conducting.

2) The gap of all the calculated DNA models (1,2) as it is the case in all minimal basis *ab initio* calculations is too large (10-12 eV). According to the experience, accumulated in our Laboratory in the case of polyacetylene (11), if one applies a better basis set and introduces correlated one-electron bands with the aid of the electronic polaron model (11), the gap decreases nearly to the experimental value.

(In the case of alternating transpolyacetylene from 8.3 to 3.0 eV, while the experimental value is 2.0 eV.) Further in the case of polyethylene using a double ζ + polarization functions basis set and applying the electonic polaron model (11) in the Moller-Plesset approximation Suhai (12) has obtained a quasi particle gap of 10.3 eV (the experimental value is 8.8 eV), while the corresponding HF gape is 13.4 eV.

The generalized electronic polaron model (11) used in these calculations applies Takeuti's electronic polaron theory (23) according to which one can define quasi particle (QP) energy levels in the conduction and valence band, respectively.

$$\epsilon_c^{QP}(k_c) = E^{(N+1)}(k_c) - E^{(N)} = A(k_c) , \tag{8a}$$

$$\epsilon_v^{QP}(k_v) = E^{(N)} - E^{(N-1)}(k_v) = -I(k_v) . \tag{8b}$$

Here $E^{(N+1)}(k_c)$ is the total energy per unit cell of the chain with an extra electron in the conduction band at level k_c, $E^{(N)}$ is the total energy per unit cell of the ground state and $E^{(N-1)}(k_v)$ stands for the total energy per unit cell of the system with one electron missing from the level k_v of the valence band. Per definitionem $A(k_c)$ is the electron affinity (the energy gained by putting the extra electron into the conduction band at level k_c) and $I(k_v)$ the ionization potential (the energy needed to ionize an electron from level k_v of the valence band), respectively.

One can approximate the exact total energies $E = E_{HF} + E_{corr}$ as

$$E + E_{HF} + E_2 \tag{8}$$

where E_2 is the correlation correction in the approximation of the second order of the Møller-Plesset many body perturbation theory (MP2) (6). Substituting (8) and the expressions (Koopmans' theorem)

$$\epsilon_c^{HF}(k_c) = E_{HF}^{(N+1)}(k_c) - E_{HF}^{(N)} \tag{9a}$$

$$\epsilon_v^{HF}(k_v) = E_{HF}^{(N)} - E_{HF}^{(N-1)}(k_v) \tag{9b}$$

into equ.-s (8a) and (8b) one obtains

$$\epsilon_c^{QP}(k_c) = \epsilon_c^{HF}(k_c) + E_2^{(N+1)}(k_c) - E_2^{(N)} \tag{10a}$$

$$\epsilon_v^{QP}(k_v) = \epsilon_v^{HF}(k_v) + E_2^{(N)} - E_2^{(N-1)}(k_v) \tag{10b}$$

Substituting into Equ.-s (10) the corresponding MP2 expression

$$E_2 = - \sum_{I,J,A,B} \frac{|\langle IJ||AB\rangle|^2}{\epsilon_A + \epsilon_B - \epsilon_I - \epsilon_J} \tag{11}$$

(where the summations over the composite indices I,J, etc. means summations over band indices (i,j, etc.), k-values and spins) we can write (11)

$$E_2^{(N)} = \sum_I \sum_J{}' \epsilon_{IJ}^{(N)} = \sum_{I\neq V} \sum_{J\neq V}{}' \epsilon_{IJ}^{(N)} + \sum_{I\neq V} \epsilon_{IV}^{(N)} \tag{12a}$$

$$E_2^{(N-1)}(k_v) = \sum_{I\neq V} \sum_{J\neq V}{}' \epsilon_{IJ}^{(N-1)V} \tag{12b}$$

$$E_2^{(N+1)}(k_c) = \sum_I^{(c)} \sum_J^{(c)}{}' \epsilon_{IJ}^{(N+1)C} = \sum_I^{(c)} \sum_J^{(c)}{}' \epsilon_{IJ}^{(N+1)C} + \sum_I \epsilon_{IJ}^{(N+1)C} \tag{12c}$$

Here the prime at the summation over J means that the states I = J have to be excluded, the pair correlation contributions $\epsilon_{IJ}^{(N)}$ are defined as

$$\epsilon_{IJ}^{(N)} = - \sum_{AB} \frac{|\langle IJ||AB\rangle|^2}{\epsilon_A + \epsilon_B - \epsilon_I - \epsilon_J} \tag{13}$$

V stands for the state v, k_v and C for the state c, k_c, respectively. Further in Equ. (12c) the summations over I and J contain also (with the exception of the last sum of the right hand side of (12c)) the extra occupied conduction bands state c. Substituting Equ.-s (12) with the definition (13) into Equ.-s (10) one obtains after some calculation the final expressions

$$\epsilon_c^{QP}(k_c) = \epsilon_c^{HF}(k_c) + \Sigma_c^{(N+1)}(e) + \Sigma_c^{(N+1)}(h) , \tag{14a}$$

$$\epsilon_v^{QP}(k_v) = \epsilon_v^{HF}(k_v) + \Sigma_v^{(N)}(e) + \Sigma_v^{(N)}(h) . \tag{14b}$$

Here the electron and hole self-energies Σ are defined as:

$$\Sigma_c^{(N+1)}(e) = \sum_I \epsilon_{IC}^{(N+1)c} \tag{15a}$$

$$\Sigma_c^{(N+1)}(h) = \sum_I^{(c)} \sum_J^{(c)}{}' \epsilon_{IJ}^{(N+1)c} - \sum_I \sum_J{}' \epsilon_{IJ}^{(N)} \tag{15b}$$

$$\Sigma_v^{(N)}(e) = \sum_{I\neq V} \epsilon_{IV}^{(N)} \tag{15c}$$

$$\Sigma_v^{(N)}(h) = - \sum_{I\neq V} \sum_{J\neq V}{}' (\epsilon_{IJ}^{(N)} - \epsilon_{IJ}^{(N-1)V}) \tag{15d}$$

From Equ.-s (15) it is easy to see the physical meaning of the self energies: $\Sigma_c^{(N+1)}(e)$

describes the new pair correlations formed between the extra electron in state C and all the other electrons and $\Sigma_c^{(N+1)}(h)$ gives the reduction of the pair correlations due to the fact that no scattering of the electrons into these newly occupied states is possible. Similarly $\Sigma_v^{(N)}(e)$ expresses the increase of pair correlations caused by the new scattering possibility to the empty state V and $\Sigma_v^N(h)$ gives pair correlations between the other electrons and the one in the state V. From Equ.-s (15) and from this discussion follows that $\Sigma_c^{(N+1)}(e)$ and $\Sigma_v^{(N)}(e)$ are negative while $\Sigma_c^{(N+1)}(h)$ and $\Sigma_v^{(N)}(h)$ are positive. As the detailed derivation shows it (11)

$$|\Sigma_c^{(N+1)}(e)| > |\Sigma_c^{(N+1)}(h)| \tag{16a}$$

but

$$|\Sigma_v^{(N)}(e)| < |\Sigma_v^{(N)}(h)| \tag{16b}$$

Therefore the quasi particle gap is always smaller than the HF one. Further considerations (11) indicate also that the band widths of the conduction and valence bands, respectively, are smaller in the QP (correlated) description , than in the HF case.

One has to point out, however, that if one compares the gaps (the energy difference of the lower limits of the conduction band and of the upper limits of the valence band) with experimental gaps, for the latter not the first electronic excitation energy should be used which due to exciton formation has a still smaller energy.

3) In the case of all the three homopolynucleotides polycytidine (1), polyadenilic acid and polythymidine (2) a charge transfer of about 0.2 e per unit cell has been found from the sugar-phospate part of the nucleotide to the nucleotide base. In the case of cytidine an STO-3G basis, in the other two cases, a different, somewhat better minimal basis has been used in a new crystal orbital program written for linear chains (24). Therefore, we can conclude that the amount of transferred charge is rather independent of the type of base present in the nucleotide, and even in the case of better basis sets, one would expect a nonnegligibile CT from the sugar to the nucleotide base.

If we analyze the band structure of a polynucleotide, we find that the bands of the composite system are rather similar in their positions (but not necessarily in their width) to the positions of the bands of the corresponding nucleotide stack and those of a sugar-phosphate chain (see, for instance, Table II of (1)). In all the three investigated cases, the valence and conduction bands of the polynucleotide are of "base stack" type, while the band below the valence band and the band above the conduction band are of "sugar-phospate chain" type. This shows that in these cases the simple HOMO (donor) $\Rightarrow$ LEMO (acceptor) picture of CT breaks down, and actually we could discover it only after we performed a "bookkeeping" of the charges of the bases and in the sugar-phosphate unit (applying Mulliken's population analysis).

This rather considerable CT in DNA has two important consequences. On the one hand, as Clementi has emphasized (25), the CT strongly changes the potential of B-DNA and with it the water and ion structure around it. On the other hand, the question can be raised whether B-DNA is not an intrinsic conductor. To answer this question, one has to investigate whether the total energy per unit cell would not be lower (larger negative value) if we had performed the band structure calculations in a way that takes away 0.2 e from the band below the valence band (the highest filled "sugar-phosphate" type band) and put this extra charge into the conduction band. If one calculates, however, the energy difference between the lower edge of the conduction band and the upper limit of the band below the valence band with the aid of a minimal basis, one finds energy differences at 13.3, 13.1, and 14.1 eV for polycytidine, polythymidine, and polyadenilic acid, respectively (1,2). Due to these rather large gaps (which are artifacts of the Hartree-Fock theory of solids, as is well known) we can rule out the possibility that in the Hartree-Fock level we can obtain in B-DNA free charge carriers and it would therefore become an intrinsic semiconductor if the base stack is periodic. (The base stacks have conduction band widths between 0.3 and 0.9 eV, while the widths of the highest filled "sugar-phosphate" type bands are between 0.1 and 0.5 eV in the three investigated polynucleotides (1,2).)

One can expect however, a completely different situation when one will be able to perform band structure calculations with a higher quality basis set (double ζ + polarization functions) and also compute correlated quasi particle energy bands for these systems using the above described version of the electronic polaron model. In this connection one should mention that a recent double ζ calculation of a cytosine stack (13) has decreased the STO-3G gap of 10.7 eV only by about 0.5 eV. Performing with this basis also an MP2 calculation (which has provided only about 50 per cent of the ground correlation energy) the quasi particle gap decreased to ~6 eV. Therefore, one can easily imagine that other improved calculations would provide for the polynucleotides a quasi-particle gap between the "sugar-phosphate" band and the conduction band of 4-5 eV instead of 13-14 eV. If this were really the case, the question could be raised again whether taking away δe charge (the amount of transferred charge had to be reinvestigated in these more sophisticated calculations) form the highest filled "sugar phosphate" quasi particle band and putting it into the quasi-particle conduction band would result in a lowering of the total energy per unit cell. If the answer to this question is affirmative, one would expect intrinsic semiconduction in B-DNA (at least in the case of a periodic base stack).

To conclude, we can say that the question of intrinsic semiconduction in B-DNA is an open one and can be settled only after much more sophisticated calculations have been performed on the polynucleotides. Though the simple Hartree-Fock theory does not allow the generation of free charge carriers in these systems, the CT from the sugar to the nucleotide bases hints at this possibility.

The Effect of Disorder on the Transport Properties of DNA

Even if DNA is an intrinsic semiconductor due to the charge transfer between the sugar units and the nucleotide bases (see previous discussion) or can be made conducting by doping, these considerations are valid for the case when the DNA stacks are periodic. Though the sugar-phosphate chain is periodic even in real, aperiodic DNA, the widths of the bands originating from it are considerably smaller (1,2) (widths between 0.1 and 0.4 eV) than those of the nucleotide base stacks (widths between 0.3 and 0.9 eV) (1,2). Therefore one would expect a smaller hole mobility in them. The electronic structure mobility would be in periodic base stacks. Further the transport of holes along the periodic sugar phosphate chain would be further perturbed by the aperiodic potential caused by the different base pairs (26), by the water and ion environment around DNA and by the polypeptide chain surrounding DNA.

Keeping in mind all these facts one can raise the question, whether there is any electronic (hole) transport in real, disordered DNA. The answer is of course not trivial. As we have seen from negative factor counting calculations (using both its simple (26) and matrix block form (28)) on DNA (described in some detail in the Volume of the previous conference (29); see also (30)), the disorder in the sequence of the nucleotide base stacks destroys the band structures of the periodic stacks introducing many peaks and gaps in the density of states (DOS) curves (see Fig.-s 1-3 of 29).

In these calculations, however, complete disorder in the sequence was assured (generated by a Monte Carlo program) though a mathematical statistical analysis (31) of many experimental nucleotide base sequences (see for instance (32)) shows certain regularities in them including a preference for the same unit several times (31). These regularities certainly act in some degree in the direction of the restauration of band structure. Further the perturbation caused by the water and ion environment can introduce extra impurity levels into the gaps of the DOS curves of a disordered stack increasing the probability of a hopping type charge transport.

Finally one has to study the relative position of the mobility edge (33) to the Fermi level. The Fermi level can be determined from the degree of internal charge transfer or/and of the amount of charge transferred by external dopant. To find out the position of the mobility edge one has to investigate with the help of a suitable (Green matrix) method (34) the localization properties of the different states. One can still expect a Bloch-type conduction in a disordered nucleotide base stack, if 1.) the Fermi level falls inside a continuous region of filled levels and 2.) if the states around the Fermi level are still delocalized (the mobility edge lies below the Fermi level).

In reality one would expect that the main charge transport mechanism in native disordered DNA is hopping, but in some regions of it (depending on the sequence) also coherent Bloch-type conduction occurs, if free charge carriers are generated in it. The important, but very complicated problem of charge transport in disordered DNA needs of course in the future still further, much more detailed investigations.

Conformational Solitons in DNA caused by Carcinogen Binding

In a previous paper (35) different possible long range effects of carcinogens were reviewed. Here we should like to elaborate on a possibility only shortly mentioned in ref. 35: long-range effects of carcinogens through formation of solitons in DNA. Let us assume that a bulky carcinogen, like the ultimate of 3,4-benzopyrene is bound to a nucleotide base. Certainly, in the neighborhood of the carcinogen binding the structure of DNA, first of all the positions of the stacked base pairs will become strongly distorted. With this conformational change a change in the electronic structure of DNA will be coupled because the electronic interaction between the stacked base pairs is strongly dependent on their relative position. In this way a non-linear change (conformational change coupled with electronic structure change) takes place at the site and neighborhood of the carcinogen binding (see Fig. 1).

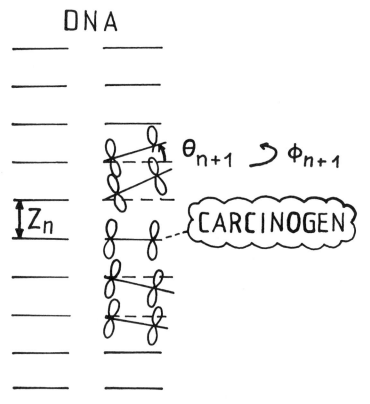

Figure 1. Distortion of a DNA stack due to the binding of a bulky carcinogen to a nucleotide base. Z_n denotes the change of the stacking distance at the n-th unit, the distortion angle θ_{n+1} is measuring the deviation of the plane of the n+1-th nucleotide base from its original position (perpendicular to the main axis of the double helix) and ϕ_{n+1} is the deviation of the latter unit from the equilibrium value of rotation around the helix axis (36° in DNA B).

In vitro a covalently bound ultimate carcinogen would remain attached to DNA for an indefinitely long time. Not this is the case, however, *in vivo* where for instance repair enzymes can remove the carcinogen within a few hours. Until the carcinogen sits at its binding site the above described non-linear change will remain localized. After the removal of the carcinogen, however, it seems rather probable that the system will not relax immediately (the original conformation will not be restored instantaneously) because for this bulky molecular constituents of DNA (together with the water and ion structure surrounding them) have to be moved. On the other hand it is well known that 1.) solitons have several orders of magnitude longer life times than simple electronic excitations and 2.) they can travel as solitary waves in an extended system (36). Therefore, one can postulate that after the removal of a carcinogen from DNA the non-linear (but previously local) change caused by its binding can travel through rather large distances along the chain (see Fig. 2) causing a long-range effect which may effect a larger segment of DNA and its interaction with a protein molecule.

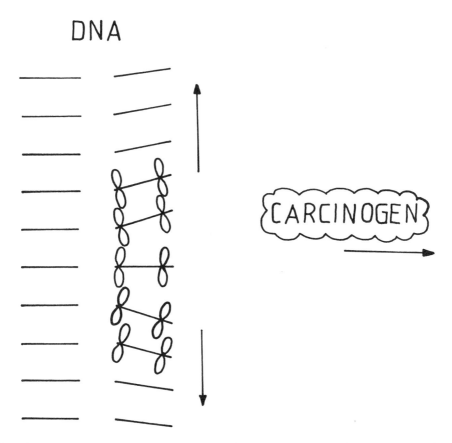

Figure 2. After the removal of the carcinogen the originally pinned down non-linear change in the DNA stack starts to travel either upwards or downwards.

To test this hypothesis one can start by writing down a Hamiltonian for the soliton. Generalizing the theory of Su, Schrieffer and Heeger (SSH) (37) which was developed for the non-linear change caused by a kink in polyacetylene, we can write

$$\tilde{H} = \tilde{H}_{el} + \tilde{H}_{conf} + \tilde{H}_{el\text{-}conf} \tag{17}$$

Here we take as first (and rough) approximation for the description of the overlapping electrons of the stacked base $\tilde{H}$ in the form (Huckel type, or tight-binding approximation)

$$\tilde{H}_{el} = -t_o \sum_{n,s,a} (\hat{C}^+_{n+1,s,a} \hat{C}_{n,s,a} + \hat{C}^+_{n,s,a} \hat{C}_{n+1,s,a}) \tag{18}$$

where n is the site index (which base), s ($\pm 1/2$) stands for the spin and $a = 1,2$ indicates the strand in the DNA double helix. Further the $\hat{C}^+$-s and $\hat{C}$-s are creation and annihilation operators, respectively, and t_o is the hopping integral for the undistorted chain. One should point out immediately that H_{el} is uncomplete because it does not contain electron-electron interaction terms.

To describe the conformational changes of the stacked bases one has to introduce three variables (in contrary of the one variable of the polyacetylene case (37)). We can denote by $Z_{n,a}$ the shift from its equilibrium position along the Z axis (the main axis of the double helix) of the n-th base in the a-th chain (if Z_n is positive Z_{n+1} will also be positive, Z_{n-1} and Z_{n-2} will be, however, negative (see Fig. 1)) and by $\phi_{n,a}$ (following Krumhansl and Alexander (38)) the rotation of this base (again measured from its equilibrium position) in the plane perpendicular to the Z axis. Finally the angle $\theta_{n,a}$ measures the tilting of the base (thus θ is the angle between the plane of the displaced base and the plane perpendicular to the Z axis) (see Fig. 2). Assuming that the motion of the base described by these three variables can still be treated in the harmonic approximation, we can write

$$\tilde{H}_{conf} = 1/2 \sum_{n,a} [K_1(Z_{n+1,a} - Z_{n,a})^2 + K_2(\phi_{n+1,a} - \phi_{n,a})^2 +$$
$$+ K_3(\theta_{n+1,a} - \theta_{n,a})^2 + M_{n,a} (\dot{Z}^2_{n,a} + \dot{\phi}^2_{n,a} + \dot{\theta}^2_{n,a})] \tag{19}$$

Here K_1, K_2 and K_3 are the force constants belonging to the three variables Z, ϕ and θ and $M_{n,a}$ is the mass of the n-th base in the a-th strand ($M_{n,a}$ of course has only four different values).

Finally, for the coupling of the conformational change with the electronic structure change we have to introduce the modified hopping integral $\tilde{t}_{n+1,a;n,a}$ as

$$\tilde{t}_{n+1,a;n,a} = t_o - [a_1(Z_{n+1,a} - Z_{n,a}) + a_2(\phi_{n+1,a} - \phi_{n,a}) + a_3(\theta_{n+1,a} - \theta_{n,a}) \tag{20}$$

where a_1, a_2 and a_3 are the electron-base displacement (phonon) coupling constants. One should point out that Equ. (20) is a straightforward generalization of Equ. (2.2) of SSH (37). Using the expression (20) we can write for the non-linear electron-displacement coupling term

$$\tilde{H}_{\text{el-conf}} = \sum_{n,s,a} (\tilde{t}_{n+1,a:n,a} - t_o)(\hat{C}^+_{n+1,s,a}\hat{C}_{n,s,a} + \hat{C}^+_{n,s,a}\hat{C}_{n+1,s,a}) \tag{21}$$

Following SSH (37) and Krumhansl and Alexander (38), respectively, one can substitute the Hamiltonian (17) with definitions (18-21) (after taking expectation values of the electron-operators with the many electron reference function which give one) into the classical canonical equations of motion of Hamilton. For actual numerical calculations one has to choose a set K_i and α_i (i = 1,2,3) values or one can try to determine them on the basis of quantum mechanical potential surface calculations and by computing the corresponding electron-phonon matrix elements.

However, in reality, the three variables describing the change of the positions of a nucleotide basis are not independent but have to be coupled. Further an at least approximate treatment of the electron-electron interaction is indispensable. Therefore we have to rewrite our Hamiltonian (17) as

$$\hat{H} = \tilde{H} + \hat{H}' \tag{22}$$

where

$$\hat{H}' = \hat{H}'_{\text{el}} + \hat{H}'_{\text{conf}} + \hat{H}'_{\text{el-conf}} \tag{23}$$

Here

$$\hat{H}'_{\text{el}} = \sum_{n,m} \gamma_{n,m}(\hat{n}_n - 1)(\hat{n}_m - 1) \tag{24}$$

where

$$\hat{n}_m = \hat{C}^+_m\hat{C}_m$$

is the number operator at site m. Equ. (24) describes the electron-electron interaction in the well-known Pariser-Parr-Pople (PPP) approximation. Further H'_{conf} can be written as

$$H'_{\text{conf}} = 1/2 \sum_{n,a} [\beta_1(Z_{n+1,a} - Z_{n,a})(\phi_{n+1,a} - \phi_{n,a}) +$$

$$+ \beta_2(Z_{n+1,a} - Z_{n,a})(\theta_{n+1,a} - \theta_{n,a}) + \beta_3(\phi_{n+1,a} - \phi_{n,a})(\theta_{n+1,a} - \theta_{n,a})]$$

where $\beta_1, \beta_2, \beta_3$ are coupling constants between the different coordinates describing the distortion of the original position of a nucleotide base (these coupling constants could be computed also from the corresponding potential surface). Finally $H'_{\text{el-conf}}$ can be easily constructed, if one substitutes into Equ. (21) instead of $\tilde{t}_{n+1,a:n,a}$

$$t_{n+1,a;n,a} = \tilde{t}_{n+1,a;n,a} + t'_{n+1,a;n,a} \tag{25}$$

with

$$t'_{n+1,a;n,a} = -\{[\gamma_1(\phi_{n+1,a} - \phi_{n,a}) + \gamma_2(\theta_{n+1,a} - \theta_{n,a})] \times$$
$$\times (Z_{n+1,a} - Z_{n,a}) + \gamma_3(\phi_{n+1,a} - \phi_{n,a})(\theta_{n+1,a} - \theta_{n,a})\} . \tag{26}$$

The additional electron-phonon coupling constants (41) could be calculated again with the help of quantum mechanically computed electronic wave functions of the stack (band structures) using again a potential hypersurface to determine the wave functions of these coupled vibrations. Presently, however, to obtain an orientation one can treat these coupling constants as parameters substituting different values for them.

To solve the dynamical problem at least classically one can follow the numerical iterative procedure of Su and Heeger (39) and the Su (40), respectively. The adaptation of this procedure to our problem in the case of the simple SSH type Hamiltonian ($\tilde{H}$) as well as of the more complete Hamiltonian $\hat{H}$ (defined by Equ.-s (22) - (26)) is described in another paper (42).

Applying this procedure to the polyacetylene problem at 50 CH units with one kink, it turned out that only by a special choice of the parameter values K and α it was possible to obtain a true soliton solution (with an infinite life time) (43).

One should point out, however, that the simple SSH-type Hamiltonian (17) does not include couplings between the three different distortion variables of a DNA stack. Therefore the dynamical treatment of this Hamiltonian would describe three independent distortion problems which differ from each other only in the values of the parameters. For this reason the existing polyacetylene kink program can be easily used to study the time dependence of these distortions of a DNA stack.

If one wants to describe, however, the dynamic behaviour of the more realistic system defined by the Hamiltonian (22) one needs a somewhat more complicated program system. The development of this is in progress.

The solutions of the classical equations of motion will then provide the time evolution of the solitary wave generated at the site of carcinogen binding. In this way one can learn about the life time and the range of the travelling non-linear distortion caused by carcinogen binding and subsequent releasing as a function of the parameter values. Most probably with a realistic estimate of the parameter values one would not obtain a soliton in a strict sense (infinite life time) but rather probably a solitary wave (44) which has a long enough life time to travel along the stack causing long range interference with the DNA protein interactions. In this way possibly it can initiate the activation of an oncogene. Obviously, such results will have besides their physical significance a profound biological importance.

Finally, it should be mentioned that in our model we have neglected 1.) in contrary to Krumhansl and Alexander (38) the coupling of the distortions of the stacked bases with the changes in the conformation of the sugar rings (the sugar puckering), 2.) non-linear terms in the motion of the bases and the sugar rings and 3.) the coupling to the environment (water molecules and ions).

Acknowledgement

We should like to express our gratitude for the very fruitful discussions to Professors E. Clementi, J. Cizek, W. F. Forbes, M. Lax, F. Martino and P. Otto and to Drs F. Beleznay, M. Seel and S. Suhai of the different aspects of the problems described here. The financial support of the "Kraftwerk Union AG" and the "Fond der Chemischen Industrie" is gratefully acknowledged.

References and Footnotes

1. Ladik, J. and Suhai, S., *Int. J. Quant. Chem. QBS7,* 181 (1980).
2. Otto, P., Clementi, E. and Ladik, J., *J. Chem. Phys. 78,* 4547 (1983).
3. Otto, P., Ladik, J., Corongiu, G., Suhai, S., and Förner, W., *J. Chem. Phys. 77,* 5026 (1982).
4. For a review see: Clementi, E., *Computational Aspects for Large Chemical Systems,* Lecture Notes in Chemistry, Vol. 19, Springer Verlag, Berlin (1980).
5. Day, R.S. and Ladik, J., *Int. J. Quant. Chem. 21,* 17 (1982).
6. Møller, C. and Plesset, M. S., *Phys. Rev. 46,* 618 (1934).
7. Ladik, J., in *Recent Advances in the Quantum Theory of Polymers,* Eds. André, J.-M, Delhalle, J., Ladik, J., Leroy, G. and Moser, C., Springer Verlag, Berlin, p. 155 (1980).
8. Suhai, S. and Ladik, J., *J. Phys. C: Solid State Phys. A15,* 4327 (1982).
9. Cizek, J., Forner, W. and Ladik J., *Theor. Chim. Acta 64,* 107 (1983).
10. A new coupled cluster *ab initio* program which can use also localized orbitals has been recently developed in Erlangen (Förner, W., Otto, P., Cizek, J. and Ladik, J., to be published).
11. Suhai, S., *Phys. Rev. B27,* 3506 (1983).
12. Suhai, S., in *Quantum Chemistry of Polymers; Solid State Aspects,* Eds. Ladik, J. and André, J.-M., Reidel Publ. Co., Dordrecht-New York (1984) p. 101.
13. Suhai, S., to be published
14. Ladik, J., *Int. J. Quant. Chem. 23,* 1073 (1983).
15. Otto, P., Clementi, E., Ladik, J. and Martino, F., *J. Chem. Phys.* (in press).
16. Del Re, G., Ladik, J and Biczó, G., *Phys. Rev. 155,* 97 (1967); Andre, J.-M., Gouverneur, L. and Leroy, G., *Int. J. Quant. Chem. 1,* 427 (1967).
17. Ukrainski, I.I., *Theor. Chim. Acta (Berlin) 30,* 139 (1975); Blumen, A. and Merkel, C., *Phys. Stat. Sol. 83,* 425 (1977); Ladik, J., in *Excited States in Quantum Chemistry,* Eds. Niclaides, C. A., and Beck, D. R., D. Reidel Publ. Co., Dordrecht-New York (1979) p. 495.
18. Suhai, S., *J. Chem. Phys.* (submitted).
19. Hehre, W., Stewart, R. F., and Pople, J. A., *J. Chem. Phys. 1,* 2657 (1969).
20. The basis used in the integral package of IBMOL-5.
21. Arnott, S., Dover, S. P. and Wonacott, A. J., *Acta Cryst. B28,* 2192 (1969).
22. Suhai, S. and Ladik, J., *Phys. Lett. A77,* 25 (1980).
23. Takeuti, Y., *Progr. Theor. Phys. (Kyoto) 18,* 421 (1957); *Progr. Theor. Phys. (Kyoto) Suppl. 12,* 75 (1975).
24. Otto, P. and Clementi, E. to appear.
25. See for instance: Clementi, E., *Computational Aspects for Large Chemical Systems,* Lecture Notes in Chemistry, Vol. 19, Springer Verlag, Berlin (1980); Corongiu, G. and Clementi, E., *Biopolymers 20,* 551 (1981); Clementi, E., in *Structure and Dynamics of Nucleic Acids and Proteins,* Eds. Clementi, E. and Sarma, R. H., Adenine Press, New York (1983), p. 321.

26. Lax, M., private communication.
27. Dean, P., *Proc. Roy. Soc. A254*, 507 (1960); *ibid A260*, 263 (1961); Dean P., *Rev. Mod. Phys. 44*, 127 (1972); Seel, M., *Chem. Phys. 143*, 103 (1979); Seel, M. and Ladik J., unpublished results.
28. Day, R. S. and Martino, F., *Chem. Phys. Lett. 84*, 86 (1981).
29. Ladik, J. in *Structure and Dynamics of Nucleic Acids and Proteins*, Eds. Clementi, E. and Sarma, R. H., Adenine Press, New York (1983), p. 321.
30. Day, R. S. and Ladik, J., *Int. J. Quant. Chem. 21*, 17 (1982).
31. Gentleman, J. F., Shadbolt-Forbes, M. A. Hawkins, J. W., Ladik, J. and Forbes, W. F., *Mathematical Scientist* (accepted).
32. Dayhoff, M. O., Schwartz, R. M., Chen, H. R., Barker, W. C., Hunt, L. T. and Orcutt, B. C., *DNA*, Vol. 1, Mary Ann Liebert In., (1981),p. 51.
33. In a slightly disordered system only the energy levels at the band limits are localized and inside the band they are still delocalized with increasing disorder more and more states inside the band become localized and after a certain critical degree of disorder all states will be localized. One calls mobility edge that level, below which all the energy levels are localized but above which one finds delocalized states.
34. See for instance: Gazdy, B., Day, R. S., Seel, M., Martino, F. and Ladik, J., *Chem. Phys. Lett. 88*, 220 (1982).
35. Ladik, J. Suhai, S. and Seel, M., *Int. J. Quant. Chem. QBS5*, 135 (1978).
36. See for instance: Davidov, A. S. and Kislusha, N. I. *Phys. Stat. Sol. 59*, 463 (1973); Davidov, A. S., *Phys. Scripta 20*, 387 (1979).
37. Su, W. P., Schrieffer, J. R. and Heeger, A. J., *Phys. Rev. B4*, 2099, (1980).
38. Krumhansl, J. A. and Alexander, D. M. in *Structures and Dynamics of Nucleic Acids and Proteins*, Eds. Clementi, E. and Sarma, R. H., Adenine Press, New York (1983) p. 61.
39. Su, W. P. and Heeger, A. J., *Proc. Natl. Acad. Sci. 77*, 5626 (1980).
40. Su, W. P. in *Mol. Cryst. Liq. Cryst.*, Proceedings of the Int. Conf. on Low Dimensional Conductors, Boulder, Colorado, Vol. 83 (1982), Gordon and Breach Science Publ.
41. One should observe that by substituting (26) into (21) one obtains third order effects. Therefore in the course of actual calculations one perhaps can neglect this term.
42. Ladik, J. and Cizek, J., *Int. J. Quant. Chem.* (submitted).
43. Martino, F., Wang, C. L. and Ladik, J. (to be published).
44. If a solitary wave passing along a DNA stack has a lasting effect (for instance breaking hydrogen bonds between the base pairs or between the polynucleotide and a polypeptide chain) it would quickly lose its energy. Therefore, such dissipative solitary waves would have a very short life time. On the other hand the water and ion environment of DNA can easily serve as an energy reservoir slowing down in this way in a large degree the energy dissipation of the solitary wave. Thus the life time of a solitary wave travelling along a DNA stack which is imbedded in its environment can become considerably larger.